Ancient Greek Dialectic and Its Reception

Topics in Ancient Philosophy/
Themen der antiken Philosophie

───
Edited by / Herausgegeben von
Ludger Jansen, Christoph Jedan, Christof Rapp

Volume 10

Ancient Greek Dialectic and Its Reception

Edited by
Melina G. Mouzala

DE GRUYTER

ISBN 978-3-11-221386-5
e-ISBN (PDF) 978-3-11-074414-9
e-ISBN (EPUB) 978-3-11-074422-4
ISSN 2198-3100

Library of Congress Control Number: 2023932875

Bibliographic information published by the Deutsche Nationalbibliothek
The Deutsche Nationalbibliothek lists this publication in the Deutsche Nationalbibliografie; detailed bibliographic data are available on the Internet at http://dnb.dnb.de.

© 2025 Walter de Gruyter GmbH, Berlin/Boston
This volume is text- and page-identical with the hardback published in 2023.
Printing and binding: CPI books GmbH, Leck

www.degruyter.com

1922–2022
A hundred years from the catastrophe
of Asia Minor and Smyrna

Table of Contents

Preface —— XI

Abbreviations —— XIII

Introduction —— 1

Part I: Ancient Greek Dialectic in Classical Antiquity

François Renaud
Dialectic as True Rhetoric in Plato's *Gorgias* —— 37

Rafael Ferber
Socrates' "Flight into the *Logoi*": A Non-Standard Interpretation of the Founding Document of Plato's Dialectic —— 57

Claudia Marsico
Friendly Fire: Dialectic Struggles between the Megarians and Plato —— 83

Beatriz Bossi
Eros on Stage: Dialogue and Dialectic in the *Phaedrus* —— 105

Kristian Larsen
Dialectic and the Activity of the Soul when Reaching for Being and the Good in Plato's *Theaetetus* 184b3–186e12 —— 129

Melina G. Mouzala
Pursuing Self-Knowledge in Plato's *Sophist*. The Communion of the Sophistic and Socratic Dialectic in the Sixth Definition of the Sophist: A Reading Based on Proclus' Interpretation of Dialectic in the *Sophist* —— 157

Anna Pavani
On Plato's Late Dialectic: The Methods of Collection and Division —— 189

Lucas Angioni
Sophistical Demonstrations: A Class of Arguments Entangled with False Peirastic and Pseudographemata —— 211

Part II: **Reception, Interpretation, Development and Influence of Ancient Greek Dialectic in Late Antiquity and Byzantium**

Gweltaz Guyomarc'h
The Services of Dialectic: Dialectic as an Instrument for Metaphysics in Alexander of Aphrodisias —— 249

Silvia Fazzo
Aporiai with Multiple Solutions in Alexander of Aphrodisias —— 277

Inna Kupreeva
Alexander of Aphrodisias on the Principle of Non-Contradiction: The Argument "from Signification" —— 287

Ilaria L.E. Ramelli
Ancient Greek Dialectic and Its Reception in Origen of Alexandria: From Plato to Christ-Logos —— 331

Michael Griffin
Exegesis as Philosophy: Notes on Aristotelian Methods in Neoplatonic Commentary —— 371

Sarah Klitenic Wear
Syrianus and the Dialectical Cosmos —— 397

Dirk Baltzly
Proclus on Plato's Dialectic: Argument by Performance —— 413

Harold Tarrant
Elenchus and Syllogistic in Olympiodorus of Alexandria —— 427

Han Baltussen
Simplicius and Aristotle's Dialectic —— 441

Graeme Miles
Michael Psellos on Dialectic and Allegory —— 457

Index Locorum —— 469
Index Nominum —— 493
Index Rerum —— 499

Notes on the Contributors —— 521

Preface

This volume consists of the entire sum of papers presented at the International Conference entitled "Ancient Greek Dialectic and its Reception", which was held on June 15–16, 2021, under the auspices of the Research Committee of the University of Patras. The conference was organized on the occasion of the twentieth anniversary of the establishment of the Department of Philosophy of the University of Patras (1999–2019), in which I worked from 2006. It was two years earlier, in October 2019, that I took the initiative to organize this conference. But due to the condition of "force majeure" caused by the spread of the COVID-19 through Europe, the USA, and the whole world, and because of the sanitary restrictions and measures imposed by the majority of governments, including the suspension of the Universities and flights between the countries, as conference organizer, I was compelled to postpone the physical event and to proceed with its virtual implementation.

The main aim of the conference was to present different interpretive approaches to Ancient Greek Dialectic, as method, practice, activity, and generally as form of philosophical enquiry, through the investigation of its use, development, and reception by different philosophers in different periods during the history of philosophy. An equally important aim of the conference was to gain a thorough understanding of the value of Ancient Greek Dialectic and its overall contribution to the intellectual empowerment, independence, and autonomy of man with relation to the conditions of one's life and the relationships with other people, both at a personal and socio-political level, as well as the relationship with oneself.

A conference in normal circumstances is a kind of spiritual celebration and it can provide the opportunity for both fruitful communication and reflection. Furthermore, there is the hopeful expectation that something from its atmosphere will be illustrated in the volume which constitutes its output. But in the difficult conditions of COVID-19 that we have all experienced for three years now, a volume based on a vivid and stimulating conference of ancient philosophy can also be seen as a stable source of strength and hope and a kind of psychic support and healing both for those who participated and contributed to the collective work that arose from it, as well as the attendees and readers. This dimension is further reinforced, given that the topic of the volume and the conference is "Ancient Greek Dialectic and Its Reception".

Dialectic is the vehicle through which ancient Greek culture established a determinate relationship between political freedom and philosophy, while precisely those aspects of socio-political life, freedom of thought and expression, boosted its development and improved its use, as argued by Enrico Berti (1978: "Ancient Greek Dialectic as Expression of Freedom of Thought and Speech". In: *Journal of the History of Ideas* 39. No. 3, 347–370; here, p. 348). Dialectic presupposes free thought and speech (Berti 1978: p. 355) as well as the search for truth. Gregory Vlastos (1983: "The Socratic Elenchus". In: Annas, Julia (Ed.): *Oxford Studies in Ancient Philosophy.* Vol. I. Oxford: Oxford University Press, 27–58; here, p. 31) stresses that *elenchus* is first and foremost a *search*, because it is always a positive attempt to reach out for the truth, as shown by the broad range of zetetic terms that are used in the context of the Platonic dialogues. These include searching (ἐρευνῶ, διερευνῶ), inquiring (ζητῶ, ἐρωτῶ, συνερωτῶ), and investigating (σκοπῶ, διασκοπῶ, σκέπτομαι, συσκέπτομαι). Moreover, this search, which begins from general truths, step by step is transformed into a search of self which leads to care of self and introspection. From this procedure emerges the therapeutic dimension of Dialectic which can be involved in the remedy of dianoetic errors and the cure of psychic diseases. Christopher Gill (1985: "Ancient Psychotherapy". In: *Journal of the History of Ideas* 46. No. 3, 307–325; here, p. 321) stresses that "Plato, in particular, appropriates the idea that philosophy is a quasi-medicine, sometimes combining this idea with the claim that the Socratic type of dialogue is the most effective method of cure for psychic illness. The locus classicus is the opening section of *Charmides*". But this is just one aspect of Dialectic, which I wish to highlight only in terms of the benefit that contemporary man can derive from it. The readers of this volume will also have the opportunity to enjoy chapters which illustrate the many different aspects and ramifications of it, representative of different philosophers and periods, from Classical Antiquity to Late Antiquity and Byzantine philosophy.

Acknowledgments

I am grateful to the authors of this volume and several readers who commented on the texts. I am also grateful to the participants in the Patras conference, who developed many fruitful discussions and exchanged their views on a variety of subjects that were presented during the two days of sessions. I am no less grateful to the attendees of the conference who stimulated interest in and increased the enthusiasm of the participants for their research.

Melina G. Mouzala
June 2022

Abbreviations

Abbreviations for ancient Greek authors and works follow the Greek-English Dictionary by Liddell, Scott, and Jones [LSJ].

Citations of works in the chapter on Origen and citations of editions and collections of fragments and testimonia in other chapters use the following abbreviations:

Acad.	Cicero, *Academica*
Adv.Math. or AM	Sextus Empiricus, *Adversus Mathematicos*
AH	Epiphanius, *Adversus Haereses*
C.Cant.	Origen, *Commentarii in Canticum Canticorum*, Commentary on the Song of Songs
C.Io.	Origen, *Commentarii in Evangelium Iohannis*, Commentary on John
C.Matt.	Origen, *Commentarii in Evangelium Matthaei*, Commentary on Matthew
Cels.	Origen, *Contra Celsum*, Against Celsus
De vir. ill.	Jerome, *De viris illustribus*, On Illustrious People
Decr.	Athanasius, *De decretis Nicaenae Synodis*
Det.	Philo, *Quod deterius potiori insidiari soleat*, The Worse Person Usually Attacks the Better One
Deut.	Deuteronomy
Did.	Alcinous, *Didaskalikos*
Diss.	Epictetus, *Dissertationes ab Arriano digestae*
Enn.	Plotinus, *Enneades*, Enneads
Ep.	Jerome, *Epistulae*, Letters
Fat.	Cicero, *De Fato*, On Fate
FHS&G	Fortenbaugh, William W., Huby, Pamela M., Sharples, Robert W., and Gutas, Dimitri (1992; corrected edition 1993): *Theophrastus of Eresus: Sources for his Life, Writings, Thought and Influence*. 2 Volumes. Leiden and Boston: Brill.
Fr.Eph.	Origen (?), *Fragmenta in Sancti Pauli Epistulam ad Ephesios*, Fragments from the Exegesis of Paul's Letter to the Ephesians
Georg.	Virgil, *Georgica*, Georgics
H.Gen.	*Homiliae in Genesim*, Homilies on the Book of Genesis
H.Luc.	Origen, *Homiliae in Lucae Evangelium*, Homilies on the Gospel of Luke
H.Ps.	Basil of Caesarea, *Homiliae in Psalmos*, Homilies on the Psalms
HE	Eusebius, *Historia Ecclesiastica*, Church History
HE	Socrates, *Historia Ecclesiastica*, Church History
Hebr.	Epistle to the Hebrews (traditionally attributed to St Paul but probably not by him)
Hom.op.	Gregory of Nyssa, *De hominis opificio*, The Creation of the Human Being
John	the Gospel of John
In Arist. Top.	Alexander of Aphrodisias, *Commentarii in Aristotle's Topica*, Commentary on Aristotle's Topics
Isag.	Albinus, *Isagoge*, Introduction
Matt.	Matthew, Gospel of Matthew
Od.	*Odyssea*

Or.	Origen, *De oratione, On Prayer*
Nu.	Aristophanes, *Nubes*
Pan.	Gregory Thaumaturgus (?), *Oratio panegyrica in Origenem, Panegyrical Oration for Origen*
Periph.	Eriugena, *Periphyseon, On Natures*
Philoc.	*Philocalia, Anthology of Origen's Works*, composed (according to tradition) by Basil and Gregory Nazianzen
PL	*Patrologia Latina*
Plac. Hipp Plat.	Galen, *De placitis Hippocratis et Platonis, The Opinions of Hippocrates and Plato*
Protr.Mart.	Origen, *Protrepticus ad martyrium, Exhortation to Martyrdom*
Ps.	Psalm(s)
Pyrrh.Hypot.	Sextus Empiricus, *Pyrrhoniae hypotyposeis, Outlines of Pyrrhonism*
RN	Lucretius, *De rerum natura, On Nature*
Sat.	Persius, *Saturae, Satyrs*
SC	Sources Chrétiennes (series)
Sel.Ps.	Origen (?), *Selecta In Psalmos, Selected Passages on the Exegesis of the Psalms*
Simplicius, *in Ench.*	*Commentarius in Epicteti enchiridion*
SSR	*Socratis et Socraticorum Reliquiae* (collegit, disposuit, apparatibus notisque instruxit Gabriele Giannantoni, Bibliopolis)
Stoic.rep.	Plutarch, *De Stoicorum repugnantiis, The Contradictions of the Stoics*
Strom.	Clement of Alexandria, *Stromateis, Tapestry or Miscellany*
SVF	Arnim, Hans Friedrich August von (3 Vols.) and Adler, Maximilian (4 Vols.) (Eds.) ([1903–1905, 1924], 1964): *Stoicorum Veterum Fragmenta*. Leipzig: Teubner.
V.Plot.	Porphyry, *Vita Plotini*
Vid.	Ambrose of Milan, *De viduis, On Widows*
VP	Porphyry, *Vita Pythagorae*

Introduction

1 Definitions and Conceptions of Dialectic

Dialectic, also called dialectics, διαλεκτική (sc. ἐπιστήμη or τέχνη), originally means "the art of conversation or debate or, most fundamentally, the process of reasoning to obtain truth and knowledge on any topic".[1] According to other definitions of dialectic, it is "the science of conducting a philosophical dialogue (διαλέγεσθαι, to converse) by exploring the consequences of premises asserted or conceded by an interlocutor"[2] or "a method of philosophical argument that involves some sort of contradictory process between opposing ideas".[3]

The term was originally used to denote a form of logical argumentation but was later used to also designate a philosophical concept of evolution applied to diverse fields, including thought, nature and history.[4] For instance, in Kant, dialectic is the "logic of illusion" or the "logic of appearance" and it is the task of true philosophy "to reveal the places where reason transgresses its proper boundaries, producing the illusions of transcendental metaphysics".[5] In Hegel, dialectics "refers to the necessary process that makes up progress in both thought and the world" and "this process is one of overcoming the contradiction between thesis and antithesis, by means of synthesis; the synthesis in turn becomes contradicted, and the process repeats itself until final perfection is reached".[6] Karl Marx explained the change of social formations by invoking "dialectical relationships between productive forces and relations of production, between the basis and the superstructure". Friedrich Engels, after criticizing the idealistic dialectics of Kant, Schelling and Hegel, set out to prove that "materialistic dialectics provides the philosophical foundations for a

---

Note: I am deeply grateful to Professor Harold Tarrant for his comments and suggestions on the Introduction. Any errors and omissions in this section are my own responsibility.

1 Oxford Dictionary of Philosophy (3rd ed.).
2 Sedley (2015), "Dialectic", *Oxford Classical Dictionary*.
3 Maybee (2020), "Hegel's dialectics", *The Stanford Encyclopedia of Philosophy*.
4 Encyclopaedia Britannica, s.v.
5 Oxford Dictionary of Philosophy (3rd ed.). See also Bennett (1974), Loparic (1987), and Bird (2006).
6 Oxford Dictionary of Philosophy (3rd ed.). See also Forster (1993), Bencivenga (2000), and Wandschneider (2010).

comprehensive theory of development".[7] The Marxian kind of dialectics provided the framework for philosophical and social inquiry from which emerged critical theory in the first half of the 20th century. This theory, represented by German philosophers and social scientists known as the Frankfurt school, applied the Marxian dialectic to criticize social phenomena arising from contradictions within the capitalist economy.[8] Husserl and Heidegger established the foundations for a reformulation of dialectical inquiry within the frame of the phenomenological understanding of reality, and Gadamer inaugurated a new approach to dialectic entitled "philosophical hermeneutics". The latter is a dialectical method to be used for the analysis of historical and literary texts but is broadly aimed towards the understanding of reality through a different perspective.[9]

2 The Origin of Dialectic, the Presocratics, and the Sophists

The authors of this volume are concerned with dialectic in its original meaning, focusing on dialectic as the process of reasoning and arguing to obtain truth and knowledge on any topic of philosophical inquiry. Both Plato and Aristotle associate Zeno of Elea with the beginnings of dialectic. The former, implicitly in his dialogue *Parmenides*, attributes to Zeno a method of argumentation which constitutes a kind of dialectic, while Aristotle explicitly considers the Eleatic Zeno as the inventor or discoverer of dialectic.[10] In favor of this view, it can be said that Zeno practiced a kind of elenchus before the Socratic elenchus, following a procedure which consists in establishing a thesis by refuting the contradictory thesis. The refutation of a thesis was being implemented by trying to show that the thesis in question was leading to intrinsically contradictory consequences. For example, if things are many, then things would be both similar and dissimilar at the same time. This method implies the validity of the principle of non-contradiction.[11] His antinomies derived contradictory consequences from a disputed hypothesis.[12] According to one line of interpretation, Zeno was in general applying the reduction

---

[7] Hörz (1987), p. 493. The Marxian dialectic is not just an epistemological or ontological theory, offering a way to acquire knowledge, but it also necessarily involves praxis; see Dybicz-Pyles (2011), p. 303.
[8] Dybicz-Pyles (2011), pp. 303–304.
[9] Dybicz-Pyles (2011), pp. 304–313.
[10] Plato, *Parmenides* 127d–128e and 135c–136c. Aristotle, *Sophist* (Frg. 65 R³). Cf. Diog. Laertius 9.25.
[11] Berti (1978), p. 353.
[12] Sedley (2015).

of a thesis to impossibility.¹³ So, his paradoxes of divisibility and movement would be better construed as simple derivations of impossibility (ἀδύνατον) than reductio ad absurdum.¹⁴

Ancient Greek dialectic has not only the narrow sense of a philosophical method that aims at establishing the foundations of knowledge or science based on the examination of the argumentation towards an interlocutor, it also has a significant concern for ontology, for principles or for beings, for ideas and forms.¹⁵ The latter concern of dialectic places the emphasis on the need for the investigation not only of the structure of an argument and the counter-argument, not only of the structure of a dialogue, of the structure of thought and its relation to discourse, and not only of the structure of soul or self and its understanding of reality, but finally also of the structure of cosmos as well.

Given that the major problems driving ancient philosophical thought were the questions whether the principles of things are one or many and whether there is one Being or many, one of the central issues of ancient Greek dialectic is the study of the relation between the one and the many.¹⁶ Heraclitus is a philosopher who places this problem in the center of his philosophy, by thematizing it in his theory of flux and his conception of the unity of opposites. A significant proportion of the preserved fragments of his work are devoted to the theory of identity and unity of the opposites. The examples he offers are presented as empirical proofs of the fundamental theoretical principle which, at the beginning of his treatise *On Nature*, recommends to humans "to know all things as one".¹⁷ His theory of flux (for example, that one cannot step twice into the same river, for fresh water is ever flowing; and even when one steps into the same river, other and still other water will flow over one¹⁸) offers the foundations of the epistemology which, according to Aristotle, motivated the formation of the Platonic theory of Forms. No sense-perceptible thing can be grasped as the object of knowledge because it is constantly in flux. From this assumption emerges the epistemological postulate that true objects of knowledge must have features of ontological stability.¹⁹ Furthermore, in Heraclitus, the identity of a given river remains fixed although it is constantly changing. This idea alludes to the preservation of structure within a process of flux, and further to the contradistinction between a unitary form which is maintained and its ma-

---

13 McKinney (1983), p. 180. For the role of Zeno, see also Dorion (2002), pp. 200–208.
14 Mathieu (2014), pp. 393–434.
15 Fink (2012), pp. 1–2.
16 Plato, *Sophist* 253d–e and *Philebus* 15d–16d; Aristotle, *Physics* 184b 22–25 and 185b 26–186a3.
17 Lebedev (2014), p. 23.
18 Heraclitus, DKB12; DKB91; cf. DKB49a.
19 Aristotle, *Metaphysics* 1078b12–17.

terial embodiment which is constantly lost and replaced.[20] The idea that the opposites can co-exist in a unity, in the sense that they belong to the essence of a thing, can be construed as a foreshadowing of the Aristotelian theory of potentiality and theory of change. These theories presuppose "an ambivalence of essence" in the sense that a thing's very being may require the coexistence within it of opposite potentialities.[21]

This ontological approach not only endows with new ramifications the debate between monism and pluralism, which is one aspect of the central and perennial problem of dialectic, namely the search for the relation between the one and the many, but it also has an impact on this aspect of dialectic which deals with the structure of argumentation and with every single argument when one speaks, not only about the principles of things, or the opposites and the procedures of change within nature and cosmos, but also simply regarding anthropocentric matters, and finally on any matter. The Parmenidean thesis that there is a constant interrelation between being, thinking and speaking along with his denial of *ouk estin*,[22] namely the denial of negative statements about the nature or essence of things, provoke reactions that boost the development of dialectic and establish the foundations of the search for the use and function of the verb *estin* and its ontological significance.[23] According to Furley, in the sentence πολύδηριν ἔλεγχον of Frg. 7, the original meaning of ἔλεγχος, "shame" or "disgrace", had been replaced by that of "refutation"[24]. In an earlier period there had also been a neuter form of *elenchos*, τὸ ἔλεγχος, and a feminine noun, ἡ ἐλεγχείη, but their usage later declined.[25] Lesher, who has researched the semantic history of this word, notes that there is no reason to doubt that 6th- and 7th-century writers used both the neuter noun and the verb form to convey the idea of shame and disgrace.[26] He suggests that there is a literary use in the background of the word which shows that when a masculine form of *elenchos* appears, it is not surprising that its meaning is not "disgrace", but a "test" or a "contest" in which one incurs or avoids disgrace.[27] He further notes that "the idea of *elenchos* as a testing of a person's truth-

---

20 Kahn (1979), pp. 166–168.
21 Hussey (1999), pp. 97–99.
22 Parmenides, DK29 B3, B6. 1–2, B8. 7–9.
23 See Austin (2007), Kahn (2009), and Bredlow (2011).
24 Furley (1973). Cf. Furth (1968).
25 Hofmann (1950), p. 88. Homer, *Il.* XXII, 100; *Od.* XXI, 329.
26 Hesiod, *Theogony* 26; Theognis, 1011; Homer, *Il.* XI, 314–315; *Od.* XXI, 424–425. There are also the adjectives ἐλεγχής, -ές (Homer, *Il.* IV, 242) and ἐλέγχιστος (Homer, *Il.* IV, 171). See Lesher (2002), pp. 22–23.
27 Lesher (2002), p. 24. For the history of the word and the character and legacy of the Parmenidean *elenchos*, see Lesher (2002), pp. 22–35.

fulness or character becomes a common refrain in the works of fifth-century writers".[28]

The authors of this volume contribute with chapters which are concerned with *Ancient Greek Dialectic* from the period of Classical Antiquity onwards, beginning with the great philosophers whose presence and work has been identified with dialectic throughout the history of philosophy, such as Socrates and Plato, their opponents the Sophists, as they have been alluded to in Plato's work, the Megarians amongst the Socratics, as well as Aristotle. Diogenes Laertius supports Protagoras as the inventor of dialectic, the rival to Zeno's candidature for it. In Protagoras we may trace a reflection of the structure of nature, in which the opposites are predominant as principles of any change, within the structure of logos. In his philosophy, the opposites, which according to the testimony of Aristotle, were considered by all his predecessors to be predominant within the processes of nature as principles of any change,[29] undertake the primary role in the realm of logos. According to Diogenes, Protagoras was the first to say that on every issue there are two arguments opposed to each other; and these he made use of in arguing by question and answer, a practice he originated.[30] He was therefore the first to introduce the Socratic type of argument and the first to adopt in discussion the argument of Antisthenes,[31] which attempts to prove that contradiction is impossible.[32] The technique of opposed arguments, which is probably discussed in his lost work *Antilogies*, is with certainty attributed to Protagoras, but there is also some evidence that the feature of using pairs of arguments was to a certain degree characteristic of this period.[33] Representative of this technique of argument was an anonymous sophistic treatise, entitled *Dissoi Logoi* or *Dialexeis*, which was written mainly in the Doric language and is probably dated to the beginning of the 4th century.[34] The art of antilogical contradiction, the art of antilogy and disputation, was mainly employed by the sophists as part of the art of eristic, as illustrated in Plato's *Sophist* 226a. Ἀντιλογία is an ancient concept already mentioned in Herodotus, whereas the word Eristics derives from the term ἔρις (*eris*), meaning "strife" or

---

28 Lesher (2002), p. 25.
29 Aristotle, *Physics* 188a 19–27.
30 Diog. Laertius 9. 51 (trans. O'Brien 2001).
31 Prof. Harold Tarrant commented that this is not so much an argument as a "tactical ploy".
32 Diog. Laertius, 9.53; cf. O'Brien (trans.) 2001. Cf. Plato, *Euthydemus* 286c.
33 For example, see Euripides, *Antiope*, Frg. 189 (Kannicht) or the two personified logoi in Aristophanes' *Clouds*, the Just and the Unjust Argument; cf. Plutarch, *Life of Pericles* 4. 5. See Kerferd (1981), pp. 84–85.
34 Sprague (1968), Kerferd (1981), Gera (2000), and Bailey (2008).

"competition".³⁵ The latter, as depicted in both Plato's *Euthydemus* and the *Sophist* was a version of dialectic, according to which the sophists competed against one another in disputation, employing any verbal trick or even false assumptions called 'sophisms', in order to gain victory over the dialectical opponent.³⁶ The antilogical method was used in eristic exercise, in forensic training, in juridical theory, and later in skeptical inquiry.³⁷ It has also been claimed that Socrates' method of *elenchos* (dialectic refutation), to the degree that he too sought inconsistencies and contradictions in the arguments of his interlocutors, can plausibly be derived from the method of antilogic of the sophists.³⁸

3 Socrates, Plato, and the Megarians

Socrates had an idiosyncratic manner of discourse that he often called διαλέγεσθαι (conversing), which means that he would be involved in a dialogue with a particular individual and start questioning on a specific subject. So, his discourse was generated in relation to specific persons on particular occasions, where a philosophical question emerged from the very first interaction with his interlocutor.³⁹ The term διαλέγεσθαι acquired a connection with philosophy earlier than Socrates, and later the method called διαλεκτική, which was formed from the adjective διαλεκτικός, became the distinctive method of philosophy. There was a time when scholars used the term *dialectic* to denote generally the method of philosophizing in all of Plato's dialogues, both the so-called early, as well as the middle and later dialogues. But nowadays, scholars tend to call the method that Socrates is depicted as using in the early dialogues "the elenchus".⁴⁰

We can find a first illustration of Socrates' method in the *Apology* 20c–23c. There, Socrates enters into a dialogue as a questioner and aims to examine (ἐξέτασις) an interlocutor from whom he hopes to elicit knowledge while scrutinizing him during the *elenchus*.⁴¹ Perhaps the most influential characterization of the

---

35 Herodotus, 8. 77. 13. Hesiodus, *Opera et dies* 11–12. On Eristic, see also Delcomminette and Lacanche (2021).
36 Luz (2012), p. 133.
37 Luz (2012), p. 134.
38 For the affinities between these methods, see Nehamas (1990). See also Gulley (1968), p. 31, and Luz (2012), p. 134.
39 Cf. Redfield (2017), p. 129.
40 Baltzly (2012), p. 159.
41 Cf. Fink (2012), p. 3. See also Benson (2002), p. 109.

Socratic elenchus has been offered by Vlastos,[42] who in 1983 criticized Robinson, who, in his *Plato's Earlier Dialectic*, maintained that "Plato habitually thought and wrote as if all elenchus consists in reducing the thesis to a self-contradiction".[43] Vlastos offers another description, stating that "Socratic elenchus is a search for moral truth by adversary argument in which a thesis is debated only if asserted as the answerer's own belief, who is regarded as refuted if and only if the negation of his thesis is deduced from his own beliefs".[44] Furthermore, he asserts that it would be wrong to assimilate the Socratic elenchus to Zeno's dialectic, since "Zeno's refutands are unasserted counterfactuals". On the other hand, Socrates "does not debate unasserted premises, but only those asserted categorically by his interlocutor, who is not allowed to answer contrary to his real opinion".[45]

Vlastos' approach to the Socratic elenchus stimulated much discussion,[46] whereas in more recent scholarship the questions arise whether Socrates' method should be called elenctic,[47] and furthermore whether Socrates, indeed, does have a method. The latter question is the subject of a collective work published by Gary Alan Scott in 2002.[48] The second part of this book is devoted to the re-examination of Vlastos' analysis of "the Socratic elenchus" in his classic essay. Brickhouse and Smith note that Vlastos' understanding of the *elenchos* has failed to generate even a consensus, much less a universal agreement among scholars, and that several alternative accounts have been offered, although none of these has received much support. Their view is that there can be no solution to the "problem of elenchus" and no single analysis of elenctic arguments, for the simple reason that there is no such thing as "the Socratic elenchos"; rather it is an artifact of modern scholarship.[49] Carpenter and Polanski express the view that refutation is merely one function of Socrates' investigation, though a most crucial one since all of his other purposes are accomplished through refutation. These include to investigate claims of expertise, to examine the life of his interlocutor, to eliminate conceit of wisdom, to establish certain views and to stimulate the interlocutor to reorient his life, amongst others. But since Socrates has no single method of refutation or cross-ex-

---

[42] Baltzly (2012), p. 159.
[43] Robinson (1953²), p. 28. See also Vlastos (1983), p. 29.
[44] Vlastos (1983), p. 30.
[45] Vlastos (1983), p. 29. See also Vlastos (1975) on Plato's testimony concerning Zeno of Elea.
[46] Starting from Kraut (1983), the first published response to Vlastos (1983).
[47] Tarrant (2000).
[48] Scott (2002). Vlastos (1983, pp. 27–28) also believes that the 'Socrates' who speaks for Socrates in the early dialogues never uses the word μέθοδος and never discusses his method of investigation.
[49] Brickhouse and Smith (2002), pp. 146–147.

amination, we cannot have an all-embracing view of his elenctic discussions.[50] Tarrant suggests that *elenchos* was never a term used either by Plato or his later interpreters, for all of Socrates' investigations were through question and answer, but both the noun and its corresponding words were applied only to those examples of interrogation whose purpose was refutation.[51] Benson also believes that we should not suppose that whenever Socrates engages in philosophical discourse, he must be behaving elenctically. He stresses that the problem of *elenchos* requires that we understand it as a unique form of argument with unique features. He also maintains that it has a very unique constraint on premise acceptability, which he calls "the doxastic constraint": no other property is thought necessary for premise acceptability, than being believed by the interlocutor. This is a necessary and sufficient condition as well within the frame of *elenchos*, for premise acceptability.[52] But the problem is that being believed by the interlocutor is a property of the apparent refutand as well. So, as Benson emphasizes, "an individual elenctic encounter can only establish the inconsistency of the premises and the apparent refutand. It can only establish an inconsistency among the beliefs of Socrates' interlocutor" (Benson 2002, p. 105). In one of his later works on the Socratic elenchus, Benson insists that in the elenctic dialogues Plato presents us with a coherent and distinctive Socratic method which, although it is not the only method that Socrates practices, it does tend to be the predominant.[53]

Regarding Socrates' concern with definition, which Aristotle highlights as a central characteristic of Socrates' distinctive method, in recent decades the view of the priority of definitional knowledge to Socrates in the Socratic dialogues has been disputed.[54] More plausible than attributing to Socrates the commitment to a thesis of priority of definitional knowledge is to recognize that he is committed to a stronger and deeper conception of knowledge than the ordinary one, or the knowledge that false experts claim to possess.[55]

An extremely important aspect of the Socratic dialectic is self-examination as a way of reaching the truth, which is somehow in oneself. Self-examination in this

---

[50] Carpenter and Polanski (2002), pp. 89–90.
[51] Tarrant (2002), p. 63. But he confesses (Tarrant 2015, p. 165) that regardless of Plato's use of the term, it is clear that his Socrates adopts a distinctive approach to refutation that merits some technical name to describe it.
[52] Benson (2002), pp. 105 and 107. See also the citations to Benson (1987 and 2000) on p. 105, n. 11.
[53] Benson (2011), pp. 179–198.
[54] Benson 2011, pp. 193 and 195–196. For the nature of the Socratic definition, see Vlastos (1981), Benson (2000, pp. 99–111), and Wolfsdorf (2003a and 2003b). On the "ti esti question" see Politis (2012).
[55] Cf. Benson (2011), pp. 194–198.

case is equated with the struggle to achieve knowledge of oneself, not just any self but one's true self, which is revealed by the procedure of discarding false beliefs and identifying the ones that are true.[56] Knowledge of ourselves is a matter that gains priority among the aims of the Socratic dialectic, and furthermore, knowledge of ourselves proves to be a matter of knowing that ourselves are identical with our souls.[57] This recognition in turn places at the center of the Socratic search for knowledge and truth, the life-long effort for care of the soul.[58] When examining the kind of *elenchos* described in the *Sophist* (230d), Rowe puts forward the idea that it is a "challenge", not a "refutation", although in this context it happens to lead to refutation, and it is an activity rather than a method.[59] Fink aptly stresses that "to Socrates, question-and-answer dialectic is as much a form of conducting one's life as it is a certain form of conducting an argument".[60]

According to an interpretative line, if one seeks to find the strongest thread that binds the Socratic and the Platonic dialectic, then one has to acknowledge that for Plato the fundamental meaning of conducting a dialectical debate is based on the Socratic postulate of giving an account or taking one up, while conducting an examination in question-and-answer form.[61] If one claims knowledge of something then he has to give an account of it in order to prove and certify knowledge of it.[62] Dorter stresses that Plato presented his philosophy not technically but contextually, which means that there is a variety of vocabulary and terms of reference from one dialogue to another, in accordance with the question under discussion and the persons participating in the dialogue. This is the reason why, in his approach, we must have reservations about asking a global question like "what Plato means by dialectic?".[63] Whereas Robinson, in his classic work *Plato's Earlier Dialectic*, states that the word "dialectic" had a strong tendency to mean in Plato "the ideal method, whatever that may be", Benson stresses that the Greek substantive [ἡ] διαλεκτική and its cognates occur only 22 times in the Platonic corpus and only once in dialogues that are considered by Robinson to be early (*Euthd.* 290c5).

---

56 Rowe (2011), p. 203. On self-refutation, see Castagnoli (2012).
57 Cf. *Alcibiades I* 130b–131c.
58 Rowe (2011), pp. 204–205.
59 Rowe (2015), p. 150, n. 8.
60 Fink (2012), p. 3. See also Plato, *Apology* 20 c–d; 28e; *Laches* 187e–188a.
61 Cf. Fink 2012, p. 3.
62 Plato, *Phaedo* 76b–c.
63 Dorter (2006), p. 9.

Furthermore, over a third of those occurrences are concentrated within six Stephanus pages in the *Republic* (531d9, 532b4, 533c7, 534b3, 534e3, 536d6, 537c6, 537c7).[64]

In the *Cratylus* 390c, we can see that a dialectician is someone who knows how to pose questions and give answers. Also, in 398d–e it is said that the *heroes* were probably named as such because they were dialecticians, namely, able to ask questions. When their name is spoken of in the Attic dialect, the heroes turn out to be orators and askers of questions, for εἴρειν (ἔρεσθαι, ἐρωτᾶν [ask questions]) is the same as λέγειν (speak). So, the heroes are dialecticians, and the latter are persons with the ability to ask questions. In the *Meno* 75d, there is the statement that the more dialectical way to conduct a philosophical dialogue is not merely to answer what is true but also to make use of those points the questioned person acknowledges he knows. In the *Republic* 537c, a dialectician is someone who can see things synoptically. In the *Phaedrus* 265d–266c, *dialecticians* are referred to as those who practice the methods of collection and division, namely the two methodological principles or the two forms of dialectic. Collection consists in the act "of perceiving and bringing together in one idea the scattered particulars, that one may make clear by definition the particular thing which he wishes to explain" (trans. Fowler 1925). Division, on the other hand, consists in the practice "of dividing things again by classes, where the natural joints are, and not trying to break any part, after the manner of a bad carver" (trans. Fowler 1925).[65] In the *Sophist* 253d, the person who has expertise in dialectic is able to divide things by kinds and will not think that the same form is a different one or that a different form is the same. He is capable "of adequately discriminating a single form spread out all through a lot of others, each of which stands separate from the others. In addition, he can discriminate forms that are different from each other but are included within a single form that is outside them, or a single form that is connected as a unit throughout many wholes, or many forms that are completely separate from others" (trans. White 1993). And this dialectical activity is assigned only to someone who philosophizes with purity and justice. In the *Statesman* 285a–d, division and collection are described and explained through the reference to two kinds of the science of measurement. Moreover, it is said that the investigation of the statesman has been undertaken in order to make us better dialecticians about all subjects. In the *Philebus* 16c–17a, the theory that all things which are ever said to exist are sprung from one and many and have inherent in them the finite and the infinite, is presented as a tradition handed down by the ancients who were better than we

---

64 Robinson (1953), p. 70. See also Benson (2006), p. 86. More frequent is the occurrence of the substantival infinitive τὸ διαλέγεσθαι, but it is often difficult to determine when it has a technical usage.

65 For division as a method in Plato, see Fossheim (2012).

and lived nearer the gods. Immediately after this statement, Plato presents the task of any man who searches things in a dialectical way; we must not apply the idea of infinite to plurality until we have a view of its whole number between infinity and one. This is the mode of investigating, learning, and teaching one another that the gods handed down to us (trans. Fowler 1925). So, dialectic is discerning the essence of things and constitutes the recommended way of bridging the one and the many.[66] Consequently, the dialectician is able to give an account of the essence of each thing.

Robinson proposes a distinction between the early and middle dialogues, claiming that during the early period Plato gives prominence to method but not to methodology (namely theories of method), whereas during the middle he gives prominence to methodology but not to method.[67] Also, the method of *hypothesis* illustrated in *Meno* 86e–87b and in *Phaedo* 99e–100b has traditionally been distinguished from the method typically described in the *Republic* (506a–539e), although the latter also includes references to the *hypotheses* of the mathematicians and culminates with the process that reaches the unhypothetical first principle.[68] In recent Anglophone Platonic scholarship, there are two competing interpretative models, developmentalism and unitarianism.[69] Vlastos represents the most extreme version of the first model, claiming that the philosophical views expressed in the elenctic dialogues could not have been derived from the same mind as the philosophical views of the middle or classical dialogues, unless it was the mind of a schizophrenic.[70] Benson distances himself from all these kinds of approaches to the Platonic method and remains neutral between the two main models. He also agrees that the method of *hypothesis* supplants, for whatever reason, the *elenchos* as Plato's method of knowledge acquisition in the so-called middle or classical dialogues. However, he stresses that "the two sets of dialogues are not inconsistent-at least with respect to method. Indeed, the role of the *elenchos* in eliminating the

---

[66] For a concise exposition of all the relevant Platonic passages, see Dorter (2006). See also Baltzly (2012).
[67] Robinson (1953), pp. 61–62.
[68] Cf. Baltzly (2012), p. 159.
[69] See the concise presentation of the basic features of each interpretative model in Benson (2015), pp. 8–11. In parallel, there are two other competing interpretative models; there are scholars who emphasize the dialogues' dramatic creations and strategies, and those prefer to follow in understanding them the technique of literary analysis, whereas other scholars place the emphasis on the abstract argumentation of the dialogues, and those favor the technique of logical analysis. For a brief overview of those models, see Dancy (2004), p. 1; for a discussion of this opposition, see Stokes (1986), pp. 1–35. On prolepticism and perspectivism, see Gonzalez (2017).
[70] Vlastos (1991), p. 46.

false conceit of knowledge is included in the dialectical method of the *Republic*".[71] By investigating and analyzing the philosophical method in the three key middle dialogues, the *Meno*, the *Phaedo*, and the *Republic*, Benson shows that while differences in Plato's explicit recommendations of method in these three dialogues are apparent, there are certain core features that remain constant. His analysis emphasizes that what is important to realize is the continuity between the methods proposed in the *Meno*[72] and the *Phaedo*[73] and the method of dialectic in the *Republic*. According to him, in all these three dialogues dialectic consists of two fundamental processes; on the one hand, a process of identifying and drawing out the consequences of propositions known as *hypotheses*, and on the other, a process of verifying or confirming or justifying the truth of the *hypotheses*.[74] Dianoeticians, namely those who follow the dianoetic method, geometers and mathematicians, fail to practice dialectic by applying correctly the method of *hypothesis*, because in their inquiry they consider as known, as an *archē* of their investigation, what is unknown and still in need of confirmation. They disregard the unsuitability of sense-experience and the prerequisite that the confirmation of the *hypotheses* they employ can be obtained only through knowledge of the unhypothetical first principle, which is identified with the Form of the Good.[75] One's inquiry and acquisition of knowledge remains incomplete until one reaches the unhypothetical first principle of everything. So, dialectic must be considered inseparable from the Good and goodness.[76]

Diogenes Laertius states that the name "διαλεκτικοί" was given to those who occupied themselves with arguments, with extreme subtlety (περὶ τὴν τῶν λόγων τερθρείαν).[77] Those who were called "dialecticians" for engaging in dialectic, for instance the Megarians, were also called by their opponents, "eristical" or "sophistic".[78] In this volume there is also a chapter on the Megarians, who among the Socratics formed a group who were associated in a unique manner with the development of dialectic. Euclides of Megara, perhaps one of the oldest members of the

---

71 Benson (2015), p. 11, n. 29.
72 See also Karasmanis (1987), Benson (2003), Wolfsdorf (2008), Benson (2015), Iwata (2016), Judson (2017), and Ionescu (2018).
73 See also Gallop (1975), Bostock (1986), Rowe (1993), van Eck (1994), Kanayama (2000), Benson (2015), and Ferber in this volume.
74 Benson (2006), pp. 85 and 90.
75 Plato, *Republic* 510c–511d and 533b–e. Benson (2015, pp. 248–249) stresses that the procedure applied by the dianoetic and dialectic method is the same but that the difference lies in how each treats its *hypotheses*.
76 Benson (2015), p. 249. See also Dorter (2006), p. 10.
77 Diog. Laertius, 1. 17.
78 Allen (2018), pp. 31–32.

Socratic circle, was considered the founder of the Megaric school.[79] In the beginning of his section on Euclides, Diogenes Laertius states that Euclides' successors were called *Megarikoi*, then *Eristikoi*, and later *Dialektikoi*.[80] While it has been assumed that these were three successive labels applied to a single school and hence that anyone called a *Dialektikos* was really a Megarian philosopher masquerading under this name, Sedley argued that it is also possible that Diogenes described here a *diadochē* rather than a single school.[81] Ancient testimonies ascribe to Euclides and his followers Eleatic views, but some modern scholars have expressed reservations about this.[82] Generally though, the Megarians were thought to be influenced by the Eleatic view that being is one, and it is held that Euclides applied this oneness to the good and further supported that the opposite of the good cannot exist.[83]

4 Aristotle, Peripatetics, Middle Platonists, Christian Platonists

Aristotle, in his *Topics*, a treatise dedicated to the art of dialectic, examines the use of arguments in which the propositions rely upon commonly held opinions or endoxa.[84] At the beginning of his treatise, he states that its purpose is "to find a line of inquiry whereby we shall be able to reason from opinions that are generally accepted about every problem propounded to us" (*Top.* I. 1. 100a 18–20, trans. Pickard-Cambridge). According to a line of interpretation, the main difference between Plato and Aristotle with regard to dialectic can be traced in their assertions about the uses of dialectic.[85] For Aristotle, dialectic is useful for exercise, for casual encounters and with regard to the philosophical forms of scientific knowledge.[86] Two views regarding Aristotle's dialectic have been widely accepted in contemporary studies. The first is that Aristotelian dialectic was a technique of arguing from com-

---

79 On Euclides of Megara, see Boys-Stones and Rowe (2013) and Brancacci (2017).
80 Diog. Laertius, 2. 106.
81 Sedley (1977), p. 75. On the problem of the Megaric "School", see also Allen (2018).
82 See Allen (2018), p. 22, n. 13.
83 Cicero, *Academica Priora* 2. 129. See also Rankin (1983), p. 193, and Marsico in this volume.
84 These endoxic premises must express opinions that are acceptable either to all men, or to the majority, or to the wise, either to all of them or to the majority of them or the most renowned; see Aristotle, *Top.* I. 1. 100b 21–23.
85 Fink (2012), p. 13.
86 Aristotle, *Topics* I. 2. 101a 25–28. On Aristotle's dialectic, see Evans (1977), Irwin (1988), Bolton (1990), Slomkowski (1997), Spranzi (2011), pp. 11–38, Fink (2012), Weigelt (2017), and Rapp (2018).

mon beliefs, accepted opinions or reputable views. The second is that according to Aristotle, dialectic provided the way to the first principles of the sciences, or at any rate of some of them.[87] One of the most famous statements of Aristotle on dialectic is the one we see in *Top.* I.2. 101a36–b4; since it is proper for dialectic to examine, it paves the path towards the principles of all inquiries. The meaning of "principles" here is "first principles", whereas the meaning of "all inquiries" is "all specific sciences".[88] According to Aristotle, the primary function of dialectic may be to exercise, whereas exercise serves one of the most important philosophical uses of dialectic: going through ἀπορίαι.[89] Through the practice of διαπορῆσαι, namely, of working through the difficulties on either side of a philosophical or scientific topic, an intimate connection is established between dialectical method and the use of ἀπορίαι.[90] The dialectical system of *topoi* and the dialectician's expertise in formulating syllogisms, along with the ability to argue for any given thesis and raise difficulties or *aporiai*, constitute the dynamic and multifaceted framework of the Aristotelian dialectic.[91] However, in current research, there is much debate regarding the purpose and scope of the so-called "endoxic" method, while there is also skepticism about whether Aristotle really outlines or employs such a method throughout all of his texts or instead employs it only for particular topics.[92]

The second part of this volume focuses on issues related to the reception, development, and influence of ancient dialectic during Late Antiquity and the Byzantine period. Predominant in the classical tradition of dialectic is the debate between Plato and Aristotle regarding the epistemological value of the use of dialectic, an issue which has also determined the way in which the Platonic and Aristotelian tradition received and developed dialectic after Plato and Aristotle.[93] Whereas Plato considers dialectic as the supreme science through which one can reach the absolute or unhypothetical principle of all things, which is identified with the Form of the Good, Aristotle believes that such a universal discipline which

---

[87] Smith (1993), p. 335. Irwin (1988) made a distinction between a 'strong' dialectic, developed by Aristotle for the science of first philosophy and his later works such as the *Metaphysics*, and a 'weak' or 'pure' dialectic, described in the *Topics* and elsewhere. However, it has been claimed that it is false that there is a clear break between Aristotle's earlier and later works with regard to the dialectical method; see Sim (1995), p. 2.
[88] Paglieri (2014), p. 393.
[89] Fink (2012), p. 13.
[90] Rapp (2018), pp. 112–113.
[91] On the different attempts of scholars to define dialectical *topos-topoi* in Aristotle's dialectic, see Drehe (2011).
[92] See Frede (2012), Devereux (2015), DaVia (2017), Karbowski (2015 and 2019), Falcon (2019), and Griffin in this volume.

investigates all objects and all principles of science cannot be a science, strictly speaking, but only a method restricted to arguing on the basis of endoxa.[93] Gourinat notes that since the time of Plato, dialectic was considered to be a science, as we can see in the *Sophist* 253d, but Aristotle removes from it the status of a science and attributes this title only to the analytic science, as referred to in the *Rhetoric* I, 4, 1359b10. The qualification of a science is attributed again to both logic and dialectic by the Stoics, since the latter was thought to be a part of logic.[94] The Stoics consider dialectic not only as a constitutive part of philosophy but also as a true science and a significant aspect of wisdom, which can be applied both in the theoretical and practical realms.[95] For them, only the wise is *dialektikos*.[96]

Alcinous, in his *Handbook of Platonism*, representative of Middle Platonic attempts to present a summary of the doctrines of Plato or a manual of instruction in Platonism or an introduction (Εἰσαγωγή) to the teachings of Plato, devotes the fifth chapter of this work to dialectic.[97] In the introduction to the chapter he states that the fundamental purpose of dialectic, according to Plato, is firstly the examination of the essence of every thing whatsoever and then of its accidents. Alcinous explains that dialectic enquires into the nature of each thing either "from above", by means of division and definition, or "from below", by means of analysis. Furthermore, dialectic examines accidental qualities which belong to essences, either from the standpoint of individuals, by induction, or from the standpoint of universals, by syllogistic. He concludes that, logically, dialectic comprises five procedures: division, definition, analysis, and in addition induction and syllogistic.[98]

The conception of the status of dialectic as an *organon* that serves all branches of Aristotelian philosophy and especially first philosophy is a significant thesis attributed to later Peripatetics within the Aristotelian tradition.[99] Alexander of Aphrodisias, the exegete, the commentator par excellence, has a distinctive approach to Aristotle's thought, based on the assumption that the same logical, physical, and

---

[93] Aristotle, *Topics* I.1, 100a 25–30, and *Posterior Analytics* I. 6, 74b13–26 and I. 18; 81b 19–23. Cf. DaVia (2017), p. 385 and Bénatouïl (2018), pp. 13–14. While scientific inquiries involve demonstrative syllogisms, dialectical inquiries involve dialectical syllogisms which rely upon endoxa; see also Kupreeva (2018), pp. 231–232.
[94] Gourinat (2016a), p. 17. See also Crivelli (2016).
[95] Bénatouïl (2018), Gourinat (2016b, p. 378, and 2018), and Ierodiakonou (2018).
[96] Cf. Alexander of Aphrodisias, *in Top.* 1. 10–14.
[97] One must not overlook that many Aristotelian and Stoic doctrines and phrases are attributed to Plato. See Witt (1971), p. 8, and Dillon (1993), pp. xiii–xiv.
[98] Translation by Dillon (1993). Dillon (1993, p. 72) notes that the term 'dialectic' here is made to serve for the whole field of logic since the topic of logic as such was not recognized by Plato.
[99] Guyomarc'h in this volume. On dialectic in the early Peripatetics, see Kupreeva (2016) and Crivelli (2018). On dialectic as a Peripatetic method, see Baltussen (2000).

ontological principles apply to all parts of Aristotle's world.[100] An example of this tendency is Alexander's attempt to present Aristotle's metaphysics as a demonstrative science.[101] As Kupreeva notes, on Alexander's view, every science, including first philosophy, is demonstrative and definitional. Dialectic contributes towards knowledge of the first principles.[102] Alexander, in his commentary on Aristotle's *Metaphysics*, explicitly refers to the connection of the utility of two methods, the aporetic method and the method of dialectic, as illustrated in the *Topics*. He interprets the Aristotelian *aporiai* in *Metaphysics Beta* as logical and dialectical arguments from endoxic premises.[103] Alexander's *aporiai* with multiple solutions, as illustrated in Alexander's *Quaestiones*, constitute a special framework for Aristotelianism as exegetical tradition, that retains the exploratory and initially open character of *aporiai*.[104]

Origen of Alexandria, the great Christian philosopher of the Patristic period and one of the most prominent Christian Platonists, was Alexander's quasi-contemporary.[105] He was a teacher in Alexandria and in Caesaria, and he taught all the philosophical schools along with Christian philosophy-theology.[106] He wrote exegetical works on the Bible as well as other treatises and was an admirer of Plato. He also intended to establish an "orthodox" Christian Platonism, against "pagan" and "Gnostic" Platonism and non-Platonic philosophies.[107] Origen knows Plato's dialectic and attributes it to the Scripture while also employing Aristotle and the Stoics' logic-dialectics within the frame of his own zetetic method.[108]

5 Neoplatonic Philosophers and Commentators

As we have previously mentioned, there was a crucial debate between the Peripatetics and the Stoics on the question of whether logic is a part of philosophy or just an instrument (*organon*) of it. According to Alexander of Aphrodisias, those who defend the thesis that logic is a part of philosophy against the thesis that it is an *organon*, suggest that within logic there are things that are not useful, so they can-

---

100 Cerami (2016), p. 164.
101 Bonelli (2001).
102 Kupreeva (2018), pp. 232–233 and 236.
103 Alexander in *Top.* 32.12–34. 5 and in *Metaph.* 173. 27–174.4. Kupreeva (2018), pp. 239–240.
104 Fazzo in this volume; see also Guyomarc'h in this volume.
105 Ramelli (2014) has tried to show that Alexander exerted some significant influence on Origen.
106 Ramelli (2014), p. 240.
107 Ramelli (2017), p. 2.
108 Ramelli in this volume.

not belong to an instrument of philosophy. Those who defend the thesis that logic is a part of philosophy are precisely the Stoics.[109] According to Plotinus, logic is an instrument of philosophy, whereas dialectic is the precious part of philosophy. In his treatise *On Dialectic*, Plotinus seeks to establish the relation between dialectic and philosophy. To the question whether dialectic and philosophy are the same, he answers that we must not think of dialectic as the mere tool of the philosopher, because it does not consist of bare theories and rules, but deals with real beings.[110] Given that it is concerned with real beings and with what has the greatest ontological value, dialectic is not equal with other parts of philosophy, such as physics and ethics. So, one must assume that it is that part which makes the whole of philosophy worth engaging with.[111] At the beginning of his short and concise treatise on dialectic, Plotinus determines the object of dialectic, claiming that it is the art or method or discipline which will allow the soul to attain the ascent to the Good, the First Principle.[112] Dialectic, on the one hand, uses the Platonic method of division to distinguish the Forms, to discern the essence, and to work towards the primary genera that it combines in the intellectual mode (νοερῶς) until it goes through the entire intelligible world; on the other hand, it analyzes the Forms until it returns to the principle which is the starting point.[113] Dialectic is an extremely important method in Plotinus' philosophy, on the one hand because it contributes to the ascent of the soul to the Good, and on the other because it is an activity of Intellect, whereas logic is limited to the level of soul.[114]

Proclus, in the prolegomena of his Commentary on Plato's *Cratylus*, draws a sharp distinction between Platonic and Peripatetic dialectic when he discusses whether the dialogue is logical and dialectical in character. According to Proclus, whereas the Peripatetic dialectical methods are unrelated to reality,[115] which means that they are abstracted from the objects of their analysis (ψιλὰς τῶν πραγμάτων μεθόδους διαλεκτικάς), the Platonic dialectic leads us to the supreme reality, the Good. He explains that the "great Plato" knew that dialectic is suited only to those who have been completely purified in thought, educated in mathematics, pu-

---

[109] Alexander in *Prior Analytics*, 2.33–3.13. See also Gourinat (2016b), p. 378.
[110] Plotinus, *Enn.* I.3 [20] 5, 5–13. See also Gourinat (2016a), p. 18 and Gourinat (2016b), pp. 377–379. Leroux (1974) stresses the great importance of Plotinus' treatise since it contains the most explicit account of the distinction between logic and dialectic, which was a crucial subject of discussion.
[111] Plotinus, *Enn.* I.3 [20] 6, 1–7. See also Schiaparelli (2009), p. 274.
[112] Plotinus, *Enn.* I.3 [20] 1, 1–5. See also Schiaparelli (2009), p. 254.
[113] Plotinus, *Enn.* I.3 [20] 4, 12–18. See also Schiaparelli (2009), pp. 263–264 and Gourinat (2016a), p. 17.
[114] Schiaparelli (2009), pp. 254–255; see also Gourinat (2016b), pp. 378–379.
[115] Van den Berg (2008), p. 136.

rified through the virtues of the immature aspect of their characters and devoted to genuine philosophical study.[116] This method is the capstone of mathematical studies and leads us upwards to the one cause of all things, the Good. According to Proclus, Intellect is the projector of dialectic since it generates dialectic as a whole, from itself as a whole. It institutes the art of division, that of definition and the technique of deductive proof; moreover, it produces the art of analytical reduction.[117] In the first book of his Commentary on Plato's *Parmenides*, Proclus examines Parmenides' method, which is called 'exercise' (γυμνασία) in the dialogue, with regard to Plato's dialectic and Aristotle's dialectic. He then refutes the arguments of those who claim that Parmenides' method is different from Plato's dialectic. Within this framework, Proclus further analyzes the various aspects of Plato's dialectic throughout the relevant dialogues and states that dialectic as method of trying to attain genuine knowledge contains three sorts of activity: arguing on both sides of a question, expounding only the truth, and exposing error; the latter form of dialectic serves to refute false beliefs.[118]

The tendency to harmonize the philosophies of Plato and Aristotle, which is predominant throughout the history of post-Plotinian Neoplatonism, has the implication that dialectic in late Neoplatonism includes the procedures both of Platonic dialectic and Aristotelian logic.[119] The prevailing view was that the Aristotelian curriculum offers a good account of logic and its methods, especially with regard to the sensible world, whereas Plato offers a good account of the intelligible world.[120] Syrianus and Proclus introduce alongside a typical procedure of Platonic dialectic, such as the division (διαιρετική), some crucial procedures of the Aristotelian logic, such as definition (ὁριστική) and demonstration (ἀποδεικτική).[121]

Important aspects of the views on dialectic held by the Neoplatonic philosophers and commentators that are presented in this volume can be found when examining their epistemology, which is interwoven with their ontology. Longo stresses that Syrianus, Proclus' teacher, believes that human beings do not produce the axioms by induction from observation of sensible objects, but receive them from

---

[116] Proclus, in *Cratylum* II. p. 1. 10–2. 4 (Pasquali). In general outline, I follow the translation by Duvick (2007).
[117] Proclus, in *Cratylum* III. p. 2. 5–12. On Proclus' conception of the scientific dialectic, see also Bonelli (2016).
[118] Proclus in *Parmenidem* I 648.2–653.2; 653.3–654.18 (trans. Dillon and Morrow). See also Butorac (2009).
[119] Gourinat (2016a), p. 17.
[120] Griffin (2012), p. 174.
[121] Syrianus in *Metaph*.55.29–56.4. Proclus in *Parmenidem* I 653.18–30. See also Bonelli (2016), pp. 406–407.

Intellect by nature. So, the axioms' eternal existence in the soul and their derivation from Intellect justify and guarantee their necessary logical priority over the conclusions inferred from them in scientific demonstration.[122] Opsomer notes that Olympiodorus' method is characterized by the use of non-hypothetical principles, namely our infallible common notions. The common notions are fundamental in Olympiodorus' epistemology and philosophical method. We have them through our participation in Intellect and once they have been brought to the fore in maieutic conversations and critically examined, they provide principles for demonstration.[123] Simplicius' strategy, according to Baltussen, includes formalizing arguments into syllogistic form to ensure they are valid, after the example of Aristotle's logical works, *Analytics* and *Topics*. He tries to link his ardent tendency for harmonization of all of Greek philosophy to his overall philosophical agenda.[124]

Exegesis or commentary, on the one hand, establishes the relation of thought with some pre-existent form of philosophical thought that serves as a starting point of the thinking procedure, and on the other, it constitutes a creative philosophical activity. The verb ἐξ-ηγεῖσθαι and the Latin substantive *com-mentarium*, as Goulet-Cazé astutely notes, denote this kind of thought which is constructed on the grounds of a pre-existent philosophical basis and is developed in terms of a fruitful procedure of "thinking with". This "thinking with" implies that those who are commenting on a text must understand, explain, take a position on the philosophical issues raised by the author of the text and possibly succeed in surpassing his ideas.[125]

6 Michael Psellos: Dialectic in Byzantium

One chapter of the volume is devoted to one of the most preeminent scholars and philosophers of the 11th century, who contributed to the intellectual history of Byzantium in almost all areas of learning, Michael Psellos.[126] As Lankila notes, Byzantine intellectuals inherited the general Platonizing culture of Late Antiquity.[127] On the other hand, the Byzantine thinkers received, commented upon and developed ancient logic. Logic constituted a part of the *enkyklios paideia* and also a subject of scientific research; their engagement with the study of logic was deeply influenced

---

122 Longo (2010), pp. 627–628.
123 Opsomer (2010), pp. 705–706.
124 Baltussen (2010), pp. 714 and 717.
125 Goulet-Cazé (2000), pp. 6–7.
126 Jenkins (2017), pp. 447–448.
127 Lankila (2017), p. 316.

by Aristotelian logic, as illustrated in the *Organon*.[128] Erismann stresses that in this sense, studying the Byzantine history of logic is equal to the study of the relation of Byzantine thought to Aristotelianism.[129] Psellos is first and foremost known for his contribution to the revival of philosophical studies in Byzantium and the promotion of philosophy, specifically of ancient Greek thought, with his priority being the Neoplatonic works and the logical treatises of Aristotle.[130] As O'Meara notes, based on his *Chronographia*, Psellos acknowledges that science is knowledge that involves discursivity and rigorous methods of reasoning, such as can be found in mathematics, but it is surpassed by a superior form of knowledge which involves a non-discursive grasp of truths by intellect. So, although he underlines the value of rational thinking and the importance of the use of logical syllogisms, logical analysis, and demonstrations, he also makes a radical distinction between discursive thought and mystical experience.[131]

7 Structure and Content of the Volume

The present collection includes eighteen chapters, written by scholars from various continents and many different countries. The volume is divided into two sections grouped chronologically, the first presenting Classical Antiquity and the second Late Antiquity and the Byzantine period. In each section, the chapters are also placed according to a general chronological order and fall within two broad categories: a) The first deals with issues of dialectic and its evolution and development in Classical Antiquity. The use of dialectic by Socrates and Plato and the different methodological aspects of it are reflected in a transitional dialogue such as *Gorgias* as well as in some of the most important middle dialogues, such as the *Phaedo* and the *Phaedrus*. The connection between Plato's *Parmenides* and *Theaetetus* is examined through the investigation of the crucial notion of *the power of dialegesthai*. The dialectic of the Sophists is investigated both in contrast to and in communion with the Socratic dialectic in the *Sophist*. Through the examination of the methods of collection and division in the relevant Platonic dialogues, are presented different accounts of Plato's late dialectic. This section also includes a chapter on the method of dialectic as used specifically by the Megarians and the dialectic struggles between them and Plato, along with a chapter on Aristotle's treatment of sophistical arguments. b) The second includes important issues that are thematized

---

[128] Erismann (2017), p. 362.
[129] Erismann (2017), p. 363.
[130] Ierodiakonou (2011), p.789; see also Jenkins (2017), p. 448.
[131] O'Meara (2017), pp. 173–174; Jenkins (2017), p. 448; Ierodiakonou (2011), pp. 789–790.

by some of the most significant representatives of philosophical thought in Late Antiquity. The issues discussed in the second section present the views of ancient commentators of Plato and Aristotle on dialectic, including the Peripatetics, Christian Platonists and Neoplatonists philosophers, such as Alexander of Aphrodisias, Origen, Syrianus, Proclus, Olympiodorus, and Simplicius, who practice philosophy through exegesis and commentary. Within the frame of this philosophical activity, those philosophers reflect on the different aspects of dialectic of Classical Antiquity and create their own philosophical theories about the function and value of this method. Finally, one chapter discusses the reception of ancient dialectic in Byzantium, focusing on the philosophical personality of Michael Psellos.

8 Synopsis of the Contributions

The volume and its first section open with a chapter by **François Renaud** which argues that Plato's *Gorgias* is usually interpreted as a radical condemnation of rhetoric, a condemnation that is also often seen as incompatible with the apparent use of rhetorical tricks by Socrates. But in fact, true rhetoric is subtly referred to throughout the dialogue, as many ancient readers of the dialogue recognized. Renaud stresses that this noble rhetoric corrects or refutes instead of flattering the way conventional rhetoric does. Socrates reveals this true rhetoric only gradually (454e–455d, 480c–d, 503a–b, 504d, 508b–c, 516e–517a, 527b–c). Moreover, the dialectic Socrates practices coincides with true rhetoric, and he employs it in two ways: one that is strictly rational or argumentative, while the other, for lack of a proper interlocutor, appeals to the emotions with a view to mere persuasion, such as is the case with myths. As a result, the dramatic action is inseparably joined to the argumentation. Against his non-philosophical interlocutors, Socrates defends dialectic as justice by correcting them. The chapter concentrates on how Socrates, together with Plato, gradually alludes to true rhetoric and its connections with the dialectic being practiced and how true rhetoric also implicitly coincides with true politics.

Rafael Ferber contributes with a chapter dealing with the *deuteros plous*, literally "the second voyage", proverbially "the next best way"; it is discussed in Plato's *Phaedo*, the key passage being *Phd.* 99e4–100a3, which may be considered as the founding document of the Platonic dialectic. Ferber argues firstly that the "flight into the *logoi*" can have two different interpretations, a standard and a non-standard one. The issue is whether at 99e–100a Socrates means that both the student of *erga* and the student of *logoi* consider images ("the standard interpretation"), or whether the student of *logoi* does not consider images, since mere consistency of the *logoi* suffices for truth ("the non-standard interpretation"). Ferber favors the non-standard interpretation, which, to the best of his knowledge, was first anticipated by Leibniz. Secondly however, this interpretation entails *the problem of the hypothesis* or *the problem of the "unproved principle"*, one that is analogous to *"the problem of the elenchus"* (Vlastos), namely, how do we get from mere consistency of the logoi, which is only a negative test, to truth, in the sense of correspondence with reality? The preliminary solution lies in the assumption that "the truth of things is always in our soul" (*Men.* 86b1) or that a "global error" of the *logoi* is impossible. Finally, he argues that the non-standard interpretation also indicates the kernel of truth contained in the standard one, while he concludes with some remarks on the "weakness of the *logoi*".

In her chapter, **Claudia Marsico** notes that the Socratic circle included many lines that were different, amongst them the Megarians and Plato. They shared several important features, in contrast to those characterized by their materialism or hedonism, like Antisthenes or the Cyrenaics. However, the Megarians and Plato also had significant differences in their conceptions. The author reviews three aspects that illustrate this problematic relation between them. Firstly, she analyzes their disagreements regarding the foundation of beings and the possible Platonic allusion to the Megarians in the allegory of the sun in *Republic* VI. She then examines the methodological differences between them, as attested in the *Euthydemus*. Finally, she investigates the responses of the Megarians through the case of the third man argument, alluding to Stilpo's case against Plato.

Beatriz Bossi analyzes the way in which, in the *Phaedrus*, both aspects of dialectic, namely, dialogue and knowledge of a definition arrived at by a process of collection and division, play an essential complementary role as inseparable parts of the same process. The lively discussion, at the beginning, between two interlocutors under attack by erotic passion, paradoxically focuses on a criticism of passionate love. The alternative between a sick, passionate love and a cold, uncommitted relation offers such a poor choice that Socrates has to set it aside. Phaedrus feels under attack and abandoned by the move, but it is only after it has been made that Socrates is ready to listen to him. At this turning point in the dialogue, during which the daimon has a word to say to Socrates, he regains his self-confidence and is ready to introduce a new division that rescues passion by defining eros as a (second, higher order of) divine madness. In the final section, Socrates demonstrates to Phaedrus the strategies he has at hand to persuade him to move cautiously from similarity to similarity, till the opposite bank is reached: knowing how to engage properly in dialogue is an essential step in leading interlocutors towards the right definitions. As a result, the "right-hand" species of love (passionate, divine madness for knowledge) is chosen over the "left-hand" species (passionate, sick love: 265d–266c), while the sophistic, deceptive conception of it (non-passionate, uncommitted love), which does not even belong to the genus, is naturally dropped.

Kristian Larsen focuses on the dialectical activity of the soul when trying to reach for Being and the Good, as illustrated in Plato's *Theaetetus* 184b3–186e12. In a crucial passage in the *Parmenides*, Parmenides states that the power of conversation (*ten tou dialegesthai dynamin*) depends on forms (135b–c) and indicates that this power is a prerequisite for philosophy. Larsen raises the question of what implications this passage has for Plato's conception of dialectic and argues that the discussion of the thesis that knowledge is perception in the *Theaetetus*, and in particular the conclusion to this discussion found at 184b3–186e12, provides an explanation of Parmenides' claims about the power of conversation. The chapter provides a detailed interpretation of the final refutation of Theaetetus' thesis, that knowledge is perception, highlighting the way in which the dramatic features of the *Theaetetus* accentuate the argument Socrates develops in order to refute the thesis. The author argues that a central feature of the drama of the dialogue is Socrates' attempt at redirecting Theaetetus from mathematics toward dialectic and philosophy. In particular, Socrates aims to make Theaetetus realize that *being* and the *good* or the *beneficial* are things that are what they are, themselves by themselves; that they are on a par, ontologically speaking, and that reaching for them, and attempting to get to grips with them, in thought, is a requirement for knowledge in general and for dialectic in particular.

Melina G. Mouzala contributes with a chapter on Plato's *Sophist* and claims that when the fifth definition of the sophist places the emphasis on *logos*, it paves the way to the sixth definition. *Logos* is a human characteristic which brings to the fore and realizes the manifestation of all thinking and specifically of controversies and disputations in which our thought is involved and expressed. She argues that the same subject is reserved and developed in the sixth definition, which apart from the explicit discussion of purification or cathartic dialectic, actually thematizes division itself. According to Proclus' perspective of dialectic, the name 'eristic' in itself is neutral, since it only indicates the activity of controverting and raising objections, and since there is good and bad strife. She argues that the notion of communion (κοινωνία) is implicitly examined for the first time in the dialogue within the sixth definition of the sophist, where the Sophistic and the Socratic Dialectic are commingled. She aims to show that from the analysis of the crucial passage 230 b–d, we can infer that the basic characteristic of Socrates' cathartic method is self-purification of the person who implements the elenchus, and a specific emotional attitude of the person who is subjected to elenchus, which, due to its reflexive and self-referent character, leads to self-knowledge.

In her chapter, **Anna Pavani** notes that although accounts of Plato's Late Dialectic differ extensively, they almost all assume that the dialectical method remains *one and the same*. According to the standard interpretation, the *Phaedrus*, the *Sophist*, the *Statesman*, and the *Philebus* display one single dialectical method, which is first introduced in the "canonical passage" of the *Phaedrus* (265c8–266c8), where Socrates claims that "dialecticians" are those who can perform two complementary tasks, namely Collection and Division. In this chapter, Pavani challenges what she refers to as the "one-method interpretation". By means of a close textual analysis, she shows that, even within a single dialogue, we encounter different dialectical methods, namely methods that by dealing with different objects proceed in different ways in order to achieve different goals. By mostly focusing on the *Sophist* and the *Statesman*, she furthermore argues that we can reconstruct a *family* of the paradigmatic methods of Collection and Division (such as Dichotomy, Division by limbs, and taxonomy). This is just one of the family of strategies by which Plato aims to make both the interlocutors and us readers "more dialectical on any topic" (*Politicus* 285d5–6).

Lucas Angioni examines a class of sophistical arguments that are entangled both with peirastic arguments and with pseudographemata. Like peirastic arguments, these sophistical arguments are aimed at exposing false claims of scientific expertise, but by exploring a different route. Peirastic arguments observe "commons items" (*koina*) in order to detect conflicts and incompatibilities between propositions conceded by a false expert and produce a refutation in the strict sense. On the other hand, these sophistical arguments present themselves as direct "demonstrations" that falsely appear to be explanatorily appropriate to their object and, thereby, are aimed towards the debunking of the explanatory story offered by the real experts. Angioni explains that this is why he refers to them with the curious expression "sophistical demonstrations". These sophistical arguments are somehow similar to (while being also different from) pseudographemata. Both of these arguments and the pseudographemata have premises that are in general appropriate to the scientific expertise in question and even appropriate at large to the targeted conclusion, and both fail in capturing the principles appropriate to their explananda. However, these sophistical arguments are different from pseudographemata because they are purposely deceitful. They are essentially directed towards producing a false appearance of explanatory appropriateness to their explananda.

The second section of the volume, which includes issues related to the reception of Ancient Dialectic during Late Antiquity and in Byzantium, opens with a text on Alexander of Aphrodisias. **Gweltaz Guyomarc'h** contributes with a chapter on the services of dialectic, by focusing on dialectic as an instrument for metaphysics in Alexander of Aphrodisias. He notes that it is commonplace that ancient commentators worked to systematize Aristotle. In doing so, they ignored all that was exploratory and problematic in Aristotelian thought. Alexander of Aphrodisias offers a typical example of this tendency when he seeks to make metaphysics a demonstrative science. The author examines Alexander's use of the dialectical method in metaphysics, in particular in his exegesis on *Metaphysics* Beta, and in his practice of the aporetic method. Against the idea that Beta's aporiai are no longer genuine puzzles, Guyomarc'h sets out to show, on the one hand, that aporia retains, in Alexander, an authentically exploratory function and, on the other, that its use in metaphysics is not so much a matter of systematization as of the accomplishment of the *organon* status granted to dialectics. Thanks to the dialectical method, Alexander maintains this exploratory character of aporiai within a scientific approach. The author argues that if we show the heuristic role of the dialectic as an integral part of science, we will then be able to provide a stronger foundation for the idea that Alexander maintains the exploratory character of the aporiai.

Silvia Fazzo presents a chapter concerned with the very nature of Alexander of Aphrodisias' interest in *aporiai*, which will be placed in the middle of her investigation, summarizing previous scholarly achievements. Although the collection called *Quaestiones* has several kinds of elements in it, there is one which is most typical of Alexander's *aristotelizein*: *aporiai* with multiple solutions for a single problem. A reading of such texts, as if they were just simply dialectical exercises, meant to display for pedagogical reasons the different possible answers to a given problem, does not render justice of their uniqueness and interest. According to Fazzo, *aporiai* crucially represent the multi-layered way the tradition was built throughout the centuries, making Aristotelianism a universal grammar of scientific reasoning and a non-dogmatic system of thought.

Inna Kupreeva contributes to the volume with a chapter focusing on Alexander of Aphrodisias and the Principle of Non-Contradiction. Kupreeva provides an outline of Alexander's reading of Aristotle's argument which will allow the readers to see how his approach may be interpreted in the light of contemporary discussions of Aristotle's argument. The goal of her work is to present a study of Alexander's interpretation of Aristotle's argument in support of the Principle of Non-Contradiction, which is traditionally described as "argument from signification". Kupreeva discusses Alexander's interpretation of elenctic demonstration and attempts to show that Alexander develops his own version of unrestricted essentialist interpretation of Aristotle's argument, which has some philosophical merits.

Ilaria L.E. Ramelli presents a chapter on Origen, in which she considers Plato's dialectic (in turn influenced by Parmenides, notwithstanding the conflictual relation between them) as an important motif of inspiration for the Christian Platonist Origen. The meaning, function, and partition of dialectic in Plato is argued as having impacted Origen's philosophical theology a great deal. An analysis of the role of dialectic in imperial pre-Plotinian Platonism, Clement, Alexander of Aphrodisias, and Plotinus yields interesting comparisons with Origen. Ramelli points out that Origen knows Plato' dialectics, employs those of Aristotle and the Stoics logic-dialectics, and attributes Platonic dialectic to Scripture (one of the many convergences he found between the Bible and Plato). She investigates Origen's view of dialectics in the Thanksgiving Oration, also meant as a thinking attitude against psychagogy (here as well, Origen is likely to have been inspired by Plato in addition to Christ-Truth). Origen's zetetic method is an expression of dialectics. Ramelli's final reflections are devoted to dialectic's relation to philosophy/theology in Plato and Origen—the vantage point from which to explore and interpret Origen's dialectic—as well as in Origen's great admirer, Eriugena, who also linked dialectics very closely to theology. Ramelli argues that the nexus between dialectics and theology is provided by Christ-Logos.

Michael Griffin focuses on Exegesis as Philosophy, offering some insights on Aristotelian methods in Neoplatonic commentary. It is often pointed out that later ancient Mediterranean philosophers taught and practiced philosophy in a medium of textual interpretation—the exegēsis of literary and philosophical "classics", including Plato, Aristotle, and Homer, with care to preserve their "harmony" or compatibility. There is some consensus that this strategy did more than constrain the later ancient Platonists; it also led to creative philosophical and scientific conclusions. Moreover, the exercise of copying and interpreting a text constituted an important feature of the philosophical "ways of life" adopted in Late Antiquity. In this chapter, Griffin sets out to explore some of the methodological roots of that practice. He begins with several methodological remarks in Aristotle that associate philosophy with interpretation, and explores their development in the Stoics and Plotinus, before settling on examples from the Neoplatonists. Along the way, he traces several of the metaphilosophical and pedagogical assumptions that helped to motivate this posture toward "exegesis as philosophy" during Late Antiquity.

Sara Klitenic Wear presents a chapter on Syrianus' conception of dialectic. Her paper is an examination of how Syrianus understands certain dialectical statements from Plato's *Parmenides* and how he uses these statements as descriptions of the whole universe, particularly how one entity relates to another. Namely, we see that he reads "if a thing is/if it is not" to describe the relationship between the One and the intelligible universe; moreover, "in itself/in another" also speaks to the relationship between the One and the entities below the One. These two statements become the paradigm whereby Syrianus understands the relationship to other entities, such as the forms in his *Commentary on Metaphysics*, or the monads, as reported in Damascius' *Commentary on the Philebus*. Thus, these two dialectical statements in the *Parmenides* act as the lens by which Syrianus understands participation by intelligible entities; that is, they primarily follow the example of the intelligible realm's participation in the One. At the heart of the hypotheses "if a thing is/if it is not" and "in itself/in another" is a concern for preserving the unity of the One, while showing how it gives rise to an elaborate, inter-connecting universe that coheres in its cause.

Dirk Baltzly presents Proclus' conception of Plato's dialectic by focusing on the argument by performance. He notes that one of the many hard problems in Plato scholarship is how to understand the relation between the method of dialectic as it is described in the middle books of the *Republic* and the method that is called 'dialectic' and described or illustrated in dialogues such as the *Phaedrus* or the *Statesman*. In short, the question is what is the relation between Republican dialectic and the method of collection and division? This chapter looks at Proclus' description of the kinds of dialectic in his commentary on the first part of the *Parmenides*. Proclus argues for a threefold division of dialectic: the kind that argues both sides of the question; the kind that exhibits only the truth; and the kind that serves only to refute false beliefs (*in Parm.* 654.11–13). In describing the truth-exhibiting kind of dialectic, he does not so much *argue* for the unity of collection and division with Republican dialectic as to *perform* this unity. The chapter elucidates this distinction and its implications for how we should understand the activity of Platonic philosophizing in Late Antiquity.

Harold Tarrant contributes a chapter on Elenchus and Syllogistic in Olympiodorus of Alexandria. Olympiodoran elenchus has been previously treated by him in *On Alcibiades*, where it has a structural role, and in *On Gorgias*, where many of the rules for *elenchus* are set. Here Tarrant also examines its place in the treatment of the rather different Socrates of *Phaedo*, and, especially of *Meteorologica*, an Aristotelian treatise to which Socrates is irrelevant. He poses the question, can elenchus be non-Socratic, or even non-Platonic? He stresses that the vocabulary of *elenchus* is not found in *On Phaedo*, where the primary term used for an argument is *epikheirêma*. This term was also more common than talk of demonstration (*apodeixis*) or syllogism in the commentaries on *Phaedo* and *Meteorologica* alike, occurring about 12 and 15 times per 10,000 words, when 'demonstration' is rarely mentioned. The stronger the terminology the more reluctant Olympiodorus is to use it in these two seemingly late works. In the *Meteorologica*, where Aristotle does not use the terminology, Olympiodorus' commentary uses it 51 times in the first 60,000 words, but never in the last 50,000. So, it belongs only to the treatment of the first two books of the *Meteorologica*, not the last two.

Han Baltussen undertakes an analysis of Simplicius' views on Aristotle's dialectic, building on his earlier examinations of this method. He notes that Simplicius' sterling track record as commentator on Aristotle's works stands out among the Platonists and is especially worthy of attention. Although a commentary by Simplicius on Aristotle's *Topics* is not extant (and probably never existed), we can extract from his other commentaries some idea of his understanding of Aristotle's dialectical method, even if he does not seem to develop an explicit theoretical view on this point. Starting from the simple question of what Simplicius may have understood "dialectic" to mean, Baltussen moves beyond a terminological survey of *dialektikos*, and shows his specific interest in, and use of, dialectic as a methodology for the analysis of coherence in an author. He concludes that in some ways, Simplicius echoes certain mechanisms from Aristotle's dialectical procedures as a method useful for research, but it becomes clear that he will also adjust certain moves to fit his own philosophical agenda.

In the final chapter, **Graeme Miles** presents his work on the subject "Psellos on Dialectic and Allegory". Psellos, as a philosopher steeped in the Christian and non-Christian traditions, navigates his way through the tensions of this combined heritage with remarkable deftness. Two of his primary means of accomplishing this ever-fraught conciliation are dialectic and allegory. In addition to his employment of these modes of thought and discourse, Psellos on occasion offers definitions of both and descriptions of their activities. These definitions are often in response to others (e.g., *Philosophica Minora* 2.13) and draw directly (sometimes verbatim) from ancient sources, but nonetheless are consonant with and inform his practice as we can observe it elsewhere. Importantly, Psellos claims that both of these modes are applicable to any subject matter, allowing him considerable flexibility in their use in recovering aspects of pagan Platonism, as well as in elaborating Orthodox positions by means which are harmonious with his philosophical commitments. Despite this shared breadth, it appears that allegorical interpretation is able to reach further "down" the social hierarchies, as it emerges in some remarkable essays on popular language and customs which have received little scholarly attention.

In this volume, one can find aspects of the history of Ancient Greek dialectic, which shed light on its essence and the transformations in its understanding. By reading its chapters one can realize not only the epistemological character or value of dialectical reasoning, but also those features which help us to interpret the impact of dialectic on the way man is related, on the one hand to other men and the socio-political community, and on the other, to his dianoia and soul, i.e., to himself. The examination of all these aspects and of the diverse applications and modes of employment of dialectic contributes to the rediscovery of our relationship with the roots of human thought. We hope that this volume will significantly reinforce this effort.

Melina G. Mouzala

Bibliography

Allen, James (2018): "Megara and Dialectic". In: Bénatouïl, Thomas and Ierodiakonou, Katerina (Eds.): *Dialectic After Plato and Aristotle*. Cambridge: Cambridge University Press.
Austin, Scott (2007): *Parmenides and the History of Dialectic: Three Essays*. Las Vegas: Parmenides Publishing.
Bailey, Dominic T. J. (2008): "Excavating Dissoi Logoi 4". In: Inwood Brad (Ed.): *Oxford Studies in Ancient Philosophy*. Volume XXXV. Oxford: Oxford University Press, pp. 249–264.
Baltussen, Han (2000): "Peripatetic Method: Dialectic and Doxography". In: Baltussen, Han: *Theophrastus Against the Presocratics and Plato: Peripatetic Dialectic in the* De sensibus. *Philosophia Antiqua*, Vol. LXXXVI. Leiden and Boston: Brill, pp. 31–70.
Baltussen, Han (2010): "Simplicius of Cilicia". In: Gerson, Lloyd P. (Ed.): *The Cambridge History of Philosophy in Late Antiquity*. Vol. II. Cambridge: Cambridge University Press, pp. 711–732.
Baltzly, Dirk (2012): "Dialectic (*Dialektikē*)". In: Press, Gerald Alan (Ed.): *The Continuum Companion to Plato*. London and New York: Continuum, pp. 159–161.
Bénatouïl, Thomas (2018): "Introduction: Dialectics in Dialogue". In: Bénatouïl, Thomas and Ierodiakonou, Katerina (Eds.): *Dialectic After Plato and Aristotle*. Cambridge: Cambridge University Press, pp. 1–16.
Bencivenga, Hermanno (2000): *Hegel's Dialectical Logic*. New York: Oxford University Press.
Bennett, Jonathan (1974): *Kant's Dialectic*. Cambridge: Cambridge University Press.
Benson, Hugh H. (1987): "The Problem of the Elenchus Reconsidered". In: *Ancient Philosophy* 7, pp. 67–85.
Benson, Hugh H. (2000): *Socratic Wisdom: The Model of Knowledge in Plato's Early Dialogues*. New York: Oxford University Press.
Benson, Hugh H. (2002): "Problems with Socratic Method". In: Scott, Gary Alan (Ed.): *Does Socrates Have a Method? Rethinking the Elenchus in Plato's Dialogues and Beyond*. Pennsylvania: The Pennsylvania State University Press, pp. 101–113.
Benson, Hugh H. (2003): "The Method of Hypothesis in the Meno". In: *Proceedings of the Boston Area Colloquium of Ancient Philosophy* 18. No. 1, pp. 95–143.
Benson, Hugh H. (2006): "Plato's Method of Dialectic". In: Benson, Hugh H. (Ed.): *A Companion to Plato*. Malden and Oxford: Blackwell Publishing, pp. 85–99.
Benson, Hugh H. (2011): "Socratic Method". In: Morrison, Donald R. (Ed.): *The Cambridge Companion to Socrates*. Cambridge: Cambridge University Press, pp. 179–200.
Benson, Hugh H. (2015): *Clitophon's Challenge: Dialectic in Plato's Meno, Phaedo, and Republic*. Oxford: Oxford University Press.
Berti, Enrico (1978): "Ancient Greek Dialectic as Expression of Freedom of Thought and Speech". In: *Journal of the History of Ideas* 39. No. 3, pp. 347–370.
Bird, Graham (2006): *The Revolutionary Kant. A Commentary on the Critique of Pure Reason*. Chicago and La Salle: Open Court.
Blackburn, Simon (2016): *The Oxford Dictionary of Philosophy*. 3rd ed. Oxford: Oxford University Press.
Bolton, Robert (1990): "The Epistemological Basis of Aristotelian Dialectic". In: Devereux, Daniel and Pellegrin, Pierre (Eds.): *Biologie, Logique et Metaphysique chez Aristote*. Paris: CNRS, pp. 185–236.
Bonelli, Maddalena (2001): *Alessandro di Afrodisia e la metafisica come scienza dimostrativa*. Elenchos XXXV. Naples: Bibliopolis.

Bonelli, Maddalena (2016): "Proclus et la dialectique scientifique". In: Gourinat, Jean-Baptiste and Lemaire, Juliette (Eds.): *Logique et Dialectique dans l' Antiquité*. Paris: Librairie Philosophique J. Vrin, pp. 397–421.

Bostock, David (1986): *Plato's Phaedo*. Oxford: Oxford University Press.

Boys-Stones, George and Rowe, Christopher (Ed. and Trans.) (2013): *The Circle of Socrates: Readings in the First-Generation Socratics*. Indianapolis and Cambridge: Hackett.

Brancacci, Aldo (2017): "Socratism and Eleaticism in Euclides of Megara". In: Stavru, Alessandro and Moore, Christopher (Eds.): *Socrates and the Socratic Dialogue*. Leiden and Boston: Brill, pp. 161–178.

Bredlow, Luis-Andrés (2011): "Parmenides and the Grammar of Being". In: *Classical Philology* 106. No. 4, pp. 283–298.

Brickhouse, Thomas C. and Smith, Nicholas D. (2002): "The Socratic Elenchos?". In: Scott, Gary Alan (Ed.): *Does Socrates Have a Method? Rethinking the Elenchus in Plato's Dialogues and Beyond*. Pennsylvania: The Pennsylvania State University Press, pp. 145–157.

Butorac, David. D. (2009): "Proclus' Interpretation of the *Parmenides*, Dialectic and the Wandering of the Soul". In: *Dionysius* 27, pp. 33–54.

Carpenter, Michelle and Polanski, Ronald M. (2002): "Variety of Socratic Elenchi". In: Scott, Gary Alan (Ed.): *Does Socrates Have a Method? Rethinking the Elenchus in Plato's Dialogues and Beyond*. Pennsylvania: The Pennsylvania State University Press, pp. 89–100.

Castagnoli, Luca (2012): "Self-Refutation and Dialectic in Plato and Aristotle". In: Fink, Jakob L. (Ed.): *The Development of Dialectic from Plato to Aristotle*. Cambridge: Cambridge University Press, pp. 27–61.

Cerami, Cristina (2016): "Alexander of Aphrodisias". In: Falcon, Andrea (Ed.): *Brill's Companion to the Reception of Aristotle in Antiquity*. Leiden and Boston: Brill, pp. 160–179.

Crivelli, Paolo (2016): "Logic Within Stoic Philosophy". In: Gourinat, Jean-Baptiste and Lemaire, Juliette (Eds.): *Logique et Dialectique dans l' Antiquité*. Paris: Librairie Philosophique J. Vrin, pp. 303–319.

Crivelli, Paolo (2018): "Dialectic in the Early Peripatos". In: Bénatouïl, Thomas and Ierodiakonou, Katerina (Eds.): *Dialectic After Plato and Aristotle*. Cambridge: Cambridge University Press, pp. 47–81.

Dancy, Russell M. (2004): *Plato's Introduction of Forms*. Cambridge: Cambridge University Press.

DaVia, Carlo (2017): "Aristotle and the Endoxic Method". In: *Journal of the History of Philosophy* 55. No. 3, pp. 383–405.

Delcomminette, Sylvain and Lacanche, Geneviève (Eds.) (2021): *L'Éristique: Définitions, Caractérisations et Historicité. Cahiers de Philosophie Ancienne* XXVII. Bruxelles: Éditions Ousia.

Devereux, Daniel (2015): "Scientific and Ethical Methods in Aristotle's *Eudemian* and *Nicomachean Ethics*". In: Henry, Devin and Nielsen, Karen Margrethe (Eds.): *Bridging the Gap Between Aristotle's Science and Ethics*. Cambridge: Cambridge University Press, pp. 130–147.

Dillon, John (1993): *Alcinous: The Handbook of Platonism. Translated with an Introduction and Commentary*. Oxford: Clarendon Press.

Dorion, Louis-André (2002): "Aristote et l' Invention de la Dialectique". In: Canto-Sperber, Monique and Pellegrin, Pierre (Eds.): *Le Style de la Pensée. Recueil de Textes en Hommage à Jacques Brunschwig*. Paris: Les Belles Lettres, pp. 182–220.

Dorter, Kenneth (2006): *The Transformation of Plato's Republic*. Lanham: Lexington Books.

Drehe, Iovan (2011): "The Aristotelian Dialectical *Topos*". In: *Argumentum* 9. No. 2, pp. 129–139.

Duvick, Brian (2007): *Proclus: On Plato Cratylus. Ancient Commentators on Aristotle*. London: Bloomsbury.

Dybicz, Phillip and Pyles, Loretta (2011): "The Dialectic Method: A Critical and Postmodern Alternative to the Scientific Method". In: *Advances in Social Work* 12. No. 2, pp. 301–317.
Encyclopaedia Britannica (2021): "dialectic". https://www.britannica.com/topic/dialectic-logic, last accessed on November 7, 2022.
Erismann, Christophe (2017): "Logic in Byzantium". In: Kaldellis, Anthony and Siniossoglou, Niketas (Eds.): *The Cambridge Intellectual History of Byzantium*. Cambridge: Cambridge University Press, pp. 362–380.
Evans, John David Gemmill (1977): *Aristotle's Concept of Dialectic*. Cambridge: Cambridge University Press.
Falcon, Andrea (2019): "Aristotle's Method of Inquiry in *Eudemian Ethics* 1 and 2". In: Bonazzi, Mauro, Ulacco, Angela, and Forcignanò, Filippo (Eds.): *Thinking, Knowing, Acting: Epistemology and Ethics in Plato and Ancient Platonism. Brill's Plato Studies Series*, Vol. III. Leiden and Boston: Brill, pp. 186–206.
Fink, Jakob L. (2012): *The Development of Dialectic from Plato to Aristotle*. Cambridge: Cambridge University Press.
Forster, Michael N. (1993): "Hegel's Dialectical Method". In: Beiser, Frederick C (Ed.): *The Cambridge Companion to Hegel*. Cambridge: Cambridge University Press, pp. 130–170.
Fossheim, Hallvard (2012): "Division as a Method in Plato". In: Fink, Jakob L. (Ed.): *The Development of Dialectic from Plato to Aristotle*. Cambridge: Cambridge University Press, pp. 91–112.
Frede, Dorothea (2012): "The Endoxon Mystique: What Endoxa Are and What They Are Not". In: *Oxford Studies in Ancient Philosophy*. Vol. XLIII, Oxford: Oxford University Press, pp. 185–215.
Furley, David J. (1973): "Notes on Parmenides". In: Lee, Edward N., Mourelatos, Alexander P. D., and Rorty, Richard. (Eds.): *Exegesis and Argument: Studies in Greek Philosophy Presented to Gregory Vlastos*. Assen: Van Gorcum, pp. 1–15.
Furth, Montgomery (1968): "Elements of Eleatic Ontology". In: *Journal of the History of Philosophy* 6, pp. 111–132.
Gallop, David (1975): *Plato, Phaedo: Translated with Notes*. Oxford: Clarendon Press.
Gera, Deborah Levine (2000): "The Thought Experiments in the Dissoi Logoi". In: *American Journal of Philology* 121. No. 1, pp. 21–45.
Gonzalez, Francisco J. (2017): "Plato's Perspectivism". In: *Plato Journal* 16, pp. 31–48. https://doi.org/10.14195/2183-4105_16_4, last accessed November 7, 2022.
Goulet-Cazé, Marie-Odile (2000): "Avant-Propos". In: Goulet-Cazé, Marie-Odile (Ed.): *Le Commentaire Entre Tradition et Innovation. Actes du Colloque international de l' Institut des Traditions textuelles, Paris et Villejuif, 22-25 Septembre 1999*. Paris: Librairie Philosophique J. Vrin, pp. 5–12.
Gourinat, Jean-Baptiste (2016a): "Introduction". In: Gourinat, Jean-Baptiste and Lemaire, Juliette (Eds.): *Logique et Dialectique dans l' Antiquité*. Paris: Librairie Philosophique J. Vrin, pp. 7–19.
Gourinat, Jean-Baptiste (2016b): "Logique et dialectique chez Plotin: la restauration de la dialectique platonicienne". In: Gourinat, Jean-Baptiste and Lemaire, Juliette (Eds.): *Logique et Dialectique dans l' Antiquité*. Paris: Librairie Philosophique J. Vrin, pp. 363–382.
Gourinat, Jean-Baptiste (2018): "Stoic Dialectic and Its Objects". In: Bénatouïl, Thomas and Ierodiakonou, Katerina (Eds.): *Dialectic After Plato and Aristotle*. Cambridge: Cambridge University Press, pp. 134–167.
Griffin, Michael (2012): "What Has Aristotelian Dialectic to Offer a Neoplatonist? A Possible Sample of Iamblichus at Simplicius on the *Categories* 12, 10–13, 12". In: *The International Journal of the Platonic Tradition* 6, pp. 173–185.
Gulley, Norman (1968): *The Philosophy of Socrates*. London: Macmillan.
Hörz, Herbert (1987): "Dialectics and Evolution". In: *Biology and Philosophy* 2, pp. 493–508.

Hussey, Edward (1999): "Heraclitus". In: Long, Anthony Arthur (Ed.): *The Cambridge Companion to Early Greek Philosophy*. Cambridge: Cambridge University Press, pp. 88–112.

Ierodiakonou, Katerina (2011): "Michael Psellos". In: Lagerlund, Henrik (Ed.): *Encyclopedia of Medieval Philosophy*. Dordrecht: Springer, pp. 789–791.

Ierodiakonou, Katerina (2018): "Dialectic as a Subpart of Stoic Philosophy". In: Bénatouïl, Thomas and Ierodiakonou, Katerina (Eds.): *Dialectic After Plato and Aristotle*. Cambridge: Cambridge University Press, pp. 114–133.

Ionescu, Cristina (2018): "Elenchus, Recollection, and the Method of Hypothesis in the Meno". In: *Plato Journal* 17, pp. 9–29.

Irwin, Terence (1988): *Aristotle's First Principles*. Oxford: Clarendon Press.

Iwata, Nayola (2016): "Plato's Hypothetical Inquiry in the Meno". In: *British Journal for the History of Philosophy* 24. No. 2, pp. 194–214.

Jenkins, David (2017): "Michael Psellos". In: Kaldellis, Anthony and Siniossoglou, Niketas (Eds.): *The Cambridge Intellectual History of Byzantium*. Cambridge: Cambridge University Press, pp. 447–461.

Judson, Lindsay (2017): "Hypotheses in Plato's Meno". In: *Philosophical Inquiry* 41. No. 2–3, pp. 29–39.

Kahn, Charles H. (1979): *The Art and Thought of Heraclitus. An Edition of the Fragments with Translation and Commentary*. Cambridge: Cambridge University Press.

Kahn, Charles H. (2009): *Essays on Being*. Oxford: Oxford University Press.

Kanayama, Yahei (2000): "The Methodology of the Second Voyage and the Proof of the Soul's Indestructibility in Plato's Phaedo". In: *Oxford Studies in Ancient Philosophy*. Vol. XVIII, pp. 41–100.

Kannicht, Richard (Ed.) (2004): *Tragicorum Graecorum Fragmenta. Vol. V: Euripides*. Göttingen: Vandenhoeck & Ruprecht.

Karasmanis, Vassilios (1987): *The Hypothetical Method in Plato's Middle Dialogues*. Thesis: DPhil in Philosophy. Brasenose College. Oxford: University of Oxford.

Karbowski, Joseph (2015): "Endoxa, Facts, and the Starting Points of the *Nicomachean Ethics*". In: Henry, Devin and Nielsen, Karen Margrethe (Eds.): *Bridging the Gap Between Aristotle's Science and Ethics*. Cambridge: Cambridge University Press, pp. 113–129.

Karbowski, Joseph (2019): *Aristotle's Method in Ethics: Philosophy in Practice*. Cambridge: Cambridge University Press.

Kerferd, George B. (1981): *The Sophistic Movement*. Cambridge: Cambridge University Press.

Kraut, Richard (1983): "Comments on Gregory Vlastos, 'The Socratic Elenchus'". In: Annas, Julia (Ed.): *Oxford Studies in Ancient Philosophy*. Volume I. Oxford: Clarendon Press, pp. 59–70.

Kupreeva, Inna (2016): "Aristotelianism in the Second Century AD: Before Alexander of Aphrodisias". In: Falcon, Andrea (Ed.): *Brill's Companion to the Reception of Aristotle in Antiquity*. Leiden and Boston: Brill, pp. 138–159.

Kupreeva, Inna (2018): "*Aporia* and Exegesis: Alexander of Aphrodisias". In: Karamanolis, George and Politis, Vasilis (Eds.): *The Aporetic Tradition in Ancient Philosophy*. Cambridge: Cambridge University Press, pp. 228–247.

Lankila, Tuomo (2017): "The Byzantine Reception of Neoplatonism" In: Kaldellis, Anthony and Siniossoglou, Niketas (Eds.): *The Cambridge Intellectual History of Byzantium*. Cambridge: Cambridge University Press, pp. 314–324.

Lebedev, Andrei V. (2014): *The Logos of Heraclitus. A Reconstruction of His Word and Thought (With a New Critical Edition of the Fragments)*. Originally published in Russian. Saint Petersburg: "Nauca".

Leroux, Georges (1974): "Logique et dialectique chez Plotin *Ennéade* I. 3 (20)". In: *Phoenix* 28, pp. 180–192.

Lesher, James H. (2002): "Parmenidean *Elenchos*". In: Scott, Gary Alan (Ed.): *Does Socrates Have a Method? Rethinking the Elenchus in Plato's Dialogues and Beyond*. Pennsylvania: The Pennsylvania State University Press, pp. 19–35.
Longo, Angela (2010): "Syrianus". In: Gerson, Lloyd P. (Ed.): *The Cambridge History of Philosophy in Late Antiquity*, Vol. II. Cambridge: Cambridge University Press, pp. 616–629.
Loparic, Zeljko (1987): "Kant's Dialectic". In: *Noûs* 21. No 4 (Dedicated to Alberto Coffa), pp. 573–593.
Luz, Menahem (2012): "Antilogy and Eristics (Eristic)". In: Press, Gerald Alan (Ed.): *The Continuum Companion to Plato*. London and New York: Continuum, pp. 133–135.
Mathieu, Marion (2014): "Les Arguments de Zénon d' après le Parménide de Platon". In: *Dialogue* 53, pp. 393–434.
Maybee, Julie E. (2020): "Hegels' Dialectics". In: Zalta, Edward N. (Ed.): *The Stanford Encyclopedia of Philosophy* (Winter 2020 Edition). Stanford: Stanford University Press.
McKinney, Ronald H. (1983): "The Origins of Modern Dialectics". In: *Journal of the History of Ideas* 44. No. 2, pp. 179–190.
Morrow, Glenn R. and Dillon, John M. (1987): *Proclus Commentary on Plato's Parmenides*. Princeton: Princeton University Press.
Nehamas, Alexandros (1990): "Eristic, Antilogic, Sophistic, Dialectic. Plato's Demarcation of Philosophy from Sophistry". In: *History of Philosophy Quarterly* 7. No. 1, pp. 3–16.
O'Brien, Michael J. (Trans.) (2001): *Protagoras*. In: Sprague, Rosamond Kent (Ed.): *The Older Sophists*. Indianapolis and Cambridge: Hackett, pp. 3–28.
O'Meara, Dominic (2017): "Conceptions of Science in Byzantium". In: Kaldellis, Anthony and Siniossoglou, Niketas (Eds.): *The Cambridge Intellectual History of Byzantium*. Cambridge: Cambridge University Press, pp. 169–182.
Opsomer, Jan (2010): "Olympiodorus". In: Gerson, Lloyd P. (Ed.): *The Cambridge History of Philosophy in Late Antiquity*. Vol. II. Cambridge: Cambridge University Press, pp. 697–710.
Paglieri, Fabio (2014): "Accepted by Whom? On the Empirical Roots of Aristotle's Dialectic". In: *Revue Internationale de Philosophie* 270 No. 4, pp. 393–402.
Politis, Vasilis (2012): "What Is Behind the *ti esti* Question". In: Fink, Jakob L. (Ed.): *The Development of Dialectic from Plato to Aristotle*. Cambridge: Cambridge University Press, pp. 199–223.
Ramelli, Ilaria (2014): "Alexander of Aphrodisias: A Source of Origen's Philosophy?". In: *Philosophie Antique* 14, pp. 237–289.
Ramelli, Ilaria (2017): "Origen and the Platonic Tradition". In: *Religions* 8. No. 2. https://doi.org/10.3390/rel8020021, last accessed on November 7, 2022.
Rankin, David H. (1983): *Sophists, Socratics and Cynics*. Routledge Revivals. London and Canberra: Croom Helm and Totowa: Barnes and Noble Books.
Rapp, Christof (2018): "Aporia and Dialectical Method in Aristotle". In: Karamanolis, George and Politis, Vasilis (Eds.): *The Aporetic Tradition in Ancient Philosophy*. Cambridge: Cambridge University Press, pp. 112–136.
Redfield, James M. (2017): "The Origins of the Socratic Dialogue: Plato, Xenophon, and the Others". In: Stavru, Alessandro and Moore, Christopher (Eds.): *Socrates and the Socratic Dialogue*. Leiden and Boston: Brill, pp. 125–138.
Robinson, Richard (1953²): *Plato's Earlier Dialectic*. Oxford: Oxford University Press.
Rowe, Christopher (1993): "Explanation in Phaedo 99c6–102a8". In: *Oxford Studies in Ancient Philosophy*. Vol. XI. Oxford: Oxford University Press, pp. 49–69.
Rowe, Christopher (2011): "Self-Examination". In: Morrison, Donald R. (Ed.): *The Cambridge Companion to Socrates*. Cambridge: Cambridge University Press, pp. 201–214.

Rowe, Christopher (2015): "Plato, Socrates, and the *genei gennaia sophistikē* of *Sophist* 231b". In: Nails, Debra and Tarrant, Harold in Collaboration with Kajava, Mika and Salmenkivi, Eero (Eds.): *Second Sailing: Alternative Perspectives on Plato. Commentationes Humanarum Litterarum.* Vol. CXXXII. Helsinki: Societas Scientiarum Fennica, pp. 149–167.

Schiaparelli, Annamaria (2009): "Plotinus on Dialectic". In: *Archiv für Geschichte der Philosophie* 91. No 3, pp. 253–287.

Scott, Gary Alan (2002): *Does Socrates Have a Method? Rethinking the Elenchus in Plato's Dialogues and Beyond.* Pennsylvania: The Pennsylvania State University Press.

Sedley, David (1977): "Diodorus Cronus and Hellenistic Philosophy". In: *Proceedings of the Cambridge Philological Society, New Series.* No. 23 (203), pp. 74–120.

Sedley, David (2015): Dialectic. *Oxford Classical Dictionary.* Oxford: Oxford University Press.

Sim, May (1995): "Dialectic and Definition in Aristotle's Topics". In: *The Society for Ancient Greek Philosophy Newsletter.* 326. https://orb.binghamton.edu/sagp/326

Slomkowski, Paul (1997): *Aristotle's Topics. Philosophia Antiqua.* Vol. LXXIV. Leiden: Brill.

Smith, Robin (1993): "Aristotle on the Uses of Dialectic". In: *Synthese* 96. No. 3 (Logic and Metaphysics in Aristotle and Early Modern Philosophy), pp. 335–358.

Sprague, Rosamond Kent (1968): "Dissoi logoi or Dialexeis". In: *Mind* 77. No. 306, pp. 155–167.

Spranzi, Marta (2011): *The Art of Dialectic Between Dialogue and Rhetoric: The Aristotelian Tradition.* Amsterdam and Philadelphia: John Benjamins Publishing Company.

Stokes, Michael C. (1986): *Plato's Socratic Conversations: Drama and Dialectic in Three Dialogues.* Baltimore: John Hopkins University Press.

Tarrant, Harold (2000): "Naming Socratic Interrogation in the *Charmides*". In: Brisson, Luc and Robinson, Thomas M. (Eds.): *Plato: Euthydemus, Lysis, Charmides. Proceedings of the V Symposium Platonicum (Selected Papers).* Sankt Augustin: Academia Verlag, pp. 251–258.

Tarrant, Harold (2002): "*Elenchos* and *Exetasis*: Capturing the Purpose of Socratic Interrogation". In: Scott, Gary Alan (Ed.): *Does Socrates Have a Method? Rethinking the Elenchus in Plato's Dialogues and Beyond.* Pennsylvania: The Pennsylvania State University Press, pp. 61–77.

Tarrant, Harold (2015): "*Elenchus* (Cross-examination, Refutation)". In: Press, Gerald Alan (Ed.): *The Bloomsbury Companion to Plato.* London and New York: Bloomsbury, pp. 165–167.

Van Den Berg, Robbert M. (2008): *Proclus' Commentary on the Cratylus in Context: Ancient Theories of Language and Naming.* Leiden and Boston: Brill.

Van Eck, Job (1994): "Σκοπεῖν ἐν λόγοις: On Phaedo 99d–103c". In: *Ancient Philosophy* 14, pp. 21–40.

Vlastos, Gregory (1975): "Plato's Testimony Concerning Zeno of Elea". In: *The Journal of Hellenic Studies* 95, pp. 136–162.

Vlastos, Gregory (1981): *Platonic Studies.* Princeton: Princeton University Press.

Vlastos, Gregory (1983): "The Socratic Elenchus". In: Annas, Julia (Ed.): *Oxford Studies in Ancient Philosophy.* Volume I. Oxford: Clarendon Press, pp. 27–58.

Vlastos, Gregory (1991): *Ironist and Moral Philosopher.* Cambridge: Cambridge University Press.

Wandschneider, Dieter (2010): "Dialectic as the 'Self-Fulfillment' of Logic". Anthony Jensen (Trans.). In: Limnatis, Nektarios G. (Ed.): *The Dimensions of Hegel's Dialectic.* London: Continuum, pp. 31–54.

Weigelt, Charlotta (2017): "Aristotelian Dialectic as Midwifery". In: *Bochumer Philosophishes Jahrbuch für Antike und Mittelalter* 20. No. 1, pp. 18–48.

Witt, Reginald Eldred (1971): *Albinus and The History of Middle Platonism.* Amsterdam: Adolf M. Hakkert.

Wolfsdorf, David (2003a): "Understanding the' What is F?' Question". In: *Apeiron* 36, pp. 175–188.

Wolfsdorf, David (2003b): "Socrates' Pursuit of Definitions". In: *Phronesis* 48. No 4: pp. 271–312.

Wolfsdorf, David (2008): "The Method ἐξ ὑποθέσεως at Meno 86e1–87d8". In: *Phronesis* 53, pp. 35–64.

Part I: **Ancient Greek Dialectic in Classical Antiquity**

François Renaud
Dialectic as True Rhetoric in Plato's *Gorgias*

The apparent contradictions in the *Gorgias* are many. Two key examples should suffice. First, Socrates claims that rhetoric is not an art (τέχνη) but a form of flattery, indifferent to the good and devoid of any usefulness (464b–466a), and then later he refers to a true rhetoric whose purpose is the improvement of citizens (cf. 503a–b, 504d, etc.). Oddly enough, this true rhetoric resembles disciplining dialectic as he practices it, primarily as refutation, ἔλεγχος. ("Dialectic" refers here to the term διαλέγεσθαι; διαλεκτική is not used in the *Gorgias*).[1] In the second example, Socrates says that he is not a politician (473e) but then later claims that he is the only person in Athens practicing true political art (521d). These contradictions compel the reader to reflect on how the discussion and the dramatic action proceed. Yet, according to a still very common reading of the dialogue, the *Gorgias* is a complete and final condemnation of rhetoric and politics. Socrates' good rhetoric may be, in his eyes, a theoretical possibility. He challenges Callicles to name a single orator who satisfied the requirements of this rhetoric (503b). At the end of the dialogue, however, Socrates refers to Aristides the Just as an example of a model politician (526a–b).[2] What, then, is the status of these superior forms of rhetoric and politics, which he explicitly claims to practice? Laurent Pernot proposes an answer, which he unfortunately does not develop further: "It seems that there has never been an example of such an orator, unless one cares to mention Aristides the Just (526a–b) and Socrates himself (521d), who are cited from a political rather than strictly rhetorical point of view".[3]

The notion of a Socratic rhetoric raises in turn other questions. Is Socrates' use of myth and other extra-logical means part of the good rhetoric to which he refers? Are these rhetorical devices compatible with the paradox, which he defends in the dialogue, that virtue is knowledge? According to some commentators,[4] the use of

---

Note: This paper is composed of translated and revised sections of my recent book, Renaud (2022). I follow the Greek text of Dodds (1959) for the *Gorgias* and that of Burnet (1900–1907) for the rest of the Platonic corpus.

[1] This does not preclude, however, the presence of division (διαίρεσις) in the dialogue, such as in the division of the arts in 462b–465e.
[2] Admittedly, true rhetoric is not discussed in any detail, which leads Goldschmidt (1963, p. 310) to this mistaken conclusion: "le *Gorgias* n'a cure de savoir ce que serait une rhétorique 'droite'".
[3] Pernot (2005), p. 48.
[4] See, for instance, Fussi (2001), Carone (2005), and Moss (2007).

rhetorical means, such as the appeal to the emotions, implies a Platonic criticism of Socratic "intellectualism". According to others, the Socratic paradox, correctly understood, is compatible with the use of rhetorical tricks.[5] Let us take as an example the notion, crucial for the dialogue as a whole, of disciplining or punishment (κολάζειν). According to the main moral thesis Socrates defends in the *Gorgias*, the greatest evil is committing injustice, not suffering it, especially if the injustice is not punished (468e–469c, 479b–e, 508c–509c). Discipline frees from ignorance and injustice, as a bitter medicine cures an illness.[6] Socrates refers to conventional procedures such as confiscation, imprisonment, exile, and death (480c–d and, in the final myth, 524e–525a, 526e–527a). He also refers to dialectic refutation as a kind of discipline, however. In both cases, he employs the term κολάζειν, which can be translated by "chastising", "punishing" or "disciplining" (I shall translate it in most cases by "disciplining" or "disciple").[7] While in some passages, Socrates evidently means conventional punishment (such as at 480d2 in the case of an execution and likewise in the final myth), in others, he has dialectic in mind as instrument of discipline.[8] How are these two divergent uses of the term κολάζειν to be explained?

The most plausible explanation,[9] which I defend in this paper, is that of the deliberate twofold use of the conventional and philosophic meanings. This usage allows Socrates to adapt to his non-philosophic interlocutors, who at first only understand the conventional notion, in order to bring them gradually, if possible, to the dialectic conception, within the dramatic action, as he makes clear on a few occasions (475d5–e1, *cf.* 480c5–7, 505c3–4 and 521e6–8). The intention of Socrates

---

5 Cf., for example, Erler (2006), Rowe (2007), and McPherran (2012).
6 While little is said of the "greatest good" (μέγιστον ἀγαθόν, of which only four occurrences, all in 452a–d), the "greatest evil" looms large (μέγιστον κακόν, μέγιστον τῶν κακῶν, ἔσχατον κακῶν: seven occurrences), especially the two greatest evils: committing injustice and committing injustice with impunity (479d4–6, 480d5–6, 492c4–8, 511a1–3, 525b8–c1). Socrates refers frequently to medicine and justice, which are restorative arts, and seldom to gymnastic and legislation, which are preventive arts that serve to maintain the good condition of the physical and political body.
7 There are sixteen occurrences of κολάζειν in the *Gorgias*, as well as twenty-six of διδόναι δίκην (literally "giving justice", "suffering punishment"), the latter being limited, however, to two sections of the dialogue (the exchange with Polos and the final myth), where the meaning of judiciary punishment predominates.
8 The term "punishment" can be employed in both cases to translate κολάζειν in the extent to which the Platonic conception of "punishment", as distinct from the modern connotations of that term, includes a therapeutic as well as a punitive function, comparable to painful but beneficial medicine. *Cf.* Mackenzie (1981), pp. 183–184, and Shaw (2015), p. 80. As is often the case in the dialogues, terminology is not strict or systematic. I shall in most cases use the term "discipline", sometimes also "punishment" when the context requires it.
9 See Rowe (2007), pp. 143–163.

in the *Gorgias* would then be twofold: to convince his interlocutors to accept the requirements of justice understood as the justice of dialogue, in default of which he can count on the recourse of constraint (517b), that is self-control, as is the case with Callicles. Socrates characterizes discipline in the same terms as true rhetoric and true politics (504e1–3, 525b8–c1). His dialectic thus seems to be identical with true rhetoric, by which the real value of conventional rhetoric can be judged.[10] The dramatic action would thus be strictly inseparable from the argumentation: Socrates defends justice as discipline against his interlocutors primarily by disciplining them. More generally, in the *Gorgias* the interlocutors discuss while illustrating the impediments to dialogue. The primary aim of the paper is to show that *this twofold use of the notion of disciplining, at both theoretical (or thematic) and dramatic (or pragmatic) levels, applies equally to the notions of rhetoric and of politics*.[11] The two meanings or dimensions, gradually unveiled during the dialogue, are part of an overall argumentative-literary strategy which can be called one of transfer or transposition.[12]

1 True Politics

Towards the end of the dialogue, just before the final myth, Socrates makes an astonishing claim, referred to above:

> I believe that I'm one of a few Athenians—so as not to say I'm the only one, but the only one among our contemporaries—to take up the true political craft and practice the true politics (ἐπιχειρεῖν τῇ ὡς ἀληθῶς πολιτικῇ τέχνῃ καὶ πράττειν τὰ πολιτικὰ μόνος τῶν νῦν) (521d6–8; trans. Zeyl in Cooper 1997).

What does he mean by the true art of politics (ἡ ὡς ἀληθῶς πολιτικὴ τέχνη)? He already referred to the art of politics (πολιτικὴ [τέχνη]) in his classification of the true and false arts. There he defined it as the art which cares for the soul (τὴν ἐπὶ τῇ ψυχῇ, 464b3–4).[13] He explains in 521d–e the meaning of his claim as follows:

---

10 See Szlezák (1985), pp. 195.
11 See Erler (2009), p. 18.
12 This is the expression used by Diès (1972) to express an interpretation that Rowe (2012, p. 192, n. 22; *cf.* viii) also defends: "a general feature of Platonic writing", in this case "a general tendency to redefine, or transform, common-or-garden notions of things".
13 The paradoxical definition of politics as the art of caring for the soul (464b3–4), which Socrates alone practices, is more easily understood if one bears in mind that the ancient concept of politics rests on that of the city (πόλις): contrary to the modern concept of the State, it does not include a

> This is because the speeches I make on each occasion do not aim at gratification but at what's best (πρὸς τὸ βέλτιστον). They don't aim at what's most pleasant (οὐ πρὸς τὸ ἥδιστον). And because I'm not willing (οὐκ ἐθέλων) to do those clever things you recommend, I won't know what to say in court. And the same account I applied to Polus comes back to me. For I'll be judged the way a doctor would be judged by a jury of children if a pastry chef were to bring accusations against him (κρινοῦμαι γὰρ ὡς ἐν παιδίοις ἰατρὸς ἂν κρίνοιτο κατηγοροῦντος ὀψοποιοῦ) (521d8–e4; trans. Zeyl).

In this explanation he takes up the same terms he used to describe true rhetoric, namely aiming for the best, rather than for pleasure as does conventional rhetoric. Is dialectic then identical to true politics and true rhetoric? To answer this question, it is first necessary to confront some of the difficulties posed by Socrates' surprising statement. First, he said earlier that he is a stranger to politics (*cf.* οὐκ εἰμὶ τῶν πολιτικῶν, 473e6). Second, possessing the true art of politics seems incompatible with his avowal of ignorance, stated more than once in the *Gorgias* (*cf.* οὐκ οἶδα, 509a5). Socrates hardly explains the nature of true political art nor how to apply it (500e, 505a–b). Third, can he claim to possess this art if he is unable to convince Callicles and improve him (*cf.* 515a–b, 517a)?

First, it is necessary to distinguish conventional from true politics. According to Socrates, in Athenian democracy the orators-politicians fight against each other for the people's favors. He sarcastically castigates the great representatives of that type of politics, Pericles, Cimon, Miltiades, Themistocles, as mere servants of the people:

> (Soc.) No, my strange friend, I'm not criticizing (οὐδ' ἐγὼ ψέγω) these men either, insofar as they were servants of the city (διακόνους εἶναι πόλεως). I think rather that they proved to be better servants than the men of today, and more capable than they of satisfying the city's appetites (ἐπεθύμει). But the truth is that in redirecting its appetites and not giving in to them (μεταβιβάζειν τὰς ἐπιθυμίας καὶ μὴ ἐπιτρέπειν), using persuasion or constraint (πείθοντες καὶ βιαζόμενοι) to get the citizens to become better (ἀμείνους ἔσεσθαι), they were really not much different from our contemporaries. That alone is the task of a good citizen (μόνον ἔργον ἐστὶν ἀγαθοῦ πολίτου) (517b2–c2; trans. Zeyl).

Far from trying to make citizens better, these politicians exploited their prejudices, one of which is "freedom" understood as the "power" to do anything that one

---

territorial or institutional dimension; the city is the whole of human beings composing it (πόλιν δὲ τὸ τῶν τοιούτων πλῆθος) according to Aristotle's definition (*Politics* III, 1275b20–21). If the soul constitutes the true human being, as Socrates claims, then it follows that it is the souls that the political art ought to care for. *Cf.* Erler (2012), pp. 279–280. The specificity of the Socratic conception, however, lies in its private character: dialectic, even practiced in the public space of the agora, addresses to interlocutors individually; on this see Renaud 2022, Chapter II, § 7.

"wishes".¹⁴ The principal cause of corruption is the mimetic character of flattery, which is the means of avoiding being a victim of injustice or any other harm and of obtaining power and goods, both material and symbolic (511c–513c). The conflict between Athenian democracy and true politics appears unresolvable, like that between democracy and true rhetoric. These two conflicts really are one and the same, opposing divergent types of discourse and ways of life.

Yet is the idea that the political art is identical with dialectic not incompatible with Socrates' avowal of ignorance?¹⁵ This difficulty is resolved if the knowledge in question is that of the dialectic art and its rules. Socrates compares dialectic refutation to medical treatment that frees from illness (477e7–8) as well as to discipline that frees from ignorance (505c3–4). The medical and juridical analogies are interdependent; as we have seen, both aim at illustrating dialectic as discipline. By means of dialectic Socrates fulfils the function which institutional politics should but do not fulfil, namely that of improving citizens. If Socrates fails in convincing Callicles, this is because the latter violates the dialectical rules and is one of the incurable souls referred to in the final myth.

2 True Rhetoric

2.1 Gradual Unveiling

According to the classification of the arts established by Socrates (463e–466a), rhetoric is a counterfeit image of a part of politics (πολιτικῆς εἴδωλον, 463d2). He refers also, however, to a true rhetoric that cares for the soul (ἡ ἀληθινὴ ῥητορική, 517a5, cf. 503a2–9, 504d5–6). The status of rhetoric is revealed only gradually, in seven steps:

(1) 454e–455d. Socrates distinguishes at first, with Gorgias' agreement, two kinds of persuasion (δύο εἴδη πειθοῦς), one that conveys knowledge (εἰδέναι, μάθησις, ἐπιστήμη), the other mere belief (πίστις, 454e3–4). Rhetoric is the kind of persuasion used in tribunals or public assemblies (452e1–4); it "produces the

---

14 Cf. Aristotle, Politics V, 1310a32: τὸ ὅ τι ἂν βούληταί τις ποιεῖν.
15 Irwin (1979, pp. 240–241) translates ἐπιχειρεῖν by "undertake" or "attempt"; Socrates' declaration would then imply an important qualification: he undertakes or attempts to practice the true political art but does not yet possess the knowledge of its principles nor of its proper application. Shaw (2011, pp. 188–190) shows that the passage (521d6–8) allows for both interpretations ("I practice" and "I attempt") and that the larger context must be considered, after which he defends (pp. 193–195) the interpretation that Socrates attempts to practice but does not possess the political art by referring especially to 509a4–7 and 515a3–4.

persuasion that comes from being convinced, and not the persuasion that comes from teaching (πιστευτικῆς ἀλλ' οὐ διδασκαλικῆς), concerning what's just and unjust" (454e9–455a2; trans. Zeyl).

(2) 480c–d. Socrates then admits that rhetoric can be useful after all, if it fulfills a function contrary (τοὐναντίον) to that of flattery that is normally attributed to it. The useful function is that of accusing (κατηγορεῖν) and disciplining (κολάζειν): accusing oneself and one's relatives when injustice is committed (480c1-3). This function restores the health of the soul and gets rid of "the worst thing there is", injustice, together with ignorance on which it is based.

(3) 503a–b. Socrates recognizes the distinction, proposed by Callicles, between two types of rhetoric (διπλοῦν): flattery or demagogy (κολακεία, δημηγορία) and noble rhetoric, the function of which consists in "striving valiantly (διαμάχεσθαι) to say what is best, whether the audience will find it more pleasant or more unpleasant (εἴτε ἡδίω εἴτε ἀηδέστερα ἔσται τοῖς ἀκούουσι)" (503a8–9; trans. Zeyl).

It is remarkable that Callicles proposes this distinction between the two types of orators, and that Socrates passively accepts the distinction. He had asked the (binary, structuring) question of whether the orators speak with a view to the good or to pleasure. "This issue you're asking about, Callicles responds, isn't just a simple one (οὐχ ἁπλοῦν ἔτι τοῦτο ἐρωτᾷς)". Socrates' question actually requires a distinction.[16] Socrates answers: "That's good enough (ἐξαρκεῖ)". This response is surprising. It is as though Socrates had been waiting to be challenged by Callicles before taking up again the distinction he himself proposed earlier, between a good and a bad πειθώ (454e–455d). As we will see, the two distinctions are basically the same, despite the fact that the earlier one is formulated more simply in terms of knowledge and belief. The passive role played by Socrates, in this reformulated distinction, can be explained as follows. He agrees with Callicles that this other type of rhetoric is still unknown (οὐ πώποτε, 503b1), but he does not categorically deny that such a rhetoric exists (indeed he will explain its nature shortly after, at 504d5–6); he places the burden of proof on Callicles. Callicles is unable to give any examples of its existence among contemporary orators(-politicians), but mentions great politicians of the past, Themistocles, Cimon, Miltiades, and Pericles. Then follows Socrates' scathing attack, quoted above, against the four famous politicians, whom he describes as providers of "goods" rather than reformers. From that moment on, Socrates refers, in his own name, to that true, noble rhetoric.

(4) 504d–e. Immediately thereafter, Socrates himself refers to the figure of the good orator:

---

[16] Cf. 468c2–5, where Socrates requires that two questions be distinguished where Polos sees only one; *Lach.* 188c5: (Lach.) οὐχ ἁπλοῦν ἀλλὰ διπλοῦν.

> (Soc.) So this is what that skilled and good orator (ὁ ῥήτωρ ἐκεῖνος, ὁ τεχνικός τε καὶ ἀγαθός) will look to when he applies to people's souls whatever speeches he makes as well as all of his actions, and any gift he makes or any confiscation he carries out. He will always give his attention (πρὸς τοῦτο ἀεὶ τὸν νοῦν ἔχων) to how justice may come to exist in the souls of his fellow citizens and injustice be gotten rid of (ἀδικία δὲ ἀπαλλάττηται), how self-control may come to exist there and lack of discipline be gotten rid of, and how the rest of excellence may come into being there and badness may depart (504d5–e3; trans. Zeyl).

We note that the function of good rhetoric, which "gets rid of injustice" (ἀδικία ἀπαλλάττηται), corresponds exactly to the good use of rhetoric at 480c–d, namely, that of accusing and disciplining, which also eliminates injustice (ἀπαλλάττωνται ἀδικίας), "the worst thing there is", and ignorance.

(5) 508b–c. Socrates then refers explicitly to his discussion at 480c–d as follows:

> (Soc.) These consequences are all those previous things, Callicles, the ones about which you asked me whether I was speaking in earnest when I said [480c-d] that a man should be his own accuser, or his son's or his friend's, if he's done anything unjust, and should use rhetoric for that purpose (τῇ ῥητορικῇ ἐπὶ τοῦτο χρηστέον). Also, what you thought Polus was ashamed to concede is true after all, that doing what's unjust is as much worse than suffering it as it is more shameful, and that a person who is to be an orator the right way should be just and be knowledgeable in what is just (ἐπιστήμονα τῶν δικαίων), the point Polus in his turn claimed Gorgias to have agreed to out of shame (508b3–c3; trans. Zeyl, slightly modified).

Like true politics, true rhetoric rests on knowledge. "Knowledge" here is to be understood primarily as the knowledge of the dialectical rules, as expounded throughout the *Gorgias*, the most fundamental of which is the requirement of consistency in speech (logical consistency) and between speech and deeds (or moral consistency).

(6) 516e–517a. Next, Socrates establishes, in a not fully explicit manner, the link between true rhetoric and true politics, both of which he claims are unknown in Athens:

> (Soc.) So it looks as though our earlier statements were true, that we don't know any man who has proved to be good at politics in this city (ἄνδρα ἀγαθὸν γεγονότα τὰ πολιτικά). You were agreeing that none of our present-day ones [*scil.* Themistocles, Cimon, Miltiades, Pericles] has, though you said that some of those of times past had, and you gave preference to these men. But these have been shown to be on equal footing with the men of today. The result is that if these men were orators, they practiced neither the true rhetoric (οὔτε τῇ ἀληθινῇ ῥητορικῇ ἐχρῶντο)—for in that case they wouldn't have been thrown out—nor the flattering kind (οὔτε τῇ κολακικῇ) (516e9–517a6; trans. Zeyl, slightly modified).

(7) 527b–c. At the very end of the dialogue, after the final myth, Socrates comes back to the good function of rhetoric, namely accusation and disciplining (or punishing, κολάζειν):

> and that if a person proves to be bad in some respect, he's to be disciplined (κολαστέος), and that the second best thing after being just is to become just by paying one's due, by being disciplined (κολαζόμενον διδόναι δίκην); [...] and that rhetoric and every other activity (τῇ ῥητορικῇ οὕτω χρηστέον... καὶ τῇ ἄλλῃ πάσῃ πράξει) is always to be used in support of what's just (527b7–c4; trans. Zeyl, slightly modified).

On the whole, the repeated references to the good rhetoric as well as to the close links between it and the true political art, raise the following questions: is true rhetoric in the *Gorgias* the same as the philosophic rhetoric in the *Phaedrus*? Is it the instrument of true politics? And if so, does Socrates himself practice true rhetoric in the *Gorgias*?

2.2 Ancient Readings

It is instructive to look at the ancient responses to these questions. In Antiquity, the ways of understanding the relationship between philosophy and rhetoric are generally characterized by two contrary hegemonic tendencies: the dominance of rhetoric by philosophy and the dominance of philosophy by rhetoric. This antagonistic relationship does not exclude annexation nor partial appropriation of one by the other. Philosophy, notably in the case of Plato, tends to appropriate rhetoric and some of its techniques while assigning a new aim to them.

Cicero (106–43 BCE) holds the criticism of rhetoric in the *Gorgias* to be self-contradictory. In the *De oratore* (I 47) he expresses, through the mouth of Crassus, his impressions when reading the dialogue: "what impressed me most deeply about Plato in that book was, that it was when making fun of orators that he himself seemed to me to be the consummate orator (*mihi [in] oratoribus inridendis ipse esse orator summus videbatur*)" (trans. Sutton, Rachham). The *Gorgias* displays, according to him, a tension between theory and practice as Plato, and with him Socrates, reluctantly recognize the necessity of rhetoric in displaying his own rhetorical talents.[17] Aelius Aristides (c. 117–181 CE) comments on this question in a far more detailed and critical manner than does Cicero. He exploits especially what he considers a contradiction between his wholesale condemnation of rhetoric

---

[17] Ciceron, *De oratore* III, 129: *si est victor, eloquentior videlicet fuit et disertior Socrates*. On Cicero's take on the *Gorgias* and its relation to the *Phaedrus*, see Renaud (2018).

and his occasional defense of it. He opposes Plato to Plato. He cites the references to the good rhetoric in the *Gorgias* as well as the defense of rhetoric as an art (τέχνη) in the *Phaedrus*.

According to Quintilian (c. 35–95 CE), however, the *Gorgias* does not contradict the *Phaedrus*, the latter is merely more explicit and detailed on true rhetoric.[18] The *Gorgias* as refutative dialogue deals with the rhetoric of its time (*tum*) while recognizing a true rhetoric (*veram autem et honestum*). Quintilian refers to Socrates' evocation of the good orator (508c1). The Roman rhetor even conceives of his own rhetorical ideal in Platonic terms: the orator does not seek the apparent, but the true; the prize is not victory of a cause, but good conscience.[19] Socrates' argumentation itself is rhetorical to the extent to which he addresses an adversary.[20] There would not be for that reason any incompatibility between the Socratic paradox (460c) and this rhetorical practice.[21] Apuleius (c. 124–170 CE), in his *De dogmate Platonis* (Book II), takes up the distinction between two kinds or two parts (*partes*) of rhetoric: one is a science (*disciplina*) that teaches how to know the good and how to live according to justice, with a view to the science of politics; the other is the science (*sic*) of flattery (*adulandi scientia*) which finds probable arguments without employing reasoning. He links the notion of mere use (τριβή, 463b4) with persuasion (πείθειν) without knowledge or teaching (διδάσκειν, 454e–455a), which is only a shadow or image (*umbram, imaginem*) of science. According to Olympiodorus, finally, Socrates is the only politician, although he identifies rational rhetoric more with Plato than Socrates (*cf.* 9.4).[22] Olympiodorus responds to criticisms of Plato, including those of Aelius Aristides, by showing how the *Gorgias* can be read not in opposition to but in continuity with the *Phaedrus*. Socrates uses various types of discourse according to the type of soul he is dealing with, as he recommends in the *Phaedrus* (269c–272b). Olympiodorus insists on the medical analogy in the *Phaedrus* (270b–c) equally central in the *Gorgias*, including in connection with true politics (517a–b and above all 521d). In short, according to Quintilian and the Platonic tradition as attested by Apuleius and Olympiodorus, the Socrates of the *Gorgias* recognizes two rhetorics or two uses of rhetoric, one juridical or conventional and the other philosophical.

---

18 Quintilian, *Institutio oratoria* II, 15, 24–29.
19 Quintilian, *Institutio oratoria* II, 15, 27, 28 and 32 respectively.
20 Quintilian, *Institutio oratoria*. II, 15 and 28: for instance, when he speaks to Polos: *contra quem illa de simulacro et adulatione*.
21 Quintilian, *Institutio oratoria* II, 15 and 29.
22 According to *The Anonymous Prolegomena to Platonic Philosophy*, closely related to the School of Olympiodorus, Plato intends "to explain what true rhetoric is (τὴν ἀληθῆ ῥητορικὴν)" (22.40).

2.3 True Rhetoric and Dialectic

In the *Apology* (17b), Socrates defends himself against his accusers' claim he would be clever at speaking (δεινὸν λέγειν): he will speak as he is accustomed to, by telling the truth (τὸν τἀληθῆ λέγοντα). He admits that he too is an orator (ῥήτωρ) but not in their manner.[23] Yet this does not prevent him, in this exordium, from resorting to various devices of conventional rhetoric. Socrates knows these tricks, although he claims not to use them. His speech thus pretends to be non-rhetorical: he will simply be using the words that come to him haphazardly (ἐπιτυχοῦσιν). So-called improvisation is one of the *topoi* of sophistic rhetoric, as is the reversal of common opinion to create the effect of surprise and striking revelation. In short, the beginning of the *Apology* defends the new rhetoric of frankness while revealing Socrates' ability to employ conventional rhetoric at the very moment he says he is not using it.[24]

Contrary to the ancient readers, modern scholars commonly oppose the *Gorgias* and the *Phaedrus* on the ground that the *Gorgias* presents itself primarily as an attack on rhetoric, while the *Phaedrus* defends and expounds in some detail the notion of philosophic rhetoric. Scholars usually explain this contrast in terms of Plato's development. They usually suppose, moreover, that the philosophic rhetoric of the *Phaedrus* is for Plato no more than a program to be realized. This rhetoric, then, could hardly be the one practiced in the *Gorgias*. Some modern commentators, however, including Eric Robertson Dodds, consider the good rhetoric referred to in the *Gorgias* to be the same as that described in the *Phaedrus*.[25] In both dialogues, the good orator is an expert (τεχνικός, *Gorg.* 504d5, *Phdr.* 262b5), although compared to the *Phaedrus*, the *Gorgias* deals less with the rhetorical techniques than with the psychological conditions of persuasion.

What then distinguishes the two dialogues, beyond the fact that one is more explicit in regard to true rhetoric? According to Harvey Yunis (2007), the innovations in the *Phaedrus*, in comparison with Plato's previous dialogues (in the standard chronology) and his predecessors, are threefold. First, the *Phaedrus* would expand the scope of rhetoric: in addition to the political dimension proper (i.e., rhetoric as oratory), underscored in the *Gorgias*, the *Phaedrus* would include the private sphere, that is (philosophic) dialogue (261a7–9). Second, rhetoric understood as the art of "directing the soul" (ψυχαγωγία) would include psychology:

---

23 This truth-telling will be taken up by the Stoics as the only rhetoric, or "the science of speaking well" (ἐπιστήμη τοῦ εὖ λέγειν; SVF II, 293).
24 On the compatibility between the requirement of frankness and irony, see Aristotle, *Nicomachean Ethics* IV, 1124b28–1125a2, as well as *Posterior Analytics* 13, 97b14–26.
25 Dodds (1959), p. 330.

its object is the soul[26] and rests on the methodical and systematic knowledge of the different types of soul and on the different types of discourse corresponding to them (271b–272b). Third, the aim of the orator's knowledge of the subject matter is not so much the public's best interest (as in the *Gorgias*) as persuasion itself (259e–262c). As rhetoric, dialectic includes knowledge of the right moment (καιρός), that is, with whom, when, and how to speak.

The first innovation pointed out by Yunis, namely, the extension to the private sphere, requires, however, a significant qualification. It is true that the rhetoric in the *Phaedrus* encompasses the whole range of the sayable, that is, that its scope is universal (261d10–e4):

> (Socrates) Well, then, isn't the rhetorical art, taken as a whole, a way of directing the soul by means of speech (τέχνη ψυχαγωγία τις διὰ λόγων), not only in the lawcourts and on other public occasions but also in private (καὶ ἐν ἰδίοις)? (261a7–9; trans. Nehamas and Woodruff in Cooper 1997).

In the *Phaedrus*, therefore, the main subject is not conventional or forensic rhetoric, the prime object of criticism in the *Gorgias*, but philosophic rhetoric. The novelty of this rhetoric is indicated by Phaedrus' astonishment (261b2). The universal dimension of rhetoric makes it inseparable from dialectic. It is, however, incorrect to claim, as Yunis does, that this extension is absent in the *Gorgias*. At the very end of the dialogue, Socrates sums up his argument, from both the methodological and moral point of view:

> and that every form of flattery (πᾶσαν κολακείαν), both the form concerned *with oneself* (περὶ ἑαυτὸν) and that concerned with others, *whether they're few or many* (καὶ περὶ ὀλίγους καὶ περὶ πολλούς), is to be avoided, and that rhetoric and every other activity is always to be used in support of what's just (527c1–4, trans. Zeyl, emphasis added).

Regarding the question of the practice of this rhetoric, Socrates in the *Phaedrus* speaks with caution, if not with skepticism, concerning the very existence of that art ("if indeed this art exists": εἴπερ ἔστιν, 261e2). The two contradictory speeches improvised by Socrates in the first part of the *Phaedrus* demonstrate, however, his talents as orator as well as the possibility of a one-to-one encounter, in this case between the young Phaedrus and him.[27] Here, as in the *Gorgias*, the dramatic action completes the argument.

---

26 Cf. Yunis (2007), p. 84.
27 Cf. Narcy (2007), p. 955.

3 Socratic Rhetoric, Platonic Rhetoric

The noble rhetoric referred to in the *Gorgias* thus overlaps that of the *Phaedrus*. But does Socrates practice it? Most commentators believe Socrates does not practice the rhetoric of the *Phaedrus*, neither in the *Gorgias* nor elsewhere in the Platonic corpus, on the ground this would be incompatible with his avowal of ignorance. Moreover, he would not give any account of (*logos*), condition for the possession of any art (465a), limiting himself to some allusions. He would not possess that art since he fails in persuading his main interlocutor, Callicles.[28] According to other commentators, Socrates does not practice the rhetoric of belief (πιστευτική, 454a–455a), defended by Gorgias, either; that would be incompatible with the opposition, on which he insists so much, between dialectic and rhetoric. Finally, according to many, the rhetorical dimension of the Platonic dialogues would be limited to the function of mere literary reinforcement, adding nothing substantial to the argumentation nor to the nature of dialectic.[29]

Consistent with my earlier remarks about true politics, I hold that dialectic in the *Gorgias* is none other than true rhetoric. But does Socrates practice the rhetoric he teaches? Consider first the question of whether he denies teaching anything (*Apol.* 23d). That depends on what should be understood by teaching (διδάσκειν). If that term is understood as the straightforward transmission of knowledge, then dialectic is not teaching due to its questioning and maieutic function. In the *Sophist* (229b7–230d5), however, the art of refutation (ἔλεγχος), which is in its purgative function very much akin to Socratic dialectic, is presented as a form of teaching (διδασκαλική). Socrates "teaches" in making the interlocutor conscious of his own ignorance and in leading him through the art of maieutic to draw knowledge from within.[30] The appeal to the emotions is also compatible with this rhetoric. Socrates' employment of devices specific to conventional rhetoric does not imply a criticism or rejection of intellectualism (or the Socratic paradox) nor the approval of Gorgias' rhetoric.[31] As we have seen in the preceding section

---

28 Roochnik (2007), p. 79.
29 It is common to recognize, for example, the rhetorical character of the *Apology*'s exordium, referred to earlier, without however always drawing the consequences for the interpretation of Socrates' speech nor of the *Apology* as Platonic writing. Luc Brisson (1997, p. 129, n. 2), for instance, shows the extent to which Socrates (and so Plato making him speak) masters the *topoi* of judiciary rhetoric: contrary to what he says, he "*semble connaître les ficelles du métier*". Brisson does not further explore the rhetorical dimension of the *Apology*, nor the Socratic (or Platonic) irony it implies.
30 On the unity of *elenchos* and maieutic, see Renaud (2001), pp. 729–730.
31 Cf. McPherran (2012), p. 14–24, in reply to Fussi (2001).

concerning true politics, dialectic understood as rhetoric has two functions, one rigorously intellectual, such as refutation, the other extra-logical.[32] In the case of Polos and Callicles, Socrates appeals to the emotions: he tries to change their false opinions by frustrating and not by satisfying their desires, as the rhetoric of flattery does.

3.1 Rhetorical Devices

Socrates' rhetoric resides first of all in the fact that his argumentation is adapted to his interlocutor. This adaptation implies a strategic dimension, including the often progressive or gradual character of his argumentation. Socrates first seeks to focus his interlocutor's attention, then to maintain the continuity of the exchange, to shake his self-confidence, and to discipline through refutation. In the *Gorgias*, Socrates attempts to have his interlocutors admit that true power must be subjected to the constraints of justice. To this end, he deploys an argumentation that makes use of rhetorical and polemical elements, in a more striking way than in other dialogues, given his antagonistic interlocutors. His use of irony, especially in his avowal of ignorance, also fulfils strategic functions.[33] Socrates often proposes premises which he does not himself accepts, but which eventually allow him to refute his interlocutors' opinion.[34] His argumentation is sometimes elliptic, based on premises which have not been defended and which are merely accepted by the interlocutor (for instance concerning the existence of the arts, τέχναι, and the soul, ψυχή), as opposed to other occasions when the premises are defended at length. He employs fictive questions which help clarify an issue, and work to lead the exchange in a specific direction.[35] He sometimes resorts to speeches, notably in the case of myths or more generally of emotionally charged diction. These speeches belong to

---

[32] Cf. Collobert 2013, p. 115–131.
[33] Socrates appears to acquiesce to the accusation that he himself does not respect the principle of frankness (495b2). In the *Apology*, he practices judiciary *elenchos* (with the exception of the dialectical exchange with Meletus, in 24d–28a), condemned in the *Gorgias:* in the third part of his defense especially, he appeals to testimonies (πολλοὶ μάρτυρες, 32e1). Dorion (2007, pp. 88 and 89) concludes from this that "Platon confie à l'*elenchos* rhétorique, plutôt que dialectique, le soin d'assumer la partie la plus déterminante de la défense de Socrate contre les accusations de 399. [...] [S]a défense invite à penser qu'il [*scil.* Socrates] était conscient des limites de l'*elenchos* dialectique". The aim of the *Gorgias* consists precisely to reveal the strengths and limits of dialectics and therewith the insurmountable conflict between two types of discourse and two kinds of life. On this larger question see Renaud (2022), especially Chapter II, § 6.4.
[34] Compare, for example, *Gorg.* 474c and 475b–c.
[35] Cf., for instance, 451a–b, 452a–c, 455c–d, as well as Renaud (2022), Chapter I, § 2.

the same category of devices as the exposition of the division of the arts, a speech whose structure is antithetical, the language full of imagery and the harmonious assonances.³⁶ Refutation appeals to feelings of shame (αἰσχύνη), in its philosophical meaning, that is self-regarding, as opposed to conventional, other-regarding shame.³⁷ Socrates provokes Callicles' anger³⁸ which leads him to defend a radical form of hedonism.³⁹ The exchange with Callicles alternates, according to the latter's moods, between demonstration and persuasion. Socrates uses logical tactics, especially homonymy (for instance the equivocal phrase εὖ πράττειν which can mean "acting well" and "being happy": 497a3, 507c3–5). Thus, some of the argumentative strategies referred to by Aristotle in Book VIII of the *Topics* are at work in the *Gorgias*, such as hiding from the interlocutor the conclusion at which the questioner is aiming (155b23) and disguising the argument's premises (156a7–13).⁴⁰ Socrates recognizes to having behaved as a popular haranguer,⁴¹ compelled that he was by Callicles' refusal to respond. While Socrates does need his interlocutor's assent as a condition of truth (500e3–4), he nevertheless can do without if need be.⁴² He also parodies some Gorgianian figures of speech.⁴³

---

36 For example, the assonances characteristic of the beginning of his speech (464b3): δυοῖν ὄντοιν τοῖν πραγμάτοιν. Cf. Dalfen (2004), p. 241. The personification of the Laws in the *Crito* resorts to sarcasm, antithesis, rhetorical questions, and impassioned commands and appeals. The *Hippias Major* is one of the most evident cases of a massive use of irony.
37 On this transposition of the notion of shame, as well of rhetoric and politics, see Renaud (2022), *passim*.
38 For example, 490c8–d1, d6, d10, e4, 490e9–491a3.
39 Cf. Gentzler (1995), pp. 36–38.
40 Cf. Aristotle, *Sophistic Refutations* 12, 172b35–173a30. On the question of the relationship between Platonic dialectic and Aristotelian dialectic, see especially Narcy (1984), pp. 159–178.
41 519d5–6: (Soc.) ὡς ἀληθῶς δημηγορεῖν με ἠνάγκασας, ὦ Καλλίκλεις, οὐκ ἐθέλων ἀποκρίνεσθαι. According to Polos and Callicles, Socrates is ironic (εἰρωνεύῃ, 489e1), sophistic, and eristic (σοφίζῃ, 497a6; φιλόνικος, 515b5), in short, a popular haranguer (δημηγόρος, 482c5), appealing to "crowd-pleasing vulgarities (φορτικὰ καὶ δημηγορικά)" (482e3–4). The δημηγόρος at the time is the one who makes speeches (cf. *Prot.* 329a, 336b), uses doubtful devices such as the appeal to popular opinions and sentiments, in order to win the approval of the crowd (482c, 494d).
42 Cf. 505d8–e3: (Soc.) εἷς ὢν ἱκανὸς γένωμαι.
43 Such as in 467b11: ὦ λῷστε Πῶλε. Cf. Philodemus (*De vitiis* 22, 30–32) criticizes Socrates for his use of ironic epithets in 473d3: ὦ γενναῖε Πῶλε ("noble Polos") and in 494d4: ἀνδρεῖος γὰρ εἶ ("for you're a brave man"). For a general discussion of the stylistic aspect of Platonic writing, see Norden (1915), pp. 104–113, as well as Demetrius (*De elocutione* 5, 205–298), especially on allusion (ἐσχηματισμένον, 287) and the employment of terms with equivocal meanings (πολλαχῇ ἐπαμφοτερίζουσιν, 291), deliberate ambiguous uses of words, irony (εἰρωνεία) which causes perplexity (ἀπορίαν).

3.2 New Purpose

Does the use of such stratagems undermine the logical value of the argumentation?[44] Let us phrase the question differently: according to what Platonic criteria can the employment of deliberate sophisms be justified? First, as Socrates himself says, he is sometimes forced to resort to long speeches: for example, if the interlocutor does not understand, Socrates must explain his thought in the form of a speech, such as in the case of the classification of the arts. He considers the use of this method to be justified (δίκαιον, 466a2). Speeches are also necessary[45] when Callicles refuses to respond. Moreover, Socrates' argumentation is sometimes agonistic and provocative, but the victory aimed at is that of truth or at least the moral improvement of interlocutor. For dialectic is not a purely logical activity. One must distinguish between the form and the purpose of argumentation (*cf.* τοῦ ἕνεκα, 457e1). As the *Euthydemus* shows, sometimes the philosopher's ethical intention alone distinguishes him from the sophist. The overall criticism of the *Gorgias* therefore does not imply a complete rejection of judiciary or conventional rhetoric, but the redefinition of its purpose.[46] The conventional purpose (the power or success of the orator, according to Plato) becomes the liberation from ignorance and sometimes the attainment of truth (472b6) or at least the inculcation of self-control (492a–c, 508a, etc.).

4 Rhetorical Practice and Dialectic: The Final Myth

I will limit myself to myth, a key example of Socrates' rhetorical practice. The final myth offers another specimen of Socrates' rhetoric which does not flatter but disciplines. This rhetoric, as we have seen, overlaps in many important ways with dialectic. It includes private conversation but, just like dialectic, it is used in two ways: one is strictly rational or argumentative, the other appeals to the emotions and aims not at demonstration but persuasion.

The final myth can seem at first sight to defend a form of moral optimism.[47] It does evoke a world in which justice rules, but one should not consider only what the myth says, but also to whom it is addressed and to what aim. In what sense

---

44 Schofield (2000), pp. 194–195.
45 505e3: ἀναγκαιότατον; 519d6: με ἠνάγκασας.
46 Cf. Erler (2006), p. 84, and McCoy (2008), p. 136.
47 Cf. Annas (1982), p. 123.

exactly and to what extent does the myth have on the one hand dialectical value and on the other rhetorical value (in its conventional meaning)?

During the narration and explanation of the myth, Socrates addresses Callicles directly, in the vocative (ὦ Καλλίκλεις), no less than seven times, the last of which ends the dialogue (524a8, d4, 525e5, 526a4, c3, d3, 527e7). One notes, moreover, that Socrates resorts to the final myth *after* the dialectic impasse with Callicles. Since dialectic failed, Socrates hopes by means of a myth to instill mere belief in Callicles. Socrates thus implicitly recognizes the inability of dialectic to convince non-philosophers, such as Callicles, who refuse the rules of dialectic. Earlier (in 492d–494a), Socrates uses the vocabulary of belief (πίστις and πειθώ, and cognates),[48] and he tells Callicles the brief myth of the water carrier and the leaking jar. Socrates first compares desire (ἐπιθυμία) to a leaky jar which the water carriers in Hades have to refill continually (493b–c). He then makes the following remark:

> This account is on the whole a bit strange (τι ἄτοπα); but now that I've shown it to you, it does make clear what I want to persuade you (πεῖσαι) to change your mind about if I can: to choose the orderly life, the life that is adequate to and satisfied with its circumstances at any given time instead of the insatiable, undisciplined life. Do I persuade you at all (πείθω τί σε), and are you changing your mind to believe that those who are orderly are happier than those who are undisciplined, or, even if I tell you many other such stories (ἄλλα πολλὰ τοιαῦτα μυθολογῶ), will you change it none the more for that? (493c3–d3; trans. Zeyl).

Callicles once more expresses his incredulity: "You do not convince me (οὐ πείθεις), Socrates" (494a6). This rhetoric of belief (πίστις) appears to be the one Socrates identified in 454a–455a as one of the two forms of persuasion (πειθώ), opposed to that by teaching (διδάσκειν).

This use of rhetoric aims at imparting true opinions and inculcating the virtues, above all self-control. In the case of the final myth, rhetoric aims to replace Homeric poetry through stories that are both true and edifying: they are shown to be true once the λόγος in the μῦθος has been brought out and explicated. Socrates considers opinion (δόξα and πίστις in the *Gorgias*) to be a starting point towards knowledge (ἐπιστήμη). The transformation of opinion into knowledge is the ultimate goal of dialectic. That is why other dialogues, the *Phaedo* and the *Republic* for instance, seek to justify rationally what, in the *Gorgias* myth, is merely presupposed, the survival of the soul after death and the nature of justice. The judiciary reform of Zeus requires the nakedness and solitude of the soul facing judgement,

---

[48] 493c5, d1 and 494a3, 6; *cf.* 526d4, 527c5; cf. Tarrant (1990), p. 22.

in this case through the unveiling of injustice and the required punishment, an image for the disciplining of dialectic through refutation (ἔλεγχος).

The philosophical quest is itself a fight, a heroic combat: "And I call on all other people as well, as far as I can—and you especially I call on in response to your call—to this way of life, this contest (ἀγῶνα), that I hold to be worth all the other contests in this life" (526e1–4; trans. Zeyl) (526e1–4).[49] It is no longer a matter of fighting for a personal victory[50] and the glory resulting from it but rather of fighting for justice and truth, that is, dialogue and self-criticism. Socrates thus takes up the rhetorical model as contest (cf. 456c7–8) and turns it against itself, transforming it in philosophical terms into the notion of discipline as he does in the case of politics.

In short, I have defended three theses. First, in the *Gorgias*, Socrates' dialectic, conceived and practiced as discipline has two distinct functions, according to the context and the interlocutor: to refute and demonstrate on the one hand, and persuade by extra-logical means on the other. Second, the notions of discipline rhetoric and politics as conceived and employed by Socrates have a double meaning, philosophically and conventionally. And third, Socratic dialectic in the *Gorgias* coincides, in a largely implicit manner, with true rhetoric and true politics.

Bibliography

Annas, Julia (1982): "Plato's Myths of Judgement". In: *Phronesis* 27, pp. 119–143.
Aristides Aelius (1976–1980): *P. Aelii Aristidis Opera quae existant omnia*. Charles Allison Behr and Friedrich Walter Lenz (Eds.). Leiden: Brill.
Aristotle (1957): *Politica*. William David Ross (Ed.). Oxford: Clarendon Press.
Aristotle (1958): *Topica et Sophistici elenchi*. William David Ross (Ed.). Oxford: Clarendon Press.
Aristotle (1959): *Ethica Nicomachea*. Ingram Bywater (Ed.). Oxford: Clarendon Press.
Aristotle (1964): *Analytica Priora et Posteriora*. William David Ross (Ed.). Oxford: Clarendon Press.
Arnim, Hans Friedrich August von (3 Vols.) and Adler, Maximilian (4 Vols.) (Eds.) ([1903–1905, 1924], 1964): *Stoicorum Veterum Fragmenta*. Leipzig: Teubner.

---

49 *Cf. Rép.* X, 608b4–8: "Yes, for the struggle (ὁ ἀγών) to be good rather than bad is important (μέγας), Glaucon, much more important than people think. Therefore, we mustn't be tempted by honor, money, rule, or even poetry into neglecting justice and the rest of virtue (ἀμελῆσαι δικαιοσύνης τε καὶ τῆς ἄλλης ἀρετῆς)" (trans. Grube, revised by Reeve in Cooper 1997).
50 As Socrates already declares in 505e4–5, the "victory" is no longer that of having the last word but of discovering the truth: "all of us ought to be contentiously (φιλονίκως) eager to know what's true and what's false about the things we're talking about (πρὸς τὸ εἰδέναι τὸ ἀληθές τί ἐστιν περὶ ὧν λέγομεν καὶ τί ψεῦδος)". He thus subverts ironically the usual meaning of the term φιλονικία (*cf.* 457d4, e4–5, 515b6).

Brisson, Luc (1997): Platon, *Apologie de Socrate, Criton*. Translation and notes by Luc Brisson. Paris: GF-Flammarion.
Burnet, John (Ed.) (1900–1907): *Platonis opera*. 5 Vols. Oxford: Oxford University Press.
Carone, Gabriela Roxana (2005): "Socratic Rhetoric in the *Gorgias*". In: *Canadian Journal of Philosophy* 35, pp. 221–241.
Cicero (1989): *De oratore On the Orator*. Books I and II. Edward W. Sutton and Harris Rackham (Trans.). Cambridge: Harvard University Press.
Collobert, Catherine (2013): "La rhétorique au cœur de l'examen réfutatif socratique: le jeu des émotions dans le *Gorgias*". In: *Phronesis* 58, pp. 107–138.
Cooper, John M. (1997): *Plato: Complete Works. Edited, with introduction and notes by J. M. Cooper (associate editor D. S. Hutchinson)*. Indianapolis and Cambridge: Hackett.
Dalfen, Joachim (2004): Platon, *Gorgias*, In: *Platon Werke, Übersetzung und Kommentar*. Vol. III. Göttingen: Vandenhoeck & Ruprecht.
Demetrius (1902): *Demetrius on Style: The Greek Text of Demetrius'* De Elocutione. William Rhys Roberts (Ed.). Cambridge: Cambridge University Press.
Diès, Auguste (1972): "La transposition de la rhétorique" [1913]. In: Diès, Auguste: *Autour de Platon. Essai de critique et d'histoire*. Paris: Les Belles Lettres, pp. 402–432.
Dorion, Louis-André (2007): "*Elenchos* dialectique et *elenchos* rhétorique dans la défense de Socrate". In: *Antiquorum Philosophia* 1, pp. 75–90.
Erler, Michael (2006): *Platon*. München: Beck.
Erler, Michael (2009): "Platons *Politeia* und sokratische 'Politik'". In: *Latein und Griechisch in Baden-Württemberg. Deutscher Altphilologenverband. Mitteilungen des Landesverbandes B-W* 37, pp. 12–24.
Erler, Michael (2012): "Platon bei Werner Jaeger". In: Erler, Michael and Neschke-Hentschke, Ada (Eds.): *Argumenta in dialogos Platonis Teil II: Platoninterpretation und ihre Hermeneutik vom 19. bis zum 21. Jahrhundert. Akten des internationalen Kolloquiums vom 7. bis 9 Februar 2008 im Istituto Svizzero di Roma*. Basel: Schwabe, pp. 265–283.
Fussi, Allessandra (2001): "Socrates' Refutation of Gorgias". In: *Proceedings of the Boston Area Colloquium in Ancient Philosophy* 17, pp. 123–145.
Gentzler, Jyl (1995): "The Sophistic Cross-Examination of Callicles in the *Gorgias*". In: *Ancient Philosophy* 15, pp. 17–43.
Goldschmidt, Victor (1963): *Les Dialogues de Platon: Structure et méthode dialectique*. Paris: Presses Universitaires de France.
Mackenzie, Mary Margaret (1981): *Plato on Punishment*. Los Angeles: University of California Press.
McCoy, Marina (2008): *Plato on the Rhetoric of Sophists and Philosopher*. Cambridge: Cambridge University Press.
McPherran, Mark L. (2012): "Socrates' Refutation of Gorgias: *Gorgias* 447c–461b". In: Kamtekar, Rachana (Ed.): *Virtue and Happiness: Essays in Honour of Julia Annas. Oxford Studies in Ancient Philosophy*. Supplementary Volume. Oxford: Oxford University Press, pp. 13–29.
Moss, Jessica (2007): "The Doctor and the Pastry Chef: Pleasure and Persuasion in Plato's *Gorgias*". In: *Ancient Philosophy* 27, pp. 229–249.
Narcy, Michel (1984): *Le philosophe et son double. Un commentaire de l'*Euthydème *de Platon*. Paris: Vrin.
Narcy, Michel (2007): "Sokratik". In: Ueding, Gert (Ed.): *Historisches Wörterbuch der Rhetorik*. Vol. VIII. Rhet-St. Tübingen: Max Niemeyer Verlag, pp. 952–959.

Norden, Eduard (1915): *Die Antike Kunstprosa. Vom VI. Jahrhundert v. Chr. bis in die Zeit der Renaissance*. Volume I. Leipzig: Teubner.
Olympiodorus (1970): *Olympiodori in Platonem Gorgiam Commentaria*. Westernink, Leendert Gerrit (Ed.). Leipzig: Teubner. [English translation with commentary: Olympiodorus (1988): *Commentary on Plato's Gorgias*. Robin Jackson, Kimon Lycos and Harold Tarrant (Trans.). Leiden: Brill.]
Pernot, Laurent (2000): *La Rhétorique dans l'Antiquité*. Paris: Librairie générale française. [English translation: Pernot, Laurent (2005): *Rhetoric in Antiquity*, William Edward Higgins (Trans.). Washington, D.C.: The Catholic University of America Press.]
Philodemus (1911): *De vitiis*. Christian Jensen (Ed.). Leipzig, Teubner.
Plato (1959): *Gorgias*. E. R. Dodds (Trans.). Oxford: Clarendon Press.
Plato (1979): *Gorgias*. Terence Irwin (Trans.). Oxford: Oxford University Press.
Quintilian (2001): *Institutio oratoria = The Orator's Education*. Donald A. Russell (Ed. and Trans.). Cambridge: Harvard University Press.
Renaud, François (2001): "Maieutik". In: Ueding, Gert (Ed.): *Historisches Wörterbuch der Rhetorik: L-Musi*. Vol. V. Tübingen: Max Niemeyer, pp. 727–733.
Renaud, François (2018): "Cicero and the Socratic Dialogue: Between Frankness and Friendship (*Off.* I, 132-137)". In: Moore, Christopher and Stavru, Allessandro (Eds.): *Socrates and the Socratic Dialogue*. Leiden: Brill, pp. 707–726.
Renaud, François (2022): *La Justice du dialogue et ses limites: Étude du* Gorgias *de Platon*. Paris: Les Belles Lettres.
Roochnik, David (2007): "Commentary on Teloh". In: *Proceedings of the Boston Area Colloquium in Ancient Philosophy* 23, pp. 78–81.
Rowe, Christopher (2007): *Plato and the Art of Philosophical Writing*. Cambridge: Cambridge University Press.
Rowe, Christopher (2012): "The Status of the Myth of the *Gorgias*, Or: Taking Plato Seriously". In: Collobert, Catherine, Destrée, Pierre, and Gonzalez, Fransisco J. (Eds.): *Plato and Myth Studies on the Use and Status of Platonic Myths*. Leiden: Brill, pp. 187–198.
Schofield, Malcolm (2000): "*Gorgias* and *Menexenus*". In: Rowe, Christopher and Schofield, Malcolm (Eds.): *The Cambridge History of Greek and Roman Political Thought*. Cambridge: Cambridge University Press, pp. 192–199.
Shaw, J. Clerk (2011): "Socrates and the True Political Craft". In: *Classical Philology* 106, pp. 187–207.
Shaw, J. Clerk (2015): "Punishment and Psychology in Plato's *Gorgias*". In: *Polis* 32, pp. 75–95.
Szlezák, Thomas Alexander (1985): *Platon und die Schriftlichkeit der Philosophie. Interpretationen zu den frühen und mittleren Dialogen*. Berlin: de Gruyter.
Tarrant, Harold (1990): "Myth as a Tool of Persuasion in Plato". In: *Antichton* 24, pp. 19–31.
Westerink, Leendert G. and Segonds, Alain-Philippe (Eds.) (1990): *Anonymous Prolegomena to Platonic = Prolégomènes à la philosophie de Platon*. Jean Trouillard (Trans.). Paris: Les Belles Lettres.
Yunis, Harvey (2007): "Plato's Rhetoric". In: Worthington, Ian (Ed.): *Blackwell Companion to Greek Rhetoric*. London: Blackwell, pp. 75–89.

Rafael Ferber
Socrates' "Flight into the *Logoi*": A Non-Standard Interpretation of the Founding Document of Plato's Dialectic

Plato's Socrates uses the term δεύτερος πλοῦς (*Phd.* 99c9–d1) in connection with his intellectual autobiography, in the course of which he was led away from that "wisdom" (σοφία) they call "the study of nature" (φύσεως ἱστορία, *Phd.* 96a8) to instead look to "the truth of things" (τῶν ὄντων τὴν ἀλήθειαν) in the *logoi* (*Phd.* 99e6). He compares this move to a flight—a "flight into the *logoi*"—and calls this "flight into the *logoi*" the "second voyage" (δεύτερος πλοῦς, *Phd.* 99c9–d1). The decisive passage runs as follows:

[S₁] ἔδοξε δή μοι χρῆναι εἰς τοὺς λόγους καταφυγόντα ἐν ἐκείνοις σκοπεῖν τῶν ὄντων τὴν ἀλήθειαν.
[S₂] ἴσως μὲν οὖν ᾧ εἰκάζω τρόπον τινὰ οὐκ ἔοικεν·
[S₃] οὐ γὰρ πάνυ συγχωρῶ τὸν ἐν λόγοις σκοπούμενον τὰ ὄντα ἐν εἰκόσι μᾶλλον σκοπεῖν ἢ τὸν ἐν ἔργοις.[1]

Preliminary translation:
[S₁ₐ] So I decided that I must take refuge in the *logoi* and look at the truth of things in them.
[S₂] However, perhaps this image is inadequate;
[S₃] for I do not altogether admit that one who investigates things by means of the *logoi* is dealing with images more than one who looks at realities.

Initially, (1.) I will propose a non-standard interpretation of this passage, then (2.) I will proceed to address the philosophical problem raised in this passage according to this interpretation, that is, the *problem of the hypothesis* or *the problem of the "unproved principle"* before indicating (3.) the kernel of truth contained in the standard interpretation and concluding with some remarks on the "weakness of the *logoi*".

---

[1] Cf. *Phd.* 99e4–100a3, Oxford Classical Texts by Duke, E. A., Hicken, W. F., Nicoll, W. S. M., Robinson, D. B., and Strachan, J. C. G., trans. Grube, modified.

1 The standard and the non-standard interpretations

The correct meaning of the proverbial expression δεύτερος πλοῦς is suggested already by Eustathios from Thessaloniki (ca. 1110–ca. 1195), who refers to Pausanias: δεύτερος πλοῦς means "the next-best way", that is, the way adopted by those who try another method if the first does not succeed, specifically those who "try oars when the wind fails according to Pausanias" (*Eust.*, p. 1453).[2] There has been some dispute about whether this is really Plato's intended meaning and whether he is not using the expression ironically here[3]—that is, whether the second-best here is not actually the second-best, but rather the best voyage for the Platonic Socrates. But in the wake of the study by Martinelli Tempesta, who insists that the expression δεύτερος πλοῦς is proverbial and not to be confused with a metaphor,[4] there can no longer be any reasonable doubt that it refers to a second voyage

---

[2] Cf. Burnet (1911) *ad loc.*, LSJ *s.v*, and Martinelli Tempesta (2003), p. 89: "… il significato del celebre proverbio utilizzato da Platone in *Phd.* 99c–d può essere soltanto quello … di *second best*, come è suggerito inequivocabilmente da tutte le testimoninanze antiche". Tempesta (2003), pp. 123–125, also contains a useful index of the passages where the expression δεύτερος πλοῦς is used in a proverbial way; he argues at pp. 108–109 (*pace* Kanayama 2000, pp. 88–99) against the interpretation of Plb. *Hist.* 8.36.6.2 B.–W. that δεύτερος πλοῦς merely means a safer voyage that takes longer. Cf. n. 6.

[3] Cf. Burnet (1911), p. 99, *ad loc.*: "In any case, Socrates does not believe for a moment that the method he is about to describe is a *pis aller* or 'makeshift'". Cf. Gadamer (1968), p. 254: "Ein sehr ironischer Passus. Ich habe schon in meinem oben abgedruckten Buche 1931 ausgeführt, wie weit gerade die Erforschung des Seienden in den Logoi der Zugang zur Wahrheit des Seienden ist…". Gadamer seems not to see the problem: how the Socratic *logoi*—especially the hypothesis of ideas—can lead not only to consistency but also to truth, nor does he distinguish between "simple" and "complex" irony. Cf. also Thanassas (2003), p. 10: "The 'images of logoi' are the only means at our disposal for approaching the truth of beings". But the hypothesis of ideas is not an image. For the distinction between "simple' and "complex" irony, cf. Vlastos (1991), p. 31: "In 'simple' irony what is said just isn't what is meant: taken in its ordinary, commonly understood, sense that statement is simply false. In "complex" irony what is said is and isn't what is meant: its surface content is meant to be true in one sense, false in another". For Gadamer's interpretation of the *Philebus* in 1931, cf. Ferber (2010).

[4] "'Deuteros plous' is not a metaphor but just a proverb. On my view, they are two different kinds of expressions: the proverb has a single and fixed meaning that is always the same, while we can use a word or an expression in different metaphorical ways. Of course, a proverb can be used as a metaphor of something else, but it is its sole meaning that can refer to something else, which is implied by its not equivocal meaning" (Letter from 02/25/2018, quoted with the permission of the author).

not only in the chronological sense,[5] but *also* in the evaluative sense of inferiority (δευτερότης) to the first voyage. It is also clear that it is not being used ironically here,[6] as it is not used in this way in the two other occurrences in Plato (*Phlb.* 19c2–3; *Plt.* 300c2) and Aristotle (cf. *EN* 2.9, 1109a34–35; *Pol.* 3.11, 1284b19). In fact, the related comparison of the Socratic enterprise with a "raft" (*Phd.* 85d1), instead of a boat, is also not ironical (unless the irony is not "simple", but "complex").

This "second voyage" stands in contrast to the "first" one (πρῶτος). Although Socrates does not explicitly use the expression "earlier voyage" (πρότερος πλοῦς) or "first voyage" (πρῶτος πλοῦς), this implied earlier or first voyage is, in fact, reflected in his intellectual autobiography as a former student of the natural sciences (*Phd.* 95a–99d). These sciences represented a method by means of which he hoped to obtain direct access to reality, a process that ended in disappointment,[7] since it instead led to complete blindness of the soul—that is, complete ignorance (cf. *Phd.* 99e2–3)—because of its bewildering effect (cf. *Phd.* 79c7). In contrast to the "first voyage", the "second voyage" has the advantage of being safer (ἀσφαλέστερον, cf. *Lg.* 897e1–2), although it is slower and more laborious. Thus, the very notion of a second voyage implies a change in the means or method used, but not in the goal aimed, namely "to look at the truth of things" (σκοπεῖν τῶν ὄντων τὴν ἀλήθειαν, *Phd.* 99e5–6). This goal of "the second-best voyage in search of the cause" (τὸν δεύτερον πλοῦν ἐπὶ τὴν τῆς αἰτίας ζήτησιν, *Phd.* 99c9–d1) implies that the Platonic Socrates investigated the "true" (ἀληθῶς, *Phd.* 98e1) or "real" (τῷ ὄντι, *Phd.* 99b2) cause, that is, the final or second-order cause of the mechanical causes, the latter of which are mere necessary (cf. *Phd.* 99b3) or "co-causes" (συναίτια, cf. *Phd.* 98c2–e1; *Ti.* 46c7) "that would direct everything and arrange each thing in such a way as would be best" (*Phd.* 97c5–6). Thus, Socrates starts from the antinaturalistic assumption that nature has a teleological structure and that the "study of nature" ought to explain this structure, a project that was to be realised by Plato later on in the *Timaeus* (cf. *Ti.* 30a2–7).

---

5 Cf. the scholium quoted in Greene (1938), p. 14: "Since those who failed in the prior voyage (πρότερος πλοῦς) prepare the second safely, the proverb 'second voyage' is said about those who do something safely", quoted in Kanayama (2000), pp. 88–89, especially p. 89: "According to this reading, 'second' means only 'second in time' and there is no implication of the inferiority of the second voyage relative to the prior one; it rather suggests that the second voyage is better than the prior in being a safer voyage".
6 Cf. already Murphy (1936), p. 41, n. 1, and Hackforth (1955), p. 127, n. 5.
7 With Goodrich (1903), p. 382, I assume that τὰ ὄντα σκοπῶν (cf. *Phd.* 100a2) "must refer to the physical speculations previously described and condemned". For this reason, I prefer the first of the three interpretations of the "first voyage" mentioned by Kanayama (2000), p. 95, n. 112. The best overview of what belongs to Plato's intellectual history versus that of Socrates is to be found in Hackforth (1955), pp. 127–130.

But can we say in more detail what this "second-best voyage" involves? According to [S₁], it is a flight from direct perception or vision of "the things" (τὰ πράγματα, *Phd.* 99e3) to the indirect method of using *logoi*, which stand in contrast with "the things". Plato's Socrates employs here a commonly accepted way of thinking, which we also find, for example, in the *Apology* (cf. 32a4–5) and the *Seventh Letter* (cf. 343c2–3), but he then departs from the claim that the *logoi* constitute mere empty talk (ἀδολεσχία, cf. *Phd.* 70c1; *Prm.* 135d5; *Tht.* 195b10; κενεαγορία, *R.* 607b7), turning it into an interesting philosophical claim that is at odds with what is normally believed. In contrast to the common opposition between the realities themselves (τὰ ὄντα, *Phd.* 99e5)[8] and mere talk (οἱ λόγοι), Socrates claims that the things that are commonly believed to be realities are not true realities, while arguing, conversely, that paying attention to "mere talk" can lead us to the true realities. Hence, what looks like a path that leads us far away from reality turns out to be exactly the right path to reach true reality. This is—so to speak —the Socratic turn away from "the study of nature" towards what we say (διαλέγεσθαι, cf. *Phd.* 63c7–8). Although the Platonic Socrates does not yet "refer to dialectic as such"[9] in the *Phaedo* or employ the substantive "dialectical pursuit" (διαλεκτικὴ μέθοδος, *R.* 533c7),[10] he nevertheless speaks of "another form of pursuit" (ἄλλος τρόπος τῆς μεθόδου, *Phd.* 97b6–7) to find out "the reasons of each thing—why it comes into being, why it perishes, why it exists" (*Phd.* 96a9–10, trans. Rowe), namely "the art concerning the *logoi*" (ἡ περὶ τοὺς λόγους τέχνη, *Phd.* 90b7). This is the first occurrence of dialectic as an art, which is later called "the art of dialectic" (διαλεκτικῇ τέχνῃ) (*Phdr.* 276e5–6), an expression coined in an analogous way to μουσικὴ or γυμναστικὴ τέχνη (cf., e.g., *R.* 409c5–9).[11] He may have already alluded to this art in speaking of a "path" or "byway" (ἀτραπός): "It looks as if there's a byway (ἀτραπός) that'll bring us and our reasoning safely through in our search (ἐν τῇ σκέψει)" (*Phd.* 66b3–4, trans. Rowe, modified; cf. *Plt.* 258c3).[12] If it is not the path that leads directly to the goal, then the byway brings us "on to the trail

---

8 Cf. Burnet (1911), p. 99, *ad loc.*: "*ta onta* like *ta pragmata*".
9 Kahn (1996), p. 313.
10 For an interpretation, see Ferber (1989), p. 102.
11 For further remarks, cf. Müri (1944), especially p. 158. A classification of the dialectical subspecies is to be found in Gaiser (2004), p. 196. Dixsaut (2001) documents in Appendix I, pp. 345–352, "les occurrences du verbe διαλέγεσθαι dans les dialogues" and in Appendix II, pp. 353–354, "Les occurrences de διαλεκτικός, διαλεκτική, διαλεκτικόν, διαλεκτικῶς".
12 Pace Ebert (2004), p. 140, n. 16, who follows Harry (1909), I do not see in the ἀτραπός an allusion to the body which misleads us, but rather to "another form of pursuit" (ἄλλος τρόπος τῆς μεθόδου, *Phd.* 97b6–7), that is, to dialectic. On this difficult passage, cf. Burnet (1911), pp. 35–36, Dixsaut (1991), p. 332, n. 83, Trabattoni (2011), p. 41, n. 49, and Casertano (2015), pp. 292–293.

in our hunt after truth".[13] This "byway" anticipates "another form of pursuit" (ἄλλος τρόπος τῆς μεθόδου, *Phd.* 97b6–7) for which the Platonic Socrates then uses the proverbial expression "second voyage". But both expressions—"byway" and "second voyage"—indicate second-best options for reaching the goal, that is, "the true" (τὸ ἀληθές) (*Phd.* 66b7).

Socrates gives no explicit affirmative theoretical definition (by genus and difference) of this "second-best option", but primarily a negative contextual one, insofar as he distinguishes it (a) from the "first voyage" of Ionian natural philosophy on the grounds that it makes no use of sense perception—that is, it proceeds a priori—and (b) from "antilogic", that is, arguments that aim merely at contradiction (ἀντιλογικοὶ λόγοι, *Phd.* 90c1, cf. 101e2; Ar. *Nu.* 1173). Socrates defines his second-best option positively as a method of "giving an account of being" (λόγον διδόναι τοῦ εἶναι), that is, a λόγος τῆς οὐσίας by means of questions and answers (ἐρωτῶντες καὶ ἀποκρινόμενοι, *Phd.* 78d1–2).[14] Hence, the second voyage is a method involving the use of questions and answers to give an account of being or essence.

According to these negative contextual definitions, the "second voyage" should thus not be *immediately* identified with (a) the hypothetical method, (b) the theory of forms or (c) the explanation of things in terms of formal causes.[15] We may, however, ask *if* this "dialectical turn" on Socrates' part leads to (a) the hypothetical method, (b) the theory of forms or (c) the explanation of things in terms of formal causes, as Rose has argued.[16] (In my opinion, it leads indirectly to all three of them, namely, to the hypothesis of forms which explains the characteristics of things in terms of their formal causes.) Leaving aside Parmenides DK B7.8 here, then it is the first expression of what was later called "metaphysics",[17] in the sense of giving an account of invisible things of which no a posteriori experience by means of our sensory organs is possible. We may therefore also refer to this "flight" from "wisdom" (σοφία), which they call "the study of nature" (φύσεως ἱστορία, *Phd.* 96a8), as Socrates' "dialectical turn", in the sense of a "meta-physical turn".

When it comes to the meaning of the expression λόγοι in *Phd.* 100a1, which is "not easy to translate",[18] the reality is that we do not have an equivalent word in our modern European languages. Plato may be using the word at *Phd.* 100a1 in the non-technical sense of "discussions" (Grube's translation), as he does at *Phd.* 59a4.

---

13 Burnet (1911), *ad loc.* Cf. also: "It will be seen that the metaphor of the ἀτραπός gains very much when we bring it into close connexion with the hunt" (Burnet 1911, *ad loc.*).
14 Cf. Burnet (1911), *ad loc.*
15 Cf. Rose (1966), pp. 466–477, and Preus and Ferguson (1969), p. 105.
16 Rose (1966), p. 473.
17 Cf. Reale (1997), pp. 137–158.
18 Burnet (1911), p. 99, *ad loc.*

But his "definiteness of intention", to borrow an expression from Arne Naess,[19] may be more subtle in [S₁]. In fact, there have been a wide range of other translations proposed,[20] which I subdivide into non-sentential and sentential translations, with the sentential translations being further subdivided into mono-sentential, poly-sentential and ambiguous translations.

Non-sentential translations are "Begriffe", "conceptos" (Apelt/Horn/Gual) or "ideas" (Jowett). But as Burnet remarked long ago: "The term *logos* cannot possibly mean 'concept'. So far as there is any Greek word for 'concept' at this date, it is *noêma*".[21] Ambiguous translations include "rationes" (Ficino), "Gedanken" (Schleiermacher/Rufener) and "raisonnement" (Dixsaut) which do not render the linguistic aspect of *logoi* (cf. *Cra.* 431b5–c1; *Tht.* 189e4–6, 202b3–5; *Sph.* 264a8–9) and leave open whether these "rationes" or "Gedanken" or "raisonnements", when expressed in sentences, are mono- or poly-sentential. Mono-sentential translations include "definitions" (Bluck), "propositions" or "statements" (Ross/Kanayama), "postulati" (Reale) and "Grund-Sätze" (Natorp). Hence, Ross, for example, writes: "The language of 'agreement', and the fact that what Plato calls the 'strongest *logos*' is the proposition that Ideas exist, shows that *logoi* means statements or propositions".[22]

Nonetheless, with the flight into the *logoi*, Plato also looks back to the *Crito*, where he describes his Socrates as "the kind of man who listens only to the *logos* [that is, not only the proposition, but the argument] that on reflection seems best (βέλτιστος) to [him]" (*Cri.* 46b4–6). Later on, in the *Parmenides*, young Socrates' eager desire ἐπὶ τοὺς λόγους (cf. *Prm.* 135d3) also implies a zeal for *logoi*, in the sense not only of propositions, but also of arguments. In the section on μισολογία (*Phd.* 89d4), Socrates also uses the word in the sense of arguments (cf. *Phd.* 90b4.6.7.c1.4). Hence, the "mono-sentential" translations in terms of "statements" or "propositions" may be replaced with "poly-sentential" ones, not with "discussions" (Grube)—since it is possible for discussions to not contain any arguments—but rather with "arguments" (Hackforth), "theories" (Tredennick, Gallop) or "reasoned accounts" (Rowe), since theories or reasoned accounts by definition contain arguments.[23] Nevertheless, we can maintain the translation "propositions"

---

[19] Naess (1952), pp. 256 ff.
[20] For a selection, cf., e.g., Murphy (1936), p. 40, and Casertano (2015), pp. 360–362.
[21] Burnet (1914), p. 317, n. 1. Also Loriaux (1975), p. 93: "... dès 99e5, *tous logous* vise plus que de simples 'notions'".
[22] Ross (1951), p. 27.
[23] Cf. Murphy (1936), pp. 40–41: "... *logoi* are verbally contrasted with *erga*, and perhaps some word like 'theories', though it is not an exact equivalent, would bring out this contrast ..."

if, with Ebert, we translate λόγοι as "premises of theories", since premises of theories are propositions.²⁴ Thus, I will attempt to elucidate the intended meaning by translating *logoi* in the passage as follows:

[S₁ᵦ]: So I thought I must take refuge in theories and their premises, and investigate the truth of things in them.

[S₂] makes an addition and a qualification: Socrates declares the sight of the reflection of the sun in water, which is used as a comparison, to be an image (εἰκάζω) and qualifies this image as being in some sense inadequate. What is inadequate about this image? To see reality through an image suggests that one has indirect access to that reality; however, as Hackforth remarks: "[The image] is a good parallel in so far as the contrast of direct and indirect apprehension goes; but in so far as it might imply that *logoi* stand to physical objects (*erga*) in the relation of images to real things, it is misleading".²⁵

[S₃] is indeed "misleading" or "confusing".²⁶ It seems to have so far gone unnoticed that this passage has at least two different interpretations. I call these the standard interpretation²⁷ and the non-standard (or astonishing) interpretation.²⁸ In the first interpretation, Socrates pursues the parallel; in the second, he retracts it, at least in a certain sense. The issue is whether Socrates means that (a) both the student of *erga* and the student of *logoi* consider the "truth of things" in images, because *logoi* are also images (the "standard interpretation") or that (b) the student of *logoi* does not consider "the truth of things" in images, because *logoi* are not *eo ipso* images of the "truth of things", but must only be consistent (the "non-standard interpretation").

On the standard interpretation, the indirect approach is not inferior to the direct approach, because theories are also images of reality, namely "pictures in

---

24 Cf. Ebert (2004), p. 350: "Die *logoi*, die Sokrates im Auge hat, sind also offenbar Teile von Argumenten oder allgemeiner von Schlüssen, sind aber selbst keine Argumente oder Schlüsse. Sie haben den Status von Prämissen, aber nicht den von Schlüssen".
25 Hackforth (1955), p. 137, and mentioned in Frede (1999), p. 121, n. 29.
26 Gallop (1975), p. 178: "The sentence [S₃] in which Socrates qualifies his comparison of 'theories' with images (a1–2) is confusing in translation". Cf. the different translations in the appendix. But the sentence is also confusing in Greek.
27 The standard interpretation has been defended by, e.g., Gallop (1975), p. 178, Bostock (1986), pp. 157–162, Gadamer (1978), p. 254, Gonzalez (1998), pp. 188–208, Thanassas (2003), Dancy (2004), p. 295, and Costa (2017), p. 141.
28 The non-standard interpretation has been defended, e.g., by Burnet (1911), p. 99, *ad loc.:* "It is not really the case that the *logoi* are mere images of *ta onta* or *ta pragmata*". Cf. Dixsaut (1991), p. 140: "Saisir une réalité à travers un discours réflexif, ce n'est pas n'en saisir qu'une image. Au contraire, c'est l'expérience concrète qui ne livre que l'image de la chose, alors que la réflexion accède à sa réalité véritable". Cf. also Kanayama (2000), p. 47.

words" (cf. *Cra.* 431b2–c2),²⁹ just as the image of the sun in the water is an image of the real sun. *Logoi* or theories, which depict (εἰκάζειν) real things, would thus be on the same level as what Socrates later calls εἰκασία, that is, conjectures through images, or—more precisely—conjectures through images of images (cf. *R.* 511e2, 534a1–5; cf. also *R.* 598b6–8).³⁰ As the sun seen "in water or some such reflection" (*Phd.* 99d8–e1, my translation) is an image of the real phenomenon, so too would *logoi* be images of realities. Hence, the upshot of Socrates' flight into the *logoi* is that theories, as images of reality, are inadequate—even if they are "images of a higher grade than objects in the sensible world, and thus closer to Forms".³¹ Socrates would thus, in a way, anticipate Wittgenstein's picture-theory of language and thought: "The picture is a model of [phenomenal] reality" (*TLP* 2.12). As "a model of [phenomenal] reality", a picture is not an exact representation of phenomenal reality.

In the non-standard interpretation, [S₃] makes the claim that the indirect approach is nevertheless not inferior to the direct approach involving vision: the indirect approach does not use *logoi*, in the sense of images, whereas the reflection of the sun in water is an image of the real sun.

The standard interpretation, however, raises the following problems: (a) it insinuates that the Platonic Socrates treats *logoi* like images—or even εἰκασίαι—that is, conjectures through images of images; (b) it insinuates that *these* conjectures posit ideas and then (c) assumes the logical impossibility that the *logoi* which posit ideas (cf. *Phd.* 100b5) first depict what they subsequently posit; (d) it leaves open the question about what false *logoi*, which misrepresent "the truth of things", depict; and (e) nowhere *in* the *Phaedo* are "objects in the sensible world" explicitly called "images of Forms".³²

---

29 I owe this reference to *Cra.* 431b2–c2 to Costa (2017), p. 28. Costa is anticipated by Bostock (1986), p. 160, who gives the following *caveat:* "However, it is not even clear that this (rather confused) line of thought was Plato's own view at the time when he wrote the *Cratylus.* This is partly because the dialogue goes on to reject the premise that names need to be rightly framed (434c–435c), and partly because there are evidently many things in that dialogue that Plato is not very serious about, and the way the argument is extended from names to propositions may well be one of them" (quoted without footnote).
30 Cf. Ferber (1989), pp. 85–91 and pp. 111–114.
31 Gallop (1975), p. 178.
32 For anticipations of the *interpretatio difficilior*, cf. Natorp (1903), p. 156; 2004, p. 167: "For logic is not something like a mere organ or instrument with which to grasp the 'existing' objects to be found outside us; it is not merely the eye-glass that protects from blinding, in order that we may look with impunity at the externally existing being of sensible things that radiates in the sunlight, so to speak, of immediate truth in itself. This simile is defective for it is not the logical shape of being that is merely a copy ..." (trans. Politis and Connolly). Cf. Murphy (1936), p. 43: "the *logoi*

But let us (f) nevertheless assume that the standard view is right: what, then, would the "philosopher's progress" be, if "Plato's philosopher", Socrates, turns from the old method of observing facts to the new one involving *logoi* as inadequate images of facts? Would Socrates not, in this case, merely be turning from blindness to imprecision or even confusion, such that his "progress" would really be a regress—eliciting the *schadenfreude* of his enemies?

Since I am unwilling to concede this joy to his enemies, I may be forgiven for preferring the non-standard interpretation. In fact, the Socrates who turns to the *logoi* in the sense of theories and their premises is not simply looking at reflections: the criterion of consistency (συμφωνία), which *logoi* must fulfil (*Phd.* 100a5), suggests that *logoi* are not mere images or reflections of real things, in the sense of the correspondence theory of truth, but that συμφωνία or "consistency should suffice for truth".[33] In fact, we read:

> But in any case, this was my starting point: hypothesizing (ὑποθέμενος) on each occasion whatever account I judge to have the most explanatory power, I posit as true (ἀληθῆ ὄντα) whatever seems to me in tune (συμφωνεῖν) with this ... (*Phd.* 100a3–4, trans. Rowe).

Hence, in the non-standard interpretation, *logoi* as theories are not images of real things, but are posited as true even if they are only in harmony with or consistent with their premises.

In this non-standard interpretation, as in the standard interpretation, Socrates will, in the words of Shorey, "not admit that discussion is a less direct approach to truth than sense",[34] or more precisely, as Ross puts it, "not altogether admit that his method of studying things is less direct than that of the physicists ...".[35] But the non-standard interpretation gives quite a different twist to these words than the common one does: the physicists study things ἐν ἔργοις, that is, in reality.[36] Socrates studies things ἐν λόγοις, that is, in light of the premises of theories. If Socrates, as Ross claims, "will not altogether admit that his indirect method of studying things is less direct than that of the physicists", then his indirect method is no less direct than that of the physicists. If it is no less direct, then it is at least on an equal

---

are in no sense *like* the things being studied, and it becomes equally clear as we read on that the *logoi* are not *logoi of* the things. ... But surely, they are independent propositions and thoughts introduced *ab extra*". For Socrates, *logoi* (and hypotheses) are not on the same level as images or *eikasiai*; cf. *R.* 511d6–e4.

33 Vlastos (1991), p. 15.
34 Shorey (1933), p. 131.
35 Ross (1953), p. 27.
36 Robin (1950), XLIX: "L'expression *en ergois*, ... fait penser à l'*energeia* d'Aristote: *acte* qui est à la fois forme logique et réalité ; qui, à l'état pur, est Dieu même".

footing with the physicists' method of getting at the truth of things. As Kanayama argues, "[a]n enquirer who studies his objects in *logoi* studies them as directly and clearly as those who study in concrete, and what's more, without any fear of being blinded by the employment of the senses".[37]

Thus, to summarise, I attempt to elucidate the intended meaning of [S₁] in the following ways:

> [S$_{1a}$] So I decided that I must take refuge in the *logoi* and look at the truth of things in them.
>
> [S$_{1b}$]: So I thought I must take refuge in theories and their premises, and investigate the truth of things in them.
>
> [S$_{1c}$] So I thought I must take refuge in the coherence of theories with their premises, and investigate in the coherence of theories the reality of things.

2 The problem of the non-standard interpretation

The question—left unasked by Shorey, Ross, Kanayama or others—now arises: how is it possible that the indirect approach involving arguments is on an equal footing with the more direct way involving seeing, that is, how is τὸ σκοπεῖν ἐν [τοῖς] λόγοις τὰ ὄντα on an equal footing with τὸ σκοπεῖν τὰ ὄντα ἐν [τοῖς] ἔργοις? This problem is analogous to the problem that Vlastos called *"the* problem of the elenchus".[38] For our part, let us call it *the* problem of the hypothesis, that is—to borrow an expression from the pseudo-Platonic *Definitions—the* problem of the "unproved principle" (ἀρχὴ ἀναπόδεικτος, *Def.* 415b10).

In contrast to the elenchus employed by Plato's Socrates in the early dialogues, the Platonic Socrates of the *Phaedo* does not start from premises or hypotheses advanced by the interlocutor (cf., e.g., *Euthphr.* 11c4–5; *Hp.Ma.* 302c12; *Grg.* 454c4–5) to which he is not committed, but from his own premises, to which he is committed. Nevertheless, *the* "problem of the elenchus" persists in *the* problem of the hypothesis or "unproved principle". This is because all the theories can do is to arrive—in a way analogous to how Plato's Socrates and his interlocutors in the early dialogues

---

[37] Kanayama (2000), p. 47.
[38] Vlastos (1983), pp. 38–39: "[T]he question then becomes how Socrates can claim, ... to have proved that the refutand is false, when all he has established is the inconsistency of p with premises whose truth he has not undertaken to establish in that argument: they have entered the argument simply as propositions on which he and the interlocutor are agreed. This is *the* problem of the Socratic elenchus ...".

arrived at ὁμολογία (cf., e.g., *Chrm.* 157c6–7; *Ly.* 219c4; *Grg.* 487e6–7)—at συμφωνία, that is, "harmony" or "concord" (cf. *Phd.* 100a5). The meaning of the term συμφωνία or "concord" has been made more precise by Robinson, both here and at 101d5, by distinguishing "consistency" from deducibility (cf. *Prt.* 333a6–8; *Grg.* 457e1–3; *Phdr.* 270c6–7).[39]

I cannot enter into the logical problems which the translations "consistency" and "deducibility" present here,[40] but will restrict myself to making the following point concerning consistency. If a hypothesis leads to inconsistent consequences, then it is supposed to be false: "if anyone should question the hypothesis itself, you would ignore him and refuse to answer until you could consider whether its consequences were mutually consistent (συμφωνεῖ) or not (διαφωνεῖ)" (*Phd.* 101d3–5, trans. Rowe). If the consequences are not mutually consistent, then the hypothesis is false and must be rejected.

But consistency or "concord" is only a negative test of truth.[41] Hence, the *deuteros plous* also seems to be a mere negative test of truth, as was observed by Leibniz, long before Robinson, in a summary of the *Phaedo:*

> ... after establishing something like a second voyage [*secunda navigatio*] I entered another path [*aliud iter*] which, if it does not explain everything, does not tolerate that something false is said.[42]

Nevertheless, we can pose the remaining question: granting that *logoi* or theories may be consistent or harmonious like a piece of music, are they also true in the sense of corresponding to reality? Mere consistency is, for the Platonic Socrates,

---

39 Robinson (1953), p. 131.
40 Cf. Robinson (1953), pp. 126–136. But cf. also Kahn (1996), p. 316: "I suggest that the term for consequence is deliberately avoided, because Plato is here presenting the method hypothesis as more flexible and also more fruitful than logical inference. ... Whatever is incompatible with some basic feature of the model, as specified in the *hypothesis,* will be 'out of tune' *(diaphônein)* or fail to accord. But the positive relationship of 'being in accord' *(symphônein, synâidein)* is not mere consistency. It means fitting into the structure, bearing some positive relationship to the model by enriching or expanding it in some way". This point has been further developed by Bailey (2005), especially pp. 104–110, by accentuating the musical undertones of "being in accord", a point made also by Stefanini (1949), p. 258: "*Il criterio della verità, è, adunque, la legge stessa della muscia: armonia.* Ciò che resta fuori dell'euritmia universale è ad un tempo dissonante e falso".
41 Cf. Robinson (1953), pp. 135–136: "'Seeing whether the results accord', considered as a test, is merely negative. It can sometimes show that the hypothesis must be abandoned, but never that it must retained".
42 Leibniz (1980), p. 294: *[Cum ergo causas rerum ex optimi electione sumptas, neque ipse per me consequi, neque ab alio me discere posse viderem,] velut secunda navigatione instituta aliud ingressus sum iter, quod si non omnia explicet, nihil tamen patiatur dici falsum.*

not yet in itself a guarantee of truth: "if your premise (ὑπόθεσις) is something you do not really know and your conclusion and intermediate steps are a tissue (συμπέπλεκται) of things you do not really know, your reasoning may be consistent with itself (ὁμολογία),⁴³ but how can it amount to knowledge (ἐπιστήμη)?" (*R.* 533c5–6, trans. Cornford; cf. *Cra.* 436c7–d7). What is said here about mathematical *hypotheseis*, which the mathematicians lay down as "known" (*R.* 510c6) and treat as absolute or non-hypothetical assumptions, seems to me valid in an analogous way to the hypothesis of ideas (cf. *Phd.* 100b1–9). Since the hypothesis of ideas is presented as something that is much "talked about" (τὰ πολυθρύλητα, *Phd.* 100b5), it is therefore not yet established as true, even if it enjoys consensus among the interlocutors and its consequences are mutually consistent.

In fact, we find in Plato not only consensus (ὁμολογία, cf. *Grg.* 487e6–7) or consistency (συμφωνία) as a criterion of truth (*Phd.* 100a4–7), but also correspondence: "a true *logos* says that which is, and a false *logos* says that which is not" (*Cra.* 385b7–8; cf. *Sph.* 263b3–7): "The [true] statement as a whole is complex and its structure *corresponds to* the structure of the fact".⁴⁴ In the same vein, we might say that a true hypothesis as a whole is complex and that its structure *corresponds to* the structure of the facts

If the Socrates of the *Phaedo* tries to investigate "in the *logoi*" "the truth of things", by his flight into the *logoi*, he not only does "not tolerate that something false is said", but tries to reach the reality of things. Therefore, [S₃] seems to indicate that consistency is not a mere negative test of truth, but is in itself a guarantee of truth, no less than correspondence is.

Metaphorically speaking, the second sailing is no less a method of arriving at the goal—"the truth of things"—than the first sailing, just as a rowboat is no less a vehicle for reaching the final destination than a sailing boat. Or, to put it in yet another way, in dreaming—as Socrates sometimes does (cf. *Smp.* 175e2–3; *Cra.* 439c6–d; *Phd.* 60e1–61a4)—we may arrive at reality as if in a state of wakefulness, whereas in seeing with our eyes we are blinded, at least if our dreams are consistent.

To use the metaphor deployed by the Platonic Socrates in the *Republic*, by distinguishing the essence of the Good from everything else and "surviving, as if in a battle, all attacks with refutations" (ὥσπερ ἐν μάχῃ διὰ πάντων ἐλέγχων διεξιών, *R.* 534c1–2, my free translation) with a "*logos* not liable to fall" (ἀπτῶτι τῷ λόγῳ, *R.* 534c3), the projected philosopher-kings and -queens not only survive all attacks with an infallible—or at least at the end of the battle still unrefuted—*logos*, but

---

43 ὁμολογία may mean consensus or consistency. Cf. the remarks on ὁμολογουμένως (*R.* 510d2) in Ferber (1989), p. 96, where I plead for consensus.
44 Cornford (1935), p. 311.

they are "brought at last to the goal" (*R.* 540a6), namely "to lift up the eye of the soul to gaze on that which sheds light on all things" (*R.* 540a7–9), that is, "the Good itself" (*R.* 540a8–9), a "principle which is not a hypothesis" (ἀρχὴ ἀνυπόθετος, *R.* 510b7) to which the expression "something sufficient" (τι ἱκανόν, *Phd.* 101e1) may allude.⁴⁵

This is quite an astonishing claim. The question was *in principle* aptly formulated by Davidson:

> Consistency is, of course, necessary if *all* our beliefs are to be true. But there is not much comfort in mere consistency. Given that it is almost certainly the case that some of our beliefs are false (though we know not which), making our beliefs consistent with one another may as easily reduce as increase our store of knowledge.⁴⁶

In fact, the flight into the *logoi* brings with it the risk that some of the *logoi*—or even the *logos* judged to be the "strongest" (*Phd.* 100a4), in the sense of the "hardest to refute" (δυσεξελεγκτότατος, *Phd.* 85c9–d1)—are false. Now, the method through which the Platonic Socrates takes refuge in the *logoi* in the *Phaedo* is the mathematical method known from the *Meno* as the method of hypothesis (ἐξ ὑποθέσεως, *Men.* 86e3). But in the *Phaedo*, it is neither a mathematical hypothesis that is put forward (cf. *Men.* 85b7–86d2) nor the hypothesis that virtue is a science ("if virtue is a science, then it would be teachable", *Men.* 87c5–6). Rather, it is the hypothesis that ideas are, where "is" has the emphatic Parmenidean meaning of being real or really real (ὄντως ὄν).⁴⁷

> My aim is to try to show you the kind of reasons that engage me, and for that purpose I'm going to go back to those much-talked-about entities (πολυθρύλητα) of ours—starting from them, and hypothesizing that there's something that's beautiful and nothing but beautiful, in and by itself, and similarly with good, big, and all the rest. If you grant me these, and agree that they exist, my hope is, starting from them, to show you the reason for things and establish that the soul is something immortal.⁴⁸

The reasoning is roughly as follows: if the hypothesis of the ideas is true, then the soul is immortal. Not only does the theory of ideas depend on a hypothesis, but the final proof of the immortality of the soul also depends on the hypothesis of the ideas. That is, the final proof depends on the hypothesis of the ideas

---

45 Cf. Gallop (1975), pp. 190–191. This is nevertheless a debatable issue; cf. Verdenius (1958), p. 231, and Ferber (1989), p. 100.
46 Davidson (2005), p. 223.
47 On the influence of Parmenides in the *Phaedo*, cf. Hackforth (1955), pp. 84–85.
48 *Phd.* 100b1–9 (trans. Rowe).

(*Phd.* 100b7–9), while the hypothesis of ideas depends on a hypothesis or premise.⁴⁹ Hence, the immortality of the soul is, like the theory of ideas, "something necessary because of a hypothesis" (ἐξ ὑποθέσεως ἀναγκαῖον; cf. Arist. *PA*. 1.1, 642a9).

But in the short time remaining before his death—"as long as there is still daylight" (*Phd.* 89c7–8)—the Platonic Socrates cannot do what the Platonic Parmenides will later do in the *Parmenides*, i.e., consider the consequences of the negations of his hypothesis, namely "if that same thing is hypothesized (ὑποτίθεσθαι) not to be" (*Prm.* 136a1–2). What are the consequences if "the beautiful, the good and every such reality" (*Phd.* 76d8–9) are hypothesised not to be? In fact, concerning the pre-existence of the soul, Socrates assumes that if these realities do not exist, then this argument would be altogether futile (cf. *Phd.* 76e4–5).

And *even if* Socrates were to prove that these realities exist, he would not have the "five" years (*R.* 539e2) needed to ascend, with Simmias and Kebes, the upward path to "the Good itself" (*R.* 540a8–9), a "principle which is not a hypothesis" (ἀρχὴ ἀνυπόθετος, *R.* 510b7) to which the expression "something sufficient" (τι ἱκανόν, *Phd.* 101e1) *may* allude.

But without this time-consuming "exercise" (cf. *Prm.* 135c8, 135d4–7, 136c5), how does the Platonic Socrates know that his hypothesis of the individual ideas is not false *as* a hypothesis (cf. Arist. *APr.* 62b12–20)—just as other hypotheses he has advanced have been proven false (cf. Arist. *Pol.* 2, 1261a16, 1263b29–31)— or, even worse, that it is not mere idle talk (ἀδολεσχία, cf. *Phd.* 70c1; *Prm.* 135d5; *Tht.* 195b10; κενεαγορία, *R.* 607b7), as is commonly assumed?

In fact, Plato's first interpreter, Aristotle, would go on to say that the πολυθρύλητα—the Platonic ideas—are τερετίσματα (*APo.* 1.22, 83a33), that is, mere twittering, and that to speak of ideas as paradigms and of participation is κενολογεῖν, "idle talk" (cf. *Metaph.* 1.9, 991a21–22). With this critique, Aristotle is clearly referring to the δεύτερος πλοῦς of the *Phaedo* (cf. *Metaph.* 1.9, 991b3–7). Already his remark that "... [Plato's] introduction of the Forms (ἡ τῶν εἰδῶν εἰσαγωγή), was due to his inquiry in the *logoi* (τὴν ἐν τοῖς λόγοις ἐγένετο σκέψιν), for the earlier did not partake in dialectic ..." (*Metaph.* 1.6, 987b31–33) is "pretty clearly a reminiscence" of *Phd.* 99e5–100a4.⁵⁰ In *De generatione et corruptione*, Aristotle explicitly attributes the theory not to Plato, but to the Platonic Socrates, that is, to "the Socrates in the *Phaedo*" (ὁ ἐν τῷ Φαίδωνι Σωκράτης, *GC* 2.9, 335b10–14; cf. *Pol.* 2, 1261a6).

The answer Plato's Socrates gives to "*the* problem of the Socratic elenchus" is that, like Meno's slave, we have true opinions hidden in us (cf. *Men.* 81a–d, 85b–

---

49 Cf. Sedley (2018), pp. 210–220.
50 Cf. Ross (1924), Vol. I, p. 171, and now the careful article by Delcomminette (2015).

86b), because we are "fallen souls", for "the truth of things is always in our soul" (*Men.* 86b1), meaning that a Cartesian *dubitatio de omnibus* or a global error "in our soul" is impossible. Similarly, the inhabitants of the Platonic cave are not ensnared in a global error concerning moral matters either, but rather see "shadows of the just" (*R.* 517d8–9). Davidson thus writes in his article "Plato's Philosopher":

> [T]he assumption is that, *in moral matters, everyone has true beliefs which he cannot abandon and which entail the negations of his false beliefs*. It follows from this assumption that all the beliefs in a consistent set of beliefs are true, so a method like the elenchus which weeds out inconsistencies will in the end leave nothing standing but truths.[51]

In the same vein, the Platonic Socrates *could* have said in the *Phaedo*, evoking the *Meno*, that everyone has hidden true beliefs about the universals like "the equal" (cf. *Phd.* 74a5–75a3). Although "the equal" seems to belong to the *metaxy* between Ideas and sense phenomena (cf. Arist. *Metaph.* A6 987b14–18),[52] since Socrates also uses the plural forms "the equals themselves" (αὐτὰ τὰ ἴσα, *Phd.* 74c1) and "the three" (τὰ τρία, *Phd.* 104e1), the "equal" is nevertheless a universal like "the beautiful, the good and every such reality" (*Phd.* 76d8–9). But at the end of the day this hypothesis will remain true because an examination of it would leave realism about universals like the equal and "the beautiful, the good and every such reality" as the only viable option for these universals. Through the δεύτερος πλοῦς we arrive at reality, just as we do through the πρῶτος πλοῦς, because we have inside ourselves true opinions about the universals—which entail negations of false opinions and which cannot be shaken or are elenchus-resistant—but which must nonetheless be made explicit by cross-examination.

Metaphorically, we can give the answer in the following way: the rowboat contains within itself a sail, which can be hoisted—that is, by τὸ σκοπεῖν ἐν λόγοις, we arrive at the ἀλήθεια τῶν ὄντων. Or, to use another metaphor, our soul as the "place of ideas" (τόπος εἰδῶν, Arist. *De an.* 3.4, 429a27–28)[53] is a mirror of the truth, but must, in its incarnated form, be purified from its hidden contradictions by an examination of the *logoi* until it can reflect the unveiled truth.

---

51 Davidson (2005), p. 229.
52 Cf. Wippern (1970), pp. 276–277: "Die Lösung dieser Aporie kann von Platon aus gesehen nur darin liegen, dass jedenfalls die Seele und die scheinbar nur der Exemplifikation dienenden partikulären 'Ideen' der *monas dyas, trias* oder der *pemptas* eine Art *Zwischenstellung* zwischen dem Reich der absoluten eidê selbst und der sinnlich wahrnehmbaren Dinge innehaben". Cf. also Wippern (1970), p. 277, n. 14, Schiller (1967), pp. 57–58, and Ross (1924), p. 194, *ad loc.*
53 Cf. Ferber (2007), p. 183.

Once again, in the same vein, Socrates could say: everyone has hidden true beliefs about his soul and its destiny after death, for example that the soul brings life (φέρουσα ζωήν, cf. *Phd.* 105d3–4; *Cra.* 399d11–e2; *Lg.* 895c11–12) "whenever it exists" (ὅτανπερ ᾖ, *Phd.* 103e5),[54] which entails the negation of his false beliefs, for example that the soul dies with the body. The hypothesis of the immortality of the soul will remain true in the end because an examination of this hypothesis would leave it as the only viable, elenchus-resistant option.

3 The kernel of truth in the standard interpetation

Only now are we at last able to point out the kernel of truth in the "standard interpretation", according to which the student of *logoi* considers the "truth of things" in images: *logoi* or theories may become images of "the truth of things" in the sense of the correspondence theory only when purified from their hidden contradictions. But in that case, they are no longer εἰκασίαι, that is, conjectures through images, or through images of images (cf. *R.* 511e2, 534a1–5; cf. also *R.* 598b6–8), but rather they express justified true beliefs which say that which is.

I use the expression "justified true beliefs", not "knowledge", because there is a *caveat* in the *Phaedo*: as long as our soul is embodied, in the best case we may come as near as possible or "very near" (ἐγγύτατα, *Phd.* 65e4, 67a3; cf. ὁμοιότατον, 80b3 with ἐγγύς τι τούτου, 80b10) to the truth, but it remains at a "distance" from the truth—a distance caused by our corporeality. There is a distance between pure knowledge "of anything", which would imply the complete consistency of our belief system, and the closest-possible approach to this knowledge in life, a gap that cannot be bridged by a "byway" or "shortcut" (ἀτραπός, *Phd.* 66b4):

> [If] it's impossible to get pure knowledge of anything in the company of the body, then one or the other of two things must hold: either knowledge can't be acquired, anywhere, or it can be, but only when we're dead; because that's when the soul will be alone by itself, apart from the body, and not until then.[55]

If this principle is applied to the soul, it is impossible to acquire *pure* knowledge of the soul and its immortality in the company of the body—that is, in this life—al-

---

[54] For the remark "whenever it exists" (ὅτανπερ ᾖ, *Phd.* 103e5), cf. the neglected, but pertinent, comments by Wippern (1970), pp. 273–274.
[55] *Phd.* 66e4–67a2 (trans. Rowe).

though it may be possible to attain different degrees of approximation, depending on the progressive separation of the soul from the body (cf. *Phd.* 67a2–3).[56] Only after death—after our *excarnation*—will we not only *believe*, but also really *know* that we are immortal, if, paradoxically speaking, we are still alive after death.

This limit of the δεύτερος πλοῦς may also be alluded to in the *Philebus:* "but while it is a great thing for the wise man to know everything, the second-best voyage (δεύτερος πλοῦς) is not to ignore oneself, it seems to me" (*Phlb.* 19c1–3, trans. Frede, modified)—"to know everything" may be an ironical allusion in the "simple" sense to the "wise man" Anaxagoras and his "first voyage", which led Socrates to complete ignorance (cf. *Phd.* 99e2–3).

This lack of self-knowledge is mentioned again in the *Seventh Letter:* "I know that certain others have also written on the same matters; but who they are they themselves do not know" (*Ep.* 7.341a5–7, trans. Morrow).[57] This implies also that these "certain others", that is, the writers of the "so-called unwritten doctrines" too, like Aristotle, Speusippus, Xenocrates, Heraclides, Hestiaeus,[58] did not attempt the δεύτερος πλοῦς.

At the same time, the human impossibility of arriving at pure knowledge also seems to hold for Plato right up to the *Seventh Letter* because the four means of knowledge at our disposal—ἓν μὲν ὄνομα, δεύτερον δὲ λόγος, τὸ δὲ τρίτον εἴδωλον, τέταρτον δὲ ἐπιστήμη (*Ep.* 7.342b1–2)—are not able to grasp the essence, but only the quality or "the vague, general likeness"[59] of "the fifth", that is, the postulated "object itself, the knowable and truly real being" (ὅ τε καὶ ἀληθῶς ἐστιν ὄν) (*Ep.* VII.342a7–b1),[60] or the Platonic idea. Although we do not find the verb διαλέγεσθαι (*Phd.* 63c7–8) or the noun "dialectical pursuit" (διαλεκτικὴ μέθοδος (*R.* 533c7) in the *Seventh Letter*,[61] the *Seventh Letter* alludes to the method mentioned in the *Phaedo* of "giving an account of being" (λόγον διδόναι τοῦ εἶναι), that is, the method of giving a λόγος τῆς οὐσίας by means of questions and answers (cf. *Phd.* 78d1–2) by mentioning twice the process of questioning and answering (cf. *Ep.* VII. 343c8–d1, 344b6). In a kind of reformulation of the δεύτερος πλοῦς in the digression (*Ep.* 7,

---

56 Cf. Fine (2016), pp. 563–564.
57 Cf. Ferber (2007), p. 42 and p. 133, n. 3.
58 Cf. Novotny (1930), pp. 213–217, especially pp. 215–216.
59 See Bluck (1949), p. 507, in his review of Boas (1948).
60 Agamben's (1999, p. 32) emendation of *ho* through *di' ho* with reference to Ficino's translation "*quintum vero oportet ipsum ponere quo quid est cognocibile, id est quod agnosci potest, atque vere existit*" seems to me unnecessary and the replacement of "the thing itself" by "that *by which* the object is known, its own *knowability and truth*" tautological.
61 Cf. Dixsaut (2001), pp. 353–354.

344b3–c1),⁶² we find the reason indicated in the formula "because of the weakness of *logoi* or arguments" (διὰ τὸ τῶν λόγων ἀσθενές, *Ep.* 7, 343a1)⁶³—a corollary of the "human weakness" (ἀνθρωπίνη ἀσθένεια, *Phd.* 107b1; cf. *Plt.* 278c9–d6, *Lg.* 853e10– 854a1) caused by the *in*carnation of our souls. This weakness caused by our incarnation implies that even the human *nous* only comes as "near as possible" (ἐγγύτατα) (*Ep.* VII. 342d4) "in kinship and similarity" to the "fifth". Thus, in the *Philebus*, Socrates and Protarchus are only able to capture the idea or the essence (of the Good, cf. 342a4), with three characteristics or qualities (343b8–c2) which intend "no less" (342e3) than to cover the "fifth" (342e2), that is, essence. As the light of the one sun is broken into three *parhelia*, so the one Good appears to "us" (*Phlb.* 64e5) as if in three qualities: an "aesthetic" one, beauty; a relational one, symmetry; and an ontological one, truth.⁶⁴

In Kantian terms, as *homo phaenomenon*, even the dialectician is not able to grasp and communicate the "the thing in itself" (τὸ πρᾶγμα αὐτό, *Ep.* VII. 341c7), because the means of knowledge, intuition and categories give only the appearance of "the thing in itself".

Although there is in the *Seventh Letter*, in distinction to Kant, an "illumination" (ἐξέλαμψις) "of reason *and* understanding *if* one goes to the limit of human power" (ἐξέλαμψε φρόνησις περὶ ἕκαστον καὶ νοῦς, συντείνων τι μάλιστ' εἰς δύναμιν ἀνθρωπίνην, *Ep.* VII. 344b8–c1), this "illumination" also admits of degrees⁶⁵ and is not the faculty of an *ex*carnated νοῦς, but rather the *activity* or awakening of the "sleeping" *in*carnated νοῦς which comes, to repeat, only as "near as possible" (ἐγγύτατα) (*Ep.* VII.342d4) "in kinship and similarity" to the "fifth".⁶⁶

---

62 That the digression in the Seventh Letter contains a reformulation of the δεύτερος πλοῦς has been suggested by Forcignanò (2020), pp. 41–42. For an interpretation of the digression, cf. Ferber (2007), pp. 51–66 and 106–120, as well as Forcigagnò (2020), pp. 38–41.
63 Cf. Ferber (2007), especially pp. 56–66 and pp. 106–120. The interpretation of Burnyeat (2015), pp. 121–132, does not take into account this discussion. Cf. now inter alia the critique of Burnyeat by Szlezák (2017), pp. 311–323, especially pp. 318–320.
64 Cf. Ferber (2020), pp. 177–183, especially p. 183.
65 Cf. Ferber (2007), pp. 111–112, and Forcignanò (2020), pp. 45–46.
66 Pace Szlezák, (2021), p. 191: "Die Beschreibung und Deutung der Schwäche der logoi macht verständlich, dass die gesuchte Erkenntnis des on, des wahrhaft Seienden einer jeden Sache (341a1, b8), also ihrer Idee, zwar nicht ohne die vier Mittel—sie sind unentbehrlich: 342d8–e2—, aber doch irgendwie gegen sie oder trotz ihrer erreicht wird". Szlezák seems to ignore that only an excarnated *nous* can reach this knowledge, but an incarnated *nous* can come only as near as possible, cf. especially Ferber (2007), especially p. 110, and long ago Stefanini (1949), pp. XLVI-LVIII, especially p. LV: "*La verosimiglianza platonica non è apparenza di verità, ma approssimazione alla verità*".

In any case, just as we must distinguish between *acquaintance as such* and *knowledge by acquaintance*,[67] we must also distinguish between *illumination as such* and *knowledge by illumination*.[68] Illumination as such (in the rational rather than mystical sense of the *Seventh Letter*[69]) signifies direct experience without linguistic symbols (if and so far as it is possible in certain *Grenzsituationen* for human beings). Knowledge by illumination is, however, propositional and affected by the "weakness of the *logoi*". But the "illumination" of "reason and understanding if one goes to the limit of human power" implies that Plato does not speak of a non-propositional *illumination as such*, but of *knowledge by illumination*. The *logoi*, or arguments, used to formulate such knowlege-claims, are either sound or unsound—that is, they start from true propositions and contain valid deductions or not—and they can be replaced by other propositions, just as Plato's "Theory of Ideas" is formulated in different ways in the dialogues,[70] and the unwritten "Theory of the Principles" has been handed down to us in different words.[71]

This *körperliche Verdüsterung* of our soul—to borrow an expression from old Goethe[72]—may in fact remain true even in the "so-called unwritten doctrines" (Arist. *Ph*. 209b14–15) *if* old Plato did, in fact, say there: "Not only the happy *(eutychounta)* but also the proving *(apodeiknynta)* human being (cf. *Phd*. 77a5, b2–3, c2–6, d4, 87a4–c4, 105e8) must remember that he is a human being".[73] To remember that we are human beings means also to remember our "human weakness" and mortality, in light of which achieving complete consistency in our incarnated *logoi* is at the very least difficult, if not, as for the Platonic Socrates, impossible, to reach—at least for most ephemeral *incarnated* human beings "participating only to a small extent in truth" (*Lg*. 804b3–4).

In fact, with the metaphor of a "raft" (σχεδία, sc. ναῦς), "literally, 'improvised boat'"[74] with which one must sail through life (*Phd*. 85d1–2), Plato's Socrates of the *Phaedo* indicates not only the fragile instrumental character of the flight into the *logoi* as a *Hilfskonstruktion*, but also of the hypothesis of ideas—a *Hilfskonstruktion* which even for the Plato of the *Timaeus* is not a self-evident axiom, but "something necessary because of a hypothesis" (ἐξ ὑποθέσεως ἀναγκαῖον, cf. Arist. *PA*. 1.1, 642a9): *if* we distinguish between true belief and knowledge, then we must also ac-

---

67 Cf. Feigl (1967), p. 37.
68 The distinction is made neither by Szlezák (2021), p. 191, nor by Forcignano (2020), pp. 42–47.
69 Cf. Ferber (2007), especially pp. 99–100.
70 Cf., e.g., the short summary in Baltes and Lakmann (2005), pp. 2–6.
71 Cf. Van der Wielen (1941), pp. 178–179.
72 Talks with Eckermann (March 11, 1828).
73 Gaiser (1963), p. 455, Testimonium 11.
74 Kanayama (2000), p. 92.

cept ideas[75]—at least if we may hear in the voice of the Platonic Timaeus also old Plato's voice: "so here's how I cast my own vote" (*Tim.* 51d3).

A raft is not a stable vehicle like a sailing boat or a rowboat, although, like the raft of Odysseus, to which Simmias possibly alludes (cf. *Phd.* 85d1), it can also have sails (ἱστία) (cf. Hom. *Od.* 5.259–261). A second voyage on a raft with oars *and* at least *one* sail (ἱστίον) capable of being hoisted may also be an apt metaphor for the Socratic δεύτερος πλοῦς in the *Phaedo*. But for all its instability, a raft with oars and *one* sail is still a better way than swimming without the "raft" of a hypothesis through the troubled water, the πόντος ἀτρύγετος of the γένεσις and φθορά of our lives.[76]

Bibliography

Agamben, Giorgio (1999): "The Thing Itself". In: Agamben, Giorgio: *Potentialities, Collected Essays in Philosophy*. Daniel Heller-Roazen (Ed. and Trans.). Stanford: University Press, pp. 27–47.
Apelt, Otto (1928): *Platon, Phaidon oder über die Unsterblichkeit der Seele, übersetzt und erläutert v. O. Apelt*. 3rd ed. Leipzig: Felix Meiner.
Bailey, Dominic T. J. (2005): "Logic and Music in Plato's *Phaedo*". In: *Phronesis* 50, pp. 95–115.
Baltes, Matthias and Lakmann, Marie-Luise (2005): "Idea (dottrina delle idee)". In: Fronterotta, Francesco and Leszl, Walter (Eds.): *Eidos—Idea. Platone, Aristotele e la tradizione platonica*. In: *International Plato Studies* 21 (Sankt Augustin), pp. 1–24.
Bluck, Richard S. (1949): "Plato's Biography: The Seventh Letter". In: *The Philosophical Review* 58, pp. 503–509.
Boas, George (1948): "Fact and Legend in the Biography of Plato". In: *The Philosophical Review* 57, pp. 439–457.
Bostock, David (1986): *Plato's Phaedo*. Oxford: Oxford University Press.
Burnet, John (1911): *Plato's Phaedo*. Oxford: Clarendon Press.
Burnet, John (1914): *Greek Philosophy, Vol. I: Thales to Plato*. London: Macmillan and Co.
Burnyeat, Myles and Frede, Michael (2015): *The Pseudoplatonic Seventh Letter*. Dominic Scott (Ed.). Oxford: Oxford University Press.
Casertano, Giovanni (2015): *Fedone, o dell'anima. Dramma etico in tre atti*. Naples: Paolo Loffredo.
Cornford, Francis M. (1935): *Plato's Theory of Knowledge. The Theaetetus and the Sophist of Plato*. London: Kegan Paul.
Costa, Ivana (2017): "Conoscere attraverso immagini nel *Fedone*". In: Eustacchi, Francesca and Migliori, Maurizio (Eds.): *Per la rinascita di un pensiero critico contemporaneo. Il contributo degli antichi*. Milan: Mimesis Edizioni, pp. 137–148.
Dancy, Russell M. (2004): *Plato's Introduction of Forms*. Cambridge: Cambridge University Press.

---

75 Cf. Ferber (1998), p. 438.
76 An earlier version of this article appeared in the first two sections of "Second Sailing towards Immortality and God: On Plato's Phaedo 99e4–100a3, with an Outlook on Descartes' Meditations, AT VII, 67". *Mnemosyne* 74. No. 3, 2021, pp. 371–400. I thank Chad Jorgenson for some helpful remarks.

Davidson, Donald (2005): "Plato's Philosopher". In: Davidson, Donald: *Truth, Language, and History*. Oxford: Oxford University Press, pp. 223–240. [First published: Davidson, Donald (1985): "Plato's Philosopher". In: *The London Review of Books* 7. No. 14, pp. 15–17; reproduced as: Davidson, Donald (1990): "Plato's Philosopher". In: Davidson, Donald: *Plato's Philebus*. New York and London: Garland, pp. 1–15.]

Delcomminette, Sylvain (2015): "Aristote et le *Phédon*". In: Delcomminette, Sylvain, D'Hoine, Pieter, and Gavray, Marc Antoine (Eds.): *Ancient Readings of Plato's Phaedo*. Leiden and Boston: Brill, pp. 19–36.

Denyer, Nicholas (2007): "The *Phaedo*'s Final Argument". In: Scott, Dominic (Ed.): *Maieusis. Essays in Ancient Philosophy in Honour of Myles Burnyeat*. Oxford: Oxford University Press, pp. 87–97.

Dixsaut, Monique (1991): *Platon. Phédon. Traduction nouvelle, introduction et notes*. Paris: GF-Flammarion.

Dixsaut, Monique (2001): *Métamorphoses de la dialectique dans les dialogues de Platon*. Paris: Vrin.

Ebert, Theodor (2004): *Platon, Phaidon. Übersetzung und Kommentar*. Göttingen: Vandenhoeck & Ruprecht.

Feigl, Herbert (1967): *The mental and the physical: The essay and a postscript*. Minneapolis: University of Minnesota Press.

Ferber, Rafael (1989): *Platos Idee des Guten*. Zweite, durchgesehene und erweiterte Auflage. St. Augustin: Academia Verlag.

Ferber, Rafael (1998): "'Auf diese Weise nun gebe ich selbst meine Stimme ab'. Einige Bemerkungen zu Platons später Ideenlehre unter besonderer Berücksichtigung des *Timaios*". In: *Gymnasium* 105, pp. 419–444. [Reproduced with modifications in: Rafael Ferber (2020): *Platonische Aufsätze*. Berlin and Boston: de Gruyter, pp. 239–237.]

Ferber, Rafael (2003): *Philosophische Grundbegriffe 2: Mensch, Bewusstsein, Leib und Seele, Willensfreiheit, Tod*. Munich: Verlag C.H. Beck.

Ferber, Rafael (2007): *Warum hat Platon die 'ungeschriebene Lehre' nicht geschrieben?* Munich: Verlag C.H. Beck. [2nd enlarged edition of Ferber, Rafael (1991): *Die Unwissenheit des Philosophen oder Warum hat Plato die 'ungeschriebene Lehre' nicht geschrieben?* St. Augustin: Academia Verlag.]

Ferber, Rafael (2010): *"The Origins of Objectivity in Communal Discussion.* Einige Bemerkungen zu Gadamers und Davidsons Interpretationen des Philebos". In: Renaud, François and Gill, Christopher (Eds.): *Hermeneutic Philosophy and Plato. Gadamer's Response to the Philebus. Studies in Ancient Philosophy* 10. St. Augustin: Academia Verlag, pp. 211–242. [Reproduced in: Rafael Ferber (2020): *Platonische Aufsätze*. Berlin and Boston: de Gruyter, pp. 313–346.]

Ferber, Rafael (2020): *Platonische Aufsätze*. In: Daub, Susanne, Erler, Michael, Gall, Dorothea, Koehen, Ludwig, and Zintzen, Clamens (Eds.): *Beiträge zur Altertumskunde* 386. Berlin and Boston: de Gruyter.

Fine, Gail (2016): "The 'Two Worlds' Theory in the *Phaedo*'. In: *British Journal for the History of Philosophy* 24, pp. 557–572.

Forcignanò, Filippo (2018): "Experiences Without Self-justification. The 'Sticks and Stones' argument in the *Phaedo*". In: Cornelli, Gabriele, Robinson, Thomas, and Bravo, Francisco (Eds.): *Plato's Phaedo. Selected Papers from the Eleventh Symposium Platonicum*. Baden-Baden: Academia Verlag, pp. 249–254.

Forcignanò, Filippo (2020): *Platone, Settima Lettera, Introduczione, traduzione e commento*. Rome: Carocci Editore.

Frede, D. (1999). *Platons Phaidon. Der Traum von der Unsterblichkeit der Seele*, Darmstadt: Wissenschaftliche Buchgesellschaft.

Gadamer, Hans-Georg (1968): "Amicus Plato magis amica veritas". In: Gadamer, Hans-Georg (Ed.): *Platons dialektische Ethik und andere Studien zur antiken Philosophie*. Hamburg: Felix Meiner, pp. 249–268.
Gaiser, Konrad (1968): *Platons ungeschriebene Lehre. Studien zur systematischen und e-schichtlichen Begründung der Wissenschaften in der Platonischen Schule*. 2nd ed. Stuttgart: Klett-Cotta.
Gaiser, Konrad (2004): "Platonische Dialektik—damals und heute". In: Szlezák, Thomas A. and Stanzel, KarlHeinz (Eds.): *Gesammelte Schriften. International Plato Studies* 19. Sankt Augustin: Academia Verlag, pp. 177–203.
Gallop, David (1975): *Plato, Phaedo*. Oxford: Clarendon Press.
Gonzalez, Francisco J. (1998): *Dialectic and Dialogue. Plato's Practice of Philosophical Inquiry*. Evanston, IL: Northwestern University Press.
Goodrich, W. J. (1903): "On *Phaedo* 96a–102a and on the *Deuteros Plous* 99d". In: *CR* 17, pp. 381–384.
Greene, William C. (1938): *Scholia Platonica*. Oxford: Oxford University Press.
Grube, Georges M. A. (1997): "*Phaedo*". In: Cooper, John M. (Ed.): *Plato, Complete Works*. Georges M.A. Grube (Trans.). Indianapolis and Cambridge: Hackett Publishing Company, pp. 49–100.
Grünwald, Eugen (1910): "Simmias und Kebes in Platons *Phaidon*". In: *Zeitschrift für das Gymnasialwesen* 64, pp. 258–263.
Hackforth, Reginald (1955): *Plato's Phaedo*. Cambridge: Cambridge University Press.
Horn, Christoph (2011): "Kritik der bisherigen Naturforschung und die Ideentheorie 95a–102a, Platon, Phaidon". In: Müller, Jörn (Ed.): *Platon, Phaidon*. Berlin: Akademie Verlag, pp. 127–142.
Kahn, Charles H. (1996): *Plato and the Socratic Dialogue. The Philosophical Use of a Literary Form*. Cambridge: Cambridge University Press.
Kanayama, Yahei (2000): "The Methodology of the Second Voyage and the Proof of the Soul's Indestructibility in Plato's *Phaedo*". In: *Oxford Studies in Ancient Philosophy* 18, pp. 41–100.
Leibniz, Gottfried Wilhelm (1980): "*Platonis Phaedo Contractus* (März 1676)". In: Leibniz, Gottfried Wilhelm: *Sämtliche Schriften und Briefe*. Akademie der Wissenschaften der DDR (Ed.). Series 6. *Philosophische Schriften*, Vol. III. Berlin: Akademie Verlag, pp. 284–297.
Loriaux, Robert (1975): *Le Phédon de Platon. Commentaire et Traduction, II, 84b–118a*. Namur: Presses Universitaire de Namur.
Martinelli Tempesta, Stefano (2003): "Sul significato di *deuteros plous* nel *Fedone* di Platone". In: Bonazzi, Mauro and Trabattoni, Franco (Eds.): *Platone e la tradizione platonica. Studi di fi-losofia antica. Quaderni di Acme* 58. Istituto Editoriale Universitario, pp. 89–125.
Murphy, Neville R. (1936): "The *Deuteros Plous* in the *Phaedo*". In: *CQ* 30, pp. 40–47.
Naess, Arne (1952): "Toward a Theory of Interpretation and Preciseness". In: Linsky, Leonardo (Ed.): *Semantic and the Philosophy of Language*. Champaign: University of Illinois Press, pp. 248–269.
Natorp, Paul (1903): *Platos Ideenlehre. Eine Einführung in den Idealismus*. Leipzig: Verlag der Dürr'schen Buchhandlung.
Natorp, Paul (2004): *Plato's Theory of Ideas. An Introduction to Idealism*. Vasilis Politis (Ed.). Vasilis Politis and John Connolly (Trans.). *International Plato Studies* 18. St. Augustin: Academia Verlag.
Novotny, Franciscus (1930): *Platonis Epistulae commentariis illustratae*. Brno: Filosofická Fakulta.
Preus, Mary and Ferguson, John (1969): "A Clue to the *Deuteros Plous*". In: *Arethusa* 2, pp. 104–107.
Reale, Giovanni (1984, 1997): *Per una nuova interpretazione di Platone, Rilettura della metafisica dei grandi dialoghi alla luce delle 'Dottrine non scritte'*. 20th ed. Milan: Vita e Pensiero.
Robin, Léon (1950): *Platon, Œuvres completes*. Vol. I. Paris: Les Belles Lettres.
Robinson, Richard (1953): *Plato's Earlier Dialectic*. Oxford: Clarendon Press.
Rose, Lynne E. (1966): "The *Deuteros Plous* in Plato's *Phaedo*". In: *The Monist* 50, pp. 464–473.

Ross, William David (1924): *Aristotle's Metaphysics*. Vol. I. Oxford: Clarendon Press.
Ross, William David (1951): *Plato's Theory of Ideas*. Oxford: Clarendon Press.
Rowe, Christopher (2010): *Plato. The Last Days of Socrates: Euthyphro, Apology, Crito, Phaedo*. London: Penguin Books.
Sedley, David (2018): "The Phaedo's Final Proof of Immortality". In: Cornelli, Gabriele, Robinson, Thomas M. and Bravo, Francisco (Eds.): *Plato's Phaedo. Selected Papers from the Eleventh Symposium Platonicum*. Baden Baden: Academia, pp. 210–220.
Shorey, Paul (1933): *What Plato Said, Abridged Edition*. Chicago: University of Chicago Press.
Stefanini, Luigi (1932, 1949): *Platone*. Vol. I. 2nd enlarged ed. Padua: Cedam.
Szlezák, Thomas A. (2017): "Review of: Burnyeat, M., and Frede, M. 2015". In: Scott, Dominic (Ed.): *The Pseudo-Platonic Seventh Letter. Gnomon* 89, pp. 311–323.
Szlezák, Thomas A. (2021): *Platon. Meisterdenker der Antike*. Munich: C.H. Beck.
Taylor, Alfred Edward (1936): *Plato. The Man and His Work*. 2nd ed. New York: The Dial Press.
Thanassas, Panagiotis (2003): "Logos and Forms in *Phaedo* 96a–102a". In: *Bochumer Philosophisches Jahrbuch für Antike und Mittelalter* 8, pp. 1–19.
Trabattoni, Franco (2011): *Platone, Fedone*. Stefano Martinelli Tempesta (Trans.). Turin: Giulio Einaudi Editore S.P.A.
Van Der Wielen, Willem (1941): *De Ideegetallen van Plato*. Amsterdam: D. B. Centen's Uitgevers-Maatschappij N.V.
Verdenius, Willem Jacob (1958): "Notes on Plato's *Phaedo*". In: *Mnemosyne* 11, pp. 193–243.
Vlastos, Gregory (1983): "The Socratic Elenchus". In: Annas, Julia (Ed.): *Oxford Studies in Ancient Philosophy*. Volume I. Oxford: Clarendon Press, pp. 27–58. [Reproduced in Fine, Gail (Ed.) (1999): *Plato 1. Metaphysics and Epistemology*. Oxford: Oxford University Press, pp. 36–63.]
Vlastos, Gregory (1991): *Socrates. Ironist and Moral Philosopher*. Cambridge: Cambridge University Press.
Wippern, Jürgen (1970): "Seele und Zahl in Platons *Phaidon*". In: Albrecht, Michael von and Heck, Eberhard (Eds.): *SILVAE. Festschrift für Ernst Zinn zum 60. Geburtstag*. Tübingen: M. Niemeyer, pp. 272–288.

Appendix: Translations of Phd. 99e4–100a3

Quapropter operae pretium esse censui, ut ad rationes confugerem, atque in illis rerum veritatem considerarem. Forte vero nostra haec similitudo non omni ex parte congruat. Non enim prorsus assentior, eum, qui res in rationibus contemplatur, in imaginibus aspicere potius, quam qui in operibus intuetur. (Marsilio Ficino)

Sondern mich dünkte, ich müsse zu den Gedanken meine Zuflucht nehmen und in diesen das wahre Wesen der Dinge anschauen. Doch vielleicht ähnelt das Bild auf gewisse Weise nicht so, wie ich es aufgestellt habe. Denn das möchte ich gar nicht zugeben, dass, wer das Seiende in Gedanken betrachtet, es mehr in Bildern betrachtet, als wer in den Dingen. (F.D. Schleiermacher)

Es erschien mir demnach notwendig, zu den Begriffen meine Zuflucht zu nehmen und an ihrer Hand das wahre Wesen der Dinge zu erforschen. Vielleicht trifft mein Vergleich nicht ganz zu; denn ich leugne auf das bestimmteste, dass der, welcher die Dinge begrifflich betrachtet, sich in höherem Grade einer bildlichen Betrachtungsweise bediene als der, welcher sich unmittelbar an die gegebenen Dinge wendet. (O. Apelt)

Es schien mir daher nötig zu sein, meine Zuflucht zu den Argumenten zu nehmen und in ihnen die Realität des Seienden zu untersuchen. Vielleicht ist aber mein Vergleich in gewissem Sinne unpassend: Denn ich will gar nicht zugeben, dass jemand, der das Seiende in Argumenten untersucht, dabei eher in Bildern untersucht als derjenige, der es in der Wirklichkeit untersucht. (Th. Ebert)

So I thought I should take refuge in theories, and study in them the truth of the things that are. Perhaps my comparison is, in a certain way, inept; as I don't at all admit that one who examines in theories the things that are is any more studying them in images than one who examines them in concrete. (D. Gallop)

So I decided that I should take refuge in theories and arguments and look into the truth of things in them. Now maybe in a way it does not resemble what I'm comparing it to. For I don't at all accept that someone who, when studying things, does so in theories and arguments, is looking into them in images any more than someone who does so in facts. (A. Long and D. Sedley)

Il me sembla dès lors indispensable de me réfugier du côté des idées et de chercher à voir en elles la vérité des choses. Peut-être, il est vrai, ma comparaison en un sens n'est-elle point exacte, car je ne conviens pas sans réserve que l'observation idéale des choses nous les fasse envisager en image, plutôt que ne fait une expérience effective. (L. Robin)

Voici alors ce qu'il me sembla devoir faire: me réfugier du côté des raisonnements, et, à l'intérieur de ces raisonnements, examiner la vérité des êtres. Il se peut d'ailleurs que, dans un sens, ma comparaison ne soit pas ressemblante: car je n'accorde pas du tout que lorsque l'on examine les êtres à l'intérieur d'un raisonnement, on ait plus affaire à leur images que lorsqu'on les examine dans des expériences directes. (M. Dixsaut)

E mi parve necessario rifugiarmi nei concetti, e considerare in essi la realtà delle cose esistenti. Sebbene forse, in certo senso, la similitudine non si addice. Perché io non posso ammettere che chi considera le cose nei loro concetti le veda in imagine più di chi le consideri nella loro realtà. (M. Valmigli)

Perciò ritenni di dovermi rifugiare in certi postulati e considerare in questi la verità delle cose che sono. Forse il paragone che ora ti ho fatto in un certo senso non calza, giacché io non ammetto di certo che chi considera le cose alla luce di questi postulati le consideri in immagini più di chi le considera nella realtà. (G. Reale)

Juzgué, pues, que era necesario refugiarme en las proposiciones y buscar en ellas la verdad de las cosas; por cierto que la comparación de que me sirvo no me parece exacta, porque no convengo de ningún modo que quien examina las cosas en las proposiciones las examina en imágenes más que quien lo hace en los hechos. (C. Eggers Lan)

Opiné, pues, que era preciso refugiarme en los conceptos para examinar en ellos la verdad real. Ahora bien, quizás eso a lo que lo comparo no es apropiado en cierto sentido. Porque no estoy muy de acuerdo en que el que examina la realidad en los conceptos la contemple más en imágenes, que el que la examina en los hechos. (C. Garcia Gual)

Claudia Marsico
Friendly Fire: Dialectic Struggles between the Megarians and Plato

The uneven textual history and the consequent fragmentary status of the relevant works often conspire against the study of the Socratic philosophies. However, despite the speculative nature of any exploration of this kind, the attempt to better understand the philosophical exchanges of that period should not be omitted. This is the only way to proceed if we seek to attain a broad historiographical view oriented to the full development of ancient thought and not only to isolated authors. We know through doxography that the original audience did not perceive an essential qualitative difference among the Socratic philosophies. Hence, the dissimilar luck regarding text preservation does not respond to ancient selection criteria but to diverse processes and phenomena far removed from them.

The Socratics formed a curious group with plenty of quarrels and antagonisms. They developed an innovative textual format, the σωκρατικοὶ λόγοι, which underlines their connection. Still, they were also divided by deep theoretical and vital distances, in line with the testimonies about their intense polemics. To think that these polemics did not correlate with their texts seems unimaginable. It is true that because of the very format of the dialogue with Socrates as the main character, the frame is located some decades before their times and does not represent the discussions of the 4th century BCE directly. However, we often find members of the Socratic circle in action. Xenophon's Socratic works, for instance, that account for Antisthenes and Aristippus' behavior, or Plato's *Phaedo*, even if Phaedo is just a rapporteur, are instances of this phenomenon. Then, although they are not in the foreground, the arguments do not lack allusions to the hot topics that divided the members of this prolific group.

In the framework of the Socratic circle, the Megarians and Plato shared essential features. They were far from their materialist or hedonist fellows, like Antisthenes or the Cyrenaics. However, their conception of philosophy also had significant differences in line with their diverse views about dialectic. In this work, we will review three aspects that illustrate this problematic relationship in the case of Plato and the Megarians. First, we will explore their metaphysical disagreements and the possible Platonic allusion to the Megarians in Plato's allegories of the sun, the line and the cave in *Republic*, VI-VII. Then, we will focus on the methodological differences attested in the *Euthydemus*. Finally, taking the opposite view, we will study the Megarian rejection of participation, alluding to Stilpo's case against Plato's approach. The exploration of this friendly fire will help better comprehend the variants of classical dialectic that appeared within the Socratic circle, which are relevant for two main reasons. On the one hand, they show the curious phenomenon of a manifold group of intellectuals with a common origin but novel and different ideas rooted in different conceptions of dialectic as a philosophical method. On the other hand, they are incredibly influential notions that leave their polemical imprint on Western philosophy.

1 The Sun, Son of the Good

The allegories in the *Republic* are one of the best known and discussed passages in the overall corpus, to the point that sometimes they are presented as a synthesis of Plato's philosophy. The allegory of the sun offers the metaphysical foundations of the approach, the line adds the dialectic steps to grasp them, and the cave develops the practical dimension.[1] The association between the sun and the Good in all of them is crucial. For this motive, it is striking that the rest of the corpus does not take up this notion, as would be expected according to its alleged significance. Why does the Form of the Good vanish beyond these passages?

We can offer a plausible explanation: these analyses are strongly affected by the controversial discussion with other lines of thought. Indeed, the internal interpretation of Plato's texts, rich in itself, often obscures that they were forged in an agonistic context inside and outside the Socratic group. If so, it is coherent to find a particular emphasis on aspects that become placed in the background or disappear entirely in other settings. In these cases, what is relegated to a secondary place is not the topic itself but the discussion that triggered it. As a result, this development becomes redundant and exits the stage.

Hence, this could have happened in this case. But who were the adversaries in the allegories of the *Republic*? We want to stress the plausibility of this interpretation, pointing out that the passages could have been forged as a reaction against the Megarians. The testimonies inform us about Euclid's closeness to Plato. He received the group in Megara during the obscure times after Socrates' death,[2] and Plato presented some of his dialogues there.[3] Furthermore, they seem to have

---

[1] The whole passage was analyzed in Divenosa and Marsico (2011) from the perspective of its structure and relevance within Plato's philosophy. This different approach stresses its value beyond this realm as a hint of the metaphysical differences between Plato and the Megarians that are at the basis of their dialectical disagreements.
[2] See, for instance, Diogenes Laertius, 2.106 and 3.6 (SSR, II.A.5).
[3] See *Socratic Letter 14*, 2 (SSR, II.A.8).

shared the rejection of materialist views, like those of Antisthenes. They were also against hedonistic solutions, like those of Aristippus, and both preferred a model of continence that distanced them from dubious characters like Aeschines and his scandals.[4]

In fact, Euclid and Plato shared ontological views on the existence of a stable realm that put both of them on the trail of Eleatism. Cicero does not hesitate to draw this connection.[5] In the line of traditional doxography that think of schools, he claims that the Megarian group derives from the Eleatic one, according to the well-known succession that begins with Xenophanes, continues with Parmenides and Zeno and then with Euclid and his followers. The same view, taken from Aristocles of Messene, is shared by Eusebius,[6] and Diogenes Laertius reiterates it in the text from which Euclid's theoretical position is often inferred.

Diogenes Laertius claims in that passage that Euclid "practised Parmenidian ideas (τὰ Παρμενίδεια μετεχειρίζετο)", saying that "the good was one thing called by many names (ἕν τὸ ἀγαθὸν ἀπεφαίνετο πολλοῖς ὀνόμασιν καλούμενον): sometimes wisdom, sometimes god, at other times intellect, etc. What was opposite to the good he eliminated (τὰ δ'ἀντικείμενα τῷ ἀγαθῷ ἀνῄρει), saying that it did not exist (μὴ εἶναι φάσκων)".[7] The passage mentions a practice (μεταχειρίζειν), suggesting that this approach is not limited to theory. It implies a peculiar activity associated with τὰ Παρμενίδεια, a notion that involves Parmenides' beliefs but exceeds this sphere. It could entail accommodating these ideas, possibly including some kind of recognition of the new context within recent discussions[8] or a reference to the Eleatic ideas in general.

However, the studies oriented toward understanding the mechanisms of ancient historiography often cast doubts on its exegetical schemes. Cherniss' works about Aristotle's procedures are a good example.[9] In the case of the Megarians, two critical lines were combined. On the one hand, von Fritz rejected the ancient sources' information and denied the Eleatic influence over Euclid and claimed that his thought could be explained on a pure Socratic basis.[10] On the other, the very category of Eleaticism was questioned and considered an invention motivated by

---

[4] On the beginning of the historiography about the Socratic circle, see Illarraga and Marsico (2021).
[5] Cicero, *Acad.*, II.42.129 (SSR, II.A.31).
[6] *Prep. Ev.* 14.17.1 (SSR, II.O.26).
[7] Diogenes Laertius, II,106. (*SSR* II.A.30).
[8] About the role of Melissus as an author of a reshape or Parmenides' views, see Rossetti (2020).
[9] Cherniss (1935).
[10] Von Fritz 1931, and in this line, Giannantoni (1990) and Gardella (2015).

a didactic generalization forged by Plato in the *Sophist*.[11] Thus, the doubts about the historiographical status of Eleaticism based on the absence of a "formal" school led to reject the possibility of theoretical links with the Megarians.

I have already suggested some arguments to limit these negative statements in other works,[12] underlining that considering Socratism and Eleaticism as excluding options is unjustified. It could be noted that, even if these links are rejected, they reappear in other forms, as in the case of Döring's position. He accepts von Fritz's analysis but adds that the connection with Parmenides' ideas was already present in Socrates himself, who was the origin of the bond.[13] However, why should this operation be attributed to Socrates, considering that the determination of his doctrines is even more complex and lacks explicit references in the sources? Beyond this issue, the suggestion is productive for our investigation because it calls attention to the broader presence of Eleatic ideas in the Socratic environment. This is consistent with some recent lines that seek the origin of the notion of Eleaticism. For instance, Cordero underlines the role of Plato's parricide in the *Sophist*, emphasizing that the attempt to refute Parmenides is rooted in discussions with contemporary philosophers, especially Antisthenes.[14] On his part, Bredlow has claimed that the forge could have happened within the Megarian environment, highlighting the circulation of these ideas in the whole group.[15] These perspectives do not support the rejection of the Eleatic influence on the Megarians, coinciding with the judgement of the ancient sources. Hence, although the notion of Eleaticism is far from unambiguous,[16] there is no reason to remove it from the horizon of the Megarian theories.

This aspect is relevant for our approach because among the arguments to reject the influence of Eleatic ideas on the Megarians we find a curious idea. It is said that Euclid was not concerned with being but with goodness, adding that this view is similar and consistent with the metaphysical account presented by Plato in the allegories of the sun, the line and the cave in the *Republic* where the Form of the

---

11 See, e.g., Cordero (1991).
12 Marsico (2011 and 2022a). See also Brancacci (2018).
13 Döring (1972) p. 87.
14 Cordero (2005) pp. 202–204.
15 Bredlow (2011).
16 See, for instance, Capizzi (1975), Rossetti (2020), and the volumes edited by Galgano and Cherubin (2020). Despite this aspect, many interpretations keep following von Fritz's thesis, that emphasized the Socratic origin of the Megaric views denying the Eleatic root. Recent analyses have called this approach into question since it is not necessary to postulate a contradiction between both influences. Even more, the Eleatic impact was relevant not only among the Megarians, but also among other Socratic lines. See Brancacci (2018) and Marsico (2013 and 2022a).

Good is at the forefront.[17] That is, supposing that Euclid's doctrines should be explained on pure Socratic grounds, Plato's allegories are taken as a model of a Socratic view about the Good as a principle. In this way, this suggestion sheds light on the link between both philosophies, although it does not provide clues to understand their relation fully. We attempt to do that in what follows.

Let us turn back to Diogenes' testimony. As we saw, he attributes to Euclid the thesis that "the good was one thing called by many names, sometimes wisdom, sometimes god, at other times intellect, etc.", adding that "what was opposite to the good he eliminated, saying that it did not exist".[18] Some "deflationist" readings of Euclid's views often have a bias toward language, considering the ontological references dialectical tools to formulate paradoxes.[19] However, the passage contains references to things, traits of things, judgements about their existence or non-existence, and the curious list of names related to the good points to the hearth of ancient ontological discussions. Indeed, the list presents a strange combination. It mentions two concepts linked to gnoseological realm, i.e. wisdom and intellect, together with the notion of God. What is the basis of this mix?

Although this connection has not been duly explored, the combination sounds very familiar. Indeed, these notions have been studied by Aristotle in *Met.* XII, where he presents his theory of the unmoved mover and offers arguments about its existence, its traits, and its nature, i.e., the cause of the universe as its ontological principle. This approach appeals to the notion of God, claiming that the prime unmoved mover is God, so that being and God become connected.[20] Moreover, in 1074b34, he characterizes its activity as νόησις νοήσεως, adding intellect to the equation. Finally, in *Met.* XII.10, when Aristotle claims in 1075a11–15 that the universe has the good as an army has its general, the Good appears as separated and autonomous.[21]

These passages have traditionally been linked with the Platonic horizon. Indeed, they were the basis for the discussions about the evolution of Aristotle's thought, as seen in Jaeger's studies and the following polemics.[22] The resemblance with Plato's arguments in *Laws*, X about the soul as a cosmic self-mover and the *Republic*'s allegories concerning the Form of Good are undeniable and well-studied. But we are interested now in a broader dimension that points to the belief in νοῦς as a divine principle ruling the cosmos within Presocratic philosophy.

---

17 See Gardella (2014, 31), who synthesizes this view.
18 Diogenes Laertius, II,106 (*SSR* II.A.30).
19 Villar (2021) pp. 99–115.
20 *Met.* XII.1072b.
21 See Menn (1992).
22 See Jaeger (1923), von Arnim (1931), and Düring (1966).

The best-known case is undoubtedly that of Anaxagoras, but it is not the only one. If so, in this framework, Euclid's peculiarity was related to the indication that these notions were names of an elemental entity, the Good.

However, some interpretations claim that all these notions are solely ethical, and νοῦς should be understood in all those cases as the virtue of rationality. Guided by this view, Menn goes as far as to suggest that Diogenes' testimony about Euclid reinforces this idea because the inclusion of the term φρόνησις would imply that God and νοῦς also must be comprehended in this way. However, as we have said, removing the ontological dimension from the Megarian approach leads to a partial view. It is worth noting that, according to Menn, Xenophanes' notion of God does not fit well in this reduction to the ethical order.[23] Therefore, even if we agree that this explanation is adequate in some cases, it does not seem to apply to the philosophies usually associated with Eleaticism and much less to the Megarians. On the contrary, the testimony about Euclid connects these discussions primarily oriented to ontology by exploring the notions of being, good, God and intellect.

Hence, Euclid's approach contributed to this dialogue with a theory that puts the unity of the good as the central thesis, based on the association between unity and being. It means that there is a level of stable unity, of which we have intellectual but not perceptual experience. For this reason, this step involves a transformation of ontology into "agathology". We can be sure about the existence of this realm and its pure features characterized by "goodness". Moreover, as in Parmenides' metaphysical exercise, the starting point was surely the pure notion of being. It proves to have no defects, and in this sense, it is wholly good. And as being has no opposite, in this pure view, neither the good has. However, we think of opposites, drawing upon the negation of being and the negation of the good. In this view, these are products of the same compromise with a non-metaphysical approach. Following an Eleatic inspiration, Euclid eliminated the opposite of the good as a concept: non-good is not, and the good includes everything.

With this background, we can recognize several similarities with Plato's perspective. In both cases, the existence of a stable level beyond the sensible realm is stated.[24] But, if those features are so common to both lines, what difference brings them into conflict? Indeed, the strong Eleatic framework supported by the Megarians entails the rejection of participation. This choice results in meta-

---

23 Menn (1995) p. 25.
24 We cannot analyze this aspect here, but it is worth noting that the connection is reinforced if we pay attention to the discussion about the unity of virtue, which is a problem in Plato's dialogues and reflects Euclid's thesis about the Good-One that spreads in different phenomena found in our everyday experience.

physical models that are structurally similar but that diverge profoundly regarding the conditions an entity must fulfill to be known, which produces very different dialectical models. Indeed, Plato's dialectic is constructive and points to identifying the mechanisms to gain knowledge. On the contrary, Euclid's dialectic is negative. For this reason, the ancient sources associate it with skepticism: maybe there is a real, stable level, and we can grasp it through a specific metaphysical exercise, but it is not connected with our everyday experience. No significant contents related to our realm are attainable by us.

Considering these differences, it is not surprising that Plato's works contain traces of arguments oriented to confront the Megarian position at this very point. As we said, this is especially clear in the case of the allegories of the *Republic*, which bear distinct hints of this collision. The three allegories point to stress the continuity between the Good and the other levels and show a progression that can be interpreted as an amendment of the Megarian beliefs. In the allegory of the sun, the talk requires characterizing the Good because it appears as the objective of philosophical knowledge. As among the Megarians, the argument begins with the assumption that ontology becomes agathology because knowing reality is knowing the Good, although it is not directly accessible. For this reason, Plato's Socrates declares himself incapable of describing it. However, it is not entirely out of reach, so he offers in return an account about "the son" of the Good in VI.506e.

The description of the relation between the Good and the tangible things is also compatible with the Megarian view: "we predicate 'to be' of many beautiful things and many good things, saying of them severally that they are, and so define them in our speech" (πολλὰ καλά ... καὶ πολλὰ ἀγαθὰ καὶ ἕκαστα οὕτως εἶναί φαμέν τε καὶ διορίζομεν τῷ λόγῳ),[25] and "we speak of a self-beautiful and of a good that is only and merely good (καὶ αὐτὸ δὴ καλὸν καὶ αὐτὸ ἀγαθόν), and so, in the case of all the things that we then posited as many, we turn about and posit each as a single idea or aspect (πάλιν αὖ κατ' ἰδέαν μίαν ἑκάστου ὡς μιᾶς οὔσης τιθέντες), assuming it to be a unity and call it that which each really is (ὃ ἔστιν' ἕκαστον προσαγορεύομεν)".[26] Reversing Euclid's perspective, Plato accounts for the relationship between one and many beginning from the multiple things to emphasize the synthesis we certainly grasp. Indeed, Plato offers a reversal of Parmenides and Euclid's metaphysical argument. The sage of Elea and his Megarian follower invited us to think of the pure notion of being or good avoiding multiplicity. Conversely, Plato focuses on multiplicity, to note that behind that diversity we grasp at the

---

25 Plato, *Republic*, VI.507b.
26 Plato, *Republic*, VI.507b.

same time something unique. But the beginning from multiplicity stresses the connection between both levels.

On this basis, two distinct tiers are distinguished. Instead of trying to grasp the pure notion of good, the allegory of the line looks at the sensible sphere. It is justified because it is pretty challenging to focus on the Form of Good directly. Still, it is worth remembering that this is what Parmenides' goddess does, and that is also the procedure Euclid seems to have followed. On the contrary, appealing to the obstacles to grasp the Form of Good, Plato chooses as the starting point the sensible realm. He identifies who perceives, what is perceived, and a third element that makes possible the connection. In the case of vision, light is necessary to see visible things.[27] Hence, the sun is the cause of the vision.[28]

The second step will be to state that, on the intelligible level, the Good behaves in the same manner regarding intelligence and the intelligible things.[29] In the end, the sun is the "son" of the Good, a kinship that emphasizes a strong link between them. We should not forget that the horizon of mythology and its narrative explanations are not too far, and genealogy is within them a central principle to account for phenomena. Thus, Plato establishes an analogy between levels that allows him to make inferences about the intelligible realm, which is impossible from the Megarian perspective. Even more, in this way, he identifies an element that belongs to both dimensions. Intelligence operates at the intelligible level to grasp intelligible objects on the condition that the Good reveals them. Still, intelligence does not belong purely to the eidetic realm, given that it is embodied as part of human beings. Hence, we, human beings, are instances of participation. The Megarians fail because they are quite attached to Parmenides' approach and ignore that they should begin paying attention to their situation in the sensible realm. This feature contradicts their approach and its claim of disengagement of ontological levels. In this sense, the example constitutes an anti-Megarian argument.

The reasoning moves towards the knowability of the Good: "you must conceive it as being the cause of knowledge, and of truth in so far as known" (αἰτίαν δ' ἐπιστήμης οὖσαν καὶ ἀληθείας, ὡς γιγνωσκομένης μὲν διανοοῦ).[30] If this were not so, the Good should cause products contrary to its nature. At the same time, like its cause, it shows features of superiority, which leads to affirming that the objects take their reality from the Good. Hence, the Good is not reality "but still transcends reality in dignity and surpassing power" (ἀλλ' ἔτι ἐπέκεινα τῆς οὐσίας πρεσβείᾳ καὶ

---

27 Plato, *Republic*, VI.508a.
28 Plato, *Republic*, VI.508b.
29 Plato, *Republic*, VI.508c.
30 Plato, *Republic*, VI.508e.

δυνάμει ὑπερέχοντος).³¹ The feature of ἐπέκεινα τῆς οὐσίας suggests closeness with Megarian ideas, but this link stresses that this distance does not imply an abyss between levels. If so, the approach would become questionable. Indeed, the insightful views of Ferber regarding this controversial passage and its contradictory issues could be supported by this Megarian horizon.³²

At this point, Glaucon burst into laughter, saying that it had "a miraculous superiority (δαιμονίας ὑπερβολῆς)".³³ Why is this so funny? It is worth remembering that the interventions of Socrates' companions in Plato's dialogues used to be not merely ornamental. For instance, in *Republic* II, the description of the healthy *polis*, reaching quasi-poetic tones, is interrupted when in 372d Glaucon says it was "a city of pigs". Most likely, it is veiled criticism against Antisthenes' natural communities that had the cyclops as models.³⁴ In this case, the outburst is also significant and calls attention to the Megaric claim about the unattainable Good. If what is real is so distant, it risks being lost and useless to understand the surrounding world, as happens precisely in the Megarian approach. Glaucon's laugh suggests that we should not exaggerate. We must recognize the difference between the visible and the intelligible level, without separating them to such an extent that the connection becomes impossible. Hence, the allegory of the sun has the primary role of showing that both levels are connected every time we use our understanding, the same as the sun is engaged every time we see something.

The same objective is central in the allegory of the line, where the scheme describes the metaphysical continuity between levels and the mechanism to grasp the Form of the Good. If so, every attempt of disconnection is cancelled. Indeed, the reasoning reinforces the elements of mediation and connection. Both levels are subdivided into interrelated parts. The images imitate objects, and the imitated objects operate as images of *dianoia*, the inferior part of the intelligible segment, as said in 510b. As we have seen, in the allegory of the sun human intellective acts reveal an instance of connection between levels. The same happens in the allegory of the line in the case of discursive thought. It represents the use of the sciences, in which the hypotheses are oriented to conclusions related to the sensible realm.

The example offered in 510c deals with mathematics, but the same holds for any of the other τέχναι, which have a set of axioms that regulate each disciplinary field. Hypotheses are also the element that unites this segment with the superior part. In the segment of νόησις, hypotheses are oriented towards a principle, but the hypothetical character of each statement is explicit. On this point, it is

---

31 Plato, *Republic*, VI.509b.
32 See Ferber (2015) and Ferber and Damschen (2015).
33 Plato, *Republic*, VI.509c.
34 Porphyry, *Sch. Odd.* IX.106 (*SSR*, V.A.189). See Marsico (2013c) and Illarraga (2012).

worth noting that the description assumes the metaphysical distance through the introduction of the notions of ἐπιβάσεις and ὁρμαί.³⁵ Ferber underlines that both terms have a military connotation and translates them as *Angriff* and *Ansturm*, that is, "attack" and "onslaught".³⁶ This cleverly indicates the effort the Platonic philosopher must make to reach the goal. At the same time, the traditional translations that focus on the final stages toward the Form of the Good are also appropriate. However, they are usually understood as equivalent terms, although they are not.

The passage says that dialectic operates with the hypotheses as ladder rungs (ἐπιβάσεις) but also as springboards (ὁρμαί) up to the non-hypothetical.³⁷ The combination of both terms is striking since the ladder rungs suggest the idea of continuous progress, whereas the springboards imply a jump. Climbing a ladder to reach the finish, counting on a fixed structure, direction and destination, requiring homogeneous efforts, is not the same as using a springboard, which implies a jump, with a much higher risk rate of many things going wrong. Do we access the Form of the Good through ladder rungs or springboards?

Indeed, the idea of a jump justifies the usual interpretations regarding an insight.³⁸ However, this discontinuity co-exists with the emphasis on continuity described in the descent, in the framework of a process "only of pure ideas moving on through ideas to ideas and ending with ideas".³⁹ Hence, if there is a jump, it does not imply a metaphysical gap associated with an abstruse gnoseological mechanism but the conclusion of a continuous path in which, at most, the difficulty is stressed. The terms ἐπιβάσεις and ὁρμαί are united by τε καί, i.e., strong conjunction that involves the combination of both aspects. Therefore, if a Megarian philosopher wants to deny the connection between sensible and intelligible, a Platonic philosopher can oppose the allegory of the line.

Then, in the same tone, the allegory of the cave includes an experiential account of the diverse levels of the line. If there are still doubts, the tale reports how a person passes through these levels. The climb can be abrupt, but no abysses are closing the path. Only mental blockages hide the way out as a possibility.⁴⁰ In this case, as in the previous analysis, the starting point is the sensible realm, em-

---

35 Plato, *Republic*, VI.511b.
36 Ferber (2015), p. 100.
37 Plato, *Republic*, VI.511b.
38 See, for instance, Fine (1990). Even if this is not our main topic here, it is worth noting that coherentism is not the best way to comprehend Plato's approach. Coherence is certainly an important element, but it is the methodological basis to reach adequation.
39 Plato, *Republic*, VI.511c.
40 On the allegory of the cave, see Inverso and Marsico (2011), about the interpretation of the passage from a phenomenological perspective.

phasizing the phantasmatic situation of defective knowledge of most people. To highlight that they ignore the surrounding world, which because of its link with the real and intelligible realm could make them comprehend the actual structure of everything, Plato deprives them of the command of the body. They are tied and forced to focus on images and discourses about images investing in them all their time. This is the synthesis of social and political life.

However, to describe the philosophical practice, he follows the steps of the freed prisoner. He can look at the surrounding world and understand the spurious origin of the bubble in which the prisoners leave. Hence, he sees things as they are and discovers a sheer but walkable way to the surface until he exits the cave. This exit corresponds to the change of segment in the allegory of the line. The freed prisoner finds there the models copied by the mobile statues that produced the shadows he saw at the bottom of the cave. This frame of "images of images" stresses the risk of "copy overload". By allowing the mimetic varieties to multiply, we are faced with an experience comparable to that of Christopher Nolan's film, *Inception* (2010), where dreams within dreams end up erasing the wakefulness parameter. Plato suggests it is easy to lose oneself in the recursiveness of copies of copies, images of images, and stop looking for reality. We end up missing the point of reference and take as a model what comes to us in stories. However, when we observe the political consequences, what we see is closer to a nightmare than a happy dream. It is convenient, then, to stay awake and look at the world, and to do this, it is necessary to limit the autonomy of the images. Thus, the efforts and pains of the ascent are justified when reality appears before the prisoner's eyes, and he can grasp a referral point.

Moreover, the moment corresponding to the jump in the allegory of the line is also essential in the account of the cave. It is the point at which the freed prisoner sees the sun. So, the qualitative leap is the consequence of the process of philosophical reflection, and it does not suppose a cut, however powerful it may be the experience of seeing the sun. The iteration of the sun's image closes the circle of the three allegories. It implies that they can be comprehended, in an integrated manner, as an answer to those who think of the Good as an unattainable element, necessarily beyond human experience. In this way, the Megarian metaphysics and its ambivalence of positing an eidetic realm but considering it impossible to grasp are called into question. This belief is substituted by an account in which continuity predominates beyond the many obstacles of the human condition. If so, the absence of this issue in other dialogues does not imply that it is a secondary topic but a highly controversial account against adversaries that contemporary readers could recognize easily.

2 Philosophy, Eristic and Paradoxes

The difference between Platonic and Megarian philosophers in the metaphysical field implies a different comprehension of dialectics. As we have already seen in Diogenes Laertius, II.106, according to the Megarians, the hierarchy and unity of being cannot be expressed, given that we tend to separate it into different dimensions: wisdom, God, intellect, etc. We have analyzed the passage, but it is useful to stress that the list is not exhaustive. Then, it does not pretend to be systematic. It is open and points to the possibility of dispersion, which underlines the impossibility of guaranteed knowledge.

Following the inspiration of Parmenides' verses regarding language, the Megarians and the Socratic lines in general often called into question the scope of language. For instance, Aristippus criticized its tendency to use verbs in the active voice because they involve the world's transcendence. He claimed that it would be better to have a language with the middle voice because it would show adequately that we only operate with the content of our πάθη.[41] On his part, Antisthenes considered that language has total correspondence with reality since it is based on names and not on statements (the reason why we must not search for definitions), and the error does not exist.[42] From this view, according to his *Sathon*, Plato's dialectic was mistaken about the nature of language. In this context, as we have seen, the Megarians denounce that language cannot reflect reality because of a structural mismatch. Against these options, Plato chooses in the *Phaedo* the refuge in the λόγοι, that under certain conditions and associated with the correct use of hypotheses can produce reliable statements that are not radically different from the common language, in the sense that they do not require substantive changes. There is no confinement in the πάθη, rejection of the grammatical structure, nor condemnations of the way we comprehend language.

From this perspective, despite the elements that unite them in other aspects, the Megarians are the furthest away from Plato's view. At the practical level, the Cyrenaic idea of correcting language is an argument to exhibit their main theses and not a practical exercise, whereas Antisthenes uses statements to apply his method of investigation of names. On the contrary, the Megarians deny any possibility of success in searching for knowledge. They transform philosophy into a cautionary practice against those who believe that reflection can lead to successful conclusions.

---

[41] Plutarch, *Against Colotes* 24.1120d–e (SSR, IV.A.211). On this point, see Tsouna (1998), Zilioli (2012), and Lampe (2015).
[42] Aristotle, *Met.*, VIII.3.1043b4–32 (*SSR*, V.A.150). See Brancacci (1990) and Marsico (2022a).

A relevant testimony about Euclid transmitted by Diogenes Laertius claims: "Euclid rejected demonstrations not in the premises but in the conclusion".[43] This idea points out to argumentation. It involves a rejection of the link between premises and conclusion. To show the weakness of logic because of its linguistic basis, he created arguments oriented against any statement setting aside its role in the original formulation. As a result, it was possible to isolate any statement, apply a ready-made refutation, and then consider the whole argument rebutted. This mechanism is visible in Eubulides' paradoxes. It is said that "besides many arguments related to dialectic, he formulated the Liar, the Hidden, the Electra, the Veiled, the Sorites, the Horned and the Bald".[44] All of them attack vital notions associated with ontology and epistemology: truth, knowledge, existence, and relational concepts, like many and few, to call them into question.

Plato probably has this feature in mind when he mentions the eristics at the end of his own methodological account in *Phaedo*'s passage about the hypothetical method. After indicating the need "to reach one point which is adequate" (ἐπί τι ἱκανὸν ἔλθοις),[45] he adds: "You would not mix things up, as eristics do, in talking about the beginning and its consequences, if you wished to discover something real". The basic problem lies in their attitude: "perhaps not one of them thinks or cares in the least about these things. They are so clever that they succeed in being well pleased with themselves even when they mix everything up".[46] The portrait of those who invite to mix premises and conclusion probably refers to Euclid and his rebuttal of the arguments taking the conclusion as a starting point.

Furthermore, Diogenes' testimony says that Euclid "also rejected the reasoning by comparison, claiming that it is established from similar or dissimilar things. If it emerges from similar things, it is necessary to consider them more than those similar to them, whereas the comparison is inappropriate if it emerges from dissimilar things".[47] Within this mechanism, analogies are excluded. They are present in many arguments, which jeopardize Plato's dialectic. Indeed, the allegories of the sun, the line and the cave are nonsense from the Megarian perspective. The

---

43 DL, II.107 (SSR, II.A.34).
44 Diogenes Laertius, II.108 (SSR, II.B.13). The Liar ("when I lie and I say I lie, do I lie or say the truth?"; Aulus Gelius, 28.2.10), the Veiled (You do not know that who bears a veil, but you know him when he pull it off, then you know and do not know the same man), the Sorites (when does something leave being few to become many?), and the Horned ("you have that which you have not lost; you did not lose horns; then, you have horns"; Seneca, *Letters*, 49.8). The Bald is a case of Sorites with hair, as an example.
45 Plato, *Phaedo*, 101e.
46 Plato, *Phaedo*, 101e.
47 Diogenes Laertius, II.107 (SSR, II.A.34).

tone and spirit of these mechanisms are clear in the same testimony, where Diogenes quotes Timon of Phlius, who mocks Euclid saying that he "instilled in the Megarians the wrath of discussion".

But what concrete problem related to dialectic did Plato notice in the Megarian views? We have an excellent example in the *Euthydemus* to explain the distance between Plato's and Euclid's dialectical models. The identification of Euthydemus and Dionysodorus as spokespersons of the Megarians is entirely plausible.[48] As we saw, these characters pursue exercises of refutation wholly oriented to the destruction of knowledge. In the same line, they are presented as experts in a fighting style similar to wrestling. The scene is located in a gym, and the whole set resembles a spontaneous show with many fans and a curious audience. In a sense, Socrates and Crito seem to be commenting on somewhat quirky wrestling combat with three rounds (275d–278e; 283a–286b and 293b–303a), the times the brothers shoot arguments against their interlocutors.

The dialogue portrays Socrates stressing the need to train young people in virtue and invite them to learn. In return, he receives arguments that take all the current opinions about knowledge to absurdity (275d–277c). To clarify the request that has been evidently distorted, Socrates offers a protreptic based on a dialogue with all the essential features of Plato's dialectic: a genuine commitment to exploring hypotheses with a partner equally vigilant on the quality of the argument (279a–282e). Again, given that the notion of reality appears as a correlate of knowledge, the brothers strive to show their ambiguities, and Socrates and Ctesippus' remarks only unleash the storm of arguments about falsehood (283a–286b). As an answer to Socrates' insistence on collaborative dialogue (288a–293a), Euthydemus and Dionysodorus persist in taking statements at the chance. Some of them are the conclusions of Socrates' arguments, and they refute them, ignoring the way or the reason that led to them (293b–303a).

The whole sketch reminds us of Euclid's method and aims. The Megarian instruments to refute the interlocutor that we can infer from the extant texts are consistent with the overall approach in the Euthydemus: rigid models of questioning, insistence on "yes" or "no" answers, and special attention to issues related to ontology and gnoseology, as in the case of Eubulides' paradoxes. From this view, the work structure portrays the Megarian dialectic style interpreting it as a pure game. Whereas the Platonic philosopher tries to apply cooperative dialogue, the Megarian philosopher gets the fun picking statements and discrediting them with fancy

---

[48] See Dorion (2000), Villar (2020). See also Inverso and Marsico (2012), in which we claim that the Megarians were the main target of Plato's criticism in the *Euthydemus*. These developments are on the horizon of this analysis of the rejection of eristics in the framework of the collision between both dialectical models.

tricks. Plato's negative assessment of this course focuses on the attitude of those who dialogue without taking seriously what the other says, without ever leaving a glib tone. Of course, from the Megarian perspective, this is an effective strategy oriented to stress the weakness of intellectual endeavors, plenty of vices and contradictions, and hence useless tools to build knowledge.

According to this attitude, the brothers boast that "whatever he answers, he will be rebutted" (275e), and they refuse to contribute to Socrates' protreptic discourses. Hence, it is not a philosophy ready to educate, and it only produces what in *Republic* is compared with puppies chewing arguments.[49] Even more, it is dangerous. At the beginning of the dialogue, the young Ctesippus gets offended because of the ludicrous conclusions of the paradoxes. Still, as it moves forward, he learns the mechanism and uses it against the brothers. It is an infectious vice that spreads fast and, under its harmless appearance, undermines our beliefs in knowledge.

This point is reinforced in the final dialogue between Crito and a nameless character, very angry with Socrates' attitude. Often and rightly, this figure has been linked to Isocrates. At the beginning of his *Against the Sophists*, he accuses the Socratics of being cheaters, and in his *To Demonicus*, 3–4, he blames those who "only promote arguments" for his worthlessness.[50] The inclusion of this corollary emphasizes that it is impossible to detect the difference between Plato's and Euclid's dialectical models from a certain distance. If so, also Plato's method seems to be mere eristic and falls victim to the criticism of those who reject this kind of practice. To counterbalance this effect, Plato portrays Socrates rejecting the eristic approach in order to underline the distance by emphasizing their distinct aims and methods and showing the mistake in Isocrates' criticism and all those who fail to recognize the difference.

3 Paradoxes of Participation: Men, Vegetables and Figs

Let us now examine some testimonies about the collision between Plato's and Euclid's dialectic from the Megarian perspective. We have mentioned Plato's attacks and defenses against the Megarian views, but the reverse direction was also present. The Megarians are the source of several assertions that question Plato's theory of Forms, primarily pointing against participation. In another work, we have

---

49 Plato, *Republic*, VII.539a.
50 See Michelini (2000) and Villar (2020 and 2021).

suggested that the arguments in the first part of the Parmenides are a testimony of these kinds of conflict, and among them, we can find the third man argument, in which we are interested now.[51]

The testimony of Alexander of Aphrodisias reports distinct variants, one of them attributed to Eudemus of Rhodes, another to some sophists, the third to Polyxenus, and the latter to Aristotle.[52] He connects the first and the fourth based on the regression produced when we say that the things predicated in common of substances are in the proper sense such things, and similar things are similar because they participate in the same thing. Alexander mentions that Eudemus dealt with this issue in his *On Expression*, which suggests that he probably intends to clarify logic misunderstandings, as Aristotle does in *Sophistical Refutations*, 22, when he refers to the third man argument.[53]

Eudemus and Aristotle seem to have detected a problem in Plato's theory of Forms, as inferred from the first chapter of the *Categories*. He distinguishes between homonymy and synonymy, indicating that the first notion makes assumptions about the likeness between the sensible and the intelligible realms that allows using the same words to refer to both levels. For this reason, he offers the case of the term "living" (*zoion*) said of the man and said of the portrait, the first in the sense of "alive", and the second in the sense of "true to life".[54] In this case, we can detect both senses. However, can we guarantee that when we say "beautiful" about the Form of Beauty and about a beautiful thing, its image, there is not an unbridgeable gap? In contrast, Aristotle tries to build his system upon synonymy to avoid such risks, always operating on the same ontological level: we can be sure that "living" said of the man and said of the ox have the same meaning. Even when homonymy comes back in his famous discussion about being in *Met.* IV, it can be managed because it is not the pillar of the whole approach. In fact, homonymy is on the very basis of the regressive version of the third man argument, given that it is produced when we use the same predicate in two different ontological levels. This feature is closely related to participation and the discussion between Plato and the Megarian philosophers we are considering.

Indeed, if these regressive arguments examined in the Lyceum come from previous discussions, it exceeds our goals in this paper. However, it is worth noting that they were probably part of the anti-Platonic weapons we have found among Megarian theses. Our information about the non-regressive variants sug-

---

[51] See Marsico (2022), oriented to explore the references to the Megarians in Plato's *Parmenides*.
[52] See Fine (1994) and Gardella (2015).
[53] Fortenbaugh (2002), pp. 69–70.
[54] See Ildefonse and Lallot (2002), especially the introduction.

gests this origin. Alexander mentions the sophists as authors of the second variant and Polyxenus as a promoter of the third one. The category of "sophist" is quite broad, and it often refers to the Megarians, mainly because of their skeptical tendencies.[55] On his part, Polyxenus was associated with this group. The *Platonic letter XIII*, 360b–c reports that he was Bryson's friend. We know Bryson as one of Euclid's followers that fought against the ideas related to ὀρθότης ὀνομάτων, offering counter-intuitive thesis like the inexistence of obscene words.

The second version claims that when we say "the man walks", we are not talking about the Form because it does not move, not the particular, because "we do not know who he is", so it should have to do with a third man (84.7–16). According to Polyxenus, "if man exists by participation and community with the Idea, that is man-himself, then there must be some man who has his existence in relation to the Idea. But this is not man-himself (who is the idea), nor is it any particular man. So it remains that some third man has his existence in relation to the Idea".[56]

The meaning of both versions is better comprehended when we compare them with Stilpo's argument against Plato's position reported by Diogenes Laertius. In another work, we have discussed the philological details of this passage,[57] which can be read as follows: "He claimed that he who says 'man' says no one because he does not mean this man or that. Why should he mean the one more than the other? Therefore, neither does he mean this individual man. Again, 'vegetable' is not what is shown to me, for vegetables existed ten thousand years ago. Therefore, this is not vegetable".[58]

Stilpo applies to the field of language what Polyxenus states in the ontological realm. Semantics collapses if the metaphysical levels are separated, and there is no participation. The universal concepts lack a reference for two fundamental reasons. On the one hand, they have no lexical indicators of particularization. As a result, strictly speaking, they are not suitable for recognizing individuals. On the other, they are perdurable, which is incompatible with the nature of the individual they supposedly name. If there is no fundamental similarity between particular and universal, given that, according to Euclid's central thesis, we experience multiplicity and not unity, and it is impossible to bridge the gap, then the correlation

---

[55] It is used, for instance, alluding to Bryson. About other characterizations, see Marsico (2013a, *ad loc*).
[56] Alexander of Aphrodisias, *In Met.*, 84.16–21 (SSR, II.T.10).
[57] Marsico (2013a and 2013b). The latter is focused on Stilpo's attack against Plato's theory of Forms from the metaphysical point of view, which is complemented here with the dialectical aspects of this disagreement.
[58] Diogenes Laertius II.119 (SSR, II.O.27).

between language and reality is not possible. 'Man' does not refer to this man nor 'vegetable' to this vegetable.

Stilpo's argument about vegetables is behind one of his playful anecdotes, reported by Diogenes Laertius as a kind of "third fig argument". He says: "And once when Crates held out a fig to him when putting a question, he took the fig and ate it. Upon which the other exclaimed: 'O Heracles, I have lost the fig', and Stilpo remarked: 'Not only that but your question as well, for which the fig was payment in advance".[59] In a horizon of robust theoretical exchanges, Stilpo talks with Crates of Thebes, who was a Cynic. For that reason, he was reticent to accept the Megarian views, considering them *en bloc* as logicist gimmicks. Crates prioritizes common sense and appeals to an everyday object. The context has not been preserved, but it seems to have constituted a critical review of Stilpo's Megarian position. This procedure is similar to Diogenes of Sinope's refutation of movement through a walk.[60] It becomes interrupted because Stilpo takes the fig away, and with it, Crates' attempt to underline the worldly origin of every notion disappears. The exchange is quite informative about the attitude of these adversaries. Still, it keeps silent regarding the content of Crates' question, which was supported on the fig as a warrant or concrete proof.

About this issue, Muller claims that the question the Cynic fails in asking is: "Is this a fig?", trying to force Stilpo to establish a relation between the universal concept and the object.[61] Crates would be trying to force him to establish a link between the universal concept and the object. Hence, the argument could have been as follows: "if this is a fig, then the name 'fig' names an individual object". It would be equivalent to the example of the vegetables, in this case with a fruit, which strengthens the parallel. In this way, the argument shows that the name "fig" applies to "this fig" beyond its ephemeral and unstable character.

Stilpo's reaction of eating the fig, making it disappear, points to his view about the evanescent nature of the particular sensible object.[62] If so, it is inappropriate to link it with a stable concept. If the fig disappears, the name still has meaning, which implies that there is between them an unstable relation based on the distance that separates names and things. In this sense, the question "Is this a fig?", which depended on the particular fig that Crates had in his hand, reveals itself as inadequate. It is inadequate not only because its deictic features are lost, given that the fig has vanished, but because the objective of the argument—that

---

59 Diogenes Laertius II.118 (SSR, II.O.6).
60 Diogenes Laertius VI.39.
61 Muller (1985, *ad loc.*).
62 See Marsico (2013b).

is, to link the name "fig" with "this fig"—becomes destroyed, since the meaning of "fig" is keep unscathed even if the particular object, i.e., its concrete basis, has faded away.

Hence, we are in the face of Polyxenus' argument: "'vegetable' is not what is shown to me, for vegetable existed ten thousand years ago. Therefore, this is not vegetable".[63] In the same way, "fig" is not what is shown to me, for "fig" keeps existing, although Stilpo has eaten that particular fig. If so, what is named must be a third fig hidden among the veils of a fuzzy ontology. This argument was built as an anti-Platonic instrument, and it is on the horizon of the quarrels between Plato and the Megarians that we have found in the previous sections.

4 Conclusion

Plato's works were written in a hectic environment, with plenty of theoretical exchanges between diverse philosophies. Because of that, an exclusively internal interpretation of these texts is at risk of losing sight of the dialogical dimension behind them. The case of the Megarians is relevant, given that the traces of their developments can be inferred in several Platonic texts. On the one hand, we have paid attention to the case of the allegories of the central books in the *Republic*, often considered an outline of Plato's metaphysical position. In our view, these passages should be treated with greater caution since they include polemical aims. Against the Megarian denial of participation, the allegories reinforce this aspect, showing the connection between metaphysical levels from a different perspective. At the same time, this approach allows comprehending the significant fading of the Form of the Good in other works of Plato, given that the notion kept being important and Plato remained concerned about this issue, as is evident from the lecture on the Good, in which he presented the thesis of the unity of the Good in front of a dissenting and disappointed audience. Who knows if there were also his Megarian fellows among the upset and angry attendees?

At the same time, Plato tries to state the distance between the dialectical practice in both models. Although they could seem similar to those who observed at a certain distance, he invited not to confuse eristic with cooperative dialogue, given that the objective of denying every kind of knowledge is not comparable to a positive epistemic project. On the contrary, the Megarians attacked the theory of Forms because it was an instance of undue reliance on access to the stable realm. As a whole, these cases are a strong testimony about the active discussion

---

63 Diogenes Laertius, II.119.

within the Socratic group, even if they were intellectuals united by fraternal bonds. Indeed, in the controversial framework of the Socratic circle, which was plenty of frictions and enmities, doxography suggests that Plato and Euclid were close. The shared points remained, even if this link became diluted in later generations. However, that personal relationship and the beliefs that united them did not impede several discharges of friendly fire.

Bibliography

Arnim, Hans von (1931): *Die Entstechung der Gotteslehre des Aristoteles*. Vienna: Hölder-Pichier-Tempsky.
Brancacci, Aldo (1990): Oikeios logos. *La filosofía del linguaggio di Antistene*. Naples: Bibliopolis.
Brancacci, Aldo (2018): "Socratism and Eleaticism in Euclides of Megara". In: Stavru, Alessandro and Moore, Christopher (Eds.): *Socrates and the Socratic Dialogue*. Leiden: Brill, pp. 161–178.
Bredlow, Luis (2011): "Platón y la invención de la escuela de Elea (*Sof*. 242d)". In: *Convivium* 24, pp. 25–42.
Capizzi, Antonio (1975): *La porta di Parmenide: due saggi per una nuova lettura del poema*. Roma: Edizioni dell'Ateneo.
Chame, Santiago (2015): "La relación entre la sofística tradicional y la escuela megárica en torno a sus parámetros argumentativos: análisis comparativo del tratado *Sobre el no ser* y el *Eutidemo*". In: *Avatares Filosóficos* 2, pp. 195–205.
Cherniss, Harold (1935): *Aristotle's Criticism of Presocratic Philosophy*. New York: Octagon Books.
Cordero, Néstor (1991): "L'invention de l'école éléatique (Platon, *Sophiste* 242d)". In: Aubenque, Pierre (Ed.): *Études sur le* Sophiste *de Platon*. Naples: Bibliopolis.
Cordero, Néstor (2005): *Siendo se es*. Buenos Aires: Biblos.
Divenosa, Marisa and Marsico, Claudia (2011): *Alegorías del sol, la línea y la caverna*. Buenos Aires: Losada.
Döring, Klaus (1972): *Die Megariker. Kommentierte Sammlung der Testimonien*. Amsterdam: Grüner.
Dorion, Louis-André (1995): *Aristote, Les refutations sophistiques. Introduction, traduction et commentaire*. Paris: Vrin.
Dorion, Louis-André (2000): "Eythydème et Dionysodore sont-ils des Mégariques?". In: Robinson, Thomas and Brisson, Luc (Eds.): *Proceedings of the V Symposium Platonicum*. Sankt Augustin: Academia Verlag, pp. 35–50.
Düring, Ingemar (1966): *Aristoteles. Darstellung und Interpretation seines Denkens*. Heidelberg: C. Winter Universitätsverlag.
Ferber, Rafael (2015): *Platos Idee des Guten*. Sankt Augustin: Academia Verlag.
Ferber, Rafael and Damschen, Gregor (2015): "Is the Idea of Good Beyond Being? Plato's *epekeina tes ousias* Revisited (*Republic* 6, 509b8–10)". In: Nails, Debra and Tarrant, Harold in Collaboration with Kajava, Mika and Salmenkivi, Eero (Eds.): *Second Sailing: Alternative Perspectives on Plato*. Helsinki: Societas Scientiarum Fennica, pp. 197–203.
Fine, Gail (1990): "Knowledge and Belief in Republic V-VII". In: Everson, Stephen (Ed.): *Epistemology*. Cambridge: Cambridge University Press, pp. 85–115.
Fine, Gail (1994): *On Ideas: Aristotle's Criticism of Plato's Theory of Forms*. New York: Oxford University Press.

Fortenbaugh, William (2002): *Eudemus of Rhodes*. New Brunswick: Transactions Publishers.
Fritz, Kurt von (1931): "Megariker". In: *Paulys Realencyclopäedie der classischen Altertumswissenschaft*, Supplementband V. Stuttgart: Druckenmüller, columns 707–724.
Galgano, Nicola et Cherubin, Rose (Ed.) (2020): *Parmenides Eon 1 and Parmenides Eon 2*. Anais de *Filosofia Clássica* 14. No 27 and No 28.
Gardella, Mariana (2014): "Euclides de Mégara, Filósofo Socrático". In: *Ágora* 33. No. 2, pp. 19–37.
Gardella, Mariana (2015): *Las críticas de los filósofos megáricos a la ontología platónica*. Buenos Aires: Rhesis.
Gardella, Mariana (2016): "*Ánthropos peripateî*: los argumentos del tercer hombre megáricos". *Elenchos* 37, pp. 69–94.
Giannantoni, Gabriele (1990): *Socratis et Socraticorum Reliquiae*. Naples: Bibliopolis.
Ildefonse, Frederique and Lallot, Jean (2002): *Aristote, Catégories*. Paris: Seuil.
Illarraga, Rodrigo (2012): "Utopía ciclópea, utopía de cerdos. Una reconstrucción del pensamiento político de Antístenes". In: Miseri, Lucas (Ed.): *Estado, cultura y desarrollo: entre la utopía y la crítica*. Mar del Plata: EUNMdP, pp. 59–65.
Illarraga, Rodrigo and Marsico, Claudia (2021): "Behind the Witness' voice: Xenophon as a Source of the Socratic Philosophies". Paper presented at the Conference *Xenophon 2021*, Liverpool.
Inverso, Hernán and Marsico, Claudia (2011): "Girar con todo el cuerpo hacia lo que aparece: teoría y corporalidad en el símil platónico de la caverna desde una matriz fenomenológica". In: *El hilo de la fábula* 6, pp. 197–208.
Inverso, Hernán and Marsico, Claudia (2012): *Platón, Eutidemo. Introducción, traducción y notas*. Buenos Aires: Losada.
Jaeger, Werner (1923): *Aristoteles: Grundlegung einer Geschichte seiner Entwicklung*. Berlin: Weidmann.
Marsico, Claudia (2011): "Megaric philosophy: between Socrates' stamp and the ghost of Parmenides". In: Cordero, Néstor (Ed.): *Parmenides, venerable and awesome*. Paradise: Parmenides Publishing, pp. 353–362.
Marsico, Claudia (2013a): *Filósofos socráticos, Testimonios y fragmentos I. Megáricos y Cirenaicos*. Buenos Aires: Losada.
Marsico, Claudia (2013b): "'Ni el hombre es blanco ni el caballo corre'. Argumentos antiplatónicos en Estilpón de Mégara". In: *Méthexis* 25, pp. 9–33.
Marsico, Claudia (2013c): "Sobre los cerdos. Aspectos de la phúsis en Antístenes". In: Bieda, Esteban and Marsico, Claudia (Eds.): *Expresar la* phúsis. *Conceptualizaciones antiguas sobre la naturaleza*. Buenos Aires: UNSAMEdita, pp. 120–138.
Marsico, Claudia (2014): "Encrucijadas dialécticas: *élenchos*, dispositivos antierísticos y filosofía megárica en las *Refutaciones sofísticas*". In: *Archaí* 14, pp. 137–148.
Marsico, Claudia (2020): "Aristippus' legacy. On the Development and Central Tenets of the Cyrenaic Philosophy". In: *Journal of Classical Studies* 21, pp. 23–46.
Marsico, Claudia (2021): "Oddyseus is not a liar. Ancient Semantic Analysis in Antisthenes' investigation of names". In: *Journal of Classical Studies* 23, pp. 45–60.
Marsico, Claudia (2022a): "From love to parricide. Parmenides among the Socratics". In: Volpe, Enrico (Ed.): *La via dell'essere. Studi sulla ricezione di Parmenide nell'antichità*. Sankt Augustin: Academia Verlag, pp. 131–150.
Marsico, Claudia (2022b): "Intra-Socratic Polemics: The *Parmenides* as Part of an Anti-Megaric Programme". In: Brisson, Luc, Mace, Arnaud, and Renaut, Olivier (Eds.): *Plato's* Parmenides. *Selected Papers of the Twelfth Symposium Platonicum*. Sankt Augustin: Academia Verlag, pp. 133–146.

Menn, Stephen (1992): "Aristotle and Plato on God as Nous and as the Good". In: *The Review of Metaphysics* 45. No. 3, pp. 543–573.
Menn, Stephen (1995): *Plato on God as Nous.* Carbondale: Southern Illinois University Press.
Michelini. Ann (2000): *"Socrates plays the buffoon: Cautionary protreptic in* Euthydemus". In: *American Journal of Philology* 121. No. 4, pp. 509–535.
Muller, Robert (1985): *Les mégariques.* Paris: Vrin.
Rossetti, Livio (2020): *Verso la filosofía. Nuove prospettive su Parmenide, Zenone e Melisso.* Sankt Augustin: Academia Verlag.
Tsouna, Voula (1998): *The Epistemology of the Cyrenaic School.* Cambridge: Cambridge University Press.
Villar, Francisco (2018): "Los argumentos dialécticos de Eubúlides de Mileto: el Mentiroso, el Sorites, el Velado y el Cornudo". In: *Factótum* 19, pp. 48–59.
Villar, Francisco (2020): "Isócrates y el crítico anónimo del *Eutidemo* de Platón". In: *Agora* 39. No. 2, pp. 169–191.
Villar, Francisco (2021): *El Eutidemo de Platón como respuesta a las críticas de Isócrates al círculo socrático de Contra los sofistas y Encomio de Helena.* Buenos Aires: Universidad de Buenos Aires. Doctoral Thesis.
Zilioli, Ugo (2012): *The Cyrenaics.* London: Routledge.

Beatriz Bossi
Eros on Stage: Dialogue and Dialectic in the *Phaedrus*

1 What is the Relation between Dialectic as "Engaging in Dialogue" and as "the Highest Stage of Our Cognition"?

Dialectic is defined as a method (*Republic* 533c7), a science (511c5), and a capacity to engage in dialogue (511b4; 532d8; 533a8; 537d5). As Plato's explanations of dialectic are notoriously brief and likely to have been intended to be so, scholars have developed different interpretations of it. Notomi (2004) introduces the problematic relation of dialogue and dialectic as follows:

> Dialogue and dialectic look so different (or even opposite) that scholars tend to separate the two and ascribe them to different stages in the development of Plato's philosophy. It is often supposed that dialogue is the Socratic method, reproduced mainly in the earlier [aporetic] dialogues, while dialectic, presented in the middle and later dialogues, is the originally Platonic concept of philosophy [that reaches the truth],[1] which has little to do with actual dialogue.[2]

Notomi suggests that we should reject the assumption that knowledge is a system of doctrines or the cognitive state of possessing them but take it as "the soul's ability and activity of conducting the kind of discourse (*logos*) that allows it to deal properly with the forms".[3] Since dialectic is described as the supreme intellectual *dynamis*, he claims that Plato takes knowledge as an ability or process of inquiry (510b; 532a; 533c; 476d; 477b–478b), and focuses on dialectic as "that which *logos* itself grasps by the ability to engage in dialogue" (511b). He also suggests that Plato probably coined the word διαλεκτικός "out of the common verb διαλέγεσθαι (to engage in dialogue) and gave it more technical connotations".[4]

Note: I am deeply grateful to Prof. Thomas M. Robinson for his suggestions and his checking of the English version of this paper.

1 The dialectician is the one who captures the definition of the essence of each thing: Ἦ καὶ διαλεκτικὸν καλεῖς τὸν λόγον ἑκάστου λαμβάνοντα τῆς οὐσίας (*Republic*, 534b3–4).
2 Notomi (2004), p. 2.
3 Notomi (2004), p. 1.
4 Notomi (2004), p. 2.

https://doi.org/10.1515/9783110744149-008

There is, as it happens, a revealing passage in Xenophon,[5] in which he makes explicit "these more technical connotations" by showing how Socrates himself derived the verb διαλέγεσθαι (to arrive at truth by discussion) from διαλέγειν (to sort out, select, divide):

> (...) only the self-controlled have power to consider the things that matter most, and, sorting them out after their kind by word and deed, to prefer the good and reject the evil. And thus, he said, men become supremely good and happy and very skilled in discussion. The very word "to dialogue", according to him, owes its name to the practice of meeting together for common deliberation, *sorting things after their kind*, and therefore one should be ready and prepared for this and be zealous for it; for it makes men become excellent, the best leaders and the best skilled in dialogue: τοῖς ἐγκρατέσι μόνοις ἔξεστι σκοπεῖν τὰ κράτιστα τῶν πραγμάτων, καὶ λόγῳ καὶ ἔργῳ διαλέγοντας κατὰ γένη τὰ μὲν ἀγαθὰ προαιρεῖσθαι, τῶν δὲ κακῶν ἀπέχεσθαι. καὶ οὕτως ἔφη ἀρίστους τε καὶ εὐδαιμονεστάτους ἄνδρας γίγνεσθαι καὶ διαλέγεσθαι δυνατωτάτους· ἔφη δὲ καὶ τὸ διαλέγεσθαι ὀνομασθῆναι ἐκ τοῦ συνιόντας κοινῇ βουλεύεσθαι διαλέγοντας κατὰ γένη τὰ πράγματα. δεῖν οὖν πειρᾶσθαι ὅτι μάλιστα πρὸς τοῦτο ἑαυτὸν ἕτοιμον παρασκευάζειν καὶ τούτου μάλιστα ἐπιμελεῖσθαι· ἐκ τούτου γὰρ γίγνεσθαι ἄνδρας ἀρίστους τε καὶ ἡγεμονικωτάτους καὶ διαλεκτικωτάτους (*Memorabilia*, IV, V, 11–12).

Indeed, as Notomi concludes, dialectic is connected to the ability to question and answer (531d–e; 534d);[6] is carried out through refutation (534c); and is defined as the knowledge of how to conduct a dialogue (511b) to grasp the definition of the essence of each thing (534b). Thus, it seems that knowing how to argue cannot be separated from knowing what a thing is.

I should like to defend the view that this is precisely what we learn in the *Phaedrus:* a) that Socrates knows how to engage in dialogue with Phaedrus (despite their emotional drives) and also how to use arguments to persuade him, since he himself knows what eros is from the very beginning, and b) that, as Xenophon maintains, only those who are self-controlled can sort things out according to their kind, so that they are able to prefer the good and reject the evil.

Dialectic implies examining the species of both difference and identity in nature: τί εἶδος τὸ τῆς ἑτέρας τε καὶ τῆς αὐτῆς φύσεως (*Republic*, 454b6–7), knowing how to "render and exact an account (δοῦναί τε καὶ ἀποδέξασθαι λόγον, 531e4–5)" of any opinions offered in the discussion, and concluding with a synoptic view (σύνοψις, 537c1; c7).[7]

---

5 I am grateful to Prof. Tomás Calvo Martínez for this reference.
6 At *Cratylus* 390c 10–11, Socrates also asks Hermogenes whether he would not call someone who asks and answers questions a dialectician.
7 Dialectic is the only method that travels this road, doing away with hypotheses and proceeding to the first principle itself: ἡ διαλεκτικὴ μέθοδος μόνη ταύτῃ πορεύεται, τὰς ὑποθέσεις ἀναιροῦσα, ἐπ' αὐτὴν τὴν ἀρχὴν ἵνα βεβαιώσηται (533c7–d1). According to Vallejo, Annas (1982, pp. 282–284), R.

Gonzalez[8] rejects treating dialectic as a purely formal method of constructing arguments in total abstraction from the content of Plato's philosophy. For him the exchange of questions and answers as actually practiced in the dialogues is not accidental to the results. He regards dialectic in the "aporetic" dialogues as positive and constructive, despite the lack of propositional results,[9] and he claims that "rather than seeing dialectic as a mere tool used in system building, Plato clearly understands the very content of philosophy to be dialectical".[10]

In this paper I shall focus on the intellectual difficulties Socrates faces in the first part of the dialogue in view of his own passion for Phaedrus, and on the peculiar "rubbing up and down" of the first, poor conceptions of love as a necessary step towards the "awakening" of true knowledge in Socrates, who has an affinity with truth, and is used to a life exploring the issue at hand.[11] In a sense, the *Phaedrus* can be interpreted as a journey from sick and dependent passionate love for beautiful bodies towards divine and liberating passionate love for beautiful young boys "in conjunction with philosophy", something not just "defined" by Socrates but also "enacted" by him in relation to Phaedrus, to show him how to become a good lover.

Robinson (1953, p. 173), and Szlezák (2003, p. 187) have similar views about the tension between deduction and intuition in this process. Though it is unclear whether the ascent proceeds through logical premises, or is due to abstract synopsis, Vallejo ("El enigma de la dialéctica", a paper presented at the IV Congress of the SIFG, UCM, April 25–27, 2022) thinks it cannot constitute an abrupt miraculous leap, since the dialectician must justify somehow the un-hypothetical principle that constitutes the foundation of the hypotheses (534c1–2). However, he finds that dialectic presents us with an enigma, since the process of justification of the hypotheses is rendered unclear by the metaphorical language Plato uses.

8 Gonzalez (1998), pp. 2–6.
9 Gonzalez (1998), p. 10.
10 Gonzalez (1998), p. 11.
11 Here, I am using the expressions and concepts concerning dialectic and dialogue that are used by the author of the *Seventh Letter*: "Only barely, when the [three], i.e., names, propositions, and along with them appearances and perceptions, are rubbed against each other, each of them being refuted through well-meaning [non-adversarial] refutations in a process of questioning and answering without envy, will wisdom along with insight commence to cast its light in an effort at the very limits of human possibility" (344b3–c1).

2 Why Did Socrates Not Refute Phaedrus in the First Part of the Dialogue?

The first part of the *Phaedrus* (227–242) is quite like the early aporetic dialogues in the sense that, apart from some Socratic clues that go unnoticed by Phaedrus, there is no true definition of love but rather a sophistic defense of unpassionate love (by Lysias) followed by an almost completely imitative logos (by Socrates). However, as was noticed above, this first dramatic scene, in which passion is rejected in both speeches, is paradoxically full of passion and competitive jealousy in both characters. Socrates must reach Phaedrus where he is, intoxicated with sophistic views on the benefits of unpassionate love, and this makes refutation and real progress impossible.

Only after the daimon forbids Socrates to cross the river and abandon his interlocutor are we introduced to a new start, in which Socrates is ready to say something on his own, and Phaedrus is ready to listen to him. After defining the right genus of love as "divine madness", Socrates gives it back its passionate feature. While the beautiful παλινῳδία gives back his honor to the god Ἔρως, it also gives back his dignity to Phaedrus, in the sense that he can imagine himself as a soul divinely inspired to love not just beautiful speeches but wisdom itself.

In the third part, Socrates recalls the clues that enabled them to engage in positive dialogue. After grasping the genus of love, namely, "divine madness", the essential step was choosing the "right-hand" species of it, namely, passion for the young such as to take them to the realization of their best potentiality, while rejecting the "left-hand" one that consists of merely possessive, sterile passion. The right-hand species will also admit of at least two hierarchical kinds, the highest being love between followers of Zeus (philosophers) and at a lower-level love between followers of Ares (the military class).

One may wonder why the process took so long, and why Socrates could not refute Phaedrus' definition of love (as an unpassionate and uncommitted drive) by asking him the proper questions. Why was the first conversation between them a frustrating though necessary step? My hypothesis is that neither Socrates nor Phaedrus was ready for the task, due to the manipulative, biased relationship between them. It was presumably Socrates' intellectual frustration which provoked his desire to leave the scene, and it took a warning from his daimon to return him to dialectical mode. However, the presentation to the reader/listener of Lysias' text and Socrates' mimetic discourse served to expose the weaknesses of both kinds of eros, passionate and unpassionate, and allowed Socrates to awaken to the true being of eros as a special type of divine madness, and to dare to manifest this openly to Phaedrus.

One of the presuppositions of the author of the *Seventh Letter* is that dialectic gives rise to knowledge of ideal reality in someone possessed of an ideal nature (343e2–3). This affinity is a correspondence between the goodness of one's soul and the goodness of the object of knowledge. Gonzalez[12] observes that, in the *Laches*, the *Charmides*, the *Euthydemus*, and the *Meno*, Socrates exhibits, respectively, the courage, temperance, wisdom, and virtue which he seeks to define. If knowledge is to be gained through dialectic, it must be found in one's soul. The soul, he suggests, could not even seek the true nature of virtue if it did not already "divine" it in its defective instantiations. In the *Phaedrus* (249e) we learn that "every human soul has by nature seen all beings" though in varying degrees. If a soul is neither defective nor corrupted, it will have within itself a knowledge of what is.

While the Socrates of the first scene of the *Phaedrus* remains intoxicated by Phaedrus' poison, he cannot operate as a refuter or a dialectician. Once he gets rid of this manipulative situation, he is in a position to "divine" what real love is, to communicate his findings to his interlocutor, and to "love Phaedrus with philosophy". This way, their emotional drives, their dialectical findings, and their ability to engage in fruitful dialogue prove to be interconnected.

It is worth noticing that, in a much-debated passage in the *Sophist*, before introducing the sixth definition (which I shall consider below) Plato divides the vices of the soul (πονηρίαι) into two species: ignorance, which is a sort of "deformity", and cowardice, lack of moderation, and injustice, which are to be taken as a "disease". A corrective art (κολαστική) is propounded to punish the vices, while refutation is claimed to be the most effective and pleasing liberation from the incapacity to learn (ἀμαθία), since mere admonition is not enough to cure it (228e–230d).

It seems to me that the dramatic action in the *Phaedrus* offers a perfect example of these "deformities" and "diseases" alongside their respective cures. Phaedrus' incapacity to learn at the beginning of the dialogue is due to his biased love for Lysias' speech and for Lysias himself. On the other hand, Socrates' incapacity to refute him is due to his own intoxication with Phaedrus and the divinities of the place. The only way to "save" Phaedrus is to give him the opportunity to liberate himself from his obsessive, blind desire to please his beloved, i.e., from his "lack of moderation", to put it in the wording of the *Sophist*. This is precisely what Socrates manages to do when he "punishes" him by refusing to defend unpassionate love and by threatening to leave him alone (since he seems to have guessed that admonition is useless). At this point, Phaedrus re-acts: he regains self-control and is ready to listen. Only then can he accept Socrates' new technique of collection and division and be seduced by the beauty of the palinode, which enhances his

---

12 Gonzalez (1998), pp. 269–270.

view of his own soul. At the end, Socrates will have a genuine dialogue with Phaedrus, and will be able to teach him how to differentiate ignoble, sophistic rhetoric from one based on dialectical skills.

3 What Makes the Difference between the Dialectician, the Sophist, and the Noble Sophist?

At the beginning of the *Sophist*, Socrates says that it is not easy to recognize the class of the philosopher, for philosophers, in the eyes of the ignorant, appear disguised in all sorts of shapes: sometimes they come disguised as statesmen, sometimes as sophists, and on occasion they can give some people the impression that they are altogether *mad* (216c4–5; d1–2).[13] As opposites are to be understood together, grasping the philosopher implies being able to distinguish him from the sophist.

In a previous paper,[14] I attempted to collect their similarities. Both the sophist and the philosopher: 1) are said to share an interest in the same topics (225b13–c 9); 2) are experts in the arts of contradiction and refutation (225a–c); 3) are ready to strive hard to capture their prey: they are ἀγωνισταί (261a–c); 4) try to "capture" the attention of the young (222aff.); 5) make use of irony and rhetoric; 6) operate in the realm of the imitative; and, last but not least, 7) may to some extent "deceive" their interlocutors (240d).

However, though both the philosopher and the sophist may be ready to deceive their interlocutors, they do so with different purposes: while the sophist "sells ignorance for knowledge", the philosopher may need to hide his views at the beginning and prepare his interlocutor gradually, in hopes of setting him on a different path (*Phaedrus*, 262d). Though both may make use of rhetoric, the sophist usually does so to simulate that he knows something (268aff.), while the philosopher may do so to simulate that he does *not* know something, on occasions when the interlocutor is not ready to listen...

Both the noble rhetorician and the sophistic rhetorician proceed gradually from one similarity to the next, the former basing himself on dialectic and seeking

---

[13] All translations of the *Sophist* by Nicholas White in Cooper, John and Hutchinson, Douglas S. (Eds.) (1998): *Plato, Complete Works*, slightly modified.
[14] "Back to the sixth definition: Why noble? Why sophists? (*Sophist* 228a–231b)", Vrin, forthcoming.

persuasion through knowledge, the latter with no care for knowledge at all.[15] Between these two contrasting figures, Plato also introduces, in the *Sophist*, the controversial figure of the "noble sophist", and then sets this hybrid figure in context.[16] The Stranger adds that "some teachers" (i.e., "noble" sophists) are persuaded that a soul will not obtain any advantage from any learning that is offered to it "until someone shames it by refuting it, removes the opinions that interfere with learning, and exhibits it as cleansed, and believing that it knows only the things that it does know and nothing more" (230d1–4). The Stranger declares that he feels afraid to call such people "sophists" because he does not want "to pay them too high an honour" (231a1–3).[17] When Theaetetus suggests that there is a similarity between a sophist and these people, the Stranger complains that one should not confuse the wildest wolf with the gentlest dog and recommends that Theaetetus be "especially careful about similarities", since the type they are talking about is "very slippery". However, he lets his metaphorical description of them stand (231a7–8), and he concludes that

> within education, according to the way the discussion has now turned, the refutation of the empty conceit in one's own wisdom is nothing other than noble sophistry: τῆς δὲ παιδευτικῆς ὁ περὶ τὴν μάταιον δοξοσοφίαν γιγνόμενος ἔλεγχος ἐν τῷ νῦν λόγῳ παραφανέντι μηδὲν ἄλλ' ἡμῖν εἶναι λεγέσθω πλὴν ἡ γένει γενναία σοφιστική (231b5–8).

The Stranger's reluctance to regard sophists as a whole as "purifiers" is consistent with Plato's description of them as those who are not able to distinguish between "false" and "true" opinions.[18] Yet the Stranger is ready to admit the existence of a peculiar kind of sophist whose art could be regarded as "noble" (literally: "true-born"), and this sophist will necessarily distinguish between ignorance and knowl-

---

[15] The philosopher expects to make the interlocutor acquire, on his own, either true opinions or dialectical knowledge. If the process is successful, the interlocutor is expected to become wiser, pursue excellence, and even manage to change his character for the better. When it comes to recognizing his own limitations, the philosopher is ready to admit that he does not know what he is searching for (*Charmides*, 165b5–c2); to submit his theses to refutation; and even to believe that being refuted is advantageous if, thereby, a true conclusion is reached (*Gorgias*, 458a2–b).

[16] The Stranger talks of two methods of teaching: traditional admonition and cross-examination, a division adopted by those who think that "lack of learning is always involuntary, and that if someone thinks he is wise, he will never be willing to learn anything about what he thinks he is clever at. Admonition is a lot of work but does not do much good" (230a6–9).

[17] The sentence is ambiguous. I follow Kerferd (1954) here: Plato does not want to give the sophists the honour of being purifiers. For a contrary view see Cornford, *ad loc.*

[18] At most, they would replace their disciples' former views with new opinions of their own because they might turn out to be *better* in the sense of "more convenient" or "more useful" than their previous ones.

edge. It is easy to understand why the task of such sophists is "noble" thanks to their powers of refutation (230d8), but the difficult point to grasp is why they are called "sophists" rather than "philosophers". Interpreters find an allusion here to the art of Socrates. Indeed, Plato seems to fluctuate between describing Socrates as the ideal philosopher or the wisest man in town and occasionally allowing himself to give the impression that his Socrates, in the heat of a discussion, while attempting to purify souls through refutation, wants to win the case at all costs (consider Callicles' impressions in the *Gorgias*). So, while Socrates fits into the category of "philosopher" in terms of the nobility of his objectives, he might also fit into the category of "sophist" in terms of the sophistic nature of some of his methods.

4 Does Socrates Play the Noble Sophist in the *Phaedrus*?

I suggest that Socrates in the *Phaedrus* does not incarnate the hybrid "noble sophist" because he does not seem to deliberately use sophistic methods to persuade Phaedrus, though he does admit to having misled him. True, he cannot do his job as a refuter straightforwardly because his own passion for Phaedrus gets in the way, and, apart from a few clues that he offers, he is not in a position to engage in a dialogue that would expose the weaknesses of Lysias' thesis.

Aware of his own personal limitations, Socrates moves cautiously. On the one hand, he feels impelled to satisfy Phaedrus' request about an imitative speech that would defend the thesis that the beloved should not please the passionate lover. On the other hand, he provides clues, in passing, that go unnoticed: the non-lover is an impostor (237b), and there are two principles that rule us: the one is our inborn desire for pleasure (ἔρως), the other our acquired judgement that what is to be pursued is that which is best (237d–e). However, despite his disclosure about the real drives of the "non-lover" and his appeal to a civilized principle of reason, his own passionate drives together with his perception of Phaedrus' emotional demands preclude him from cross-examining him. This Socrates does not seem to be the self-confident, ironical master of discussion and refutation that we find in the *Gorgias* or the *Republic*. For one thing, he is conflicted. He claims he has not managed to get to know who he is, and he holds deep-seated doubts about his own capacities. More precisely, he wonders whether he has become a beast more convoluted and more swollen (πολυπλοκώτερον καὶ μᾶλλον ἐπιτεθυμμένον) than Typhoon or a more peaceful and simpler creature, to whom a divine and clear lot (ἀτύφου μοίρας) "has been given by nature" (230 a 3–6). This much more "human" Socrates, who turns out to be under the influence of irrational forces, projects these into a wild atmosphere and makes them responsible for his half-yielding to the wishes of Phaedrus. However, he does not use sophistic methods to persuade him: he simply cannot purify Phaedrus because he must first purify himself from his own doubts... Indeed, contrary to Phaedrus' expectations, he turns out to be able to refuse to argue in defense of the non-lover, and to decide to cross the river. This "punishing" corrective step puts an end to any potential games of jealousy and rivalry, and places Phaedrus in a disposition to listen to him.

Only after the beautiful palinode has had a converting effect on Phaedrus' soul does Socrates find the suitable moment to start cross-examining him, to get him to grasp the difference between ἔρως understood as a human sickness of passion and ἔρως as a divine madness for knowledge on the one hand, and noble and ignoble rhetoric on the other. Though the method that both the sophist and the dialectician use is the same, namely, going step by step from similarity to similarity, there is a big difference: he who does not know what each thing is can never acquire the art of leading others, and releasing them from what is only opinion. Moreover, he could not even escape being himself deceived (262b5–8).

Indeed, Socrates claims that the "art" of the speaker who does not know the truth and chases opinions instead is likely to be some ridiculous thing, not an art at all (262c1–3). He then picks the two speeches delivered by Lysias and himself and declares that "by some chance, the two speeches do, as it seems, contain an example of the way in which someone who knows the truth can toy with his audience and mislead them" (262c10–d2). It seems clear that Socrates knew (or suspected) what love is while not being able to guide Phaedrus properly, but one may wonder what kind of truth Lysias knew while misleading Phaedrus. He is likely to have known that passionate lovers may be overwhelming and selfish, and he used this knowledge to mislead Phaedrus into favoring unpassionate love.

Socrates asks Phaedrus whether he had a "definition" of ἔρως from the start, and Phaedrus agrees that he did. However, he seems to have collected common opinions rather than providing a true "definition" by collection and division: ἔρως is introduced as a natural desire for pleasure (237d6–9) and as a strong force that overcomes rational opinion of what is right (238b7–c4). But Socrates seems to be focused on method rather than content. He proves that Lysias writes artlessly, since he does not define the terms to be used in the discussion and also starts from the very end (264a), while the various parts of his speech appear to have been thrown together at random (264b).

Since Socrates is not able to persuade Phaedrus about the implicitly perverse features of the dispassionate "non-lover", one could object that his cautious attitude plays no effective part in the dramatic action. However, one should notice that Phaedrus' obsession is a hindrance to his being cross-examined. When passion interferes, I suggest, direct confrontation or mere admonition are equally counterproductive. The route of caution and mimesis seems to be the only one to follow at first, since, to begin a dialogue, both interlocutors need to start from a common plateau, however apparent.

After the corrective tentative abandonment of Phaedrus, however, the game is over. Passion is controlled somehow, and the intellectual process can begin. Love gains a higher status. With the palinode and the concept of peculiar "divine madness" serving as the strategy to make Phaedrus feel that he is a soul capable of the highest summits, he can expect to behave as a real lover who would take care of his beloved by leading him to philosophy, rather than as a non-lover who should expect pleasure for nothing in turn.

5 Ambiguity in the Speech of Lysias

Lysias' speech is characterized by conceptions tending in opposite directions. The listener will have as many clues in favor of a disinterested, self-controlled friendship as there are in favor of a discreet demand for sexual satisfaction. From the start, the speech is ambiguous. The non-lover declares that he strives for *their* convenience: "I've told you how convenient it would be *for us*, in my opinion, if this worked out", ὡς νομίζω συμφέρειν ἡμῖν γενομένων, while openly claiming for *his* gratification: "I don't think I should lose the chance to get *what I am asking for*, merely because I don't happen to be in love with you", ἀξιῶ δὲ μὴ διὰ τοῦτο ἀτυχῆσαι ὧν δέομαι, ὅτι οὐκ ἐραστὴς ὤν σου τυγχάνω (230e7–231a2).

The non-lover also claims that he will put his heart into whatever he thinks *will give pleasure* to the boy (231b7); he is a master of himself because he "has not been overwhelmed by love" and declares that *if he persuades the boy*, he will, first, give him his time "with no care for present pleasure" but he will "plan instead for the advantage that is to come" (233b6–c1). Furthermore, non-lovers claim that they are friends before they achieve their goal (233a1–2), and that a boy "can expect to become a *better person*" if he is won over by a non-lover, rather than by lovers, who praise their beloved far beyond what is best (233a). The boy should grant his favors to non-lovers, and "to those who are best able to return them" (233e), because "in that larger pool" he will have better chances of finding someone who deserves his friendship (231d–e). Non-lovers will also not be criticized for bad behavior (232b; 234b) and, not being jealous, will not isolate the boy from family and friends (232b–d). And they "will share his goods with him when he is older", remain "steady friends their whole lives" and "will prove their virtue when the bloom of youth has faded" (234a–b). In a word, a secret affair is easier to keep private, causes no harm, and should work to the benefit of both sides (234c).

However, the non-lover, like the lover, even when described as "virtuous", does not really care for either the boy's soul or his body, despite the large list of merits he assigns himself: moderation, prudence, modesty, secrecy, and forgiveness. These rhetorical resources mask self-interest[19] and sexual desire. Though praise may do great harm (240a–b), the appeal to this reason turns out to be superficial, for it seems to imply that the non-lover's *lack* of praise is the cause of making the boy a better person.[20] Indeed, he seems to disregard the care of the boy in every relevant respect, since he feels free from any commitment to his education.[21]

Socrates cannot unmask Lysias straightforwardly because Phaedrus is not only delighted with Lysias' professional speech, which he regards as "paradoxical and clever" (227c) and superb (234c), he is also convinced that Lysias is the most talented writer of the day.[22] The paradox is that he seems to be "mad" with love for Lysias' "non-erotic" speech. Most interpreters see it as simply utilitarian: the non-lover expects to get gratification, just because he is persuaded that he *deserves* it or is worthy of receiving it, due to his sober self-control, which he calls "virtue".

6 The Efforts of Socrates

After reading Lysias' speech, Phaedrus is eager to know how it struck his interlocutor. Socrates finds it "inspired" (δαιμονίως (234 d 1), and declares himself to be "in ecstasy" while watching Phaedrus' radiant face and sharing his *Bacchic frenzy* (συνεβάκχευσα (234d5). When Phaedrus disregards this as a joke, and puts pressure on Socrates, he says he only paid attention to the style, and comments in passing that he suspects that not even Lysias would be satisfied, as though he thought

---

19 Griswold (1985, p. 49) comments: "The non-lover's self-interest is in fact as poor a basis for friendship as the lover's lust".
20 Hackforth (1985, p. 31) rightly observes that the argument adduces no positive action by the non-lover but merely his abstention from the lover's praise and flattery.
21 Rosen (1968, p. 8) claims that "the non-lover desires gratification as an objectified commodity, independent of the personality of the boy, who is to him not a beloved but a reified unit in the free market economy, whose wares are subject to the laws of supply and demand" in the context of his "technically competent selfishness".
22 Hackforth (1985, p. 26) comments that Phaedrus' admiration is almost exclusively on account of its form, as its content is difficult to reconcile with his own encomium on Ἔρως in the *Symposium* 180b. Rosen (1968, pp. 5–6) points out that Phaedrus' love for speeches is passive, for he wishes that he could memorize Lysias' speech, not that he could write his own.

that "he did not have many things to say", and were just "showing off, trying to demonstrate that he can say the same thing in two different ways" (235a).

These comments prove that though Socrates might have been distracted by Phaedrus' shining joy, he realized nonetheless how poor it was in form and content. After disagreeing about the quality of the speech and talking of "having his breast full", Socrates feels that he can say "different and not worse things" than Lysias: ἕτερα μὴ χείρω (235 c 5–6), but he also claims that he is aware of his own ignorance (235c–d).[23] Despite his modesty, Socrates seems to be playing the competitive partner to gain Phaedrus' preference.

However, his interlocutor challenges him to make a better speech, neither shorter *nor different* from the one he has read (235d6–7). Socrates reacts by commenting that Phaedrus is "good as gold" to think that he is claiming that Lysias failed in absolutely every respect and that he can make a speech that is "different" on every point from his (235e). As Phaedrus takes Socrates' words at face value, Socrates claims that he was only criticizing his beloved "to tease him", and that he should not take it seriously, for he does not think he could match the product of his "wisdom" with a more intricate speech (236b5–8). Socrates' criticism is ironical; he masks his comments with humor so as not to annoy Phaedrus.

7 Who is Who?

It is worth noticing that Socrates refers to Lysias as Phaedrus' *beloved*, which means that there is an inverse roleplay in his speech, for while the speaker assumes the role of "the non-lover", in real life he is "the beloved", who apparently should expect to become the "non-beloved". So Lysias, I suggest, by playing the part of the "non-lover", is perhaps indicating to Phaedrus the way he would like to be treated by his lover. His speech shows that he clearly wants to get rid of overwhelming, obsessive, and demanding lovers who finally change their mind and forget their promises. By contrast, he might expect to give his favors to a non-lover because of his self-mastery, forgiveness, and flexibility, and also the secrecy of the relationship. The proposal must obviously have been regarded by Phaedrus with enormous pleasure, for he can expect to be gratified without a demand for much care for his non-beloved, while being praised as an excellent person.

---

[23] Griswold (1985, p. 53) observes that "Socrates claims no responsibility for what he says (it derives from another author), how he says it (the divinities are the cause), or why he says it (he is compelled to do so)".

Should we take the "non-beloved" to be pretending when he claims that the priority of the non-lover is not pleasure but friendship? Not necessarily; the emphasis of the relationship is placed, in several passages, on a lengthy, strong friendship (233c–d). By creating the profile of the "non-lover", by praising the non-lover's self-control and virtue, by expecting that he will restrain his immediate erotic desire, and by focusing the relation on advantageous friendship, the non-beloved (namely, Lysias) expects to be able to exert some influence on his lover (namely, Phaedrus), presumably to inhibit his (disgusting) appetites.[24] Lysias seems to be fed up with passionate, forgetful lovers and would rather have a generous, pleasant, and sensible friend. However, since he cannot eliminate sexual gratification from the relationship, he mentions it here and there, but attempts to downplay it as much as possible by praising self-mastery and concentrating on advantageous friendship. If this reading is plausible, it becomes clearer why Lysias' speech must be ambiguous and cunning while also possibly containing "some truth", as Socrates seems to suggest at the end of the dialogue when reconsidering the speech.

One may wonder why Socrates refuses to imitate the speech about a rational, sober relationship that Lysias promotes. He himself gives us the answer: he will refuse to praise the non-beloved's self-mastery because he thinks that his supposed temperance is only a pretense, and of no value. Indeed, only philosophers who have seen the form of Beauty can passionately "love the young with philosophy", in a divine, erotic way and in genuine mastery of their appetites.[25] In this fancy-dress ball,[26] nobody plays the part he should, but the very opposite one: Lysias, who is Phaedrus' beloved, plays the "non-lover"; Phaedrus, who is Lysias' lover, plays Socrates' beloved;[27] and Socrates, the universal lover, plays Lysias' cautious rival. In this way Plato creates a puzzling game on stage, played out in a highly in-

---

24 This type of relation is not reciprocal. Sex is something the beloved would not enjoy at all, if we are to believe Socrates' description of the effects of the "disgusting" lover on the boy (240b5–7). Socrates wonders what comfort or pleasure the lover will give him during the time they spend together, as it will be disgusting in the extreme for him to see the face of that older man who has lost his looks; to be constantly, and against his will, watched over and guarded with suspicion throughout; and to hear excessive praise or insults (240d4–e7).
25 In the *Symposium* (212a), only those who have ascended the ladder of love to its topmost rung and have contemplated Beauty itself are the ones in a position to become genuinely virtuous.
26 Socrates regards almost everything to have been spoken "in play" (265c8). In the *Symposium*, too, Alcibiades, who is expected to be Socrates' beloved, plays the jealous lover.
27 I agree with Hackforth (1985, p. 53, n. 1) that Socrates' veiled suggestion that Phaedrus himself is the boy, and Phaedrus' acceptance of this, are nothing more than playful.

tense atmosphere of jealousy and erotic passion,[28] while the speeches themselves run counter to passionate love.

8 What is Socrates' Purpose?

Delighted with the perspective of having Socrates at his disposal, Phaedrus forces him to speak, for he knows that, despite his reluctance, he is willing to do so. Socrates complains about his being ridiculous, "improvising on the same topic as a seasoned professional" (236d). While pretending to be merely imitating Lysias' speech against passionate love, Socrates offers some important clues to unmasking the non-lover's intentions. However, though he declares at the beginning that the non-lover is a pretender, at the end, he concedes that the boy should please the non-lover rather than the lover, and this shows the embarrassment he feels at his situation along with his desire to please Phaedrus.

Socrates calls for the aid of the Μοῦσες, but this turns out to be misleading, since the inspired speech will be disowned as false, and will require a recantation.[29] Socrates wraps up his head in expectation of "galloping through the words" to reach the end quickly, so as not to feel embarrassed by having to look at Phaedrus and lose the thread of the argument (237a). He protects himself from the influence of Phaedrus' gaze, while he attributes his shame to his fear of making a fool of himself, being a layman (ἰδιώτης) not a good poet like Lysias (236d4–5). Socrates claims that his speech against passionate ἔρως aims at making Phaedrus admire Lysias even more: if he previously thought him "wise", he would now seem even more so (237a–b). Phaedrus is apparently immune to Socratic irony.

Socrates produces a story that is a partial imitation of that of Lysias, which is not called a "λόγος" here but a "μῦθος" (237a9; 241e8 and 243a4) while in other places the same discourse is called a *logos* (241d3; 242e3; 243c2; 244a1; 264e7; 265c6; d7; e3; 266a3). He also calls the palinode both a *mythos* (253c7; 265c1) and a *logos* (252b2; 265b8; c6, d7, e3, 264e7 and 266a3). But Lysias' speech is never called a "μῦθος". In my opinion, the reason for this is that his speech reveals his real intentions. Now, if "μῦθος" and "λόγος" are not mutually exclusive in relation to Socra-

---

[28] For an emphasis on the eroticism of the situation, and Socrates' resisting the sexuality of his response to the beautiful youngster, see Tejera (1992), p. 291.
[29] On the other hand, it should be noticed that Socrates' invocation of the Muses, who represent beauty and pleasure (258d–259e), suggests, according to Yunis (2011, p. 112), "the erotic mode of discourse that becomes Socrates' concern later".

tes' first discourse, why call it a 'μῦθος" in the first place? Because, I suggest, it is likely to show that the story is fictitious:

Now there was once upon a time a boy, or rather a youth, of great beauty: and he had many lovers. Among these was one of peculiar craftiness, *who was as much in love with the boy as the others, but had made him believe that he was not in love*: ὃς οὐδενὸς ἧττον ἐρῶν ἐπεπείκει τὸν παῖδα ὡς οὐκ ἐρῴη. And once, in pressing his suit to him, he tried to persuade him that he ought to give his favors to a man who did not love him rather than to one who did (237 b 2–5).

Phaedrus is not able to realize the message of Socrates' story is symbolic because he is caught up in the trap of his beloved sophist. Why does Socrates claim that the non-lover is a concealed lover? How can he possibly guess that the non-lover is hiding a passion for the boy?[30] He is likely to be making fun of Lysias' thesis, since it is clear that "we know, that non-lovers have an appetite for those who are beautiful": ὅτι μὲν οὖν δὴ ἐπιθυμίατις ὁ ἔρως, ἅπαντι δῆλον: ὅτι δ'αὖ καὶ μὴ ἐρῶντες ἐπιθυμοῦσι τῶν καλῶν, ἴσμεν (237d3–5).

By censuring both characters equally Socrates gets rid of a sophistic rival in the theoretical arena and a possible rival in real life.[31] He is likely to have grasped the essential consequence of the deceit, namely, that the gratification of the non-lover by the "non-beloved" would be gratuitous, even if the speech of Lysias does promise friendship and virtue. Socrates' message seems to be that, despite appearances, both the lover and the non-lover would be equally egocentric in remaining indifferent to the boy.

9 When Does Socrates' Method Become Successful?

Socrates does in fact introduce a methodological procedure that was absent from Lysias' discourse, and implies both universality and a theoretical approach, since "the majority do not know the nature of things", and because of this, they cannot

---

[30] Taylor (1955, p. 303) observes that by putting his speech in the mouth of the concealed lover, Socrates safeguards his own character by abstaining from defending a morally disgraceful thesis. In similar mode, Calvo (1992, p. 51) argues: "By resorting to a tale, Plato wants us to grasp that it is not Socrates, but his character who argues for the Lysianic thesis". I do not find this argument fully convincing, for in that case the palinode, which is also called a "myth", would be equally alien to Socrates.

[31] Tejera (1992, p. 292) finds the speech sophistic and deceitful because the suitor is just another lover who will do the passion-guided things he pretends to be warning against. Calvo finds it self-accusatory (1992, p. 56).

reach an agreement on the matter they are investigating and wind up in conflict with themselves and each other (237b7–c5). Accordingly, the first thing is to get to know what ἔρως is, what power it has, and whether it brings profit or damage with it. However, as noticed above, Socrates introduces common opinions, namely, that eros is appetite (237d6–9) and that when it takes command in us and drags us without reasoning toward pleasure, it provokes ὕβρις (237e2–238a2). Though he admits the possibility that ἔρως can be moderated by rational opinion, ends up reducing it, in this part of the dialogue, to a "force" (ῥώμη), from which, he claims, it takes its name ἔρως (238b7–c4).

After defining love in its vulgar form, Socrates has the impression of having suffered from a "divine passion" (238c6); of having been captured by the "madness of the Nymphs"; and of feeling "on the edge of speaking in dithyrambs" (238d1–3). He finds himself in a Dionysian setting, and Phaedrus is made responsible for his state. Socrates hopes that "perhaps the attack may be averted", as he declares that it is up to the god (238d5–7).[32] He hopes he will come round to himself, as if he were aware that he is talking of love in an unphilosophical way,[33] and wishes to place his hope in the god's help (Plato might be making him adumbrate of the daimon's prohibition to cross the river).

Socrates starts his discourse anew:

> My friend, just as I was about to cross the river, the familiar divine sign came to me which, whenever it occurs, holds me back from something I am about to do. I thought I heard a voice coming from this very spot, forbidding me to leave until I made atonement for some offense against the gods. In effect, you see, I am a diviner (μάντις), and though I am not particularly good at it ... I am still good enough for my own purposes. I recognize my offense clearly now. In fact, the soul too, my friend, is itself a sort of diviner; that's why, a while ago during my speech, I was disturbed by a very uneasy feeling, as Ibycus puts it, that "for offending the gods I am honoured by men". But now I understand exactly what my offense has been (242b8–d2).

---

32 At 244c2, Socrates links divination with madness through the etymological connection between *mantike* and *manike*. Since he regards himself as a diviner, one might be tempted to identify this god with Apollo, "the god that inspires prophecy" and gives sound guidance to many people, rather than Dionysus. However, Dionysus is also included on the list of gods that provoke a certain divine madness, namely, the mystic madness (265b3–4) that provokes relief from present hardships. Furthermore, at the end of the dialogue, Socrates proposes to pray to the deities of the place before leaving, starting with their "beloved Pan" and all the other gods of this place: ὦ φίλε Πάν τε καὶ ἄλλοι ὅσοι τῇδε θεοί (279b8). This indicates that Socrates is grateful to all the gods and does not despise any.

33 Hackforth (1985, p. 47) remarks that "we see here the rationalist becoming gradually conscious of an influence which he feels to be irrational, or supra-rational, half unwilling to yield".

At the end of the dialogue, after Socrates has gained self-confidence, he is ready to recall that a reasonable man should not please his fellow slaves, except as a side effect, but rather his good masters, the gods (273e9–274a2). Thus, the dramatic action indicates that the way towards that dialectical τέχνη which leads to true philosophical definitions is deeply connected to the emotional and religious dispositions that Socrates has as a "diviner".

The apparent paradox is that Socrates 1) felt "divine madness" for Phaedrus under the influence of the wild Nymphs and the gods of the place, 2) hoped to be able to "reject the attack", 3) divined an offense against Eros in his former speech precisely because they were interpreting eros as mere "vulgar egocentric passion", and 4) finally defined eros as "divine madness" … again.

If Lysias' speech can be regarded as coming and going in opposite directions, Socrates' seems to do the same, unless we are able to discriminate different kinds of divine madness, depending on the personality and attributes of the gods that bestow it upon human beings. In my view, a plausible solution to this apparent paradox is to take the Dionysian divinities in the first scene as responsible for sending "divine intoxicating madness" for the beauty of bodies (let us remember that Socrates covers his head to protect himself from looking at Phaedrus' face), while in the palinode, Zeus' followers are inspired with "divine madness" for knowledge of the Forms… and the desire to provoke ἀνάμνησις of real knowledge in beautifully disposed embodied souls, together with the desire to care tenderly for them in every respect.

> I agree with Politis[34] that dialectic, understood as an ability or power,
> (…) includes not only knowing essences and Forms and the search for this knowledge, but also the soul's liberation and reorientation (as described in the Cave); and that, for Plato, this liberation and reorientation is part of that which enables us to search for essences and Forms.

In this sense, I take the first scene of the *Phaedrus* as a dialogical, mimetic exercise that in the end will provoke an emotional and intellectual liberation from mad love as a grave sickness (238e4–5). The dialectical power of "reorientation" will place Socrates on the right track to start anew the search for the essence of love, which will turn out to be defined as divine madness…of a new kind.[35]

Indeed, the misfortunes of the boy under the pressure of a mad lover, who "is ruled by appetite and is a slave to pleasure", will make him weaker and inferior to

---

34 Vasilis Politis (2022), p. 193.
35 No matter how long and difficult the way is, Socrates says, they should make their detour for the sake of important noble goals, for it is noble that he who strives after what is noble will also bear whatever he has to (273a8–b1).

himself, keep him "away from the good company of anyone who would make a better man of him", and "from what would most improve his mind, namely, divine philosophy", "in ignorance and in total dependence on himself" without family and friends; in disgusting company; and feeling the misery he would of necessity be driven into (238e–239b; 239c4–5; 239c–240a; 240b5–7; 240c4–6).[36]

On the contrary, the good fortune of the boy who is guided by a passionate lover ruled by divine madness for knowledge will make him become who he is —a follower of Zeus and a happy philosopher who is beloved by somebody who, in addition to caring for his body with tenderness (and without abuse), will also take care of his soul and help him develop his emotional and intellectual potential.[37]

10 The Sophist and the Philosopher Use the Same Method but Have Different Goals

In this section, I shall focus on Socrates' final analysis of the former stages of the dialogue that took them from deviation from the right path to a return to it. Though almost everything seems to Socrates to have been spoken "in play", among all the things that have been said "by chance", he thinks it would be gratifying if somebody could grasp with *techne* two kinds (δυοῖν εἰδοῖν) of procedure (265c8–d1). The first one consists in "seeing together things that are scattered about everywhere and collecting them into one kind" (εἰς μίαν ἰδέαν), so that by defining each thing one can make clear what is to be taught (265d3–5), while the second one consists in being able "to cut up each kind according to its species along its natural joints, and not to try to break any part, as a bad butcher might do" (265e1–3). And then, if it is a simple thing, to enquire what power of acting or of being acted upon it possesses, and if it has many forms, to consider the power of each one of these to act or be acted upon (270c10–d7). The method described should be applied to the nature of the soul, using that "technical" rhetoric

---

[36] When his passion fades and he regains self-control, the lover feels ashamed to tell the truth to the boy, abandons his promises, and runs away. The beloved must chase after him, angry and cursing (241b). Hackforth (1985, pp. 47–48) indicates that the terms *noûs* and *sophrosyne* attributed to the lover who comes to his senses, are likely to be used in their *vulgar* meaning, since the mere dying of passion cannot automatically involve moral goodness and wise cognition. (For a consideration of spurious moderation and courage see *Phaedo*, 68e.)
[37] I agree with Vlad (2022, p. 253) that the second speech will focus on the relationships with the gods and that some dialectical knowledge is reached through divination.

which, far from remaining on a mere empirical, artless level, should, by discourses and training, impart to the soul the convictions and virtues desired (270b4–9).

After describing the method, Socrates takes the example of Thrasymachus (or anyone else who is "a serious teacher of the art of rhetoric"), whose skillful procedure may look very similar to the one he has described: "He will first describe the soul with perfect accuracy, and enable us to understand whether it is one and homogeneous or, like the body, takes many forms, for this is what we call explaining its nature" (271a5–8).

Socrates proceeds gradually and cautiously so as not to annoy Phaedrus, who apparently does not grasp the irony implied.

Secondly, Socrates adds, Thrasymachus will consider the soul's power of acting and being acted upon by certain things, and thirdly he will classify the kinds of speech and soul there are, along with the ways in which various souls are affected by speeches, and he will show why one type of soul is necessarily persuaded by a certain type of speech while another remains necessarily unconvinced (271a10–b5). He will also be able, when meeting someone, to discern that he is of a particular character, and will know whether to apply one kind of speech rather than another to persuade him to favor a certain activity or belief. In addition, he will know when to talk and when to remain silent, and when to appeal to "pity or exaggeration" or any other of the kinds of speech he has learnt at school (272a–b). However skillful, this type of rhetorician, Socrates claims, would only care about what is convincing to and accepted by the crowd, and would believe that he has no need to know the truth (272d–273b).

These passages indicate that the difference between the dialectician and the sophistic rhetorician does not lie in the use of the method of collection and division, or in the analysis of the soul's power of acting and being acted upon, or in various psychological strategies employed, but in which of the two is interested in inquiring into the truth.

Socrates concludes that no one will ever possess the *techne* of speaking unless he acquires the ability to enumerate the sorts of characters to be found in any audience, to divide everything *according to its kinds* and to grasp each thing firmly by means of one form. He also comments that such a man should aim to please the gods rather than men, no matter how difficult the task is (273d–274a). At this stage, after describing the way dialectic operates, Socrates dares to rebuke Phaedrus, because his interlocutor seems to be focusing on who the speaker is and where he comes from, instead of on whether his words are true or not (275c1–2).

Phaedrus accepts the criticism, and Socrates introduces another kind of discourse that is "by nature better and *more capable*" than mere writing: the one that is written down, with knowledge, in the soul of a learner, which can *defend itself* and knows for whom it should speak and for whom it should remain silent (276a5–7). Socrates explains:

> One who uses the *techne* of dialectic chooses a fitting soul and plants and sows within it discourses accompanied by knowledge, discourses capable of helping themselves as well as the man who planted them, which are not barren but produce a seed from which more discourses grow in the character of others. Such discourses make the seed forever immortal and render those who have it as happy as any human being can be: τις τῇ διαλεκτικῇ τέχνῃ χρώμενος, λαβὼν ψυχὴν προσήκουσαν, φυτεύῃ τε καὶ σπείρῃ μετ' ἐπιστήμης λόγους, οἳ ἑαυτοῖς τῷ τε φυτεύσαντι βοηθεῖν ἱκανοὶ καὶ οὐχὶ ἄκαρποι ἀλλὰ ἔχοντες σπέρμα, ὅθεν ἄλλοι ἐν ἄλλοις ἤθεσι φυόμενοι τοῦτ' ἀεὶ ἀθάνατον παρέχειν ἱκανοί, καὶ τὸν ἔχοντα εὐδαιμονεῖν ποιοῦντες εἰς ὅσον ἀνθρώπῳ δυνατὸν μάλιστα (276e5–277a4).

Socrates insists on the need for "seeing the truth" concerning the subject about which one speaks or writes and being able to define each thing in itself and to divide it into its species until one reaches something indivisible. Once one understands the nature of the soul, following this method, one needs to discover which kind of speech is appropriate for each kind of soul, and to operate accordingly for purposes of teaching or persuading someone (277b–c).

Finally, Socrates is ready to tell Phaedrus the whole truth about Lysias or anyone else who, believing that some law or political document he is writing "embodies clear knowledge of lasting importance", is in fact ignorant of what is just and unjust and what is good and bad.[38] And this ignorance deserves reproach, even if the crowd praises it (277d6–e3).

On the contrary, Socrates and Phaedrus should pray to become like those who are able to write out in human souls what it is to be just, noble, and good, because only in those descriptions lies perfection, and only they are worth attending to (278a–b).

Socrates concludes that they must announce that if someone has composed his writings "with knowledge of the truth", is able to support them by discussion of that what he has written when they are challenged, and has the power to show by his own speech that his written words are of little worth, to call him a lover of wisdom—a philosopher—would fit him (278c–d). For the capacity to support one's views, to offer grounds for them, while striving for wisdom, is what distinguishes the philosopher from the rest.

---

[38] Politis (2022, p. 208) observes that "many people are under the illusion that they are thinking in a good and dialectical way whereas in actual fact, they are thinking in a bad and eristic way. This illusion is compatible with one's using words the way a dialectician would: the only way of securing against this is one's viewing (ἐπισκοπεῖν) the subject of one's words in a certain Form-orientated way".

It is clear from the passages quoted above that wisdom does not come automatically from the act of collecting and dividing, not only because the sophists also use this method, but because it is implied in every process of thinking and speaking (266b3–c1). Rather, it seems to come from a different, somewhat mysterious source ("by chance"? by demonic inspiration?) that is related to the knowledge of what is just, noble, and good, and this in turn is based on the type of soul that one is, and one's ability to keep striving for knowledge and excellence. Collection and division may be right or wrong, so Socrates would not attribute "the art" of dialectic to those who collect and divide without knowledge.

11 Conclusions

In the written word, as in Plato's *Phaedrus*, there is necessarily much that is playful. Dialogue is a necessary step to dialectic, even when the route apparently takes us nowhere. Socrates manages to get to know Phaedrus' emotional drives and, more importantly, to guess what kind of speech will not persuade him at all. True, his cautious and rather imitative speech, based on collecting and dividing common opinions, produces no effect on his interlocutor. However, once he can distance himself and listen to his daimon, he regains his best self and proceeds as a teacher. His second speech manages to describe the complexity of the soul, to collect different kinds of eros into the "right-hand" and the "left-hand", to divide the former into two species (namely, divine madness for knowledge, i.e., for Beauty itself, and divine madness for bodily beauty) and to subdivide the first one into a top class (love among mature philosophers and young, promising followers of Zeus) and a lower class (love among the military and young followers of Ares).

However, these better results, while presented from an employment of the method of collection and division, are not drawn from that method,[39] but from inspired divination, as the powerful metaphors of the palinode seem to suggest. The

---

[39] I agree with Larsen (2020) that we should carefully distinguish between dialectic in a general sense, as an ability necessary for speaking and thinking, and its more specific meaning, taken as the capacity "to look to one and to multiplicity as they are in nature" (266b5–6). More importantly, those who are not dialecticians also collect and divide, but they fail to do so correctly (Larsen 2020, p. 115). Socrates' speeches may look similar to sophistical procedures, but what distinguishes them are substantial suppositions concerning the nature of eros and the soul which support the arguments of the second speech. "These suppositions may be said to motivate and come to expression in the divisions found in the second speech, but they do not result from them" (Larsen 2020, p. 109). In a word, "only in combination with other considerations and insights, collection and divisions may help us to arrive at definitions" (Larsen 2020, p. 131).

dialogue has shown us that the *appropriate* use of the dialectical method of collection and division is, somehow, independent of the methodical procedure itself; its source is rather related to that vital "divine" experience of striving for excellence and wisdom that allowed Socrates to overcome his own low passionate drives.

When regarded as a gift from God, madness is the source of the best things we have (244a). Deprived of that divine passion, Socrates would perhaps have behaved just as a reasonable friend.[40] Through the beauty of the palinode, Phaedrus is likely to have implicitly understood that he is not to behave as a moderate friend who would share his material goods with his non-beloved, while remaining indifferent to the peculiarities of his soul, personality, and education. He must have felt fascinated to learn that his soul is called to an erotic, divine destiny which could guide his life, permeate his loving relationships, and prompt him to stimulate his beloved to become the best version of himself. And not only Phaedrus, but, more importantly, readers of all times...

Only after this "revelation" takes place does the authentic dialogue begin, in which Socrates will behave like the true lover who will show Phaedrus his own persuasive weapons while taking care of his emotional drives, and in this way will be able to clear up his doubts about himself. Socrates will regard his first speech as "neither healthy nor true" (242e5–243a1); and indeed it was so, not only because it reduced ἔρως to its vulgar definition as "unavoidable" appetite, but because he yielded to Phaedrus up to the point of repeating Lysias' conclusion.

Though inconsistent in this respect, however, the speech was not completely fruitless[41] since it led to a new stage of "divination". Together with his exposure to the madness of the divinities of the place, to Phaedrus' demands, and to his own erotic drives, Socrates also displayed his resistance, in the dramatic action, by offering certain clues, by covering his head, by leaving the speech "in the middle", and threatening to cross the river. In this way he managed to make Phaedrus stop manipulating him, and he got him ready to listen to a palinode which would transform his perspective.

Neither a wolf who would devour Phaedrus nor a lamb who would please him at all costs, Socrates attempted a strategy to counteract Lysias' sophistic hook, by identifying the non-lover with a concealed lover. Though in the end his concession

---

[40] The view that the non-lover's gratification without erotic excess must be "licit" and "good" in Socrates'/Plato's eyes, (Trabattoni 2011, pp. 293 and 296) seems to me quite controversial, though Lysias may find it so. Neither moderation nor reasonable calculation is enough to make the gratification good, as Socrates claims at 244a. Hackforth (1985, p. 53) interprets Socrates' refusal in the same way: "on the hypothesis that the boy must yield to one or the other, he ought rather (μᾶλλον) to yield to the non-lover than to the lover", but, he adds, this hypothesis is not Socrates' own.
[41] Calvo (1992, p. 56) does not find the tale impious or blasphemous.

to Phaedrus took him to an inconsistent conclusion, he somehow managed to purify himself (243a) and make Phaedrus admire the palinode. By passionately falling in love with divine knowledge, i.e., with the vision of the Forms, the philosopher will be capable of playing the "noble lover" who attempts to guide his "beloved" towards wisdom. What is to be praised after the palinode is loving the boy passionately with self-mastery, which, in this context, means educating him in all respects without abusing him.

In my view, the success of the magnificent visions of the palinode and the subsequent dialogue would not have been obtained[42] without the dramatic strategies of the first round, because the process of persuasion includes certain steps. When passion is involved, cross-examination is neither fruitful nor possible. Apparent imitation and partial consent are useful steps to get to know where the interlocutor stands. A dead end was required to provoke a new start, one which involved proceeding gradually, from similarity to similarity, until the opposite shore was reached.[43]

Bibliography

Annas, Julia (1982): *An Introduction to Plato's Republic.* Oxford: Oxford University Press.
Bossi, Beatriz (2015a): "Plato's *Phaedrus:* A Play inside the Play". In: Cornelli, Gabriele (Ed.): *Plato's Styles and Characters.* Berlin: de Gruyter, pp. 263–278.
Bossi, Beatriz (2015b): "El *Fedro* de Platón: un ejercicio de buena retórica engañosa". In: *Anales del Seminario de Historia de la Filosofía* 32. No. 2., pp. 345–369.
Calvo Martínez, Tomás (1992): "Socrates's First Speech in the *Phaedrus* and Plato's Criticism of Rhetoric". In: Rossetti, Livio: *Understanding the Phaedrus, Proceedings of the II Symposium Platonicum.* Sankt Augustin: Academia Verlag, pp. 47–60.
Cornford, Francis Macdonald (1935): *Plato's Theory of Knowledge.* London: Kegan Paul, Trench, Trubner and Co.

---

42 Certainly, many Platonic dialogues start from agreed definitions that are subsequently found to be incoherent. Truth cannot be reduced to convention, but without making the attempt to find some point that could be agreed on, dialogue is impossible. In the last part of the dialogue, recollection of noetic visions will ground the "mutual agreements" on a different basis.
43 I agree with Trabattoni (2011, p. 304) that, despite being well built (clear, organic, and structured) Socrates' first speech is wrong because of the dialectical mistake at the beginning of it, when unity and multiplicity are not related in the proper way. However, in my view, the "mistake" about eros was due to a "necessary" concession to Phaedrus, without which he would hardly have stayed to listen to the palinode. The *Phaedrus* displays an artful touch, where every step is carefully measured, not only with regard to philosophical content, but also with regard to the dramatic strategies that need to be taken into account if persuasion is to be given a chance.

Gonzalez, Francisco J. (1998): *Dialectic and Dialogue, Plato's Practice of Philosophical Inquiry*. Evanston: Northwestern University Press.

Griswold, Charles L. (1986): *Self-Knowledge in Plato's* Phaedrus. New Haven and London: Yale University Press.

Hackforth, Reginald (1952): *Plato's* Phaedrus, *Translated with an Introduction and Commentary*. Cambridge: Cambridge University Press.

Kerferd, George (1954): "Plato's Noble Art of Sophistry". In: *Classical Quarterly*, 4. No. 1–2, pp. 84–90.

Larsen, Jens Kristian (2020): "What are Collections and Divisions Good for? A Reconsideration of Plato's *Phaedrus*". In: *Ancient Philosophy* 40, pp.107–133.

Nehamas, Alexander and Woodruff, Paul (1995): *Plato, Phaedrus*. Indianapolis: Hackett Publishing.

Notomi, Noburu (2004): "Socratic Dialogue and Platonic Dialectic: How the Soul Knows in the *Republic*". In: *Plato Journal, International Plato Society* 4, pp. 1–7.

Politis, Vasilis (2022): "Dialectic and the ability to orientate ourselves, *Republic* V-VII". In: Larsen, Jens Kristian, Haraldsen, Vivil Valvik, and Vlasits, Justin (Eds.): *New Perspectives on Platonic Dialectic, A Philosophy of Inquiry*. New York: Routledge, pp. 193–212.

Robinson, Richard (1953): *Plato's Earlier Dialectic*. London: Oxford University Press.

Rosen, Stanley (1968): "The Non-Lover in Plato's *Phaedrus*". In: *The Society for Ancient Greek Philosophy Newsletter* 66. https://orb.binghamton.edu/sagp/66.

Szlezák, Thomas (2003): *La* Repubblica *di Platone. I libri centrali*. Brescia: Morcelliana.

Taylor, Alfred Edward (1955): *Plato, the Man and his Work*. London: Methuen.

Tejera, Víctor (1992): "The *Phaedrus*, Part I: A Poetic Drama". In: Rossetti, Livio (Ed.): *Understanding the Phaedrus, Proceedings of the II Symposium Platonicum*. Sankt Augustin: Academia Verlag, pp. 290–295.

Trabattoni, Franco (2011): "Un'interpretazione 'platonica' del primo discorso di Socrate nel *Fedro*". In: Casertano, Giovanni (Ed.): *Il Fedro di Platone*. Naples: Loffredo, pp. 285–305.

Vlad, Marilena (2022): "Dialectic as Philosophical Divination in Plato's *Phaedrus*". In: Larsen, Jens Kristian, Haraldsen, Vivil Valvik, and Vlasits, Justin (Eds.): *New Perspectives on Platonic Dialectic. A Philosophy of Inquiry*. New York: Routledge, pp. 249–263.

Yunis, Harvey (2011): *Plato, Phaedrus, Introduction and Commentary*. Cambridge: Cambridge University Press.

Kristian Larsen
Dialectic and the Activity of the Soul when Reaching for Being and the Good in Plato's *Theaetetus* 184b3–186e12

The science of dialectic (*dialektikē*) is crucial to Plato's conception of philosophy. Indeed, it may be argued that philosophers, according to Plato, are characterized especially by their mastery of dialectic. It is, nevertheless, a matter of controversy how Plato conceives of this science. In fact, many critics, following in the footsteps of Julius Stenzel and Richard Robinson,[1] have claimed that Plato advocated a number of different conceptions of dialectic and, in particular, that his supposed later dialogues are characterized by radical changes in the way the procedures of and the objects proper to dialectic are understood.

Central to the controversies concerning these alleged changes in Plato's understanding of dialectic are his *Parmenides* and the *Theaetetus*. While these dialogues are not, generally, considered to belong among the later dialogues, many regard them as transitional, being the first that allegedly expound a number of doctrinal changes concerned with his conception of forms and of dialectic that are characteristic of the later Plato. Thus, critics have argued that the *Parmenides* is essentially a critique of the transcendent forms characteristic of the so-called middle period dialogues, and that the inquiry into knowledge found in the *Theaetetus*, in contrast to related inquiries found in, for instance, the *Phaedo* and the *Republic*, is characterized by its lack of reference to transcendent forms and a more explicit interest in the world revealed to us through perception; some also argue that it foreshadows a new conception of dialectical definitions introduced in full only in the *Sophist*.[2]

A passage that seems particularly important for coming to grips with Plato's supposed later conception of dialectic, or indeed with his conception of dialectic more generally, is *Parmenides* 135b5–c6. Here Parmenides explains that, despite

---

[1] See Stenzel (1917), Robinson (1941), and Robinson (1953).
[2] See especially Stenzel (1917), where it is argued that the *Parmenides* heralds a radical change in Plato's way of conceptualizing forms while the *Sophist* contains solutions to the *aporiai* set out in the *Parmenides*, partly by introducing a new conception of *logos ousias* that is already foreshadowed in the *Theaetetus*; this line of interpretation is in many respects followed by Mary Louise Gill (2012) and Charles H. Kahn (2013); Gregory Vlastos (1954), in contrast, argued that the *Parmenides* is a dialogue that documents Plato's honest perplexity. On the *Theaetetus*, see also Robinson (1950).

the many problems he has raised in the previous part of the dialogue, should one not allow that there are forms for each of the things that are, one would not have anything toward which one might turn one's thought (*dianoia*), and one would thereby utterly destroy the power of conversation (*tēn tou dialegesthai dynamin*); and this, he further suggests, would be a serious problem for philosophy.

While it is a matter of some controversy what conversation (*dialegesthai*) means in this passage,[3] it seems safe to say that the expression "the power of conversation", taken as a whole, has a semi-technical sense and means something like a power inherent in discourse that, in order to be realized, requires that we engage in conversation for the purpose of achieving knowledge and not merely for having a casual conversation with each other. That Parmenides means to highlight a power that inheres in conversation that may be realized only when conversation is engaged in specifically for the purpose of achieving knowledge can be seen both from Parmenides' suggestion that the power of conversation will be destroyed if we deprive thought (*dianoia*) its proper objects (for it is difficult to see why a power that should characterize mere casual conversation should depend on us positing objects towards which thought may be directed) as well as from his suggestion that philosophy itself will be threatened if this power is destroyed. The upshot is, I suggest, that if one denies that there are forms in some sense at least, one destroys the power of discourse, understood as a knowledge-directed activity, by abolishing the entities toward which we must direct our mind if we are to unfold this power, and that this in turn makes philosophy impossible since philosophy is, essentially, a specific unfolding of the power of such knowledge-directed discourse; and, if we are allowed to suppose that the power of conversation for the philosopher will come to expression as dialectic, dialectic being, after all, the science concerned with conversation, we may say that Parmenides is also suggesting that dialectic will become impossible if there are no forms.

But while few will deny the importance of this passage for Plato's conception of dialectic, its significance is notoriously controversial. Is Parmenides recommending that we posit transcendent forms? Or is he only recommending that we posit forms in some other sense?[4] What reasons are there, more specifically, for claiming that the power of conversation will be destroyed if there are no forms? And what is the power of conversation precisely? The *Parmenides* itself seems to offer few clues as to how we should answer the first two questions and even fewer as to the last two.

---

3 For a recent discussion, see Gill (2012), p. 18, n. 1.
4 Allen (1997), p. 111 defends the former view, while Gill (2012), pp. 9–10 and 34–39, defends the latter.

At the beginning of the dialogue Socrates does suggest that we need to posit separate forms for certain things, at least, such as likeness, one and many, the just, the beautiful, and the good (see 130b3–10) if we are to avoid Zeno's so-called paradoxes, and it may be argued that the assumption that there are forms is meant to be a metaphysical solution to a problem raised by a metaphysical doctrine, that of Parmenides and Zeno, namely the problem that rational speech about the things revealed to us through sense-perception becomes impossible.[5] If this suggestion is along the right lines, it may perhaps also provide an explanation for why Parmenides twice commends Socrates for his zeal for speech (*hē hormē epi tous logous*; 130b1, 135d3): positing forms somehow secures a foundation for the basic human activity of speaking. Nevertheless, the dialogue itself does not provide any further clues as to why forms are necessary for the existence of the power of *dialegesthai* or what that power is, more specifically. Rather than explaining to Socrates why the denial that there are forms should lead to the destruction of this power, Parmenides simply states that Socrates is "only too aware of that sort of consequence" (135c2–3). The dialogue thus seems to suggest that this point should be obvious to the reader, either because it is self-evident, or because it is explained elsewhere. Since it is hardly self-evident that forms are called for if we are to account for the power inherent in speech, it seems more reasonable to assume that Plato means to indicate that reasons for Parmenides' claim can be found elsewhere.

In this chapter, I argue that the discussion of the thesis that knowledge is perception from the *Theaetetus*, and in particular the conclusion of this argument at 184b3–186e12, provides a partial explanation of Parmenides' claim that the power of conversation depends on forms and that the *Theaetetus* for that reason is important for understanding Plato's conception of dialectic. There are two reasons why I turn to the *Theaetetus* for this purpose, rather than to other dialogues that also provide apparent explanations for why the power of conversation depends on forms.[6] First, Plato himself clearly connects the arguments of the *Theaetetus* with those of the *Parmenides*, both at a dramatic and at an argumentative level. Second, critics who believe that Plato substantially revised his conceptions of forms and dialectic will be reluctant to accept explanations for Parmenides' claim that rely on dialogues now commonly believed to precede the *Parmenides*. But they generally regard the *Theaetetus* as being close to the *Parmenides* both

---

5 See Allen (1997), p. 85. See also Larsen (2022), pp. 185–186.
6 Vasilis Politis (2021) argues that *Republic* VII, 523–525 provides an explanation of Parmenides' claim at *Parmenides* 135b5–c3. My argument in this chapter is not that *Theaetetus* 184b3–186e12 provides the *only* explanation of Parmenides' claim, but that it provides *an* explanation. In fact, I believe that *Theaetetus* 184b3–186e12 in many respects parallels the argument from *Republic* 523–525, but it lies beyond the scope of the present chapter to explore this connection further.

in time of composition and in content; thus, if an explanation for Parmenides' claim can be found in the *Theaetetus*, it will be more difficult for critics who adopt a developmentalist approach to Plato to dismiss it outright as irrelevant for understanding the claim concerning the power of conversation found in the *Parmenides*.

1 The Question Concerning Dialectic in the *Theaetetus*

Before I turn to *Theaetetus* 184b3–186e12, it is necessary to make a few remarks about some dramatic and argumentative aspects of the dialogue that are relevant to my overall argument. First, the *Theaetetus* contains a direct reference to the encounter between the young Socrates and Parmenides, which is dramatized in the dialogue *Parmenides* (*Tht.* 183e7–184a2). Direct references between dialogues are exceedingly rare in Plato,[7] and for that reason we are entitled to infer that Plato intended that his readers think about the *Parmenides*, and the arguments made there, when reading the *Theaetetus*. This dramatic link between the dialogues is further accentuated by the fact that the *Parmenides* contains Plato's earliest depiction of Socrates, dramatically speaking, while the *Theaetetus* contains one of the latest, the dialogue taking place a few weeks before Socrates' death. The two dialogues depict the beginning and the end of Socrates' philosophical career, as it were, and indicate that this career is somehow bound up with Parmenides.

The suggestion that the two dialogues are closely connected is further corroborated at the argumentative level. The inquiry into knowledge central to the *Theaetetus* sets out from the denial that anything is one thing, itself by itself (*auto kath' hauto*; 152d2–3), whereas the inquiry of the *Parmenides* sets out from the assumption made by the young Socrates that we must posit forms, for certain things at least, themselves by themselves (*auto kath' hauto*; 128e6–7). The first part of the *Parmenides*, it may be argued, explores what problems follow from assuming that we must posit some things that are what they are, themselves by themselves, namely forms, while the first part of the *Theaetetus* explores what problems follow if one denies that there are any things that are what they are, themselves by themselves. Together, they may therefore be said to exemplify the hypothetical mode of

---

7 The *Sophist* also contains a reference to the *Parmenides* (217c5–7). In addition, the beginning of the *Timaeus* rehearses positions defended in the *Republic*; and the *Phaedo* contains a possible reference to the *Meno* (at 73a7–b2). However, both of these latter two references are less clear than those in the *Theaetetus* and *Sophist*.

investigation recommended to the young Socrates by Parmenides in the *Parmenides* as an exercise in dialectic (135e8–136a2). In the light of these connections between the dialogues, there is every reason to think that the *Theaetetus* is highly relevant for coming to grips with Plato's conception of dialectic.

Nevertheless, the *Theaetetus* is first and foremost an inquiry into the nature of knowledge, and it could be objected that this inquiry is directed at knowledge *simpliciter* and, thus, has little to say about dialectic. For, in this dialogue Socrates asks what characterizes any kind of knowledge in so far as it is knowledge and, in contrast to the *Republic*, for instance, he seems to disregard questions about what kinds of knowledge there may be, and what the objects of knowledge are (see in particular 146e7–8 and contrast with, e.g., *R.* 509d1–511e5).

The dramatic setting of the dialogue, however, suggests that Plato's attitude in the *Theaetetus* toward the question whether there are different kinds of knowledge may be more complex than it at first appears. The inquiry into knowledge is carried out in a conversation between Socrates, a philosopher who describes himself as a lover of *logos* (146a6; see also *Phdr.* 236e5) and connects his own activity as an intellectual midwife with the "whole business of conversation" (*hē tou dialegesthai pragmateia*; 161e6–7), presumably thereby meaning the business of knowledge-directed conversation central also to dialectic, and two mathematicians with rather different attitudes toward conversation-based inquiry. Theodorus, the older mathematician, is reluctant to enter into philosophical conversation (162a4–b5); he states from the beginning of the dialogue that he is unaccustomed to the "dialect" of Socrates, that is, the type of discussions Socrates usually engages in (146b3), and he later explains that he rather quickly turned away from "bare" or "abstract arguments" (*psiloi logoi*; 165a1–3), presumably thereby meaning pure dialectical discussions.[8] Theaetetus, on the other hand, admits from the beginning of the dialogue to having an interest in the type of questions Socrates poses (148e1–3), he proves easy to engage in the inquiry, he never grows tired of or angry at Socrates' questions, and he is praised by Socrates on several occasions for the good qualities of his soul (155d1–7, 185e3–5), a praise that is also reflected in the short introduction to the dialogue where Eucleides and Terpsion praise Theaetetus for being a real gentleman (142b7–c1). At the same time, both mathematicians are depicted as being somehow attracted to Protagoras: Theodorus admits to being a friend of the sophist (161b9–10, 162a4–5, 183b), and Socrates at one point even describes Protagoras as Theodorus' teacher (179a10), while Theaetetus admits to hav-

---

[8] In both the *Phaedrus* 262c8 and the *Symposium* 215c7, this expression is used to refer to the kind of speeches Socrates typically engages in.

ing read Protagoras' book "Truth" several times (152a5) and finds Protagoras' teaching impressive, at least at the beginning of the inquiry (153a4, 157d10–12).

Thus, even if the inquiry aims at articulating a unified conception of knowledge, the drama of the dialogue points to three rather different conceptions of knowledge and its importance for human life: the expert knowledge possessed by Theodorus and Theaetetus, the self-proclaimed wisdom of Protagoras that seeks to associate itself with this kind of expert knowledge, and the type of expertise and wisdom Socrates claims for himself and describes as intellectual midwifery (149a4–151d6, 161e4–5, 210c4–d2) and connects with knowledge of one's own knowledge and ignorance (210b11–c4, see also 187a1–6 and 154d8–e5). As I shall argue below, part of the drama of the *Theaetetus* consists in the fact that Socrates, through the dialogue, seeks to turn Theaetetus from sophistry and "mere" mathematics toward philosophy and dialectic. This drama revolves, in part, around Theaetetus' conception of being and benefit or the good, two things central also to the young Socrates' first steps into the realm of dialectic depicted in the *Parmenides*.

2 The Passage 184b3–186e12 and Its Connection to the Previous Part of the Dialogue

The discussion of the thesis that knowledge is perception found in *Theaetetus* (running from 151d7 to 186e12) contains one of the most celebrated and influential arguments against identifying knowledge (*epistēmē*)[9] with perception. At the end of this long and complex discussion, we find a passage that seems to contain a separate and independent refutation of the thesis, namely 184b3–186e12, that I shall refer to as the "final refutation passage". In this passage Socrates asks Theaetetus whether certain properties common to the things we perceive, such as their being and non-being, their similarity and dissimilarity, can be perceived through the

---

9 *Epistēmē* is usually translated as "knowledge". In other dialogues, however, including the *Sophist* and the *Statesman*, *technē* and *epistēmē* are used interchangeably and the examples of *epistēmai* Theaetetus presents as his first candidate answer in the *Theaetetus* at 146c7–d3 includes *skytotomikē te kai hai tōn allōn dēmiourgōn technai*. Theaetetus' initial answer in fact demonstrates that, when Socrates raises the question what knowledge is, Theaetetus is thinking first and foremost about knowledge in the sense of "expert knowledge", that is, the knowledge a practitioner of a specific profession or craft possesses in contrast to the layman, the "grasp" of a subject that you may learn from a skilled practitioner. As I will argue below, this aspect of *epistēmē* comes to the fore in a number of arguments in the later discussion, including the passage 184b3–186e12, and it is part of the reason why the notion that knowledge is perception is dismissed. This aspect of *epistēmē* is important to bear in mind.

senses (185a4–d5)[10] and Theaetetus states that they cannot and suggests that it is the soul, rather, itself by itself (*autē di' hautēs*), that examines (*episkopein*) these properties (185d6–e2). On the basis of some further discussion of the activity of the soul when it examines such properties and the concession that it is impossible to obtain truth without grasping being or to obtain knowledge without truth, Theaetetus finally concedes that knowledge cannot be perception (186e4–8). Many critics will agree that the final refutation passage where this conclusion is established is a high point in the dialogue and possibly also in the history of philosophy.[11] At the same time, the passage appears to raise as many questions as it purports to answer.

One major question is what purpose the final refutation passage serves within the wider argument that runs from 151d7 to 186e12. For Socrates has already reached the conclusion, at 182d8–e6, that if everything is always changing in every way, we are not entitled to call a particular instance of perception "seeing more than we call it not seeing, or any other sort of perception",[12] since we would thereby impose an unwarranted stability upon it. This conclusion may seem to render Theaetetus' proposal that knowledge is perception empty or meaningless and therefore refuted, since that definition, at least as it is interpreted by Socrates, depends on the ontological assumption mentioned above, that nothing is one thing, itself by itself. What Socrates has established immediately prior to the final refutation passage is, then, that the assumption about being supporting Theaetetus' proposed definition, which Socrates connects with Heraclitus, renders that very definition meaningless.[13] But if it is already clear that Theaetetus' answer is untenable, why is the final refutation passage needed?

---

[10] The Greek text used is that of the *Platonis Opera, Tomus I* (ed. E. A. Duke, W. F. Hicken, W. S. M. Nicoll, D. B. Robinson, and J. C. G. Strachan).

[11] The reader may compare the somewhat backhanded praise of Plato's "achievement" towards the end of Myles Burnyeat's essay "Plato on the Grammar of Perceiving" (1976, p. 50) with Martin Heidegger's more generous, if also very critical, remarks on the importance of the passage from his 1934 lecture on the *Theaetetus* (1997, pp. 175–176 and 181–182).

[12] All translations of the *Theaetetus* are taken from Rowe (2015); at times, I have made minor adjustments to the translation in order to make a particular point.

[13] Here I follow Burnyeat's so-called B-reading, according to which the extreme implications of this ontology spelled out at 182d8–e6 is already implied in the initial formulation of that theory (Burnyeat 1990, p. 46). Catherine Rowett argues that the passage 183b7–c3 is where Theaetetus' first proposal "is declared non-viable, and left to die" (Rowett 2018, p. 173), and that the passage 184b3–186e12, rather than constituting a conclusion to the first section of the dialogue, "forms a transition between Theaetetus' first proposed definition... and his second" (Rowett 2018, p. 197).

A further, closely related question is what ontological status, if any,[14] Socrates accords the common properties (*ta koina*), which he introduces in this passage, properties such as being and the good that the refutation contained in the passage revolves around. Particularly controversial issues are what being, that is, *ousia*, means and what reasons the interlocutors have for claiming that truth, and therefore knowledge, depends on grasping *ousia*. Is the argument simply meant to establish that "to know is to judge, or is a species of judging"[15] and that, since "a true judgement" will involve "the verb 'to be'",[16] knowledge will depend, only in this limited sense, on grasping being?[17] Or is Socrates aiming to show that knowledge, including the knowledge characteristic of the philosopher, dialectic, is somehow dependent on grasping the being of the things we seek to know? The former is by now the majority view, and it may seem to be corroborated by the way being is first introduced in the final refutation passage, a point to which I shall return below.

However that may be, if one looks at the larger context within which the final refutation passage is embedded, that is, the long discussion of Theaetetus' thesis that knowledge is perception and its connection to Protagoras and Heraclitus, a case can be made for the claim that the conception of being is clarified only gradually within the final refutation passage and that the full significance of the term can be understood only if interpreted in the light of the final section of this passage.[18] It is the process through which Theaetetus, with Socrates' help, arrives at a better understanding of being and its connection to the good and beneficial within the final refutation passage that I wish to highlight, since it is central to my claim that this passage is important for understanding Plato's conception of dialectic. I therefore begin by discussing the connection between the final refutation passage and its wider context before I turn to the question how we should understand the concept of *ousia* in the passage in section III below.

---

[14] John Cooper claims that "it nowhere matters to his argument what their [that is, the common properties] metaphysical status is" (Cooper 1970, p. 138, n. 19).
[15] Ryle (1939), p. 317.
[16] Burnyeat (1976), p. 49.
[17] Cooper (1970), p. 131, n. 12 objects to this general line of interpretation, correctly pointing out that "the principle of selection for the κοινά is not their implication in all judgements, but their applicability to objects of different senses. So the supposed special position of at least some of them as regards the power of judgement", including being, "is not Plato's reason for illustrating the independent activity of the mind by judgements involving them".
[18] On this point, see the interpretation of the passage by David Sedley (2004, pp. 109–110), an interpretation to which this chapter is indebted. Cooper, in contrast, explicitly denies that οὐσία can mean nature (Cooper 1970, p. 137, n. 18), as does Burnyeat (1976, p. 45).

How to understand the connection between the final refutation passage and the section of the dialogue that precedes it is a matter of controversy. Myles Burnyeat, voicing the perspective of his so-called B-reading, famously argued that what we find at the end of the previous part (at 183b7–c3) is the conclusion to a prolonged indirect refutation of Theaetetus' thesis that knowledge is perception, where "the thesis under consideration supplies the materials for its own refutation".[19] More precisely he argued that the Protagorean measure doctrine and the Heraclitean ontology that underpins it, the main subjects of the inquiry that ends at 183b7–c3, "provide sufficient conditions for Theaetetus' definition to come out correct" and that the argument Socrates presents "down to 160e" seeks to establish "that they are the only sufficient conditions which could reasonably be devised", wherefore they are also necessary.[20] If this is correct, then Theaetetus' thesis is self-defeating since the Heraclitean ontology, at least as it is interpreted by Socrates, renders the thesis itself meaningless. The final refutation passage, in contrast, according to Burnyeat, contains a direct refutation, that is, a refutation that "proceeds, for the first time in the dialogue, from premises which Plato himself accepts as true".[21] This suggestion does provide an explanation of the differences between what appear to be two distinct refutations of the same thesis. The question remains, however, why Plato chose to provide them in the first place. Would one not suffice? If the thesis is indeed self-defeating, what more needs to be said? And why should Socrates wish to provide a refutation that builds on premises Plato accepts?

David Sedley, on the other hand, has argued that the refutation culminating at 183b7–c3 refutes Theaetetus' first answer "only in the Heraclitean form in which he [that is, Socrates] has developed and criticized it", that is, only in the form where perception is understood in the context of an ontology that maintains that there are no things that are what they are, themselves by themselves. He further suggests that it "remains a theoretical possibility that Theaetetus' definition ... could be defended on some other basis". What the final refutation passage then provides is a refutation that demonstrates that "Theaetetus' definition is inherently faulty, regardless of any perceptual theory one might adopt".[22] This interpretation may seem to square better than Burnyeat's does with a caveat that Socrates makes at 183c3, namely, that Theaetetus' thesis that knowledge is perception can now be dismissed "unless Theaetetus here has something else to say".[23] Nevertheless, it seems

---

19 Burnyeat (1990), p. 53.
20 Burnyeat (1990), p. 10.
21 Burnyeat (1990), p. 53. See also Heidegger (1988), p. 150.
22 Sedley (2004), p. 105.
23 On this passage, see also Rowett (2018), pp. 173–174.

to face a similar problem also. For if the final refutation passage provides a refutation of the thesis that knowledge is perception, irrespective of one's theory of perception, we are left to wonder why Plato bothered to write the long discussion from 152a1 to 183c3.

The problem facing both accounts is, then, that the dialogue's argumentative structure becomes questionable if we cannot point to any deeper connections between the final refutation passage and the previous criticism of Protagoras' measure doctrine and the Heraclitean ontology Socrates associates with it. If a link can be established between the two refutations that will enable us to read the whole passage 152a1–186e12 as a continuous refutation of the thesis that knowledge is perception, it will provide us with a reading of the dialogue more generous to Plato, and one that is also preferable because Socrates, at the end of this passage (at 186e11–12), clearly indicates that this is how he understands his prolonged inquiry into Theaetetus' first proposed answer. Moreover, Theaetetus' definition, as interpreted by Socrates, that he appears to refute at 183b7–c3, rests on an assumption about being that explicitly denies what Socrates himself argues is a necessary assumption as a young man in the *Parmenides*—that there are some things that are what they are, themselves by themselves. It is therefore fair to assume that the older Socrates, when linking Theaetetus' definition with Protagoras' measure doctrine and the Heraclitean ontology and criticizing both, is also attempting to teach Theaetetus an important lesson about being and the necessity of positing at least certain things as being what they are, themselves by themselves, if dialectic and philosophy are to be possible. If this suggestion is correct, it likewise speaks in favor of finding a link between the discussion that runs from 152a1 to 183c3 and the final refutation contained in the passage 184b3–186e12 that will allow us to see this lesson reflected also in the final refutation passage.

In an influential paper from 1970, John Cooper suggested, while discussing what the expression "to grasp being" can mean in the final refutation passage, that the "refutation of Protagoras earlier in the dialogue [at 177c6–179b9] seems to offer a clue".[24] While I do not accept Cooper's ultimate conclusions about Plato's conception of knowledge and being in the *Theaetetus*, I think he is right in suggesting that the previous arguments against Protagoras that culminate at 177c6–179b9 offer clues for deciding what the expression "to grasp being" means when used in the final refutation passage[25]—and indeed for understanding the full implication

---

24 Cooper (1970), p. 141.
25 Scholars who point to the connection between Theaetetus' claim at 186a10–b1 and the argument directed against Protagoras at 177c6–179b9 include Campbell (1881), p. 143, n. to line 4, Cornford (1935), p. 107, n. 1, McDowell (1973), pp. 190–191, and Sedley (2004), p. 109. As far as I know, only

of that passage for Plato's conception of knowledge and dialectic. In what follows I argue that we are not faced with two distinct refutations in the passage from 152a1 to 186e12 but with two complementary steps in one prolonged refutation. Step one (152a1–183c3) contains a criticism of, especially, Protagorean relativism and the Heraclitean ontology underpinning it, while step two (184b3–186e12) builds upon and elaborates on the arguments articulated in step one, in particular the arguments directed against Protagoras, and contains a general conclusion about knowledge important also for understanding Plato's conception of dialectic. To demonstrate how these two steps are connected, it is helpful to begin by looking at a short section of the final refutation passage, namely 186a2–186b1, and to consider how it is related to the previous refutation of Protagoras.

In the argument leading up to this section, running from 184b3 to 186a1, Socrates and Theaetetus establish that certain properties common to the things we perceive cannot themselves be objects of perception but are rather things that the soul examines or inspects (*episkopein*), itself by itself (*autē di' hautēs*). The text runs as follows:

> [T1] Socrates: So to which of the two sets of things do you assign being (*tēn ousian*)? This is what is most constantly present in all cases.
>
> Theaetetus: I myself count it among the things that the soul reaches out to, itself by itself.
>
> Socrates: The like, too, and the unlike, the same and the different?
>
> Theaetetus: Yes.
>
> Socrates: What about beautiful and ugly, good and bad?
>
> Theaetetus: These too, it seems to me, are more than anything things whose being (*ousian*) the soul examines (*skopeisthai*) in relation to one another (*pros allēla*), reckoning up in itself (*analogizomenē en heautē*) past and present in comparison with future (186a2–b1).

The fact that Theaetetus here, when considering where to place "beautiful and ugly, good and bad", brings up the activity of "reckoning up" past and present in relation to the future indicates that he has the argument directed against Protagoras in the previous part of the dialogue, an argument that culminates at 177c6–179b9, in mind. For this argument was centered precisely on the notions of the good and the beneficial in relation to the future.

To see the full importance of that argument for the drama of the *Theaetetus*, however, it is important to note that the term "good" (as well as "beautiful") was introduced already at an earlier stage in the *Theaetetus* and has been an implicit

---

Cooper and Sedley make something of the fact that what is at stake in the earlier argument is the question of expertise.

theme prior to the refutation at 177c6–179b9. When Theaetetus initially suggests that knowledge is perception, Socrates associates this suggestion with Protagoras and the doctrine that all the things we are accustomed to say *are* in fact *become* through the interaction of two types of movement (see 152c8–d6); on this basis he then asks Theaetetus, at 157d7–9, if he likes "the proposal that neither good, nor beautiful, nor any of the other things that were on our list just now *is* at all, but is rather always in a process of *coming to be*". As Francis M. Cornford observed, at this point in the dialogue Theaetetus "apparently feels no qualms when Socrates slips in the words 'good' and 'beautiful', as if these qualities were on the same footing with 'hot' or 'white' or 'large'",[26] qualities that each of us, according to the Protagorean measure doctrine, is a measure of. Indeed, from the perspective of what Parmenides grants to the young Socrates in the *Parmenides*, we may conclude that, since he accepts that these things have no being on their own, he is granting an ontological assumption that ultimately renders the power of conversation, philosophy, and dialectic impossible.

Now, there is a further and simpler reason why it may be thought problematic to assimilate the good, in particular, to perceptible qualities and the beneficial that goes together with the good (for what is beneficial for something is beneficial because it helps realize the good of that thing),[27] at least on the assumption that each of us is an adequate measure of such qualities. For this assimilation dissolves an important basis for distinguishing the expert from the layman, a distinction Theaetetus and Theodorus should be interested in maintaining, no matter what they are inclined to think about dialectic, given that they themselves are practitioners of a recognized expertise. This consequence is gradually brought to light by Socrates in the ensuing discussion (see, e.g., 161d2–e3) until it finally becomes an explicit theme in a defense Socrates delivers on behalf of Protagoras (see 166d4–7). What seems to motivate this defense is, precisely, the problem that the notion of being an expert or wise about something appears meaningless if all things, including what is good and beneficial, are to each of us the way they appear to us. For what sets the expert or the wise person apart from the layman more than anything else, it may be argued, is the expert's ability to predict what *will be* good or beneficial to someone, an ability the layman lacks, even if the layman may still be the best judge of what appears good or beneficial to him or her at any given moment. This is the basic point of the refutation of Protagoras at 177c6–179b9.

---

[26] Cornford (1935), p. 51.
[27] I here leave the beautiful out of the discussion; it may be noted that a case could be made for connection the beautiful or fine with the notion of benefit as well, in so far as a beautiful or fine practice, for instance, is something that will help something realize its proper virtue, thereby making it good.

My aim here is not to discuss in any detail Socrates' defense on behalf of Protagoras or the various steps in the refutation of Protagoras' doctrine that follow.[28] Important for my purpose here is only that the argument against Protagoras at 177c6–179b9 sets out from the question concerning the expert's knowledge of the good and beneficial. Socrates here suggests that, while some people may be willing to follow Protagoras in claiming that, for instance, justice has no more than a kind of borne along being (*tēn pheromenēn ousian*; 177c6) and that whatever a city posits as just is just for that city as long as it so posits it, they will not be willing to make the same claim concerning the things that are good and "insist that whatever a city lays down as beneficial for herself, because it thinks it is so, is actually beneficial for as long as she so lays it down" (177c6–d5).

At this point, Socrates thus explicitly seeks to exclude the good and the beneficial from the realm of perpetual change to which the notion "good" was apparently condemned at 157c7–9. In fact, he proceeds to suggest that a city is aiming (*stochazesthai*) at this very thing, that is, what is good or beneficial, whatever it might call it, while making its laws, attempting to make them as beneficial as possible for itself, even while many a city misses (*diamartanein*) this target (177e4–178a3). That Socrates here describes legislating as an activity where one aims at or looks toward (*blepein*; 177e7) what is good but risks missing the target further underlines the idea that the good or beneficial is something stable, a thing that is not perpetually changing and relative to how each of us perceives it.

That Socrates at first focuses on the act of legislating and the question of what is good for cities is possibly a result of the fact that he, in his previous defense of Protagoras, had emphasized the role of the expert in devising legislation. In order to consolidate the conclusion that the good and beneficial are objective matters that one may misunderstand, however, he proceeds to widen the perspective considerably. It will be easier for everyone to agree to this conclusion, he suggests, if one asks about the entire class (*eidos*) to which the beneficial happens to belong (178a5–7), the class of things that is concerned with the future (178a7–8).

To explain more clearly what class of things he has in mind, Socrates then mentions a number of arts or types of expertise, asking whether it will be the expert or the layman who will be able to judge correctly what will prove good or beneficial in the future: is it, for instance, the doctor or the patient that is the better judge of whether or not the patient will catch a fever and feel sick in the future (178c2–7), or the gymnastic trainer or the musical expert that is the better judge of whether or not something will appear to be in tune at a later point (178d4–6)?

---

28 For further discussion of this matter, see Burnyeat (1990), pp. 22–28.

The general point Socrates wishes to establish concerning the good and beneficial is, then, that, at least when it comes to the question what *will* be good and beneficial in the future, it is the expert, and not the layman, who is the better judge. Relativism, or the notion that each of us is a measure of how things are, seems very difficult to reconcile with our trust in experts. As we shall see, it is also impossible to reconcile with the particular type of conversation Socrates engages in, that is, dialectical conversation aimed at seeking knowledge of the most important matters in human life.

What the argument does not establish, however, is what would have to be true about reality for expert knowledge and dialectic to be possible, since the Heraclitean ontology that underpins Theaetetus' definition, or at least the Protagorean interpretation Socrates has presented of it, is still in play. Once this ontology is dismissed, however, at 182d8–e6, further progression is possible on the way toward an answer to the questions what knowledge is, how it may be said to depend on conversation, and how we must conceive of the reality toward which knowledge, both that of the expert and the dialectician, is directed. These are the questions that the claim advanced by Theaetetus in T1 (at 186a10–b1) picks up on.

3 The Implication of Theaetetus' Concession Concerning the Good

Thus far, I have argued that Theaetetus, when claiming in T1 that the good and the bad and the beautiful and the ugly belong to the "things whose being (*tēn ousian*) the soul examines (*skopeisthai*) in relation to one another (*pros allēla*), reckoning up in itself (*analogizomenē en heautē[i]*) past and present in comparison with future" (186a10–b1), is recalling Socrates' earlier argument against Protagoras. I have also indicated that this claim demonstrates an important change in Theaetetus' understanding of the good (as well as of the beautiful): he no longer accepts that it has no being, itself by itself, as he was inclined to do at the beginning of the inquiry. For to claim that things such as the good and the beautiful have a being that the soul may examine is, it may be argued, to accept what was suggested as an objection to Protagoras, that they have something like a nature on their own that one may inquire into and come to understand to a lesser or greater degree.

Assuming that the view that the good and the bad and the beautiful and the ugly each has a being of its own that one may inquire into is central to Socratic inquiry,[29] it may seem reasonable to assume that Socrates is pleased with Theaetetus' answer. Indeed, as noted above, Plato depicts a younger Socrates in the *Parmenides* as setting out from a closely related view (at 130b7–9), according to which both unity, the just, the beautiful, and the good are things that are, themselves by themselves, and a Parmenides who concedes (at 135b5–c6) that we need to posit forms, in some sense at least, for those things Socrates mentions if the power inherent in conversation (*tēn tou dialegesthai dynamin*) and philosophy are to be possible. In the light of these observations, it may be argued that Theaetetus in T1 is conceding a premise central to the young Socrates' and to Parmenides' conception of conversation aimed at truth and of Plato's conception of dialectic more generally. If mathematicians differ from dialecticians in that the former only dream about being, as Socrates suggests in the *Republic*, whereas the latter can be said to be awake given their knowledge about the nature of things such as the good, the beautiful, and the just (see 533b5–c3 with 476c1–d2), it may in fact seem that Socrates is helping Theaetetus to wake up.

Some critics, however, would deny that this is the implication of Theaetetus' concessions in T1. John McDowell, for instance, has suggested that Theaetetus, by bringing up the argument directed at Protagoras, may be "in danger of missing the point of the present section",[30] and that Socrates' response to Theaetetus' concession—that is, "hold it there" (186b2)—is meant to highlight this problem. For, McDowell argues, by bringing up the earlier argument about future benefit Theaetetus risks "taking the present passage [that is, T1 and context] as merely adding some further questions, e.g., those about being" to those about the future, thereby missing Socrates' general point in the passage we have been considering. That point, according to McDowell, is that not even when it comes to our present perceptions can we say that perception is knowledge. Even for a trivial judgement such as "this wind is cold" we rely on being, a "concept" that is not available to us in perception. McDowell is not alone in making this suggestion; in fact, on this score, his view is in agreement with interpretations of the passage advanced by Gilbert Ryle,[31] Myles Burnyeat,[32] and a number of subsequent critics.

Two things seem to speak in favor of the reading advanced by these critics. First, when Socrates initially brings up being for discussion prior to T1, namely at 185a9 and 185c5–6, he seems to have something rather trivial in mind, namely the mere fact that, concerning two things, we may say that they both are. And

---

[29] Readers skeptical of the suggestion that Plato's Socrates would accept that the bad and the ugly have a being or nature of their own may consult *Republic* 476a1–8.
[30] McDowell (1973), p. 190, n. on 186b2–10; see also Cornford (1935), p. 107, n. 1.
[31] See Ryle (1939).
[32] Burnyeat (1976 and 1990).

when the notion of *ousia* is used at 185c9 and then repeated in T1 at 186a2, it seems to be with the same meaning. From this point of view, it is not difficult to see why many critics reject the idea that *ousia* anywhere in the passage 184b3–186e12 could mean something like the "nature" of something that someone possessing expertise, in contrast to a layman, understands, insisting rather, in the words of Cooper, that it "throughout the passage ... seems to mean (something like) the existence of this or that".[33] Second, as a conclusion to the argument advanced in the final refutation passage as a whole, Socrates suggests that they should seek knowledge, or an answer to the question what it is, not in perception but in whatever it is the soul does when it is occupied with, or takes trouble with (*pragmateuesthai*), the things that are, itself by itself (187a3–6). And this, Theaetetus suggests, is the activity of opining or forming judgements (*doxazein*). This suggestion seems to sit well with Burnyeat's claim that in the final refutation passage Socrates' primary aim is to distinguish "perception and judgement in a way that effectively denies to the senses the judgmental function they had in the *Republic*",[34] a function on which knowledge, according to Theaetetus' suggestion at 187a7–8, depends.

There are, nevertheless, reasons to reject McDowell's suggested reading of Socrates' remark at 186b2—that is, "hold it there"—as well as the widespread view that *ousia* cannot mean something like nature or essence in the later part of the final refutation passage (namely at 186a11, 186b6–7 and 186c3). Let us look first at Socrates' remark at 186b2 and the exchange between him and Theaetetus that follows. Here is the text:

> [T2] Socrates: Hold it there. It's through touch that we will perceive the hardness of what is hard, and similarly the softness of what is soft—right?
>
> Theaetetus: Yes.
>
> Socrates: Whereas what our soul tries to judge by itself, going close up to them and comparing them with each other, is their being, and that they are,[35] their oppositeness to one another, and again the being of their oppositeness?

---

[33] Cooper (1970), p. 137, n. 18. Burnyeat objects to the suggestion that the passage is concerned with "the 'is' of existence, since its negation 'is not' is later singled out as something that is also true of both the colour and the sound" (1976, p. 44) and suggests that what is contrasted here is the simple perception of a color and the thought that it is something.

[34] Burnyeat (1976), p. 36.

[35] Rowe translates *tēn de ge ousian kai hoti eston* at 186b6 as "their being, namely that they are", and the *kai* may indeed be read epexegetically. But since Theaetetus has just stated that the soul examines the being of the good and the bad in relation to each other, I find it more natural to suppose that Socrates here follows up on Theaetetus' suggestion and makes a distinction between the being of hardness and softness and the fact they are.

Theaetetus: Certainly, yes (186b2–10).

While it is true that Socrates in T2 focuses primarily on being, and leaves the good and the other properties mentioned in T1 out of the discussion for the moment, the expression "hold it there" (*eche dē*; 186b2), taken on its own, could well be regarded as a way of emphasizing the importance of what Theaetetus has just conceded in T1—that the good and the bad and the beautiful and the ugly belong in the category of properties that the soul examines on its own—rather than as a warning against bringing them up in the present part of the inquiry. For this is certainly an important concession, given Theaetetus' earlier acceptance of a Protagorean view of things. The importance of this becomes all the clearer if we remember that Protagoras is described throughout the preceding parts of the dialogue as a friend, and even as a teacher, of Theodorus (161b9–10, 162a4–5, 164e4–7, 171c8–9, 179a10), that is, of Theaetetus' teacher, and that it is indicated that Theaetetus is somewhat attracted to the position of Protagoras. Part of the drama of the argument in the *Theaetetus* is surely that two mathematicians, who undoubtedly possess genuine, expert knowledge, find the teachings of Protagoras attractive, teachings that render the very notion of expert knowledge they rely on, as well as their field of study, it may be argued, devoid of meaning.[36] As I have argued, it is precisely the concepts of the good and beneficial that Socrates highlights in order to make Theodorus realize that Protagoras' position is incompatible with the notion that the expert and the layman differ as regards their ability to judge the future correctly.

To appreciate this aspect of the drama of the dialogue fully, one also needs to recognize that the previous refutation of Protagoras that culminates at 183b7–c3 is carried out exclusively in conversation with Theodorus. At 169a6–b4, a reluctant Theodorus is finally forced into participating in the dialectical exchange, "up until the point when" the interlocutors are in a position to decide whether Theodorus should, "after all ... be the measure when it comes to geometrical figures" (169a1–3), that is, whether it is an expert such as Theodorus, or rather the layman, who is the better judge of questions pertaining to geometry. And he remains Soc-

---

36 On this point, see especially Gadamer (1982), pp. 295–299; see also Howland (1998), pp. 56–57, and Larsen (2019), pp. 7–12. On my view, the fact that both mathematicians seem unaware of the ontological and epistemological implications of the possibility of expert knowledge does not speak against the claim that Socrates uses the fact that some people possess expert knowledge as an argument against the suggestion that knowledge is perception; as Gadamer points out, in order to possess expert knowledge one need not necessarily be able to answer the question what knowledge is, just as one does not need to possess expert knowledge in a particular field in order to be able to answer the question what knowledge is. I thank Catherine Rowett for pointing out this potential objection.

rates' sole interlocutor up till 183c3, where Protagoras is finally dismissed, to Theodorus' relief who—even though he by that time has become rather engaged in the whole inquiry (see 177c3–5 and 182b6–8)—remarks that he now "too" has to "be released from the role of respondent, according to our agreement, which was that I should continue until the matter of Protagoras' thesis was settled" (183c4–7).

It is in order to highlight the dramatic importance of the fact that Theaetetus now, on his own, recognizes the significance of what was established concerning expert knowledge in the exchange between Socrates and his teacher Theodorus, I suggest, that Plato presents us with what may at first sight look like two independent refutations of the same thesis but which I argue are rather two interconnected steps. In one sense, I believe, Catherine Rowett is correct in claiming that Theaetetus' thesis is "declared non-viable" already at 183b7–c3.[37] In another sense, however, the thesis is still up for discussion, both because the fact that it is "non-viable" has not yet been brought fully home to Theaetetus, and, more importantly, because the full implications of the refutation of Protagoras and Heraclitus for the way being, the good and beneficial, and knowledge-directed conversation should be understood, has not yet been made clear. This is precisely what the conversation with Theaetetus, contained in the final refutation passage, is meant to do. It is thus a new step in a general refutation of the thesis that knowledge is perception that builds on, and elaborates upon, the previous step.

This suggestion, that it is in order to highlight the importance of what Theaetetus concedes concerning the good and the beautiful that Socrates tells him to "hold it there", not in order to warn him against missing a general point concerning being, is further corroborated by the general conclusion Socrates draws, just after T2:

> [T3] Socrates: There will be some things, then, that human beings and animals alike are naturally able to perceive as soon as they are born, namely those things the experience of which extends through the body to the soul; whereas calculations about these, *both as regards their being and as regards their benefit (ta de peri toutōn analogismata pros te ousian kai ōpheleian)*, come, to the people to whom they do come, only with difficulty, late on, and after much trouble and education (186b11–c5; my emphasis).

Here, Socrates is very far from suggesting that Theaetetus should forget about benefit and the good in order to concentrate solely on being, the supposedly most important concept for understanding our capacity to make judgements. He is rather encouraging him to see benefit, or the good, as on a par with being.

---

37 Rowett (2018), p. 173.

The question remains how *ousia* should be understood in the final refutation passage as a whole. It can hardly be denied that the expression, when first introduced at 185c9, seems metaphysically innocent. As already mentioned, it seems to refer to the fact that something we perceive can be said to be, perhaps in the simple sense that it is *something or other*, rather than in the sense that it exists or has a specific nature, since *ousia* is here paired with *to mē einai* which, in this context, seems to mean that something that exists may, in another sense, be said not to be, presumable some specific thing, a point emphasized by Burnyeat.[38] But does this give us reason to conclude that the final refutation as a whole offers no "defence … of the idea that knowledge, let alone truth, presupposes a grasp of being in the specific sense of existence, reality, or essence" and that, "if any such narrow notion of being were intended, the argument would be left to limp on an unargued assumption of the first magnitude"?[39] Not necessarily. For an argument for such an idea, or at the very least the outline of an argument, is presented precisely in Theaetetus' claim in T1 and the short discussion that follows it in T2 and T3.

To make it plausible that this is so, a few additional remarks about the structure of the argument leading up to T1, presented in the passage 184b4–186a1, are called for. Here I take for granted a point made by Cooper, namely that Socrates' primary aim in that passage is to draw a simple, if important distinction between two types of activities the soul may be engaged in, a "perceptual use of the mind, in which it operates through the medium of the bodily senses" and "a further and higher use, in which the mind works independently of the body and the senses".[40] The latter is the activity the soul is engaged in when it aims to examine or inspect the common properties. As I see it, the argument presented from 184b4 up to 186a1 is intended to make one point, and one point only, namely that, when it comes to the properties that are common to several things we perceive that cannot themselves be perceived, the soul is able to engage in an activity, first described as a kind of thinking (*dianoein*) at 185a4 and later, at 185e2, as an examination or inspection (*episkopein*), that differs significantly from the activity it is engaged in when it perceives through the bodily organs. What Socrates then ensures in the first half of T1 (186a2–8) is that Theaetetus agrees that being, likeness and unlikeness, and identity and otherness, belong among such common properties, that is,

---

38 Burnyeat (1976), p. 44. But see Rowett (2018), p. 215, who objects to Burnyeat's line of interpretation.
39 Burnyeat (1976), p. 45.
40 Cooper (1970), p. 127.

properties that the soul, as Theaetetus puts it at 186a4–5, reaches out to (*eporegesthai*), itself by itself.[41]

Important as the first stage (184b3–186a1) of the argument contained in the final refutation passage may be, it does not attempt to answer the question what the activity of the soul when reaching out for the common properties consists in or how we are to conceive of the properties themselves. This, I urge, is precisely what the passage 186a2–e12 is meant to help clarify and it is, for that reason, the most important part of the entire passage. As already mentioned, it is addressed to Theaetetus, not to Theodorus; and indeed, Theaetetus stands sorely in need of such clarification if he is to realize his full potential for philosophy hinted at throughout the dialogue. For, being rather young and also a mathematician, he still has little understanding of being and the activity through which it is cognized,[42] a fact reflected also in his earlier enthusiasm for the flux doctrine and the idea that knowledge might be perception. Early in the dialogue, Theodorus states that certain guardians (*epitropoi tines*) appear to have destroyed Theaetetus' property (144d1–3) after his father died, a remark that may be seen as an expression of Platonic irony; for the word used for property is also the word for being, *ousia*, and the guardians who have taken responsibility for Theaetetus' education are, presumably, first and foremost Theodorus and, through him, Protagoras, two teachers who have not helped Theaetetus reach a better understanding of being at all.[43] It is this problem that Socrates seeks to rectify and, in doing so, he is directing Theaetetus away from mathematics toward dialectic.

---

[41] At 187a7–8, Theaetetus suggests that the activity the soul engages in when it occupies itself with the things that are should be called "to form an opinion" (*doxazein*), and, at 187b5–6, that knowledge may happen to be true opinion (*alēthēs doxa*). That thinking and the act of forming an opinion amount, more or less, to the same thing is later suggested by Socrates himself (at 189e6–190a7), and this is also suggested by the Eleatic visitor in the *Sophist* (at 264a8–b4), a point I return to below. From that perspective, it seems unproblematic to use the term *doxazein* to describe the activity of thinking about, or preoccupying oneself with, the things that are. By suggesting that knowledge may be identical with a true opinion, however, Theaetetus shifts the focus from the *activity* of thinking, or of forming a belief, to the *product* of this activity. This, I believe, is part of the reason why the inquiry of the *Theaetetus* ultimately fails.

[42] As regards the distinction between dialectic and mathematics, in addition to the passages already mentioned above (on page 143), the reader may wish to consult Socrates' description of the mathematical disciplines in *Republic* 510c2–511d5 and 533a10–c6.

[43] On this point, see Howland (1998), p. 54. As I see it, the *Theaetetus* and the *Sophist* can be read as a continuous attempt to redirect Theaetetus from mathematics to dialectic, and to the question what being is and how and why knowledge depends on it, and the passage 186a2–e12 may be read as one crucial, but not final, step in this process. For more on this point in relation to the *Sophist*, see Larsen (2015), pp. 310–311.

Let us now take a closer look at Theaetetus' answer in T1 to Socrates' question, what he thinks about beautiful and ugly, good and bad, and consider more carefully what his answer and the ensuing discussion in T2 and T3 tell us about being, the good or benefit, and other common properties and the activity of the soul that is directed toward them. Theaetetus' answer is that, as regards the beautiful and the ugly, the good and the bad, the soul examines (*skopeisthai*) their being (*ousia*) in relation to one another, while calculating (*analogizesthai*), in itself, past and present in comparison with the future. He thus suggests that the soul in fact performs two distinct, if also closely related, activities when it comes to these matters. It *examines* their being in relation to each other and, while doing so, it *calculates* "past and present in comparison with future".

The activity of calculating, David Sedley argues, to my mind persuasively, "suggests a reference to the work of the expert", namely the work that was up for discussion in the 177c6–179b9 passage where Protagoras was refuted, which in turn suggests that Theaetetus now "seems to have bestowed on [being] a richer profile" in comparison to that found in the discussion running up to 186a2,[44] according to which being means something like the nature or essence of the subject matters an expert is concerned with. At the very least, we may rule out the suggestion that he has a general concept of being, underlying all kinds of judgment, in mind here. For whatever he means precisely by "calculation", it is aimed at providing knowledge of the being of four specific terms, the beautiful and the ugly, the good and the bad, and the fact that Theaetetus states that the soul, when trying to come to grips with their being, compares them with each other, clearly indicates that the being of one of these, say the good, is different from the being of the others.

To highlight the importance of Theaetetus' answer, Socrates then proceeds to point out, in T2, that it is through touch that we perceive the hardness of what is hard and the softness of what is soft (186b2–4). However, as regards hardness and softness,[45] when it comes

i. to their being (*ousia*),
ii. to the fact that they are (*eston*),[46]
iii. to their being opposite to each other (*ten enantiotēta pros allēlō*), and
iv. to the being of oppositeness (*tēn ousian au tēs enantiotētos*),

the soul itself tries (*peira[i]n*) to judge (*krinein*) them, going over (*epaneinai*) and comparing (*symballein*) them to each other (186b6–9). While the finer points of Soc-

---

44 Sedley (2004), p. 110.
45 For this reading, see Campbell (1881), p. 143, n. 11.
46 See n. 35 above.

rates' distinction between perceiving hardness and softness and trying to judge their being may seem less clear than we might wish them to be, his emphasis on *ousia* in both i and iv seems to me to point in the same general direction as Theaetetus' earlier claim, bestowing upon being, to use the words of Sedley, a somewhat richer profile in comparison with the preceding parts of the argument.

To see this point more clearly, let us note that i, ii, iii, and iv are all described as matters that the soul attempts to judge on its own, that is, without having recourse to perception. Neither the being of hardness and softness, the fact that they are, that they are opposite, or the being of oppositeness is something we may perceive. At the same time, the fact that Socrates mentions *being* (*ousia*) twice, first referring to the being of hardness and softness and then to the being of oppositeness, indicates the importance of being for intellectual inquiry. We may judge that two things, such as hardness and softness, are opposite, just as we may judge that they are two. But we may also ask ourselves what the being of hardness, softness, and oppositeness are.[47]

What Socrates is pointing to in T2 is then, I suggest, that it is possible to note *that* two perceptible things are and also *that* they are opposite to each other, but that we may also, once we notice this, think that there is a further question, namely *what* hardness and softness are, and *what* the being of oppositeness is. To notice *that* they are does not necessarily require much effort, while answering the question *what* they are, I take it, will require real effort, effort that may be described as a kind of calculation. For, in deciding what they are, one would be attempting to answer *ti esti* questions, while in noticing that they are, one would not, even if noticing that they are may be the natural starting point for attempting to answer the *ti esti* questions.

This emphasis on the soul's attempt to decide what the being of certain things are, and not just that they are, that they may be opposite to each other etc., is followed up, finally, in the suggestion Socrates makes in T3, as a conclusion to the argument set out in T1 and T2.

What is particularly striking about T3 is that Socrates there singles out calculations about *being* and *benefit*. This is striking because Theaetetus has already conceded that the soul examines, itself by itself, quite a number of things, such as being, likeness, unlikeness, identity, otherness, the beautiful, the ugly, the good, and the bad (186a2–b1). Had Socrates' point been to emphasize the difference be-

---

[47] Here, it may be argued, we find a clear parallel to the 523–525 passage from the *Republic* mentioned above in n. 6 that according to Vasilis Politis helps explain Parmenides' claim that the power of conversation will be destroyed if we do not posit at least certain forms. Exploring the connection between the *Republic* and the *Theaetetus* further in this regard lies, however, beyond the scope of the present chapter.

tween mere perception, deprived of conceptual content, and judgements that rely on general concepts, as many critics following Gilbert Ryle have suggested, the reference to benefit would make little sense, as would the claim that the calculations about being and benefit only come with difficulty, at length, and through hard work and education.[48] If the main point Socrates wishes to make concerns the difficulty of understanding the being of something in the sense of its *nature* or *essence*, however, and what is good or beneficial, it makes perfect sense to suggest that calculations about being and benefit require serious effort. For one of the hallmarks of the expert is, as I have argued became clear in the previous part of the *Theaetetus*, to be able to judge correctly what will prove beneficial, and this ability depends, it may be suggested, on knowing what the object the expert is concerned with is, essentially, and what its proper good is.

If this reading is along the right lines, we may conclude that the full significance of Theaetetus' concession about being at 185d6–e2, that there are certain "things" common to what we perceive, including being, that the soul examines itself by itself, does not become clear before T3. To put this point differently, while *ousia* is already introduced in the discussion at 185c8 and at first seems to signify the mere fact that something we perceive is (or perhaps is *something*), the later discussion makes clear that recognizing that something is (or is something) is not the end point of that activity of the soul that we may call thinking, but is rather the starting point. Once we recognize *that* something is, we may in some cases also need to proceed to ask *what* it is and try to settle that question, at least if we are inclined to seek knowledge and believe that the question what that particular something is stands in need of further inquiry in order to be settled properly.

4 Knowledge, *ti esti* Questions, and Dialectic

Having consolidated the conclusion that there are a number of "properties" or "things", including being and the good or beneficial, that we cannot perceive but can only reach toward through thinking, Socrates finally asks Theaetetus (at 186c7) whether it is possible to reach, or "hit upon" (*tygchanein*), truth (*alētheia*) if one does not reach being. Theaetetus agrees that it is not, and Socrates asks how someone who has not hit upon the truth about something could have knowledge of it (*pote toutou epistēmōn estai;* 186c9–10). Again, Theaetetus concedes that this is not possible. The conclusion of the entire inquiry that started when Theae-

---

[48] The reader may compare Michael Frede's suggestion about the latter point (Frede 1987, pp. 7–8) with that made by David Sedley (2004, p. 110).

tetus first suggested that knowledge might be perception is, therefore, that knowledge is to be found, not in our experiences (*en tois pathēmasin*), but only in our reasoning about these (*en tō[i] peri ekeinōn syllogismō[i]*), since it is possible to grasp (*hapsasthai*) being and truth only in reasoning, not in our experiences (186d2–6). In addition to wrapping up the whole discussion of Theaetetus' initial thesis, these final steps of the refutation also add two important points: i) truth depends upon hitting being, and ii) knowledge depends on truth.

Now, this connection between knowledge and being was already hinted at at the beginning of the inquiry, at 152c5–6, where Socrates suggested that perception, on the interpretation he was suggesting at that point, would be unerring (*apseudes*) and always about being, just as we are inclined to think that knowledge will be. As I have argued in the previous two sections, however, the final refutation passage provides only a gradual clarification of being, by picking up certain elements from the previous discussion of Protagoras' measure doctrine, that imbues being with a richer or ontologically more loaded sense than it had at the beginning of the inquiry. If this argument is along the right lines, we are entitled to infer that the idea that knowledge depends on being mentioned already at the beginning of the inquiry is also given a more precise meaning in the final refutation passage. Further, this passage also gives the connection between being and truth a more precise meaning. At 152c5–6, Socrates suggested that knowledge is unerring and about being but did not offer any explanation for these claims, while the final steps in the final refutation passage does offer an argument for the claim that knowledge depends on being, namely, that we cannot be said to know something unless we reach or hit upon truth about this something, and that we cannot obtain such truth if we do not hit upon, or grasp, the being of that something we wish to know.

Without pressing the point too far, I would like to suggest that truth, in the concluding steps of the final refutation passages, means, more or less, what at least the earlier Heidegger claimed truth originally meant to the Greeks, namely un-concealment.[49] Put differently, Socrates' argument, as I understand it, is that knowledge is the result of a specific activity, called thinking or calculation, when that activity succeeds in grasping the being of something in the sense of revealing it to us as it is. Truth is not so much a function of our judgements as it is a function of being in the sense that we only obtain truth if our thinking is of such a quality that it manages to uncover or reveal the being of what we inquire into or think about. If this is correct, the point Socrates is making is not that we only have knowledge if our judge-

---

[49] See Heidegger (1992), pp. 15–17, and Heidegger (1988), pp. 10–19. On this point, see also Rowett (2018), pp. 43–45.

ments correspond to matters-of-fact, but that we only have knowledge if we grasp the being, that is, if we are able to answer the *ti esti* question, of the subject matter we claim to know.

But even granted that Plato is making these points about truth, being, and knowledge that I argue he is, the reader may still wonder what all of this has to say about dialectic. For, it may be objected, what I have established is at best a general point about knowledge, not a specific point about philosophy or dialectic. To address this worry, let me conclude this chapter with some observations about Parmenides' claim about the power of conversation at 135b5–c6 in the *Parmenides* and its connection with the *Theaetetus*.

At the beginning of the chapter, I argued that, even if it may be slightly unclear what Parmenides means by conversation (*dialegesthai*) in the 135b5–c6 passage, the expression "the power of conversation" has a quasi-technical sense and means something like "the power inherent in conversation, engaged in for the purpose of achieving knowledge". The claim he makes is that this power will be destroyed if we deny that there are forms, and that this in turn will prove fatal for philosophy. Should we, then, follow those critics who translate the expression *tēn tou dialegesthai dynamin* as "the power of dialectic"[50] and conclude that the point of the passage from the *Parmenides* is to indicate what is required, ontologically speaking, if philosophy and dialectic is to be possible? I wish to resist this suggestion.

First, when Socrates introduces forms at the beginning of the *Parmenides*, he does not do so solely for the purpose of philosophy. He does so primarily in order to avoid certain consequences that follow from Zeno's paradoxes that would prove fatal for our ability to speak in a reasonable manner about the world we perceive. Positing forms, the young Socrates seems to suggest, is necessary if any knowledge of this world is to be possible.

Second, the fact that Parmenides states that it is the power of conversation, of *dialegesthai*, not the science of conversation, i.e., *dialektikē*, that will be destroyed if one does not allow that there are forms, suggests that he is making a much more sweeping claim than he would by claiming that philosophy and dialectic will become impossible. I would like to suggest that Parmenides is indicating that our ability to engage in conversation has a specific power inherent in it, namely a power to obtain knowledge, that may be unfolded more or less successfully, and may be used to obtain knowledge of different things, and that this power, however it is unfolded, depends on our *dianoia* or thought having forms toward which it may be directed. It is not just philosophy, but knowledge generally, that stands

---

50 See, for instance, Gill (2012), p. 18, n. 1.

in need of forms—even if it is philosophy, it may be argued, that will help make this clear to us, when we inquire into knowledge and being, and even if philosophy, or dialectic, may be regarded as the highest unfolding of this power. Indeed, the word Plato uses for the science peculiar to philosophy, *dialektikē*, can be translated as the science of conversation and it may be argued that Plato when using this term means to suggest that the philosopher, as dialectician, is the one who understands fully what the power of conversation is because he or she does not merely rely on it, as do other people who possess knowledge, but also inquires into and seeks to master it in full.

Is such a conception of conversation and the power inherent in it reflected also in the *Theaetetus*? The answer, I think, is yes. At 161d7–162a2, after having developed Theaetetus' thesis that knowledge is perception by connecting it to the Protagorean measure doctrine and the Heraclitean flux-ontology, Socrates makes the following statement about Protagoras' doctrine, on the supposition that it is true: "How can anyone be justified in supposing him qualified to teach others, for large fees, and the rest of us more ignorant than him, so that we need to go to him to be taught? ... I say nothing about my side of things, and how ridiculous it makes *me*, if this theory is correct—me and my art of midwifery, presumably along with the whole business of conversation (*sympasa hē tou dialegesthai pragmateia*). Examining the things that appear to and are believed by one another, and trying to refute them, when each person's appearances and beliefs are correct—isn't this just an inordinately drawn-out piece of tomfoolery, if Protagoras' 'truth' is true ... ?".

The point Socrates makes here is, then, that, if Protagoras' teaching is true, a teaching Socrates claims is grounded in the Heraclitean doctrine that denies that anything is what it is, itself by itself, the whole business of conversation is rendered pointless. There can be little doubt that Socrates means to include dialectic as part of that business—the point of dialectical inquiry being, among other things, to test what each of us believes while aiming for knowledge about the subject matter that is inquired into. But it seems clear to me that he also means to include types of expert knowledge more generally under this heading. For the development of any kind of expert knowledge, it may be argued, depends also on this kind of truth-seeking conversation where false opinions are substituted with correct ones, and it is this fact, and the fact that this is only possible if one does not accept Protagoras' doctrine and the ontology that underpins it, that Socrates seeks to make Theodorus and Theaetetus realize in the *Theaetetus*. Thinking (*to dianoeisthai*) is, as Socrates later suggests to Theaetetus, nothing other than conversing (*dialegesthai*) with oneself, asking oneself questions and answering them, affirming one thing and denying another (189e4–190a2). Such thinking is not the privilege of philosophers, I submit, but common to everyone who aims to achieve knowledge about anything. It is the kind of thinking that may be carried out in conjunction with other people, in dialogue, of which Socrates is a profound defender and on which our scientific community depends.

The reader may still worry whether I end up suggesting that the *Theaetetus* has nothing to say about philosophy and dialectic as a specifically philosophical kind of expertise, but only about the activity of thinking or conversing that philosophy, as well as other kinds of expertise, depend on. I mean to suggest no such thing. During his conversation with Theodorus, Socrates at one point (at 175b7–175d2) explains that what characterizes philosophers, in contrast to politically inclined people, is their relentless inquiry into matters such as justice and injustice, kingship, and human happiness, for the purpose of deciding what they are. These are the matters that Socrates elsewhere describes as the greatest matters (see *Apo* 22d7–8; see also *Soph.* 230e1–4), of which knowledge is required if we are to live flourishing lives, and that Plato depicts Socrates as interested in already as a young man in the *Parmenides* where he insists that we need to posit forms for them (see 130a7–9). When Theaetetus, in the passage I have analyzed in this chapter, comes to see that benefit or the good is on a par with being and that both being and benefit are stable objects of inquiry that it requires real effort to acquire knowledge about, he is, I submit, on his way to becoming a real dialectician. For the inquiry into such matters, and the attempt to reach clarity about them, is what dialectic is truly about according to Plato. In this regard, at least, I doubt that Plato ever changed his conception of dialectic.

Acknowledgements

This chapter benefited from being presented in an early version at the conference "Ancient Greek Dialectic and its Reception" at the University of Patras. I thank Melina Mouzala for inviting me to participate in the conference. I also thank Catherine Rowett and Justin Vlasits for helpful comments on and fruitful criticism of a later version of the chapter. Special thanks are due to Peter Larsen and Vivil Valvik Haraldsen, who read and commented on several versions.

Bibliography

Allen, Reginald E. (1997): *Plato's Parmenides: Revised Edition.* New Haven: Yale University Press.
Burnyeat, Myles (1976): "Plato on the Grammar of Perceiving". In: *The Classical Quarterly* 26. No. 1, pp. 29–51.
Burnyeat, Myles (1990): *The Theaetetus of Plato, with a translation of Plato's Theaetetus by M. J. Levett, revised by Myles Burnyeat.* Indianapolis: Hackett Publishing Company.
Campbell, Lewis (1881): *The Theaetetus of Plato.* Oxford: Oxford University Press.

Cooper, John M. (1970): "Plato on Sense-Perception and Knowledge (*Theaetetus* 184–186)". In: *Phronesis* 15. No. 2, pp. 123–146.
Cornford, Francis MacDonald (1935): *Plato's Theory of Knowledge*. London: Kegan Paul.
Frede, Michael (1987): "Observations on Perception in Plato's Later Dialogues". In: Frede, Michael: *Essays in Ancient Philosophy*. Oxford: Oxford University Press, pp. 3–8.
Gadamer, Hans-Georg (1982): "Mathematik und Dialektik bei Plato". In: Gadamer, Hans-Georg: *Gesammelte Werke—Band 7: Griechische Philosophie III, Plato im Dialog*. Tübingen: Mohr Siebeck, pp. 290–312.
Gill, Mary Louise (2012): *Philosophos*. Oxford: Oxford University Press.
Heidegger, Martin (1988): *Gesamtausgabe II. Abteilung: Vorlesungen 1923–1944, Band 34: Vom Wesen der Wahrheit: Zu Platons Höhlengleichnes und Theätet*. Frankfurt am Main: Vittorio Klosterman.
Heidegger, Martin (1992): *Gesamtausgabe II. Abteilung: Vorlesungen 1919–1944, Band 19: Platon: Sophistes*. Frankfurt am Main: Vittorio Klosterman.
Howland, Jacob (1998): *The Paradox of Political Philosophy*. Lanham: Rowman & Littlefield.
Kahn, Charles H. (2013): *Plato and the Post-Socratic Dialogue*. Cambridge: Cambridge University Press.
Larsen, Jens Kristian (2015): "The Virtue of Power: The Gigantomachia in Plato's *Sophist* 245e6–249d5 Revisited". In: *New Yearbook for Phenomenology and Phenomenological Philosophy* 13, pp. 306–317.
Larsen, Jens Kristian (2019): "Measuring Humans Against Gods: On the Digression of Plato's *Theaetetus*". In: *Archiv für Geschichte der Philosophie* 101. No. 1, pp. 1–29.
Larsen, Jens Kristian (2022): "On common forms and dialectical inquiry in Plato' *Parmenides*". In: Brisson, Luc, Macé, Arnaud, and Renaut, Olivier (Eds.): *Plato's Parmenides. Selected Papers of the Twelfth Symposium Platonicum*. Baden-Baden: Academia Verlag, pp. 183–191.
McDowell, John (1976): *Plato: Theaetetus. Clarendon Plato Series*. Oxford: Oxford University Press.
Politis, Vasilis (2021): *Plato's Essentialism: Reinterpreting the Theory of Forms*. Cambridge: Cambridge University Press.
Robinson, Richard (1941): *Plato's Earlier Dialectic*. 1st Edition. Ithaca: Cornell University Press.
Robinson, Richard (1950): "Forms and Error in Plato's *Theaetetus*". In: *The Philosophical Review* 59. No. 1, pp. 3–30.
Robinson, Richard (1953): *Plato's Earlier Dialectic*. 2nd Edition. Oxford: Oxford University Press.
Rowe, Christopher (2015): *Plato: Theaetetus and Sophist*. Cambridge: Cambridge University Press.
Rowett, Catherine (2018): *Knowledge and Truth in Plato: Stepping Past the Shadow of Socrates*. Oxford: Oxford University Press.
Ryle, Gilbert (1939): "Plato's *Parmenides* (II.)". In: *Mind* 48. No. 191, pp. 302–325.
Sedley, David (2004): *The Midwife of Platonism: Text and Subtext in Plato's* Theaetetus. Oxford: Oxford University Press.
Stenzel, Julius (1917): *Studien zur Entwicklung der platonischen Dialektik von Sokrates zu Aristoteles*. Breslau: Trewendt & Granier.
Vlastos, Gregory (1954): "The Third Man Argument in the *Parmenides*". In: *Philosophical Review* 63, pp. 319–349.
Zekl, Hans Günther (1965): *Platon: Parmenides*. Hamburg: Felix Meiner.

Melina G. Mouzala
Pursuing Self-Knowledge in Plato's *Sophist*. The Communion of the Sophistic and Socratic Dialectic in the Sixth Definition of the Sophist: A Reading Based on Proclus' Interpretation of Dialectic in the *Sophist*

1 The Problem of Coherence between the Sixth and the Previous Definitions

Heidegger, in his analysis of Plato's *Sophist*,[1] points out that "the sixth definition of the sophist always struck commentators as a consideration lying outside the framework of the previous definitions" mainly because "they were at a loss to see how this definition could be brought into the framework of the dichotomies".[2] He also believes that "if one understands the preparatory definitions to be connected through Plato's supposed concern with building a conceptual pyramid, then indeed it will be difficult to fit the sixth definition among the others".[3] If we were ready to adopt this view in an unexamined way, the consequence would be to take for granted a gap or a discontinuity between the first five definitions and the sixth. The aim of this paper is to draw a line of interpretation that claims that when the fifth definition places the emphasis on *logos* (λόγος), it paves the way to the sixth; *logos* is a human characteristic which brings to the fore and realizes the manifestation of all thinking and specifically of controversies and disputations in which our thought is involved and expressed. The same subject is reserved and developed in the sixth definition. Moreover, my aim is to show that the sixth definition, apart from the explicit discussion of purification or cathartic dialectic, actually thematizes division itself. Based on Proclus' interpretation of eristic in the *Sophist*, I will show that the notion of communion (κοινωνία) is implicitly examined for the first time in the dialogue within the sixth definition of the sophist, where the Sophistic and the Socratic dialectic are commingled. Finally, I maintain that from the analysis of the crucial passage 230 b–d, we can infer

---
1 Throughout this paper, I follow the translation by White (1993) as a general outline.
2 Heidegger (1997), p. 245. Cf. Narcy (2013), p. 70.
3 Heidegger (1997), p. 245.

that the basic characteristic of Socrates' cathartic method is a specific emotional attitude of the person who is subjected to elenchus, which due to its reflexive and self-referent character, leads to self-knowledge. This kind of self-knowledge is a way of self-recovery or self-recollection that also proves to have a collective or non-individual character, since the same emotional attitude, in cooperation with the cohesive and therapeutic intervention of the unificatory logos, binds again the person who is subjected to elenchus with the latent commonality of an intersubjective wisdom that has been forgotten.

Heidegger aptly remarks that we must not ignore the inner concatenation or the interconnection of the definitions till the sixth definition, because the individual definitions bring to the fore definite objective characteristics graspable in the sophist as he ultimately appears to us.[4] Notomi notes that these definitions represent particular features or aspects of the sophist's art, rather than individual figures of historical sophists.[5] I agree that it is necessary to assume a supposed universality within these particular features of the sophist's art and activity, since we speak of definitions. However, in my view, exactly because these features or aspects are presented as definitions, one must also suppose that they are grasped through inductive reasoning (ἐπαγωγή), so initially they are observed in the behavior of individual figures including the historical sophists. Cornford[6] notes that some have held that all the divisions define one class of historical persons from different approaches, and even that all the definitions are "adequate". However, he stresses that a fatal objection to this view is that there never existed any class of persons who could be characterized by the sixth definition as well as by the first five and the seventh, adding that the cathartic art of the sixth division was practiced by Socrates alone. Robinson,[7] when referring to the final definition of the sophist, notes that its weakness lies in what are chosen as the distinguishing features of the definition. The problem which Robinson raises regarding the last definition, and which we can recognize in the full range of negative descriptions of the sophist in this dialogue, is that Plato directs the "simple accusation of mauv-

---

4 Heidegger (1997), p. 245.
5 Notomi (1999), p. 78, n. 14.
6 Cornford (1935), p. 173. We know that the Socratic Dialectic evolves and moves itself through stages which are identified as successive—though not necessarily successful—attempts of definitory statements about the requested notion. Each stage gives a definition which is subjected to elenchus, but whatever the result of this elenchus is, the attempt is not eliminated or nullified, because something always remains which is used as a basis for further attempts. What is more, each new definition is focused on the remedy of the deficiency of the previous definition, so the interlocutors discard nothing from the previous discussion.
7 Robinson (2013), pp. 11–12.

aise foi" against all sophists. To paraphrase Robinson, to claim that each definition represents an *essential* feature of *all* sophists presupposes that the Visitor from Elea, the *dramatis personae* who is charged by Plato to undertake the task of leading this dialectical inquiry, has landed himself in the position of those who have failed to see that the supposed universality of any general proposition must be invariably the result of an inductive process, and hence, in the final analysis, is illusory. This is still true, even if as Robinson notes, we must take the phrase "all sophists" here to be understood as "all sophists except those of noble lineage".[8]

Given that all the previous five definitions of the sophist are negative descriptions and follow the framework of the dichotomies, the emergence of the sixth definition, which constitutes an evidently positive description of the sophist and *prima facie* violates the sequence of the initial dichotomies, raises the problem of its coherence with the previous definitions. We must remember that the very first dichotomy that the Visitor set out in 219 a8–d2, was the division of the totality of the arts into two categories, the art of production and the art of acquisition. All the first five definitions stem from the same original central division and, more specifically, from the branch of the art of acquisition, although they are reduced to different types or kinds of acquisition. We must remember that the whole type of arts placed under the heading of acquisition (τέχνη τις κτητική), has to do with learning (μαθηματικόν), acquiring knowledge (τὸ τῆς γνωρίσεως), commerce (χρηματιστικόν), combat (ἀγωνιστικόν), and hunting (θηρευτικόν), as the Visitor states in 219 c2–7. In 219d 4–e2, the art of acquisition is divided in two types of expertise in acquisition; the one type is mutually willing exchange (ἑκόντων πρὸς ἑκόντας μεταβλητικόν), through gifts and wages and purchase; the other type, which brings things into one's possessions by actions or words, is expertise in taking possession (χειρωτικόν). Furthermore, according to the Visitor's suggestion in 219d 9–e2, the possession-taking should be divided in two parts; that done openly, which we label *combat* (ἀγωνιστικόν), and that done secretly, which we refer to as *hunting* (θηρευτικόν).

So, the art of acquisition is the thread that binds the first five definitions. But we must note that the first definition focuses on hunting, and in turn goes back to possession-taking (χειροῦσθαι); whereas the second, third and fourth, which stem from the same line of the initial division (production-acquisition), go back to exchanging, i.e., that part of acquisition which is in contradistinction to possession-taking.[9] The second, third and fourth definitions have in common with the

---

[8] Robinson (2013), p. 12.
[9] The mutually willing exchange or exchanging part of the acquisitive art is called both ἀλλακτικόν (in the second definition, 223c 7) and μεταβλητικόν (in the original division of acquisition and

first definition, where the division applies to the case of the angler, only the fact that they still focus on someone who has an acquisitive art. But they differ from it as far as they are oriented towards exchanging instead of hunting, namely they abandon the possession-taking part of the acquisitive art. But as Heidegger notes, the fifth definition goes back to χειροῦσθαι (possession-taking) and one might think that "this definition claims the last remaining structural moment out of the framework which determines the angler and so exhausts the pregiven frame".[10] Narcy also remarks that now that the totality of the arts of acquisition has been scrutinized, one might think that the same concern for exhaustiveness demands that we look for the sophist among the arts of production. But, in his view, this expectation is dashed, since the art of discriminating is an art neither of production nor of acquisition; it is an unclassifiable art, which finds no place within the initial dichotomized division of the arts into production and acquisition, which was supposed to be exhaustive.[11] Dorter also believes that all definitions but the sixth begin with the angler definition's division of power into art and non-art, and art into production and acquisition. According to this view, in the first five definitions which all locate the sophist within the genus of acquisition rather than production, the motivation of a sophist, i.e., "what sophists value", is the starting point of the inquiry in each case. But after the establishment by the Visitor of a value-free stipulation in the sixth definition (227 a–b), all question of motivation disappears and the seventh definition begins not from the art of acquisition like the others, but from the art of production.[12]

As I noted, Heidegger aptly remarks that the fifth definition goes back to possession-taking (χειρωτικόν), which in turn goes back to acquisition, and up to this point stems from the same line of division from which the first definition also originates. But from that point onwards, these two definitions follow a different line of division because the fifth definition goes back to that part of possession-taking which is in contradistinction to hunting, i.e., combat, which means possession-taking that is done openly. In my view, acquisition continues to be discussed in the

---

in the second, third, and fourth definitions; 219d5, 224c 10; e1). In 223c 6–7, when beginning to sketch the division which leads to the second definition, the Visitor reminds us that expertise in acquisition has two parts: hunting and exchanging. We must suppose that this division is the same as that referred to in 219d 4–7; cf. White (1993), p. 9, n. 7. We note that in 223c 6–7 exchanging (ἀλλακτικόν) is presented in contradistinction to hunting. So, *hunting* comprises here all the possession-taking part of the art of acquisition, whereas that which we label *combat* is not mentioned.
10 Heidegger (1997), p. 245. See also the scheme which illustrates the genealogy of the first five definitions in Heidegger (1997), p. 210.
11 Narcy (2013), pp. 67–68.
12 Dorter (2013), p. 91.

sixth definition and the link that associates the fifth with the sixth definition is *logos*, i.e., that part of possession-taking art that is done with words, namely openly; this is the part named ἀγωνιστικόν. This bond between the two definitions may take the form of antilogic or even of eristic, but the background of any form of *logos* here is ἀγών, because the predominant philosophical subject is ἀγών in relation with *logos*. Heidegger notes that the sixth definition has a more positive descriptive character, since it immediately prepares the way for the seventh, where the positive consideration begins, but he stresses that, of course, it is not an arbitrary introduction of a new point of view, since what is taken up is the phenomenon of ἀντιλέγειν, dealt with in the fifth definition and itself already encompassing the earlier ones.[13]

2 Agōn, Antilogic, Eristic: The Genealogy of Dialectic

It is noteworthy that the fifth definition begins from the reference to that part of the possession-taking art which we label *combat*, i.e., ἀγωνιστικόν. This is a derivative originating from the word ἀγών, which is a foundation stone and a pillar notion of ancient Greek culture.[14] According to Burckhardt's analysis, while on the one hand the πόλις was the driving force in the rise and development of the individual, given the essential condition of freedom, the ἀγών was a motive power which proved capable of working upon the will and the potentialities of every individual and provoked a dawning of the competitive spirit.[15] Freedom also was the common, necessary, and essential condition for the development of dialectic and philosophy itself.[16] Burckhardt explains that since personal competitiveness was

---

[13] Heidegger (1997), pp. 245–246. Narcy (2007, p. 201) also believes that the fifth definition appears as an appropriate prelude to the sixth definition.
[14] Szlezák (2010), p. 65. Burckhardt (1999, pp. 160 and 163–165) describes the phase from the end of the Dorian migration almost to the end of the 6th century BC as the period of ἀγών or the agonal age of the history of the Greek civilization. All the various manifestations of ἀγών were part of the world depicted by Homer and we can see examples of the motif "always to be the first and outdo all the others" both in the *Odyssey* and the *Iliad*.
[15] Burckhardt (1999, p. 162) explains that this agonistical attitude was unknown to other people. Even in primitive or barbarian people, men had the chance to develop competitive activities independently of warfare, but always within the given tribe or social stratum. Whereas in the Asiatic cultures, despotism and the caste system were almost completely opposed to such activities, by custom every Greek could participate openly in certain athletic competitions.
[16] Berti (1978), pp. 347–352.

no longer to be satisfied by training directed simply towards military efficiency, the consequence of the presence of the *agonistic* spirit in the education of this society had two aspects, one linked with religion and the other with gymnastics.[17] Henceforth, many different aspects of mental activities, including conversations in philosophy and in legal procedure, bore this character of competitiveness, while ἀγών became the predominant feature of socio-political as well as cultural life.[18] The word occurs in various contexts in Plato.[19] According to Coulson, in the context of ancient Athens' agonal ethos, Plato opposed conventional Athenian ἀγών. His philosophy *"per se* was both extrinsically and intrinsically agonistic" since "it publicly competed with the eristic of sophists and others, yet taught a private way to usefully struggle for a good life of inner harmony".[20]

Heidegger stresses that for the Greeks, ἀγών properly means "contest", "competition". So, in his opinion, when the Visitor introduces ἀγωνιστική, the only part from the pregiven frame of the acquisitive and possession-taking part which remained untouched till the fifth definition, he means appropriation by means of battle, but the original determination of this battle is ἀμιλλᾶσθαι. While ἀμιλλᾶσθαι means competing with the other over something offered to both and not fighting against the other, μάχεσθαι means confrontation not *with* the other but *against* the other.[21] I believe the first feature of combat that Plato wants to highlight is that it is done openly; this feature is appropriate to and common amongst on the one hand ἀγών and on the other λόγος and λέγειν, which in this case manifest as taking-possession by words. It is also remarkable that Plato wishes to set as the starting point of this division, a broader meaning of ἀγών than that which Heidegger puts forward as its proper meaning, since the Visitor proposes to further divide combat into two parts: the first is competition (ἀμιλλητικόν) and the second is fighting (μαχητικόν; 225a). Ἀγών is transformed into μάχη (fighting) and this in turn into ἀμφισβήτησις (controversy). In 225a8–b1, the Visitor states that whereas it would be fitting and proper to give a name like violence (βιαστικόν) to this part of fighting in which one body fights another, for that in which words are pitted against words, it would be proper to call it controversy (ἀμφισβητητικόν). So, it is conspicuous that controversy is characterized by a quality of

---

17 Burckhardt (1999), p. 161.
18 Burckhardt (1999), p. 166.
19 See, for example, *Cratylus* 421d, *Alcibiades I* 119c–124a, where the word and its compound derivatives are bound with care of self and self-knowledge, *Laws* 751d, and with the meaning of wrestling of discourses, *Protagoras* 335a.
20 Coulson (2020), pp. 211–212.
21 Heidegger (1997), pp. 210–211.

fighting which is detached from violence.²² Since controversy has to be divided into two types, in the next division one must focus on the part which occurs in private discussions and is broken down into questions and answers, which we usually call disputation (ἀντιλογικόν). It must be noted that in this type of controversy the discussions are private, in contrast to the forensic type, consisting of long public speeches directed against other speeches of the same type, which deal with just and unjust matters. The emphasis here seems to be on the private nature of the discussions, a feature which removes from them the rhetoric, demonstrative or compulsory character of a public or forensic speech. But still the common ground of λόγος, namely λέγειν, ἀντιλέγειν, which underlies all the divisions of disputation, enables ἀντιλογικόν to retain the openness of the agōnistikon part of possession-taking art. The character of privacy in the discussions referred to here invokes an atmosphere of freedom from which the participants derive a great deal of benefit. Regarding this issue, we should recall the *Theaetetus* 172c–173e, where Socrates thematizes the comparison and opposition between those who have knocked about in courts and the like from their youth up and those who have been brought up in philosophy and similar pursuits. The first group seem to have been trained as slaves, the others as free men.²³ The free man always has time at his disposal to converse in peace at his leisure; he does not care how long or short the discussion may be, if only it attains the truth. The orator is always talking against time, hurried on by the clock and is not permitted to enlarge upon any subject he chooses. The adversary stands over him and is ready to read the sworn statement (διωμοσία or ἀντωμοσία) from which no deviation is allowed. Moreover, in the *Theaetetus* 173c–d, Socrates states that the leaders in philosophy from their youth up have never known the way to the ἀγορά and do not even know where the law-court

---

22 However, it is true that some titles of Protagoras' works include words borrowed from the vocabulary of wrestling. For example, let us recall Protagoras' notorious work "Truth or Refutations" (trans. O' Brien 1972); Sextus Empiricus, *Adversus Mathematicos* VII. 60. The same title can also be rendered as "Truth or Down-Throwers" or "Truth or the discourses that defeat", since the original Greek title is Ἀλήθεια ἢ Καταβάλλοντες (λόγοι), meaning that the arguments one uses within this dialectical frame aim to defeat and knock down the opponent. Also, Diogenes Laertius (*Vitae philosophorum* IX 55) testifies among others the title "On wrestling", where wrestling is used as a metaphor for intellectual debate; cf. Plato, *Sophist* 232d9–e1. Cornford (1935, pp. 175–176) notes that first, fighting is distinguished from friendly competition and then fighting in the form of verbal disputation from the violence of physical warfare. But Reid (2010, pp. 157–158) notes that even in wrestling the opponent's cooperation is required to make the activity possible, and "it is precisely my opponent's resistance that builds my own strength". He stresses that even if the objective is to throw or pin my competitor to the ground, the implementation of this activity depends on my ability, which is grounded on my pre-achieved stability and balance.
23 I partly follow the translation by Cornford (1935) and partly that by Fowler (1921).

is, or the senate-house, or any other place of public assembly. So, given that in the *Sophist* and the *Theaetetus* both ἀντιλογικόν and the conversations of the philosophers are at odds with the forensic or public type of speech, we can assume that ἀντιλογικόν already bears a character which is comparable with or akin to philosophy.

The term ἀντιλογικόν seems to allude to Protagoras, since he is the author of the notorious work in two books entitled Ἀντιλογίαι, i.e., "Contradictory Arguments". The title of Protagoras' work indisputably introduces a method which originates from it and as testified by Diogenes Laertius, "Protagoras was the first to say that on every issue there are two arguments opposed to each other; these he made use of in arguing by the method of questioning, a practice he originated".[24] However, the point of this part of the division is not the kind of ἀντιλογία introduced by Protagoras, but the method of ἀντιλέγειν implemented as κατακεκερματισμένον (chopped up) into questions and answers. The contradistinction between long discourse and this kind of dialectic which consists of questions and brief answers, is usual in several Platonic dialogues, including the *Sophist* (217c–218a).[25] Also, in *Protagoras*, it is clarified that Protagoras is able to speak at length and nobly too, and able as well to respond briefly to questions, or when he is a questioner, to wait and listen to the answer, and these are rare gifts (329b; cf. 334dff.). So, regarding this point, it would be reasonable to suggest that Protagoras follows a method which *prima facie* resembles the Socratic method.

While waiting to see what the *differentia specifica* of the sophist will be at the end of the fifth definition, we learn that disputation has two parts. The first involves controversy about contracts and is not carried on in any systematic or expert way, whereas the other part is done expertly and involves controversy about general issues, including what is just itself and what is unjust itself. The former is not named but the latter we normally call 'debating' (ἐριστικόν), says the Visitor. Cornford emphasizes the opposition between the random and artless disputes of the ordinary life, including the contracts, and the disputation conducted by the

---

[24] Diogenes Laertius IX 51=DK 80b 6a (trans. O' Brien). See also Nestle (1940), p. 289. Euripides' statement in Fragment 189 from the *Antiope*, that the man clever at speaking may make contradictory arguments out of every subject may be due to Protagoras; see Bates (1930), pp. 8–9. The impression that Protagoras is a central person from the cycle of sophists to whom Plato makes an allusion here is not arbitrary, because after the recapitulation of the definitions, in 232b–e, Theaetetus refers to Protagoras, by repeating specific terms that are used within the fifth definition, like ἀντιλογικούς (232b12), ἀμφισβητητικούς (232d2), ἀντιλογικῆς τέχνης (232e3), ἀμφισβήτησιν (232e4).
[25] Nehamas (1990, p. 6) stresses that Euthydemus and Dionysodorus are presented as practitioners and teachers of an art of question and answer, referred to by Socrates as ἐριστικὴ σοφία, which shows that this method was not an exclusive appropriation of Socrates; see Plato, *Euthydemus* 272d.

rules of art, which we call "eristic".²⁶ However, Narcy believes that the reference to controversy about what is just itself and what is unjust itself in the *Sophist* evokes the opposition raised in 175c 1–3 of the *Theaetetus*, between the man only interested in the question "what injustice have I done to you or you to me", and the man who searches for "justice and injustice in themselves". Narcy assumes that the Visitor from Elea actually thematizes the difference stated by Socrates in the *Theaetetus* 175e between the man trained in forensic oratory and the man trained in philosophy.²⁷

The final characteristic of the sophist that the fifth definition attributes to him, is that his type is precisely the money-making branch of expertise in debating, since he does not lose from his activity, but rather gains money. The one part of debating in which the person involved gets pleasure from his activity, with the result that he neglects his own livelihood, is appropriate to call *chatter* (ἀδολεσχικόν), and is conducted in a style which is unpleasant to most people who hear it. This part wastes money (χρηματοφθορικόν), while the other makes money from debates between individuals (χρηματιστικόν). We can reach the conclusion that since eristic or debating involves controversy about what is just and unjust itself, to the Visitor's eyes the sophist has the same intellectual interests and is involved in the same kinds of questions and discussions as the Platonic dialectician²⁸ or the philosopher as described by Socrates in the *Theaetetus* 172c–176a.²⁹ Both Cornford and Narcy note that in the recapitulation of all the first six definitions (231e), this type will be simply called "the eristic". But the latter notes that within the fifth division sophistry and eristic are no longer co-extensive, because eristic is of two types, of which the one deserves the title of the sophist, whereas the other is just a babbler.³⁰ The second type of eristic bears a resemblance to philosophy. The aforementioned scholars are representatives of two different lines of interpretation.

Cornford states that he does not agree with Campbell that it is Socrates who is meant by "babbler" (ἀδολέσχης), though the latter was indisputably notorious for neglecting his affairs and becoming poorer and poorer while remaining devoted to his mission;³¹ nor does he agree with Diès that the babbler is the true dialectician. Cornford's main argument is that this would make the true philosopher a species of eristic, arguing for fame or victory, although it is true that the term "babbling"

---

26 Cornford (1935), p. 176.
27 Narcy (2007), pp. 200–201.
28 Plato, *Republic* 475d1–476d6, 480a 11–12, 534b7–d1, and 537d3–7.
29 Cf. Narcy (2007), p. 201.
30 Cornford (1935), p. 176, and Narcy (2013), p. 65.
31 Plato, *Apology* 29d–31c. See Campbell (1867), p. 40, n. 8.

was applied to philosophy by its enemies and more specifically to Socratic conversation.³² He also believes that these babblers referred to here, who do not take fees, must be some followers of Socrates who can also be described as Eristics, and with little doubt these would be the Megarians.³³ On the other hand, Narcy believes that within the framework of the fifth definition, sophistry and non-sophistry both pursue the same type of discussion about the same subjects. If one compares the description of the philosopher that Socrates offers in the *Theaetetus* with the fifth definition in the *Sophist*, "it seems that philosophy and eristic are one and the same thing".³⁴ This view is somehow connected with the problem that there was a lack of general consensus among the Greeks regarding the question as to who should be considered as a sophist. In the 4th century, the word "sophist" could bear both a positive or at least a neutral and a negative meaning in parallel.³⁵ Many years after the death of Socrates, the orator Aeschines referred to him as "the sophist", while Isocrates grouped together those who occupied themselves with contentious disputations (ἔριδες) with Socrates and Plato.³⁶

---

32 Cornford (1935), p. 176. See Eupolis 352 and Aristophanes, *Clouds* 1485. See also Plato, *Phaedo* 70b–c, *Phaedrus* 270a, *Republic* 489a, *Theaetetus* 195b, *Statesman* 299b, and *Parmenides* 135d.
33 Cf. Diogenes Laertius, II. 30, 106, and 107.
34 Narcy (2013), pp. 66 and 68; cf. Narcy (2007), p. 199. He calls chattering interchangeably "(Socratic) eristic".
35 For example, if we consider two contemporaries of Plato, Xenophon and Isocrates, we can see contradictory approaches. Xenophon criticizes the sophists and mentions them in order to contrast them with Socrates (e.g., *Memorabilia* I.1.11; I.6.13–14; *Cynēgeticus*, Chapter 13). Isocrates deviates from the critical view of Xenophon and Plato (e.g., *Antidosis* 268, 235, 313; *Panēgyricus* 3; *Helenae Encomium* 9). See Notomi (1999), pp. 48–54. I agree with Notomi that although we are inclined to take this criticism as reflecting some common negative view of the sophists amongst 4th-century Athenians, it is also possible that this is the view shared mainly by the followers of Socrates or the Socratic circles, or persons who were founders and proponents of the Socratic Schools.
36 Aeschines, *In Timarchum* 173; Isocrates, *Helenae Encomium* 1. See Nehamas (1990), p. 3. Also, Isocrates used the word *sophist* with the general meaning of man of wisdom. He was also trying to ensure that the name and task of *philosophy*, with the special meaning he attached to it, would be attributed to his own educational perspective, while at the same time denying the Platonic conception of it; Isocrates, *Antidosis* 170, 235, 268, 270, 313. See Nehamas (1990), p. 4.

3 Proclus on the Socratic-Platonic Dialectic

3.1 General Remarks

In the first book of his Commentary on Plato's *Parmenides*,[37] Proclus sets out to state his opinions on the view of many interpreters of Plato's doctrine, who attempt to distinguish the dialectical method presented by Parmenides in the homonymous dialogue from the dialectical method which was highly prized by Plato.[38] In the frame of this discussion, Proclus states that dialectic, as a method of trying to attain genuine knowledge, contains three sorts of activities.[39] The first, which is suitable for young men, is useful for awakening the reason that is, as it were, asleep in them and provoking it to inquire into itself. Proclus explains that this is actually an exercise (γυμνασία), in training the eye of the soul for seeing its objects and for taking possession of its essential ideas by confronting them with their contradictories. Its advantage is that it explores not only the path that leads, as we may say, in a straight line towards the truth, but also the bypaths that lie alongside it, trying them also to see if they reveal anything trustworthy.[40] It is worth noting that, according to Proclus, the second of the three arguments of those who claim that Parmenides presents here a dialectical method different from that of Socrates, is that it is called by him an *exercise* (γυμνασία). In their opinion, this implies that it uses arguments both for and against a thesis, like Aristotle's dialectic, which the latter in teaching it states contributes to logical exercise. But Plato's dialectic as described in the dialogues, they say, leads to the highest and purest stage of knowledge and insight because it pays no attention to human opinion and uses only irrefutable knowledge since it is based on intelligible Forms, through which it advances to the very first member of the intelligible world.[41]

If the Parmenidean γυμνασία uses arguments both for and against a thesis and takes possession of its essential ideas by confronting them with their contradictories, it would be reasonable to assume that there is a common basis between this method and the Protagorean antilogic.[42] But Proclus claims that the same

---

[37] Proclus, *In Platonis Parmenidem*, I 648.2 ff. As a general outline, I follow the translation by Morrow and Dillon (1987). For Proclus as reader of Plato's *Sophist*, see Charles-Saget, Annick (1991).
[38] Plato, *Parmenides* 127d–128e; 135c–136c. See also Morrow and Dillon (1987), *Introduction* to Book I, p. 9.
[39] Proclus, *In Parm.* I 653. 5–7; cf. V 989. 10–17.
[40] Proclus, *In Parm.* I 653. 7–18.
[41] Proclus, *In Parm.* I 648. 20–649. 8.
[42] Cf. Berti (1978), pp. 352–354.

method has affinities with the Socratic method. His counter-argument against those who insist on radically differentiating the Parmenidean γυμνασία from the laws that Socrates lays down for dialectic, is that Parmenides does not use a different term or language from that which Socrates uses if one invokes the *Republic* 526b.[43] We see that Proclus solves the problem of disagreement between the Parmenidean and the Socratic dialectic by attributing to γυμνασία a broad sense, in order to include with relation to the latter case, training in the appropriate studies. Moreover, given that according to Proclus, dialectical activity is of three kinds, arguing on both sides, expounding truth, and exposing error (τὸ ψεῦδος), he states that it is the first alone that is called γυμνασία by both Parmenides and Socrates.[44] Both these men call γυμνασία the primary use of the dialectical method, namely logical exercise which includes examining both sides of the question and then in turn examining the difficulties in true beliefs or rapping them against one another and showing up any that ring unsound, while constantly exposing the interlocutor to those difficulties latent in his opinions. Proclus emphasizes that this is the method by which Socrates trains his young men, as for instance Theaetetus or Lysis. His evaluation of this kind of exercise is that it is recommended for young and ambitious persons attracted by knowledge, to strengthen them against weariness in enquiry and the tendency to give up.[45] But an equally important remark by Proclus at this point, where there is a lacuna,[46] is that all the types of dialectic are ready to hand by Socrates, when contending with sophists masquerading as experts and masters of wisdom, in order to show his adversaries where they contradict themselves. These dialectical methods are in some way cathartic of overweening self-opinion, and the result in his adversaries is that after being pounded from all sides, they may eventually be brought to a recognition of their own false pretenses.[47]

The second kind of dialectical activity, according to Proclus, is that which Socrates says unfolds before the mind the whole intelligible world, making its way from Form to Form until it reaches the very first Form of all, sometimes using anal-

---

[43] It is most probable that Proclus' intention in *In Parm.* I 652.29–653.2, is to differentiate the Parmenidean method from the Aristotelian, and present it as absolutely compatible with an aspect of the Socratic method. This aspect is manifested through the Socratic protreptic, encouraging young people to exercise themselves in the appropriate studies, e.g., the mathematics. Cf. Morrow and Dillon (1987), p. 43, n. 36.
[44] Proclus, *In Parm.* I 654.15–19.
[45] Proclus, *In Parm.* I 653. 3–5; 654. 19–34; V 989. 16–23.
[46] Proclus, *In Parm.* I 654. 34–655. 2. Cf. Morrow and Dillon (1987), p. 44, n. 38.
[47] Proclus, *In Parm.* I 655. 2–12. According to Proclus, examples of this kind of Socratic dialectic are found in the *Gorgias*, *Protagoras*, and *Republic*.

ysis, sometimes definition, now demonstrating, now dividing, both moving downwards from above and upwards from below; and after having examined in every way the whole nature of the intelligible world it reaches that which is beyond all being. These are the functions of dialectic spoken of in the *Phaedrus* and the *Sophist*.[48] The third distinct kind of dialectical activity is that which purges of double ignorance when directed against men falsely confident of their opinions. Proclus clearly identifies this kind of Dialectic with the method described within the sixth definition of the sophist in the *Sophist* 231a–d. In Proclus' analysis of this third kind, we can see the contradistinction between the philosopher and the sophist. While the philosopher is compelled to use refutation as a method of catharsis on men obsessed by their conceit of wisdom, the sophist, when engaged in refutation, was thought to assume the guise of the philosopher, like a wolf pretending to be a dog.[49]

We can understand that Proclus ascribes to the Socratic-Platonic method all the three kinds of dialectic that he recognizes. Apart from the method which proceeds by collections and divisions, divisions and definitions, as presented in the *Phaedrus* and the *Sophist*, he attaches to Socrates both the γυμνασία, namely a method which is relative to the Protagorean antilogic-since it examines both sides of a question and helps the soul to take possession of its essential ideas by confronting them with their contradictories-and the cathartic method which is identified with the sophistry of noble lineage. Proclus states that the circumstances of the dialectical process, and most of all the dialectical partners, determine which of the three kinds of dialectic will be applied in each case. When the dialectician is reasoning with himself, having to do with men who are neither adversaries needing to be trounced nor pupils needing to be exercised, he employs the highest form of dialectic, that which reveals the truth in its purity, i.e., the second kind. When he is holding forth before young men who have not been exercised through arguments, have not been trained in mathematics and thus have not been prepared for the theory of Being, he employs the first kind of dialectic, namely γυμνασία. When he has to face sophists or other men hampered by their double ignorance and incapable of receiving scientific reasoning because of their self-conceit, he employs the cathartic method, i.e., the third kind of dialectic.[50]

---

48 Proclus, *In Parm.* I 653.18–654.1. Plato, *Phaedrus* 265d–266a; *Sophist* 253b–254a. Cf. Proclus, *In. Parm.* I 649. 32–651. 9.
49 Proclus, *In Parm.* I 654. 1–14; cf. V 989. 11–14. For the reading "πειραστική" instead of "παραστατική" in I 654. 2, see Morrow and Dillon (1987), p. 44, n. 37.
50 Proclus, *In Parm.* I 655. 12–656. 14; cf. V 989. 10–23.

3.2 Proclus on Dialectic in Plato's *Sophist:* The Dual Meaning of *Eris*

On the occasion that Parmenides in *Parmenides* 135d refers to ἀδολεσχία, Proclus recalls that dialectic is commonly called "babbling" (ἀδολεσχία) and the dialecticians are called "babblers". In his reference are included quotations from comic poets who mention as a babbler not only Socrates but also Prodicus.[51] But what he stresses is that when Parmenides encourages Socrates to exercise himself in treating the arguments, he does not simply call his method "babbling", but adds "what the many call babbling". He also mentions two cases, the *Phaedo* 70b–c and the *Theaetetus* 195b–c, where Socrates shows that he is fully aware of what is generally regarded as useless and condemned by the multitude as idle talk and implies that in his earlier life the name had been applied to him in comedy. Regarding the second passage, according to Proclus' reading, what Socrates clearly calls babbling is the primordial characteristic of dialectic, namely the practice of raising difficulties, of turning the same propositions this way and that, and not being able to leave them alone. The term "babbler" was used to refer to a man who did not find it easy to stop examining the same arguments.[52] Still, when Proclus starts interpreting the classifications of the arts in the *Sophist* (219b–225d), he affirms that we even find the Visitor from Elea there, placing dialectic under the rubric of babbling.[53]

When commenting on these divisions, Proclus[54] states that we must clearly place dialectic under the acquisitive art in no other way than through competition (ἀγωνιστικόν), and in turn under verbal controversy (διὰ λόγων ἀμφισβητικόν). Furthermore, it is clear that dialectic would belong under this part of the art of disputation which proceeds privately by questions and answers. This is the art of antilogic (ἀντιλογικόν). And in a further divisional stage, dialectic will find its place under the part of antilogic which inquires into general principles that admit of controversy (e.g., about the just itself, the noble itself, and their opposites). Proclus asserts that the Visitor calls this part eristic, not meaning that it is disreputable strife or antilogy, but indicating only its activity of controverting and raising objections. By referring to Hesiod,[55] from whom the dual meaning

---

[51] Proclus, *In Parm.* I 656. 17–26; cf. V 991. 21–992. 7.
[52] Proclus, *In Parm.* I 656. 27–657. 18. For the value of dialectic considered as ἀδολεσχία, see also *In Parm.* V 990. 1–11.
[53] Proclus, *In Parm.* I 657.23–27.
[54] Proclus, *In Parm.* I 657.27–658.26.
[55] Hesiodus, *Opera et dies* 11–12: "Οὐκ ἄρα μοῦνον ἔην Ἐρίδων γένος, ἀλλ'ἐπὶ γαῖαν εἰσὶ δύω". See Burckhardt (1999), p. 165: "In Hesiod we find the agon manifested in civic and rural life, that is to

of Ἔρις originates, Proclus claims that there is a correct way of carrying on controversy, just as there is good and bad strife. There is one species of eristic that makes money and this brings in the ingenious sophist. According to him, we shall obviously place the dialectician under the other species of eristic, the one that wastes money and neglects private affairs for its insatiable interest in discussion; for he clearly does not belong in the other class, since this is sophistry.

The main conclusions we can reach from Proclus' overall approach are as follows. a) Antilogic is a common background of dialectic which in one of its types constitutes γυμνασία, namely logical exercise involving arguing both sides, i.e., both for and against a thesis, or examining both sides of a question, and taking possession of essential ideas by confronting them with their contradictories. b) When the dialectician is reasoning with himself, without the need to confront adversaries or pupils that need to be exercised in the appropriate preparatory for dialectic studies, he employs the highest form of dialectic, that which expounds and reveals truth in its purity. This contains the main techniques of diairesis, definition, demonstration, and analysis, as presented in the *Phaedrus* and the *Sophist* c) Dialectic is identified with ἀδολεσχία, which stems from verbal controversy and in turn from antilogic. This is the one kind of ἔρις which wastes money and time on examining the same propositions again and again and raising difficulties. There is a distinction between it and sophistry in that the latter is ἔρις that makes money. But independently of this distinction, there is another one between a correct and a false way of carrying on controversy, just as there is good and bad strife. d) There is a kind of dialectic which aims at exposing error. In this kind, the philosopher is compelled to use refutation as a method of catharsis on men obsessed by their conceit of wisdom. Because of their double ignorance and their self-conceit, these men are incapable of receiving scientific reasoning. e) Proclus interprets the fifth definition of the sophist as a scheme of divisions which brings to the fore the real origin and identity of dialectic.

---

say, a kind of competitiveness that formed a parallel to the aristocratic and ideal form of the agon. This is associated with his doctrine of the good and the bad *Eris* (strife), to be found at the beginning of *Works and Days*. The good *Eris* was the first to be born (while the bad was only a variant form fostering war and conflict) and Hesiod seems to find her not only in human life but also in elemental Nature, for Cronos had placed her among the very roots of the earth. It is the good *Eris* who awakens even the indolent and unskilled to industry".

3.3 Proclus on the Difference between the Maieutic and the Contentious Method of Discussion

In his comments on Plato's *Parmenides*, Proclus also highlights the main differences between midwifery (μαιεία) and verbal contest (ἀγών).[56] We can assume that principally the distinction between the Socratic dialectic and the Eristic dialectic is grounded on the same characteristics. The person who applies the midwifery causes his interlocutor to make progress and to grow always more perfect in his concepts. Whereas the end result of midwifery is the eliciting of the thought hidden within the interlocutor, the aim of a verbal contest is victory over the opponent and his reduction to complete perplexity. Rather than being overthrown by the maieutic man, the interlocutor is being benefited by him, since the maieutic man gives birth to the latter's ideas; the midwifery is performed for his improvement rather than his defeat. A major difference between the Socratic dialectic and the verbal contest is that in the latter, the interlocutor is self-protective and self-defensive in terms of his attitude towards the dialectical process since the dialectical partners are considered to be opponents.

4 The Methodological Intervention of the Sixth Definition of the Sophist: The Dual Meaning of Catharsis

What is striking in the introduction to the sixth definition of the sophist is the abrupt turn to a range of arts which the Visitor refers to, without saying whether they derive from the original division of arts into the productive and the acquisitive parts. The Visitor clarifies that all these things are kinds of dividing (διαιρετικά που ξύμπαντα 226c), since there is a single kind of expertise involved in all of them, which is the art of discrimination (διακριτικὴ τέχνη). Furthermore, he notes that this art of discrimination is of two types; in what we call discriminations the one kind separates what's worse from that which is better, and the other separates like from like. For the second kind there is not an ordinary name, but for the kind of discrimination that leaves what is better and throws away what is worse there is a name; any discrimination of this kind is called by everyone, cleansing (καθαρμός τις [226d]). It is worth noting that what Plato thematizes from the very beginning

---

56 Proclus, *In Parm.* IV 907. 8–18.

in this sixth definition, is the method of diairesis or division itself.[57] But the question is, why does Plato choose to thematize the method of division itself this time? According to a line of interpretation, the whole procedure of divisions treated in the previous five definitions can be construed of as focusing on what proves to be likeness, since the divisions made until now have only allowed us to understand that sophistry looks like the Socratic attitude and practice. Socrates is also a hunter of young people, he also engages in dialogue with them for the sake of virtue and knowledge, and his dialogical activity is also pursued through λόγοι or arguments.[58]

It is true that it was one of Plato's basic intentions in a series of dialogues, such as the *Euthydemus*, the *Cratylus*, the *Theaetetus*, and the *Sophist*, to render perfectly clear and distinct the difference between his teacher, Socrates, and himself on the one hand, and the Sophists and Antisthenes on the other. As Bonitz remarks, not only the comedies of Aristophanes, but also the speeches of Isocrates acknowledge no distinction between them, regarding them all as controversialists, "engaged in eristic" (περὶ τὰς ἔριδας διατρίβοντας).[59] It would be a reasonable expectation to suppose that this might be the case in the sixth definition, since in the fifth, the Visitor has reached the crucial point where it has been demonstrated that dialectic, as verbal controversy, stems from acquisition, from the possession-taking part of acquisition which we label ἀγωνιστικόν, from antilogic, and finally from eristic. Based on Proclus' analysis, which is grounded on the comparative examination of the relevant Platonic dialogues, we can understand that ἀδολεσχία is the name by which dialectic is commonly referred to by the multitude, namely those who consider it something useless and money-wasting. Since ἔρις, as we have seen in Proclus' interpretation, is considered to be twofold, and there is good and bad strife, sophistry is just one branch of dialectic that is distinguished from the other kinds of it—which are all encompassed within ἀδολεσχία— simply by the aim to earn money from the dialectical activity. However, earning money does not prevent someone from carrying on controversy in a correct way. Someone who earns money from debating is a sophist, but that does not mean that he is not a dialectician.

So, what the fifth definition of the sophist only achieves is to offer a graspable feature of the sophist as he ultimately appears. It does not offer a real distinction in terms of dialectic between the *dialectician* and the *dialectician who appears as a sophist*. The question is, what is the contribution of the sixth definition? In my

---

57 Cf. Klein (1977), p. 20, and Larsen (2011), p. 110.
58 Larsen (2011), pp. 109–110.
59 Bonitz (1968), p. 136, n. 29, Mouzala (2011), p. 68, and Isocrates, *In sophistas* 1–2.

view, it follows on from the fifth definition, offering a significant modification of the method of division, initially promising to strengthen it so as to be able to find the real distinction. But most importantly, it is through this definition that Plato structures his speech about division, in such a way that he achieves a harmonization between the philosophical dialogue and his method. Within its frame, the methodology of a new aspect of diaeretical dialectic, and specifically the anatomy of cathartic dialectic, is illustrated through the philosophical content of the dialogue. The point at issue comes with the description of the elenchus in 230b–e, where there is not a definition of the sophist but only an illustration of the sophistry of noble lineage.

As Heidegger aptly stresses, διακρίνειν (discriminating) is a term that has a more specific meaning than διαιρεῖν (dividing), because discriminating means not only to divide, but to distinguish from one another and to set off against one another, the things taken apart in this procedure. And this discrimination can be either separation of like from like, where things that are the same are set off against each other, or segregation, i.e., separation of the worse from the better.[60] It is obvious that in the first case one separates things that are the same and this could mean that he identifies different species within a summum genus, following the diairetical descent from the genera to the infimae species and working within a frame of a taxonomy; in this way he employs the usual method of dialectic in its diairetical aspect. Whereas in the second case, division takes an axiological sense and becomes, as Heidegger points out, an extracting of the worse from the better. Heidegger proposes that separating the better from worse has two senses here. The first one is an extracting of the worse from the better with the result that the better remains left over. A second sense of setting off emerges from the extraction of the worse from the better when the latter is made free of the former, and we call this procedure "purifying" or "cleansing". "Καταλειπούσης το βέλτιον" (226d 5–7) then has also the sense of "making free" or allowing the thing to unleash its innate proper possibilities. This "making free" has also the meaning of a "clearing away" of obstacles so that what is purified can proceed released.[61]

It is obvious that the usual diairetical branch of dialectic is not the main dialectical target here, since the Visitor states that he does not have an ordinary name for that part, and he leaves it aside. But we must note that he does not abandon it, as the methodological parenthesis in 227a–c shows. Not only regarding the sixth definition but even during the whole dialogue, we will constantly need to keep

---

[60] Heidegger (1997), p. 247.
[61] Heidegger (1997), p. 248.

in mind the contrast between this kind of discrimination which is separation of like from like and the other, which is cleansing, i.e., discrimination that leaves the better and throws away the worse. In fact, these are two stages which are complementary to one another, because cleansing follows the procedure of the separation of like from like, while separation of like from like precedes cleansing. We can deduce this from the rationale that being better or worse does not exclude being like, and on the other hand when something looks like another, it may be in parallel better or worse than that. So, we do have a separation in both of these cases, but the criterion of separation is different in each.

It has been claimed that καθαρμός represents the Platonic choice to give priority from now on to a value-oriented way of dividing.[62] But the methodological parenthesis in 227a7–b6, which is of prominent significance since it explains the rules of exercising the art of diairesis, reintroduces a neutral version of the study of this kind of separation which leads to the distinction between better and worse.[63] The meaning of this methodological comment is that when the method works with the likenesses and its primary purpose is to understand what is alike or not within a genus, then its orientation is value-neutral. I believe it is obvious that the Visitor describes a method which includes here both the inquiry of the alike or the akin, which leads to the common nature or genus of all the things which are alike or akin, and cleansing, i.e., the kind of discrimination that leaves what's better and throws away that which is worse. Cornford points out that the division should be preceded by a collection to fix upon the genus we are to divide.[64] In fact, this method to which the Visitor refers in the methodological parenthesis of passage 226a–c, incorporates all the three procedures of dialectic used in the sixth definition. These include collection, for example the recognition of the kinship between all of the different types of discrimination, then division, which is tantamount to separation of like from like, and finally cleansing, namely, discrimination or separation of the better from the worse. The latter also entails isolation, removal and expulsion of the worse, and release or liberation of the better. Collection works by investigating the kinship of different kinds (κατὰ τὴν ὁμοιότητα, τὸ συγγενές 227b), division works by investigating the lack of kinship and by discriminating like from like (ὅμοιον ἀφ' ὁμοίου 226d, τὸ μὴ συγγενές 227b), and καθαρμός by separation that leaves the better and throws away the worse. So, the new aspect of dialectic, here in the *Sophist*, is the duality of division, and the new branch of division, which is purification.

---

62 Larsen (2011), p. 111.
63 Cf. Rosen (1999), p. 119.
64 Cornford (1935), p. 186.

A significant question is, why does the Visitor insist on a value-free method, as that described within the methodological digression (227a7–b6), while giving priority to a value-oriented method such as καθαρμός?[65] I believe the answer lies in the application of this method in 227b6–c6, which constitutes a synthesis of the two dialectical tendencies, but finally is not oriented towards a value-neutral view of diairesis. As far as our aim is to show that all the capacities that are assigned to living or non-living bodies differ from the arts which are concerned with the purification of the soul (χωρὶς τῶν τῆς ψυχῆς καθάρσεων), and specifically with the purification that concerns thinking or discursive thought (καθαρμὸς τῆς διανοίας), it does not matter to our method which name would seem to be the most appropriate, just so long as it keeps the purification of the soul separate from the purification of everything else. One major difference between the two orientations is that the value-neutral method belongs to the domain of pure or formal logic, while the value-oriented method has to be evolved in the domain of soul, where given that it is divided, non-rational parts are also included. This difference is relative with the double sense of καθαρμός. Καθαρμός as a technique of division would be to proceed to more and more upgraded abstractions or to abstract on the basis of differences of value. Καθαρμός as a personal process which leads to self-knowledge presupposes involvement in the dialectical procedure and results from it, but is accomplished within the soul itself. It is the moment when each must finally come face to face with his own soul and its contradictory powers.

So, the next step of dividing is the division of purification into two types, one dealing with the soul and the other dealing with the body. However, this new division is already a kind of purification, because while earlier when focusing on the division of arts according to their likeness, the Visitor supported a value-neutral stance, he now follows one oriented towards what is better, since he moves from the inferior kinds of bodily purification to the superior ones that have to do with the soul. Furthermore, it is most probable that this shift in the subject of the division mirrors the Socratic dialectic's attempt and main concern to re-orientate and direct man's interest towards the care of the soul rather than the body (ἐπιμέλεια ψυχῆς), as illustrated primarily in the *Apology* and afterwards in many other dialogues.

---

[65] Dorter (2013), pp. 91 and 93. Dorter notes that the first five definitions all have as starting point of the divisions the motivation of the sophists, and locate them within the genus of acquisition rather than production, or a combination of the two as aggressive money-makers. The question of motivation disappears when the Visitor introduces his value-free stipulation in the sixth definition, and the seventh definition significantly begins for the first time from the art of production.

5 Dialectic, Self-Purification and Thumos

The genus to which elenchus belongs is purification of the soul, and elenchus is recognized as the principal and most important kind of purification (230d 7–8). Elenchus makes us get rid of ἀμαθία, the most difficult type of ignorance, marked off from the others and overshadowing all of them, that is reduced to ugliness and this in turn to disproportion. Given that no soul is willingly ignorant of anything (228c7–8), when a soul is ignorant, this means that it is ugly and out of proportion and this disproportion is interpreted by the Visitor in terms of motion of the soul towards truth.[66] When something that is in motion aims at a target and tries to hit it, but on every attempt passes by it and misses, we say that it is out of proportion. In passage 228c1–5, the phrase "τῆς πρὸς ἄλληλα" shows that being ignorant is psychic ugliness, because soul lacks measure and proportion with relation to truth, while it is innate to soul to aim towards it as perennial target.[67] In 228c10–d2, being ignorant, when a soul swerves aside from σύνεσιν (understanding), it is construed as being in παραφροσύνη. White notes that the latter is a quite strong word, which means "being beside oneself" or "derangement" and asserts that the point here is to claim a connection between internal disproportion and external failure.[68] According to this suggestion, we must suppose that through the external failure there is a hint at the internal disproportion. So, when the soul swerves aside from truth and understanding, this failure implies an internal psychic disproportion. But internal lack of proportion already thematizes self-knowledge, or the

---

[66] This approach is compatible with the Platonic conception of the soul as self-moving and principle of all motions, as well as with the fact that later, in 248d, knowledge is described by means of physiological terms in its dual form, as action and passion (γιγνώσκειν καὶ γιγνώσκεσθαι, ποίημα καὶ πάθος).

[67] Rowe (2015), pp. 154ff., presents an insightful interpretation of 226b–231b, according to which this passage combines Socratic intellectualism with a version of psychic partition, although any reference to the parts of the soul is absent. He also agrees that the phrase πρὸς ἄλληλα (228c 4) indicates that soul is taken here as a collection of things-e.g., beliefs, desires, anger, etc.-with a joint target, namely truth. He aptly compares three passages, 228b2–4, 228c1–5, and 230b4–5. The meaning of this parallelism is that the cause of ignorance in a soul, namely its wandering away from truth or the understanding it desires, is a lack of proportion between its elements. Then, he uses disagreement (228b4) as a link between vices and ignorance and assumes that what is implicitly said, is that all badness is properly to be understood as ignorance and has to be corrected by teaching.

[68] White (1993), p. 15, n. 12. Campbell (1867, p. 50) notes that "ignorance is a kind of deformity, and may be compared to the bodily state, in which the movements of different members are inharmonious and fail of accomplishing their end". This interpretation suggests an analogy on the basis of which we might think here of inharmonious movements of the parts of soul.

need of awareness of what causes this psychic disproportion, which is described as external failure of being in due proportion to truth. This interpretation is reinforced if we think that one possible meaning of σύνεσις is not only comprehension, but also "conscience".[69] In my view, the thematization of self-knowledge is more explicitly declared when in 230 a5–9, the part of teaching which is presented in contradistinction to elenchus is admonition. Some people argue that lack of learning is always involuntary, and that if someone thinks he is wise, he will never be willing to learn anything about what he thinks he is clever. Admonition (νουθετητικόν) is thus presented as an externally imparted knowledge which is disputable if it can penetrate the soul of those who think they are wise, since they lack willingness to learn, which presupposes the activation of self-knowledge.[70] Elenchus is the efficacious part of teaching which can help one to get rid of the belief in one's own wisdom (δόξης ἐκβολήν) and ensure the prerequisite for attainment of knowledge, i.e., self-knowledge. Double ignorance is cured only by the activation of self-knowledge, which is achieved through the elenchus.

Purification is a term with religious and medical connotations and its religious significance is prior and wider than the medical one.[71] It has been claimed that, in general, Plato speaks of purification in three different contexts. The first is the religious one and is often associated with the Orphic-Pythagorean tradition. The second is the legal one, bearing a social-political dimension, while the third is the epistemological one.[72] I will add a fourth context, the psychological, which of course is tightly bound with the epistemological. Let us keep in mind that regarding ugliness of the soul, i.e., ἀμαθία, Plato does not invoke medical purification, because ugliness of the body which corresponds to ignorance of the soul is treated by gymnas-

---

[69] Liddell-Scott-Jones Lexicon, s.v. Bernabé (2013, pp. 48–49), notes that the word ἀνόητον (228d4) has also a special significance here, adding that it is used by Plato at *Republic* 605b and *Timaeus* 30b to characterize soul's irrational part. But he connects this word rather with the *Gorgias* 493 a, claiming that it can also mean "uninitiated".
[70] Cf. Plato, *Charmides* 164e7–165a7. See also Cornford (1935), p. 179, n. 3.
[71] Bernabé (2013, pp. 41–47), in his comparative examination of the use of the term within the sixth definition and in other dialogues, notes that the most ancient usage of the word is applied to the activities of the magicians or religious practitioners, and claims that Plato is not interested in medical purification, as is clear from 227a–c, but gives the priority to the transposition of religious language. He supports that since Plato had already used religious vocabulary in previous dialogues to transpose religious concepts into philosophical tenets, in the *Sophist* he reshapes old arguments against the sophists by transposing the previously transposed language; the result is that the difference between genuine and false purifiers and initiators still lies at the base of the sixth definition.
[72] Solana (2013), pp. 75–76. Bernabé (2013, p. 47) speaks of a double sense of καθαίρειν, which can refer either to a religious purification or to refutation of a wrong idea.

tics, not by medicine as sickness or disease. Since there is an analogy between ugliness and disproportion on the one hand and ignorance on the other, there must also be an analogy between gymnastics and elenchus.[73] But this analogy has even more implications. As Cornford remarks, "Gymnastic is the parallel remedy for physical ugliness, the deformity due to lack of proportion. This is, somewhat strangely, treated as analogous to a lack of proportion or co-ordination between impulses in the soul, causing them to miss their mark".[74] So, the field where we can exercise gymnastics is the field of the contradictory impulses and powers of the soul. This is what I aim to show next.

Guthrie explains thoroughly the reasons why he believes that in 230b–e, the Visitor gives a precise and detailed description, not of the Sophistry but of the elenchus as practiced by none but Socrates himself (neither by the Megarians nor by his arrogant young followers).[75] A whole range of scholars before and after him associate the sixth definition, not with the sophist but with Socrates or those who use his method.[76] On the other hand, Kerferd traces reasons for which it is plausible that Plato could recognize as practitioners of this method the sophists and ascribe to them achievements that are similar with those of Socrates, although Trevaskis and Booth dispute his arguments and are involved in a controversy with him.[77] There is also, in the middle of this debate, a line of interpretation which I shall partly follow. As Rosen notes, the diairesis within the sixth definition, instead of producing a definition of the sophist, have arrived at a hybrid of the sophist and the philosopher.[78] This result paves the way for another interpretation; that one needs to recognize that Sophistry finally is a hybrid, and that it includes two separate tasks; the purification of beliefs that hinder understanding and the art of producing apparent knowledge.[79] It is also relative to the insightful reading of Narcy, who combines this stage of definition with the later one in which "the real sophist" is tightened through the creation of illusions with the arts of produc-

---

[73] However, Rosen (1999, p. 130) notes that the Visitor "concludes this long speech by comparing elenctic to the purging of the body by the physician (230c8ff.). Since elenctic is a kind of teaching, this contradicts the previous analogy between teaching and gymnastic". Cf. Giannopoulou (2001), pp. 112–113. She also stresses that in 230c–d, the Visitor "suggests that the purification of ignorance becomes the work not of 'psychic gymnastics' (or, at least, not only of psychic gymnastics) but of 'psychic medicine' which expels epistemic obstructions from the soul".
[74] Cornford (1935), p. 179; cf. Solana (2013), p. 77.
[75] Guthrie (1985), pp. 128–129.
[76] For an account of those scholars, see Notomi (1997), p. 65, n. 72.
[77] Kerferd (1954), Trevaskis (1955), and Booth (1956).
[78] Rosen (1999), p. 131.
[79] Solana (2013), p. 81.

tion. Narcy poses a crucial question: "Does that mean that the 'real sophist' is merely an imitator of the sophist 'faithful to his lineage'?"[80]

In my view, the passage 230b–e contains the first attempt to show an example of the notion of communion (κοινωνία) within the dialogue, but it also contains both the common ground and the crucial differences between the Socratic and the Sophistic dialectic. If we try to see the description of the elenchus in 230b–e from Proclus' perspective of dialectic as previously analyzed, we will find merging and commingling within it two kinds of dialectician. Indisputably, as Proclus recognizes, the elenchus is the kind of dialectic which purges of "double ignorance", since it is directed against men falsely confident of their opinions (230a). The refutation is a method of catharsis which the philosopher is compelled to use on men obsessed by their conceit of wisdom.[81] Proclus,[82] regarding this method, makes a distinction between the philosopher and the sophist based on two criteria; pretension and lack of self-purification on the part of the sophist, and honesty, modesty, and self-purification on the part of the philosopher. The sophist, when engaged in refutation, was thought to assume the guise of the philosopher, like a wolf pretending to be a dog,[83] as the dialogue puts it, because his own soul is unpurged. The philosopher is the dialectician who truly refutes another and not merely appears to do so, and he is truly a purger of false opinion because his own soul is already purged.

This dialectical method, which aims at exposing the error according to Proclus, is basically common between the sophist and the philosopher, and what changes is only the attitude. The philosopher does not pretend to refute and he himself has been submitted to the same elenchus, with the result that he is able to purify as being already purified. Let us recall that according to Proclus, it is the same method which Socrates uses when contending with sophists masquerading as experts and masters of wisdom, who are hampered by their double ignorance and incapable of receiving scientific reasoning because of their self-conceit.[84] But generally, this is a difference regarding only the moral stance and not the method of the di-

---

80 Narcy (2013), p. 70. Cf. Rowe (2015), p. 153.
81 Thinking one knows things one does not know is the most shameful sort of ignorance; Plato, *Apology* 29b; *Alcibiades I* 117e–118a; cf. *Symposium* 204a.
82 Proclus, *In Parm.* I 654. 1–10.
83 We can compare the reference to the similarity between wolves and dogs in the *Sophist* 231a with the *Republic* 416a–c, where there is a reference to the education of the guardians. The sophists, who pretend to be philosophers, are analogous to the wolves which are similar to dogs. The former are "savage masters" of the opinions of those who are persuaded or manipulated by them instead of being "well-meaning allies", such as the philosophers (trans. Bloom 1968).
84 Proclus, *In Parm.* I 654. 34–655. 6.

alectician *qua* dialectician. The part of the description which uses the analogy of the elenchus to medicine (230c3–d1), namely to the doctors who believe that the body cannot benefit from any food that is offered to it until what is interfering with it from inside is removed, reflects a common belief amongst all dialecticians who engage in refutation, including the sophists.[85] The evidence for this is that the phrase "τῶν προσφερομένων μαθημάτων ὄνησιν" (230c8–d1) can refer to none but the sophists since Socrates disavows possessing and offering of any knowledge and learning.[86]

The dialectical activity depicted in 230b 4–8 seems to be identified with the Socratic elenchus as we know it from its rough descriptions in the Platonic corpus.[87] Its value constitutes in that it helps the interlocutor to achieve a collecting together of his opinions and arguments during the discussion, to comparatively examine them and finally realize that they conflict each other. Confronting the opinions with their contradictories and bringing to the fore the contradictions in one's mind, not only contributes significantly to logical exercise but also stimulates and activates the internal process of self-knowledge in one's soul. Still, this is just one path towards self-knowledge. The Socratic antilogic could be simply characterized as a kind of reversal of the Sophistic or Protagorean type of antilogic, aiming at exposing the error and substituting for it something truer. The latter tries to exploit the method of arguing both for and against a thesis and extracting fruitful results from the confrontation of contradictory arguments. One must not arbitrarily deny the Protagorean type of antilogic, the achievement of stimulating the inner psychic process towards self-knowledge. Regarding the view that Protagoras' two opposite statements method may be equated with that of the sixth definition, Trevaskis states: "No doubt, if two opposite statements may be made on any subject, both of which are said to be true, a person who previously had held only one of them might be persuaded 'that he did not know what he thought he knew'".[88] But, in his opinion, this is not to say that the two methods are the same.

In my view, based on Proclus' remarks on Socrates' dialectic previously analyzed, we can infer that the difference between the Socratic and the Sophistic dialectic is not a difference of method,[89] at least regarding the thorny passage 230b–d

---

[85] Cf. Plato, *Theaetetus* 166d5–167a6.
[86] Plato, *Apology* 33a5–6; *Theaetetus* 150c4–d7. Cf. Notomi (1999), p. 66.
[87] Plato, *Apology* 23c–e; *Theaetetus* 150b–151d; 210b–c. Tarán (1985, p. 97, n. 53) suggests that the irony which most often accompanies the elenchus in the aporetic dialogues is absent from this passage because the priority is given to the ethical purpose of the elenchus, which is to be seen in its being considered as a method of education.
[88] Trevaskis (1955), p. 47. See also Kerferd (1954), pp. 88–89.
[89] Cf. Nehamas (1990), pp. 6 and 9.

of the *Sophist*, where elenchus is described. The crucial difference lies in the psychological condition and preparation of those who examine, on the one hand, and the psychological reaction of those who are being examined on the other. On the part of those who examine, the necessary condition for someone to be considered a philosopher is to have achieved self-purification; namely, to have already purged himself when applying the elenchus. In turn, the question arises whether there is within 230b–d, an effect of the Socratic dialectic that serves as a boundary which marks it off from the Sophistic dialectic, in terms of those who are submitted to elenchus. Our text says that when they become self-conscious of having opinions that conflict with each other at the same time on the same subjects in relation to the same things and in the same respects, they get angry at themselves and become calmer toward others. Anger against oneself is a reaction of the person who is cross-examined, which differentiates this description of the Socratic elenchus from others.[90] In the *Apology* 23c, Socrates states that his young followers get pleasure from listening to him when he scrutinizes people, and many times imitate him and they themselves attempt to scrutinize other people. The reaction in this case is that those who are being examined by his followers get angry with Socrates, not with them or with themselves.[91] Narcy remarks that the method described in the *Sophist* 230b–d by the Visitor while conversing with Theaetetus, is precisely the one used with relation to Theaetetus himself the day before, in the homonymous dialogue. And what is more, the Visitor re-employs the same words that Socrates had used when he had described his method as maieutic in the *Theaetetus*. Narcy quotes Robin who thinks of it as a remarkable short version of the critical method of examination which was set up in the dialogues of the first period, and then again in the *Theaetetus*.[92] In the *Theaetetus* 151c2–5, while he describes his maieutic method, Socrates invites Theaetetus not to get angry ("μὴ ἀγρίαινε") with him like a woman robbed of her first child if he will take from him a false statement like an abortion and cast it away.[93]

---

[90] Rowe (2015, p. 164) states that the suggestion that those subjected to the method become angry with themselves but less aggressive towards others, is the reverse of what Socrates says usually happens in his case, and of what actually does usually happen when we see him in action.

[91] Dorion (2012, p. 253, n. 11) also notes that whereas Socrates' interlocutors often become angry at their refuter (*Ap.* 21c–d, 21e, 22e–23a, 23c; *Grg.* 506b–c; *Tht.* 161c–d), the respondent described at 230b becomes angry at himself. He believes that this is undoubtedly an idealized description of the effects of the elenchus on the respondent.

[92] Narcy (2013), pp. 68–69.

[93] Zaks (2018, pp. 376–378) believes that the method of elenchus described at 230b–e harmonizes well not only with Socrates' method in the *Theaetetus*, but also with his method in earlier dialogues. He claims that the calming effects of the refutation stressed by the Visitor, echo Socrates' own declarations. For example, at *Tht.*210c2–4 Socrates declares that his midwifery makes people

We have seen that Proclus, in his comments on Plato's *Parmenides* (IV 907. 8–18), makes a bold distinction between midwifery and the contentious method of discussion. In a verbal contest the aim is the victory over the opponent and the reduction of him to complete perplexity, but a perplexity not in the sense of the Socratic ἀπορία. The result of midwifery is the eliciting of the thought hidden within the soul of the interlocutor. On the contrary, the participant of a contentious discussion is self-protective and defends himself in the sense of hiding himself. This means that he refuses to reveal what he has in his soul. This in turn means that he does not trust his interlocutor. The result of this psychological reaction is that he does not fulfill what Nehamas correctly considers as one of the most central requirements of the Socratic elenchus; the interlocutors must always answer with views they truly believe.[94] When someone tries to achieve victory over the opponent and is aggressive toward the interlocutor, in a sense he also tries to defend his beliefs and reserve his mental and psychological condition untouchable. When somebody tries to defend himself and preserve his *self* exactly as it is, he also reserves and maintains the errors and the deceptions which burden his self. Moreover, being aggressively defensive of oneself has the consequence of being unable to become calmer toward the others and have a fruitful exchange of opinions.

Complementary to the psychic attitude of anger against oneself is αἰσχύνη, i.e., the state of shame to which the dialectician leads the soul of the person who is submitted to elenchus (230d1). Shame seems to be a presupposition of self-purification, since a soul has to be provoked to feel shame in order that the removal of the opinions that interfere with learning may take place.[95] Self-recovery emerg-

---

calmer. See also *Grg.* 457c4–458b3. Furthermore, Zaks maintains that although the presence of ἀπορία is not explicit in the sixth definition of the sophist, the description of the elenchus implies the aporetic effects triggered by the elenctic method. In his opinion, these effects seem compatible with the calming effects of the refutation that are stressed by the Visitor.
94 Nehamas (1990), p. 10. For the relation between thumos and doxa, see Renaut (2018).
95 Candiotto (2018, pp. 576–585) notes that "shame functions as a transformative experience that is crucial for the 'therapeutic' dynamics instantiated by the *elenchus*". She believes that shame, as a psychological mechanism triggered by the *elenchus*, can result in two important outcomes: the interlocutor's cognitive transformation and the audience's purification. According to her interpretation, emotions collaborate with reason not only to purify the soul of the interlocutor, but also to convey a message to the audience, which is included in this transformative experience, since emotions play a crucial role also in the purification of the audience. On the role of shame and courage as two complementary feelings on which the psychological process of *elenchus* is based, see Fiore (2020). On the extended role of the emotions in the Socratic *elenchus*, see also Candiotto (2015). She argues that the view that Socrates propounded is the reverse of the so-called Socratic intellectualism: "emotion is the more primitive guide to the discovery of the Good since it shows the way to reach knowledge, and has the power to transfer it into our lives" (Candiotto 2015, p. 235). On the role of emotions in Plato's dialogues see Blank (1993) as well as Candiotto and Renaut (2020).

es when αἰσχύνη makes someone to be self-conscious and start believing that he knows only those things that he does know, and nothing more. So, the crucial psychic attitudes which result from the Socratic elenchus is anger against oneself, accompanied by the tendency to become calmer toward others, and shame. Thumos and the thumoeidic emotion of shame are the emotional instruments of the Socratic elenchus which, due to their reflexive and self-referent character, lead to the activation of the self-knowledge process.

It would be worthwhile to consider here both the self-referent and the social dimension of the role of thumos as part of the soul which is placed in a median position between reason and appetite. Thumos is the seat of the emotions, through which one gains access to consciousness of the internal psychic conflicts due to the contradictory impulses of soul. Moss, based on Plato's account of thumos in the *Republic*, notes that thumos' anger and shame are directed not simply towards an action or object, but rather towards a person (oneself or another) *qua* agent.[96] The reaction which the Socratic cathartic dialectic aims to bring about is an awareness of the fact that one has self-confidence while supporting contradictory views which reveal the mind's confusion. According to a line of interpretation, shame reveals what a person really believes, in opposition to what he says or even thinks he believes, bringing to the fore deep beliefs which correspond to an innate moral sense of what is good and bad.[97] Thumos may be considered the seat of the anger and it is that part of the soul which strives for mastery, victory and honor, or good reputation. So, it is also the seat of the motivations that make one orientate oneself toward the others.[98] Cooper notes that "the motivations that Plato classifies under the heading of spirit are to be understood as having their root in competitiveness and the desire for self-esteem and (as a normal presupposition of this) esteem by others".[99] The emotional mechanism of the Socratic elenchus, expressed through anger against oneself and combined with the thumoeidic emotion of shame, makes us respond to the thoughts that others have about us. Following on, it pushes us to dig deeper into our own thoughts and beliefs, while fostering introspection as a reaction to the dialectic interaction within the frame of elenchus.

This introspection, given that it takes place within a soul which under the Platonic perspective already has achieved knowledge and has seeds of this knowledge,

---

[96] Moss (2005), pp. 156–157.
[97] Moss (2005), pp. 139 and 146–147; she also refers to McKim (1988) and Kahn (1983 and 1996). Cf. Fiore (2020), p. 200.
[98] Plato, *Republic* 548c5–7; 581a 9–10. See Moss (2005), p. 155, Tarnopolsky (2015), p. 244, and Jimenez (2020).
[99] Cooper (1984), pp. 14–15. Cf. Singpurwalla (2013), p. 42.

is actually recollection. The cathartic process of dialectic is tantamount to the creation of a thread which discriminates and binds the beliefs that are worth being preserved, becoming a remedy against not only the loss of knowledge, but also the loss of self. This process of self-purification takes the form of a kind of self-recovery or self-recollection which imbues each person with self-understanding that derives from a supra-subjective and universal source of thought.[100] The same emotional attitude which we previously investigated, in cooperation and conjunction with the cohesive and therapeutic intervention of the unificatory logos, binds again the person who is subjected to elenchus with the latent commonality of an intersubjective wisdom which has been forgotten.[101]

Bibliography

Bates, William Nickerson (1930): *Euripides: A Student of Human Nature*. Philadelphia: University of Pennsylvania Press.
Bernabé, Alberto (2013): "The Sixth Definition (*Sophist* 226a–231c): Transposition of Religious Language". In: Bossi, Beatriz and Robinson, Thomas M. (Eds.): *Plato's Sophist Revisited*. Berlin and New York: de Gruyter, pp. 41–56.
Berti, Enrico (1978): "Ancient Greek Dialectic as Expression of Freedom of Thought and Speech". In: *Journal of the History of Ideas* 39. No. 3, pp. 347–370.
Blank, David L. (1993): "The Arousal of Emotion in Plato's Dialogues". In: *The Classical Quarterly* 43. No. 2, pp. 428–439.
Bonitz, Hermann (1886, 1968): *Platonische Studien*. Hildesheim: Olms.
Booth, N. B. (1956): "Plato, Sophist 231a, etc.". In: *The Classical Quarterly* 6. No. 1–2, pp. 89–90.
Burckhardt, Jacob (1999): *The Greeks and Greek Civilization*. Oswyn Murray (Ed.). Sheila Stern (Trans.). New York: St. Martin's Griffin.

---

100 Cf. Kosman (1983), p. 212: "...it is part of the nature of culture and thus of the tradition that it defines, that it is continuously forgetting itself and continuously engaged in the task of its own recovery. This dialectic of self-loss and self-recovery (which we know equally well from its ontological analogue in Plato) is what I meant earlier to refer to by the phrase 'cultural anamnesis': the tradition's recollection of itself...for according to Plato, understanding *is* recollection, and human self-interpretation-the project of reclaiming human self-understanding-is therefore fundamentally a project of self-recollection".
101 Kosman (1983) explains the way in which dialectic itself is "cathartic" in these words: "...preliminary definitions and accounts are not merely provisional and propaedeutic, but are in some important sense correct. For the process of elenchus central to the dialogues is not a process of rejection relative to a preferred logos, but a process of *catharsis*, of purifying the elements of traditional orthodox wisdom and of the common understanding as embodied to logos" (p. 209) ... "Ultimately the dialogues are exercises in the redemptive appropriation of a common wisdom" (Kosman 1983, p. 212).

Campbell, Lewis (1867): *The Sophistes and Politicus of Plato: With a Revised Text and English Notes*. Oxford: Clarendon Press.

Candiotto, Laura (2015): "Aporetic State and Extended Emotions: The Shameful Recognition of Contradictions in the Socratic Elenchus". In: *Etica and Politica/Ethics and Politics* 17. No. 2, pp. 233–248.

Candiotto, Laura (2018): "Purification Through Emotions: The Role of Shame in Plato's *Sophist* 230b4–e5". In: *Educational Philosophy and Theory* 50. No. 6–7, pp. 576–585.

Candiotto, Laura and Renaut, Olivier (Eds.) (2020): *Emotions in Plato. Brill's Plato Studies Series*. Vol. IV. Leiden and Boston: Brill.

Charles-Saget, Annick (1991): "Lire Proclus, lecteur du *Sophiste* (avec un appendice par C. Guérard: Les citations du *Sophiste* dans les oeuvres de Proclus)". In: Aubenque, Pierre and Narcy, Michel (Eds.): *Études sur le Sophiste de Platon*. Bibliopolis, pp. 475–508.

Cooper, John M. (1984): "Plato's Theory of Human Motivation". In: *History of Philosophy Quarterly* 1. No 1, pp. 3–21.

Cornford, Francis Macdonald (1935): *Plato's Theory of Knowledge: The Theaetetus and the Sophist of Plato, translated with a running commentary*. London: Kegan Paul.

Coulson, Lee M. J. (2020): "The Agōnes of Platonic Philosophy: Seeking Victory Without Triumph". In: Reid, Heather L., Ralkowski, Mark, and Zoller, Coleen P. (Eds.): *Athletics, Gymnastics, and Agon in Plato*. Sioux City, Iowa: Parnassos Press-Fonte Aretusa, pp. 211–222.

Diès, Auguste (1925): *Le Sophiste, Platon. Texte établi et traduit*. In: Plato: *Œvres completes*. Volume VIII, Part III. Paris: Les Belles Lettres.

Dorion, Louis-André (2012): "Aristotle's Definition of Elenchus in the Light of Plato's *Sophist*". In: Fink, Jacob L. (Ed.): *The Development of Dialectic from Plato to Aristotle*. Cambridge: Cambridge University Press, pp. 251–269.

Dorter, Kenneth (2013): "The Method of Division in the *Sophist*: Plato's Second *Deuteros Plous*". In: Bossi, Beatriz and Robinson, Thomas M. (Eds.): *Plato's* Sophist *Revisited*. Berlin and New York: de Gruyter, pp. 87–99.

Fiore, Giulia (2020): "Honesty, Shame, Courage: Reconsidering the Socratic Elenchus". In: *Acta Classica Universitatis Scientiarum Debreceniensis* 56, pp. 197–206.

Fowler, Harold N. (Trans.) (1921): *Plato. Theaetetus. Sophist*. Loeb Classical Library 123. Cambridge: Harvard University Press.

Giannopoulou, Zina (2001): "'The Sophistry of Noble Lineage' Revisited: Plato's 'Sophist' 226b1–231b8". In: *Illinois Classical Studies* 26, pp. 101–124.

Guthrie, William Keith Chambers (1985): *A History of Greek Philosophy: The Later Plato and the Academy*. Vol. V. Cambridge University Press.

Heidegger, Martin (1997): *Plato's Sophist*. Richard Rojcewicz and André Schuwer (Trans.). Bloomington: Indiana University Press. [Translation from Heidegger, Martin (1992): *Platon: Sophistes*, Frankfurt am Main: Vittorio Klostermann.]

Jimenez, Marta (2020): "Plato on the Role of Anger in Our Intellectual and Moral Development". In: Candiotto, Laura and Renaut, Olivier (Eds.): *Emotions in Plato. Brill's Plato Studies Series*. Vol. IV. Leiden and Boston: Brill, pp. 285–307.

Kahn, Charles H. (1983): "Drama and Dialectic in Plato's *Gorgias*". In: *Oxford Studies in Ancient Philosophy* 1, pp. 75–121.

Kahn, Charles H. (1996): *Plato and the Socratic Dialogue: The Philosophical Use of a Literary Form*. Cambridge: Cambridge University Press.

Kerferd, George B. (1954): "Plato's Noble Art of Sophistry". In: *The Classical Quarterly* 4. No 1–2, pp. 84–90.
Klein, Jacob (1977): *Plato's Trilogy: Theaetetus, the Sophist, and the Statesman*. Chicago: The University of Chicago Press.
Kosman, Aryeh L. (1983): "Charmides' First Definition: Sophrosyne as Quietness". In: Anton, John P. and Preus, Anthony (Eds.): *Essays in Ancient Greek Philosophy*. Vol. II. Albany: State University of New York Press, pp. 203–216.
Larsen, Jens Kristian (2011): *The Quarrel between Sophistry and Philosophy*. Dissertation. Copenhagen: University of Copenhagen.
McKim, Richard (1988): "Shame and Truth in Plato's *Gorgias*". In: Griswold, Charles L. (Ed.): *Platonic Writings/Platonic Readings*. University Park: Pennsylvania State University Press, pp. 34–48.
Morrow, Glenn R. and Dillon, John M. (Trans.) (1987): *Proclus, Commentary on Plato's Parmenides. With Introduction and Notes by John M. Dillon*. Princeton: Princeton University Press.
Moss, Jessica (2005): "Shame, Pleasure, and the Divided Soul". In: *Oxford Studies in Ancient Philosophy* 29, pp. 137–170.
Mouzala, Melina G. (2011): "Names and Falsity in Plato's Cratylus". In: *Journal of Classical Studies (Zbornik za klasicne studije MS)* 13, pp. 51–77.
Narcy, Michel (2007): *Dialectic with and without Socrates: On the Two Platonic Definitions of Dialectic*. In: Bosch-Veciana, Antoni and Monserrat-Molas, Josep (Eds.): *Philosophy and Dialogue: Studies on Plato's Dialogues*. Vol. I. Barcelona: Institut d'Estudis Catalans, pp.193–204.
Narcy, Michel (2013): "Remarks on the First Five Definitions of the Sophist (*Soph*. 221c–235a)". In Bossi, Beatriz and Robinson, Thomas M. (Eds.): *Plato's Sophist Revisited*. Berlin and New York: de Gruyter, pp. 57–70.
Nehamas, Alexandros (1990): "Eristic, Antilogic, Sophistic, Dialectic. Plato's Demarcation of Philosophy from Sophistry". In: *History of Philosophy Quarterly* 7. No. 1, pp. 3–16.
Nestle, Wilhelm (1940): *Vom Mythos zum Logos: Die Selbstentfaltung des griechischen Denkens von Homer bis auf die Sophistik und Sokrates*. Stuttgart: Alfred Kröner Verlag.
Notomi, Noburu (1999): *The Unity of Plato's Sophist: Between the Sophist and the Philosopher*. Cambridge: Cambridge University Press.
O'Brien, Michael J. (Trans.) (1972, 2001): "Protagoras". In: Sprague, Rosamond Kent (Ed.): *The Older Sophists*. Indianapolis and Cambridge: Hackett Publishing Company, pp. 3–28.
Reid, Heather (2010): "Wrestling with Socrates". In: *Sport, Ethics and Philosophy* 4. No. 2, pp. 157–169.
Renaut, Olivier (2018): "Thumos and doxa as intermediates in the *Republic*". In: *Plato Journal, International Plato Society*, 18, pp. 71–82.
Robinson, Thomas M. (2013): "Protagoras and the Definition of 'Sophist' in the *Sophist*". In: Bossi, Beatriz and Robinson, Thomas M. (Eds.): *Plato's Sophist Revisited*. Berlin and New York: de Gruyter, pp. 3–13.
Rosen, Stanley (1999): *Plato's Sophist: The Drama of Original and Image*. South Bend, Indiana: Carthage Reprint, St. Augustine's Press.
Rowe, Christopher (2015): "Plato, Socrates, and the *genei gennaia sophistikē* of *Sophist* 231b". In: Nails, Debra and Tarrant, Harold in Collaboration with Kajava, Mika and Salmenkivi, Eero (Eds.): *Second Sailing: Alternative Perspectives on Plato. Commentationes Humanarum Litterarum*. Vol. 132. Helsinki: Societas Scientiarum Fennica, pp. 149–167.
Singpurwalla, Rachel (2013): "Why Spirit is the Natural Ally of Reason: Spirit, Reason, and the Fine in Plato's *Republic*". In: *Oxford Studies in Ancient Philosophy* 44, pp. 41–65.

Szlezák, Thomas A. (2010), *Was Europa den Griechen verdankt, Von den Grundlagen unserer Kultur in der griechischen Antike.* Tübingen: Mohr Siebeck.
Solana, José (2013): "Socrates and 'Noble' Sophistry (*Sophist* 226b–231c)". In: Bossi, Beatriz and Robinson, Thomas M. (Eds.): *Plato's* Sophist *Revisited.* Berlin and New York: de Gruyter, pp. 71–85.
Tarán, Leonardo (1985): "Platonism and Socratic Ignorance". In: O'Meara, Dominic J. (Ed.): *Platonic Investigations.* Washington, D.C.: The Catholic University of America Press, pp. 85–109.
Tarnopolsky, Christina (2015): "Thumos and Rationality in Plato's *Republic*". In: *Global Discourse* 5. No. 2, pp. 242–257.
Trevaskis, John R. (1955): "The Sophistry of Noble Lineage (Plato, 'Sophistes 230a5–232b9')". In: *Phronesis* 1. No. 1, pp. 36–49.
White, Nicholas P. (1993): *Plato, Sophist. Translated, with Introduction and Notes.* Indianapolis and Cambridge: Hackett Publishing Company.
Zaks, Nikolas (2018): "Socratic *Elenchus* in the *Sophist*". In: *Apeiron* 51. No. 4, pp. 371–390.

Anna Pavani
On Plato's Late Dialectic: The Methods of Collection and Division

1 Introduction: Plato's Late Dialectical Method

Accounts of Plato's late dialectical method differ substantially—the method has been taken to have either Forms or particulars as its objects; to operate either analytically or taxonomically; to be also or preferably dichotomous; and to either reach or miss the alleged target, which either corresponds to or differs from a definition.[1] Despite their variety, almost all accounts assume that Plato's dialectical method remains *one and the same* across the late dialogues. According to the standard interpretation, which I refer to as the "one-method interpretation", the *Phaedrus*, the *Sophist*, the *Statesman*, and the *Philebus* present one and the same method.[2]

Even though they employ the word "method", most scholars consider Collection and Division to be informal techniques and not a "method" in the thick sense of an "ordered procedure for epistemic enrichment",[3] since both of the nec-

---

Note: An earlier version of this paper was presented at the International Conference "Ancient Greek Dialectic and its Reception" (University of Patras), June 16, 2021. I am grateful to the audience, especially Kristian Larsen and Rafael Ferber, for helpful questions and comments in response to it and to the participants in the *Kolloquium zur antiken Philosophie* (RUB) led by Barbara Sattler for comments on a prior draft.

1 What is being divided is said to be either a Form, as, e.g., Cornford (1935, p. 269) holds, or a class of particulars, as, e.g., Wedin (1987, p. 208) thinks. Mapping a Form (see Trevaskis (1967, p. 128) differs considerably from providing an exhausting and unique taxonomy of a given field. See below. As far as the procedure is concerned, Dichotomy is usually assigned a leading role and scholars like Franklin (2011, p. 1) go so far as to claim that only when Division is impeded does Plato present Division into more than two, since Dichotomy remains the most effective way of finding the natural joints (Henry 2011, p. 241). By contrast, scholars like Dorter (2013, p. 96) hold that the bisection of the *Sophist* is propaedeutic to the finer way of dividing we find in the *Statesman*. As far as the goal of Division, Aristotle's criticism, according to which Division is "something like a weak syllogism (*asthenês syllogismos*)" since it cannot prove the definition it produces (*An. Pr.* I, 31, 32–34), has proven to be very influential. Scholars have mostly followed in Aristotle's footsteps by claiming that Plato failed—although a few, like Deslauriers (1990, pp. 215–216), have argued that it is not legitimate to attribute to Plato the failure of a goal he never intended to reach.
2 See, e.g., Cherniss (1962), p. 47, n. 36.
3 Fossheim (2012), p. 101.

essary conditions for being a method, namely technical vocabulary and a set of formal rules for its proper application, are said to be missing.[4] In my deployment of the word "method", I take Plato's *methodos* more literally as the "way of inquiry".[5] Along the same lines, I take the expression "dialectical method" (ἡ διαλεκτικὴ μέθοδος at *Resp.* VII 533c7) to refer primarily to the *way* by which the interlocutors aim to answer the *ti-esti* question that is the thread running through the inquiry each dialogue depicts. Conceived in this way, the method can alternatively be seen as a "strategy", as a "(dialectical) technique", and as a "device", as suggested by the word *mêchanê*, which Socrates employs in the *Philebus* (*Phlb.* 23b7). Although (as should be obvious) Plato's *methodos* does not overlap with our modern notion of method, I shall show that both technical vocabulary and formal rules for the application of the method of Collection and Division can be found in the dialogues, albeit not in the place they are usually thought to be located.

2 The *Phaedrus:* The New Dialectical Method?

According to the standard interpretation, it is in the *Phaedrus*, the dialogue usually thought to be the earliest in the set of *Phaedrus*, *Sophist*, *Statesman*, and *Philebus*, that the new method is introduced for the first time.[6] Commenting on what is considered to be the canonical passage (*Phdr.* 265c8–266c8), with which I shall deal extensively below, Hackforth paradigmatically wrote:

> It is in this section that Plato for the first time formally expounds that philosophical method—the method of dialectic—which from now onwards become so prominent in his thought, especially in the *Sophist*, *Statesman* and *Philebus*.[7]

According to Philip, in the *Phaedrus* the features of this "important innovation" are "outlined as they would hardly be if the method were a familiar one".[8] In calling by name and allegedly *defining* the two procedures of the "new" method that will dominate the late dialogues, the *Phaedrus* is commonly said to break new ground in Plato's conception of philosophical inquiry.

Conversing with Phaedrus outside the Athenian walls, Socrates admits to being a "lover of these divisions and collections" (ἐραστής [...] τῶν διαιρέσεων καὶ συνα-

---

[4] Henry (2011), pp. 246–250.
[5] Larsen (2020), p. 559, n. 26.
[6] See, e.g., Runciman (1962), p. 63, Robin (1985), p. CLXXXIV, and Deslauriers (1990), p. 204.
[7] Hackforth (1952), p. 134. See also Robin (1985), p. CLXXXIV, and Notomi (1999), p. 5.
[8] Philip (1954), p. 335, n. 2.

γωγῶν at *Phdr.* 266b3–4) and calls those who can perform them "dialecticians" (διαλεκτικούς at *Phdr.* 266b8–c1). Collection is said to take place when

> someone brings to *one idea* things that are scattered all around by looking at them together (εἰς μίαν τε ἰδέαν συνορῶντα ἄγειν τὰ πολλαχῇ διεσπαρμένα), in order that, by determining each thing, he may make clear what he might want to teach about on each occasion (ἀεί) (*Phdr.* 265d3–5).[9]

Division, on the other hand, means,

> in turn, to be capable of cutting according to *eidē* at the joints in the manner that it is natural and not to try to break up any part, in the manner of an incompetent butcher (τὸ πάλιν κατ' εἴδη δύνασθαι διατέμνειν κατ' ἄρθρα ᾗ πέφυκεν καὶ μὴ ἐπιχειρεῖν καταγνύναι μέρος μηδέν, κακοῦ μαγείρου τρόπῳ χρώμενον) (*Phdr.* 265e1–3).[10]

The *Phaedrus* is said to be not only the earliest dialogue of the set but also the only one that provides the "instructions for use" that are allegedly missing in the other dialogues. The "prescriptive passage" of the *Phaedrus* is said to guide the "actual divisions" carried out uniformly in the *Sophist* and *Statesman* as well as in the *Philebus* (a work which I will consider only tangentially in this paper).[11] The two procedures expounded by the Socrates of the *Phaedrus* are said to be visible at work not only in the *Philebus*, where the main speaker is (again) Socrates, but also in the *Sophist* and *Statesman*. A difficulty arises in regard to the *Sophist* and *Statesman*, since it would be Socrates who provides the theoretical foundations of the "actual divisions" which are carried out in the *Sophist* and *Statesman* by another character, who has there the role of the main speaker, namely the *xenos*, at once guest and stranger from Elea. To overcome this difficulty, one may resort to the "mouthpiece paradigm", according to which, to quote Press, "the words, arguments, and apparent doctrines of the leading speaker in each dia-

---

9 Unless stated otherwise, all translations are mine.
10 As my translation shows, I take *kat'eidē* as conveying conformity ("according to the *eidē*") and *kat'arthra* as distributive ("at the joints", meaning "joint by joint"). For a different reading, see Brown (2010, p. 156, n. 14), who takes Division to be *into* forms *according to* the articulations, and Ryan (2012, p. 272), who takes the first *kata* as distributive ("form by form") and the second one as indicating conformity ("in accordance with their natural (ᾗ πέφυκεν) joints"). See furthermore Muniz and Rudebusch (2018, p. 400, n. 15), who take the second *kata* to stand in apposition to the first and suggest "*according to* forms, to cut through [a thing] *according to* joints which it has by nature".
11 The helpful distinction between "prescriptive passages" and "actual division" is due to Lane (1998), p. 14.

logue are those of Plato himself".[12] Various arguments can be presented against the "mouthpiece paradigm".[13] On my view, we do not need to solve or even to posit the problem which the "mouthpiece paradigm" aims to resolve, since we do not need to resort to the *Phaedrus*. As far as the *Sophist* and *Statesman* are concerned, in fact, what we need can be found in the triad to which the *Sophist* and *Statesman* belong.[14] More specifically, what is called for can be found in the *Sophist*

---

[12] Press (2000), p. 1. See, e.g., Lane (1998), p. 8: "I take the liberty of identifying the Eleatic Stranger's arguments with Plato's".

[13] According to this paradigm, "the exposition of Plato's thought is given entirely to an imaginary and nameless visitor from Elea" in the *Sophist* and *Statesman* (Taylor 1961, p. 5). As González (2000, p. 161) has pointed out, the assumption that the main character speaks for Plato faces a peculiar challenge in the *Sophist* and *Statesman* given that scholars are ready to assume the equation *xenos* = Plato even though they take Socrates to be Plato's mouthpiece until the *Theaetetus*. As Press (2000) explains, the "mouthpiece paradigm" has more recently come under pressure. Drawing a comparison with the case of drama, it has been more recently and, on my view, more persuasively argued that what the author wants to convey goes beyond the lines he attributes to one character, even if this character is the main one. No single character, whoever (s)he might be, but rather the dialogue as a whole is Plato's mouthpiece. For a similar view, see Delcomminette (2000), p. 17, and Trabattoni (2003), pp. 151–157.

[14] The unity of the triad *Theaetetus-Sophist-Statesman* is guaranteed by dramatic, thematic, and methodological aspects of continuity. From a dramatic point of view, the *Statesman* is meant to continue the conversation which was begun in the *Sophist*—a conversation to which the *Statesman* refers repeatedly (see, e.g., *Stm.* 258b6–7, 266d5, 284b7, 286b10, and 291c3–4). The *Statesman* opens with Socrates thanking Theodorus for having introduced Theaetetus and the *xenos* to him (*Stm.* 257a1–2); this happened in the *Sophist*. Furthermore, at the beginning of the first Division that we find in the *Statesman*, the *xenos* himself recalls the inquiry that he led in the *Sophist* (*Stm.* 258b6–7). The *xenos* is a further distinguishing feature of the diptych: The *Sophist* and the *Statesman* are the only dialogues in which this character is to be found and the only ones in which the leading role is assigned to him.

More to the point of this paper, it is stated at the outset of the *Sophist* (*Sph.* 217a4–9) and restated in the same terms at the beginning of the *Statesman* (*Stm.* 257a3–5) that the overarching aim of the discussants is to distinguish three target-terms from each other: the "sophist", the "statesman", and the "philosopher". Whereas each of the first two terms is tackled in the eponymous dialogue, the latter, which is repeatedly advertised, remains unexplored; the future search for the third term, i.e., "philosopher", first suggested at *Sph.* 217a4 is advertised at *Stm.* 257a5 and 257c1 and at *Sph.* 253e7 (for the limitations of such a future occasion, see also *Sph.* 254b4). The outset of the *Statesman*, where Theodorus claims that Socrates will be three times more thankful when the statesman and the philosopher have been discovered (*Stm.* 257a3–5), recalls explicitly the inquiry's plan exposed at the outset of the *Sophist*. In the *Sophist*, immediately before that exposition, we find the first puzzle caused by appearance, which concerns precisely the three targets sophist, statesman, and philosopher: philosophers appear to the many sometimes as statesmen, sometimes as sophists, and sometimes as being completely mad (*Sph.* 216c2–217a2). This confusion will be a constitutive part of the inquiry, since during the search for one of the two target terms,

and *Statesman*, as I shall explain below, as well as in the dialogue that serves as a sort of precursor to the *Sophist* and *Statesman*, namely the *Theaetetus*.

In the eponymous dialogue, Theaetetus tells Socrates what occurred to him and to his fellow Young Socrates during Theodorus' geometry lesson (*Tht.* 147d4–148b3).[15] It occurred to them to try (πειραθῆναι) to collect the powers, which turned out to be unlimited in number, under one term (συλλαβεῖν εἰς ἕν at *Tht.* 147d9–e1) by which they could refer to them all. It also occurred to them to divide all number into two (τὸν ἀριθμὸν πάντα δίχα διελάβομεν at *Tht.* 147e5). Even a superficial glance suffices to explain why it has been argued that the mathematical discovery is given in terms of Collection, which significantly involves naming as unifying trigger, and Division.[16] If it is true that the mathematical discovery that is attributed to Theaetetus and his fellow Young Socrates in the *Theaetetus* heralds the *kind* of philosophical enterprise that awaits them in the *Sophist* and *Statesman*, then the method introduced by the *xenos* from Elea in the *Sophist* is not completely new for the two young mathematicians. Nevertheless, the *paradeigma* of the angler at the outset of the *Sophist* shows that in spite of a certain familiarity with the method, Theaetetus needs to practice on a target that is more ready at hand—a point to which I shall turn at the end of the paper. Moreover, and more importantly for the topic at hand, if I am right in claiming that we find an embryonal employment of the method of Collection and Division already at work in the *Theaetetus*, then the alleged novelty of the canonical and prescriptive passage of the *Phaedrus* ought to be curtailed.

3 Theory and Praxis Intertwined

As explained above, the *Phaedrus* passage is said to offer the "prescriptive passages" that are allegedly missing in the *Sophist* and *Statesman*.[17] On my view, re-

---

the other ones will reappear: in searching for the sophist, the interlocutors stumble upon the philosopher (*Sph.* 253c7–9); in searching for the statesman, the sophist will reappear (*Stm.* 303b6–c7).
15 As noticed by Napolitano (2011, p. 81) in his report of their mathematical discovery, Theaetetus *always* employs the plural form "we" (ἡμῖν at *Tht.* 147c8 explained as "ἐμοί τε καὶ τῷ σῷ ὁμωνύμῳ τούτῳ Σωκράτει" at 147d7, ἡμῖν at 147e1, προσαγορεύσομεν at 147e1, διελάβομεν at 148a7, ὡρισάμεθα at 148b3) and having heard of the brilliant achievement, Socrates (the elder) praises both boys (*Tht.* 148b4).
16 For a more detailed argumentation, see also Napolitano (2011), pp. 78–79, and Negrepontis (2012), p. 5. On the passage *Tht.* 147d4–148b3, Burnyeat (1978) is still seminal.
17 The *Philebus* requires an extensive treatment that exceeds the limits of the present paper. As explained above, I refer to this dialogue only peripherally.

sorting to the *Phaedrus* to make sense of the "actual divisions" of the *Sophist* and *Statesman* is misguided for at least two reasons. On the one hand, by referring to another dialogue, we overlook that there are "instructions for use" not only in the *Statesman* but even in the *Sophist*, as I will show below. On the other hand, by overrating the prescriptive character of the *Phaedrus* passage we run into two complementary mistakes. Firstly, we fail to recognize that the *Phaedrus* passage is far less informative than usually assumed. In fact, it raises more questions than it resolves. Is the *mia idea* (only? also?) a Platonic Form?[18] What sort of objects are the "things scattered all around"? How exactly can one bring various "things scattered all around" to one *idea*? One might also wonder about the kind of cognitive process to which *synorônta* alludes. Already the "definition" of Collection (*Phdr.* 265d3–5), then, puts interpreters to the test. Yet the "definition" of Division (*Phdr.* 265e1–3) confronts interpreters with even greater challenges, since the exact meaning of the passage, beyond the superficial injunction not to proceed like a bad butcher would, is controversial.[19] Are the joints the same as the *eidê* or do they play a different role in the process of Division? Are the natural joints criteria to be followed or results to be reached within Division? How is the dialectician supposed to recognize the natural joints? Precisely because of all these open questions, one may legitimately wonder how prescriptive the so-called "prescriptive passages" are. In light of both these and the previously mentioned difficulties, it therefore seems preferable to consider the "definitions" of Collection and Division as descriptions, which leave important issues unresolved.

By overemphasizing the prescriptiveness of the *Phaedrus* passage, we are liable to a further oversight: it escapes our notice that the "canonical passage" of the *Phaedrus*, which allegedly contains only prescriptions, *does* in fact offer an "actual division". After describing Collection and Division in the passage discussed above, in fact, Socrates offers an illustration of Division just as he did before with Collection:[20]

> as just now the two speeches took the unreasoning aspect of the mind as one kind together, and just as a single body has its parts in pairs, with both members of each pair having the

---

18 According to Hackforth (1952, p. 132, n. 4), the first part of the sentence is "probably meant to include both the bringing of particulars under a Form or kind and the subsumption of a narrower Form under a wider one".

19 See FN n. 11. For the opposite view, namely that the passage on Collection causes more trouble than the passage on Division, see Dixsaut (2001), p. 126.

20 As Hayase (2016, pp. 113–114) has rightly pointed out, the passage has a well-rounded structure: after a preface on the two *eidê*, Socrates offers, first, a definition of Collection (265d3–5) followed by an illustration of Collection (265d5–7) and then a definition of Division (265d8–e3) followed by an illustration of Division (265e3–266b1).

same name, and labelled respectively left and right, so too the two speeches regarded derangement as naturally a single kind in us, and the one cut off the part on the left-hand side, then cutting it again, and not giving up until it had found among the parts a love which is, as we say, "left-handed", and abused it with full justice while the other speech led us to the parts of *mania* on the right-hand side, and discovering and exhibiting a love which shares the same name as the other, but is divine, it praised it as cause of our greatest goods (*Phdr.* 265e4–266b1, trans. Rowe, slightly modified).

The reference to an "actual division" within what is usually reduced to a "prescriptive passage" shows that theoretical reflections are intertwined with actual applications of the method, which, in turn, pave the way for further theoretical reflections. This is notably the case in the *Statesman*, where Young Socrates' mistaken setting apart of humans and beasts launches the *xenos*' famous methodological digression about "*barbaros*" and ten thousand (*Stm.* 262c8–263a3). Yet this is also the case for a few remarks scattered across the *Sophist*, which usually go unremarked. At *Sph.* 227a7–c6, the *xenos* addresses three key features of "the *methodos* of the arguments" (ἡ τῶν λόγων μέθοδος, cf. *Sph.* 227a8), namely its aim, its "axiology", and the function of names; at *Sph.* 235b8–c7, he tackles the Division's recursive splitting; at *Sph.* 264d12–265a2, Theaetetus is explicitly told that they should now attempt to "to split into two (σχίζοντες διχῇ) the *genos* set up before (τὸ προτεθὲν γένος) and to always proceed (πορεύεσθαι) along the right(-hand) part of what has been cut (κατὰ τοὐπὶ δεξιὰ ἀεὶ μέρος τοῦ τμηθέντος)" in order to reach the nature (φύσις) of the sophist.[21] The three most extensive pieces of advice we find in the *Sophist* arise *in the course of* the Divisions performed in the Outer Part of the dialogue.[22] Theory and praxis are intertwined also in the *Philebus*. There, Socrates explains that the method, of which he has always been a lover (*Phlb.* 16b5–6), is "not very difficult to clarify, but extremely difficult to use" (*Phlb.* 16c1–2). The rest of the passage with its descriptions of the method (at *Phlb.* 16c5–17a5 and at *Phlb.* 18a7–b5, respectively) and its application to sound (*Phlb.* 17a6–18d2) shows performatively that—*pace* the Socrates of the *Philebus*—both the application *and* the clarification are difficult. In all four dialogues, theory and praxis are intertwined.

Besides overturning the image of the *Phaedrus* as a dialogue which provides theoretical reflections only, (a) the "actual division" of *mania* brings us back to Socrates' Palinode, thus showing once more that actual division arises from theory, which in turn arises from previous praxis. Furthermore, (b) the division of

---

[21] Pace Brown (2010, p. 154, n.6), who takes *Sph.* 264d11–265a2 to be a description of the procedure's aim.
[22] I borrow the expression "Outer Part" from Notomi (1999, p. 27), who has aptly remarked that the two parts of which the *Sophist* consists are like the yolk and the white of an egg. The Core Part is entailed in the Outer Part, which is interrupted at 237b6 and resumed at 264b11.

mania in the *Phaedrus* points towards what I consider to be a crucial feature of the method of Collection and Division, namely its variety. Socrates distinguishes on the left-hand side—notice the terminology "right-hand" and "left-hand", which is also prominent in the *Sophist*—a *mania* "which comes about from human diseases" (*Phdr.* 265a9–11) and, on the right-hand side, a *mania* which has a divine source. The divine *mania* is divided into four parts (τέτταρα μέρη at *Phdr.* 265b2), each of which is only *now* associated with a god.[23]

It is only afterwards that Socrates "systematizes" what he addressed in the Palinode (starting at *Phdr.* 244a1).[24] There, he already distinguished the mantic, telestic, and poetic kinds of *mania*; he furthermore characterized each of these three "kinds", whereas erotic madness had only been delineated.[25] A proper systematization only *follows* the actual Division.

(b) Even if it starts off as dichotomous, the taxonomic Division of *mania* differs from the dichotomous Division as the latter is mostly carried out in the *Sophist*. Contrary to the dichotomous Divisions which always go "down the right side" (see *Sph.* 264e1), the dichotomous Division of *mania* considers not only the right but also the left side.[26] Furthermore, unlike Dichotomy as it is carried out in the *Sophist*, the Division of *mania* in the *Phaedrus* does not involve the successive rejection of the left-hand side. On the contrary, after a twofold Division, both sides, right *and left*, are named and pursued. Unlike in the *Sophist* and *Statesman*, where, as I shall explain in more detail below, the interlocutors aim to track down a given object,[27] the Division of divine *mania* displays a taxonomic character.

It has mostly escaped notice that a taxonomic Division, namely, a Division that deals with the right side as well as with the left side, also occurs in the so-called Outer Part of the *Sophist*.[28] Yet it occurs only once. Within the last Division of the dialogue, the *xenos* suggests dividing the productive *technê* not only widthwise (κατὰ πλάτος) but also lengthwise (κατὰ μῆκος) (*Sph.* 266a1–2). The result, as the *xenos* himself explains, is four parts (τέτταρα μέρη at 266a4) that are divided

---

23 See Ryan (2012), p. 270.
24 As Ryan (2012, p. 270) has rightly remarked, "Socrates feels free to interpret and enlarge upon what he said earlier, particularly in the Palinode, rather than merely recalling or reproducing it".
25 Socrates also keeps consistent in the number which get assigned to each of the four kinds of *mania*. See, e.g., the mention of the third *mania* at *Phdr.* 245a1 and of the fourth *mania* at *Phdr.* 249d4–5.
26 As noticed by Trevaskis (1967), p. 118.
27 As noticed by Deslauriers (1990), p. 208. For a similar view, see Brown (2010), p. 154, and Ionescu (2014), p. 37.
28 As aptly remarked by Brown (2010), p. 154, n. 6.

into two groups, each of which is divided into two, as Theaetetus' later recapitulation clearly shows:

> Now I understand better and I posit the two kinds of productive *technê*, each of which is divided into two, according to one cut, divine, on the one hand, and human, on the other hand; according to the other cut, a generation of the things themselves, on the one hand, and of the copies of these things, on the other hand (*Sph.* 266d5–8).

This twofold Division in width and length can be illustrated as follows (see tab. 1):

Production	Of real things (*autopoiêtikon*)	Of copies (*eidôlopoiikon*)
Divine	Divine production of real things	Divine production of copies
Examples (cf. *Sph.* 266b)	Like fire, water, the elements	Like dreams and shadows
Human	Human production of real things	Human production of copies
Examples (cf. *Sph.* 266c)	Like the construction of a house	Like the painting of a house

Tab. 1

Instead of the customary recursive cut on the right-hand side, we find a double bisection. In this sense, it is true that the Division follows a dichotomous principle. This principle, however, does not imply, as is usually thought to be the case, that the left-hand side is neglected. Rather, the contrary is the case, for both the right- and the left-hand side come under consideration. Given that, strictly speaking, the taxonomic Division also follows the principle of dividing "into two", its specificity has usually gone unnoticed. In other words, the taxonomic Division is not perceived as different from the customary dichotomous Division that we find in the *Sophist*.

4 Uniformizing Trend and Its Limitations

This oversight is not an isolated case; rather, it mirrors a general "uniformizing" trend.[29] With respect to the *Sophist*, it is generally agreed that the method of Col-

---

29 By means of contextual, linguistic, functional, and objective comparisons, Benitez (1989, p. 43)

lection and Division employed in the so-called Outer Part overlaps with the "science of free men" (cf. *Sph.* 253c7–8) thematized in the Core Part.[30] This interpretive trend can best be exemplified by Runcimas' remarks upon *Sophist* 253d–e, where the *xenos* enumerates the key capacities that distinguish the "dialectical science" (cf. *Sph.* 253d2–3). Runciman writes:

> Here the method by which the Sophist is being defined is explicitly bracketed with the method which determines the interrelation between the μέγιστα γένη. The method as a whole is summarized in the long compact sentence which runs from 253D5 to 253E2. […] Plato appears in this brief exposition to be giving a total summary of his new philosophical method.[31]

I resist this overlap on several grounds. For the sake of brevity, I shall limit myself to mentioning only one main reason here, which concerns the objects. While the search for the sophist does not need to be exclusively about Forms or to make exclusive use of Forms as criteria for dividing and collecting, I take the dialectical science, which aims at the bright region of Being (*Sph.* 254a9), to deal with Forms only.[32] With respect to the *Philebus*, the specificity of the fourfold Division of "everything that actually exists" into *apeiros*, *peras*, the mixture (σύμμειξις) of these two, and the cause (αἰτία) of this combination (*Phlb.* 23c9–d8) has not usually been appreciated. Although Socrates expressly introduces it as "a different device (μηχανή)" (cf. *Phlb.* 23b7), which does suggest a methodological shift in the passage *Phlb.* 23b5–26d10, it is usually held that

---

identifies the method in *Phlb.* 16b–19a with the method of division familiar from the *Phaedrus*, the *Sophist*, and the *Statesman*.

30 Notable exceptions to this trend are Philip (1966), p. 345, and Krohs (1998), p. 237.

31 Runciman (1962), p. 62.

32 More specifically, I take the notoriously difficult passage about the philosopher's abilities in *Sph.* 253d5–e2 to be exclusively about Forms. As is known, from a purely grammatical point of view this claim is problematic as far as the first clause is concerned (*Sph.* 253d5–6), since unlike the rest of the passage a couple of non-feminine or not securely feminine terms (πολλῶν and ἑκάστου respectively, both at *Sph.* 253d6) might invite one to think that we are dealing with the usual relationship between Forms and particulars. This reading can be resisted by suppling ἰδεῶν (of the line before) in the case of πολλῶν (d6) and *eidos* in the case of ἑκάστου κειμένου (also d6). A full-fledged defense of the philosophical reasons why I take this passage to be about Forms and Forms only goes beyond the scope of the present paper. One pertinent observation may therefore suffice: in the *Statesman* (*Stm.* 285e4–b2), the *xenos* explains that "the things now being said" (*Stm.* 286a7), which I take to also include the inquiry depicted in the *Sophist*, are "for the sake of the greatest and most valuable things, which can be shown by means of *logos* alone" (*Stm.* 286a5–6).

even if we see the fourfold division merely as innovative use of collection and division, the passage does nothing to support the notion that we are witnessing the representation of a separate method.[33]

With respect to the *Statesman*, the attempt to distinguish "Division by limbs" from dichotomous Division has typically met with some resistance, before the more recent turnaround.[34]

Moving from the level of the individual dialogue to a comparison between various dialogues, then, what I call the "one-method interpretation" seems difficult to disprove, although a few discrepancies between the *Statesman*, the *Sophist*, the *Philebus*, and the *Phaedrus* have been spotted.[35] This interpretive attitude can be illustrated by the following statement:

> Perhaps the first question which we should ask is whether we are meant by Plato to consider division as outlined in the *Phaedrus* and as practiced in the *Sophist* to be essentially one and the same method. Plato does not tell us explicitly, but, in view of the similar terminology employed in the two dialogues, it would be implausible to maintain that the two methods are to be distinguished.[36]

The main reason for challenging the "one-method interpretation" concerns its explanatory power. In subsuming all the applications of Collection and Division which we find in the dialogues, the "one-method interpretation" smooths over their significant differences and hence bends the texts to its constraints. A good example is the total neglect of the procedure of Collection, which originates from the reduction of the method of Collection and Division to the method of Division.[37] If it is considered at all, Collection is said (i) to *precede* Division, since Collection aims "to divine the generic Form which is to stand at the head of the subsequent Division"[38] or (ii) "to be completed before Division starts: once Division has started, Collection plays no more part in dialectic method".[39] (i) Expecting Collection to provide the generic term from which Division starts, we readers often over-

---

33 Fossheim (2010), p. 33. The translations of the *Philebus* are by Frede.
34 See Delcominette (2000), p. 80, Smith (2019), p. 16, and Liu (2021).
35 See, e.g., Runciman (1962, p. 62), who in considering the *Phaedrus* cautiously points out that the method is "somewhat differently" described in *Phlb.* 16c–17a, and Pellegrin (1991: 400–1), who argues for a change in the dialectical method which occurred between the *Phaedrus* and the *Sophist*. See also Dorter (2013, p. 94), who holds that the dichotomous method of Division of the *Sophist* is unique.
36 Trevaskis (1967, p. 119).
37 See, e.g., Taylor (1957), p. 377, and Koller (1961), p. 6.
38 Cornford (1935), p. 170.
39 Hackforth (1972), p. 142.

look the fact that the search for the sophist in the Outer Part of the dialogue does not start with a Collection, but rather with an "exclusive disjunction", to use Cavini's term.[40] It is the very question of whether the angler and, later and analogously, the sophist has or lacks a *technê* that properly launches the inquiry.[41] (ii) Moreover, in expecting Collection to play no further role after having launched Division, we fail to acknowledge that Collection is employed not before, but rather *within* Division itself.[42] If we were to assume that the method usually referred to as the method of Division encompasses Division only, we would fail to acknowledge that Collection is employed at all.

Starting from the passages themselves instead of constructing a model into which the passages have to fit, it is noteworthy that, even within one single dialogue from this period, we encounter different methods, that is, methods that deal with different objects, proceed in different ways, and aim at different goals. This is illustrated by the Division that proceeds widthwise and lengthwise in the *Sophist* (*Sph.* 266aff.) as well as by "Division by limbs" in the *Statesman* (*Stm.* 287cff.). Instead of a recurring bisection that sets aside at each step the side that does not lead to the target, the whole is systematically broken off into its components.[43] One after another, the limbs are distinguished. It is worth noting that the procedure just introduced has already been carried out and will be carried out right after. To put it boldly, the new kind of Division is framed by its actual employment so that we end up with the familiar sequence of employment—explanation—employment. Moreover, Division by limbs demonstrates that the *xenos* is aware of proceeding in a different way. In stating that they should divide "limb by limb" (κατὰ μέλη), "like a sacrificial animal" (οἷον ἱερεῖον), because they cannot

---

40 Cavini (1995), pp. 124–125.
41 Theaetetus is asked the following: "Will we posit that he [i.e., the angler] is someone having a *technê* or that he is someone without *technê*, but having some other capacity? (πότερον ὡς τεχνίτην αὐτὸν ἢ τινα ἄτεχνον, ἄλλην δὲ δύναμιν ἔχοντα θήσομεν; at *Sph.* 219a5–6). The two disjuncts are mutually exclusive and exhaustive: the angler has or lacks a *technê*, *tertium non datur*. By answering that "he is not *atechnos* at all" (*Sph.* 219a7) Theaetetus establishes that the angler has a *technê*. It is from the assumption that *just like the angler* also the sophist has to be posited as someone having a *technê*, which is recalled at the beginning of D1 (at *Sph.* 221c9–d2), that all Divisions take their starting point—not from an initial Collection, since even D6, which *does* begin with a Collection (*Sph.* 226b–c), implicitly assumes "having a *technê*" as its point of departure.
42 This is best exemplified by the very first Collection of the *Sophist*, which aims to secure the one side of the very first Division of the dialogue, namely that of *technê* into acquisitive and productive (*Sph.* 219a10–b2).
43 See Gill (2012), p. 191.

divide into two (*Stm.* 287c3–4), he is consciously offering an alternative.[44] The alternative character of the *xenos*' suggestion is made manifest by Young Socrates' question: by asking how they should proceed *now* (*Stm.* 287c6), Young Socrates clearly shows he has understood that they are going to proceed in a different way. The new way will differ from the one mostly applied so far, a way which Young Socrates has surely internalized by silently acknowledging the enterprise of his fellow Theaetetus all throughout the *Sophist*.

5 A Family of Methods of Collection and Division

Scholars like Gill have rightly argued that the employment of the (one) method foresees a considerable variety.[45] It has also been proposed that the various employments of the method can be best understood as members of a family.[46] Speaking of a "family" seems particularly fitting because it allows us to account for the connections within such a variety. By keeping together various degrees of similarity that cross and overlap, degrees of proximity among the various instances as well as the intrinsic possibility of expansion, the concept of "family" connects these members tightly without turning them into an undifferentiated unity.[47] As far as the members of the family are concerned, however, it seems to me more appropriate to speak of multiple methods and not of employments of what could still be one single method. In other words, we are not dealing with different employments of the same method, but with different methods altogether. To clarify adequately how each method relates to the others, insofar it comes close to another method and insofar it differs or derives from other methods, would require a more in-depth analysis than I can offer here. In what follows, I shall nevertheless attempt to reconstruct the relationships among the different methods by offering a

---

[44] In the direct aftermath, the *xenos* explains that when dichotomy is not possible, then one has to divide into the number as close as possible to two (*Stm.* 287c4–5). At the beginning of the conversation, the *xenos* explained that "it is safer (ἀσφαλέστερον) to go on cutting through the middle (διὰ μέσων) and one is more likely to hit upon *ideai*" (*Stm.* 262b6–8).

[45] Gill (2012), p. 218, n. 39. Significant differences are noticed, among others, by Pellegrin (1991), p. 398, Kahn (1996), p. 299, and Ionescu (2014), p. 37.

[46] On this issue, I have greatly profited from reading Gill's manuscript on "The Varieties of Platonic Division" (2019).

[47] I initially worked with the concept of "Familienähnlichkeit" (translated as "family resemblance" or "family likeness"), an expression Wittgenstein uses to refer to "ein kompliziertes Netz von Ähnlichkeiten, die einander übergreifen und kreuzen. Ähnlichkeiten im Großen und Kleinen" ("an intricate web of similarities that overlap and intersect") in the *Philosophical Investigations* (§ 66).

family-tree, as it were—one which is tentative and not exhaustive (since, most notably, the *Philebus* is deliberately not included).

In the family-tree, the long-lost relatives are two of the most basic functions of human thought, namely that of bringing together what is akin, on the one hand, and that of keeping it apart from the rest, on the other hand.[48] Attempts both to divide and to collect can be found from the beginning not only of human thinking but also of Plato's own work.[49] In that sense, a tree that also considers Plato's previous work, namely dialogues written before the late ones, would include as relatives at the top of the tree the methods of Collection and Division carried out in the *Euthyphro*, the *Meno*, the *Gorgias*, and the *Republic*.[50] As the diagram shows, I begin by positing as a kind of archetype Division, which includes Collection, since, as I have explained above, Collection is often carried out within the process of Division. The kind of Division that we predominantly find in the texts is dichotomous, and hence the first kind of Division is Dichotomy. The kind of Dichotomy we mostly find in the text is a recursive Dichotomy proceeding down the right-hand side and discarding the left-hand side, since the stated goal is to track down a given target without providing an exhaustive organization of the whole field. Departing from this customary Dichotomy, we also find a Dichotomy that considers the left-hand side (if we consider the development of the Divisions within the search for the sophist, where for example the first Division starts from the "discarded" side of the previous one) and a single case of double bisection (namely the above-mentioned Division widthwise and lengthwise at *Sph.* 266aff.). By exhausting the field under consideration, the latter does de facto result in a taxonomy, which seems to be the goal of the Division of *mania* in the *Phaedrus*—although in the latter passage there is a clear focus on the right-hand side, as is the case in customary Dichotomy. A more consciously taxonomic interest, one which does not limit itself

---

[48] I would like to thank Silvia Fazzo for stimulating me to think more about this issue.

[49] As is known, there is no agreement on the chronology of Plato's dialogues. For the purpose of my argument, I assume that dialogues such as the *Republic* and the *Gorgias* precede the four late dialogues *Sophist*, *Statesman*, *Phaedrus*, and *Philebus*. I am aware of the difficulty of my claim especially as far as the *Phaedrus* is concerned.

[50] An embryonic form of the method of Collection can already be found in the *Euthyphro*, the *Meno*, and the *Gorgias* (*Euth.* 5d8–e2 and 6d–e; *Men.* 71e1–72a5, 72a–74b, and 81c–d; *Gorg.* 464b–d), whereas an early version of the method of Division is already to be found in the *Gorgias* (*Gorg.* 450c7–e2; see the close affinity of this Division of arts requiring or avoiding speech with *Stm.* 258d4–e5, and 464b–466a, and see the close affinity of the fourfold Division of the care of the soul and the care of the body with *Soph.* 226e–229a). The *Republic* also hints at both procedures: Collection seems to be alluded to at 531d1–2 and 537c7, and Division seems to be at work in 532b–535a and 454a6, as aptly noticed by Ferber (1989), p. 107.

to the right-hand side, is apparent in Division by limbs, which, as I shall explain below, derives from Dichotomy (see fig. 2).

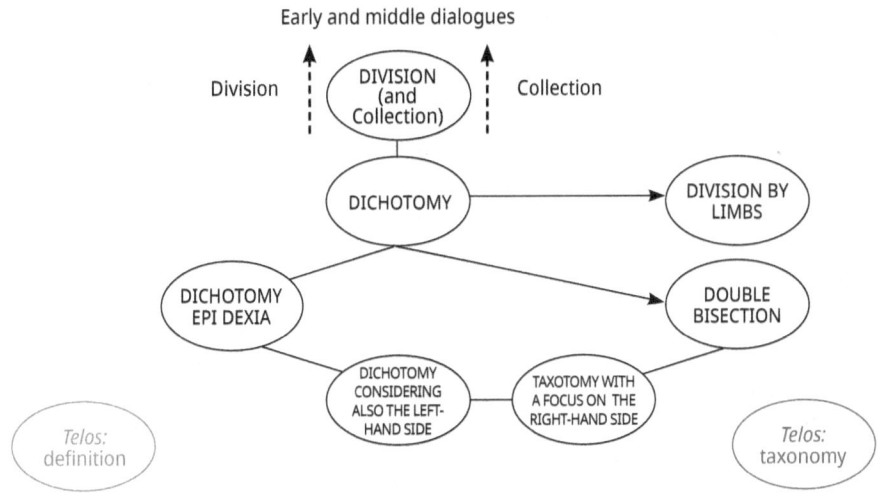

Fig. 2: The Family Tree of the Methods of Collection and Division

As the diagram shows, I take some kinds of Division to give rise to others, which thus derive from the former, as the arrows indicate. By definition, the concept of family involves expansion; such expansion accounts for the relationships among the various methods we see at work in the dialogues (as well as for further developments which, as I shall argue below, we cannot foresee yet). The fourfold taxonomic Division can be said to "conceptually" derive from Dichotomy. Yet, such a fourfold Division develops into something different from its source.[51] This is also the case for Division by limbs. In fact, Division by limbs somehow derives from Dichotomy, since the *xenos* expressly presents the necessity of dividing into the number as close as possible to two (*Stm.* 287c4–5). Division by limbs does not aim at the "uncut whole" (ἄτομον πᾶν at *Sph.* 229d5–6), which is the declared goal of Dichotomy.[52] Rather, by enumerating its *eidê* (*Stm.* 288a3 ff.), it aims to unravel the inner organization of a whole. In this sense, Division by limbs is not a pejorative deviation from the golden source; instead, it represents the "new" re-

---

[51] Liu (2021) also thinks that one method originates from the other.
[52] In translating *atomon pan* as "uncut whole", I concur with Krohs (1998), p. 241. For a different reading (i.e., "as a whole, uncut"), see Brann, Kalkavage, and Salem, *ad loc.*

sponse to a challenge which the previous kind of Division has proved unable to deal with adequately.

As the diagram shows, I take as the leading issue the question concerning the goal: What is Division for? Following Pellegrin and Brown, I distinguish between "taxonomy" and "definition".[53] In the case of taxonomy, Division aims to delineate the entire structure of a given whole, while in the case of definition, Division aims at isolating a given target. In the case of taxonomy, the classification aims to exhaust the given field, whereas in the case of definition, the search does not aim to exhaust the given field or to provide an exhaustive account of how the field is internally structured. The clarification of the goal is also fundamental for evaluating the success of any method: only by knowing the declared goal can we assess the actual outcome against it.[54] In turn, the kind of goal to be reached depends on the kind of object under investigation.[55] More basically, then, the goal of the method is intimately related to the goal of the whole discussion unfolded in the dialogue. Distinguishing the sophist from the statesman and the philosopher differs profoundly from answering the question of what love is; in the former case, we have three words that identify three items, whereas in the latter case, we have one single word that refers to what turn out to be two distinct items. In the case of "*erôs*", we need to disambiguate a word that the inquiry shows to refer both to a human and to a divine kind of *mania*, whereas in the case of "sophist", "statesman", and "philosopher", it is stated at the very beginning of the inquiry that each word picks out a different item (*Sph.* 217b1–2) and "yet it is not a small or an easy task to determine clearly what each of them [i.e., the sophist, the statesman, and the philosopher] is" (*Sph.* 217b2–4). Because of the purpose, the interlocutor is entitled to change the procedure; in turn, the purpose depends on the kind of object under investigation. The procedure depends on the goal that, in turn, depends on the object under investigation. This is so because, as the central books of the *Republic* teach, distinct objects determine the kind of grasp we

---

53 Brown (2010), p. 154, and Pellegrin (1991), pp. 398–399.
54 Remarks about the intended goal are to be found at various points during the inquiries. One of the most striking cases is probably the last remark that precedes the search for the angler in the *Sophist*. Before starting off with the search for the angler, the *xenos* claims: "I expect that he [i.e., the angler] has a *methodos* and an account not unsuitable for us, for what we want" (*Sph.* 219a1–2).
55 Even though he admits that what Socrates says of the Divine Method in the *Philebus* is strikingly similar to what he says of Collection and Division in the *Phaedrus*, *Sophist*, and *Statesman*, Hampton (1990, p. 27) claims that "the type of division one should make depends on the nature of the particular Form that is under investigation".

can have of them (*Resp.* V 477a10–478d12).⁵⁶ Each object requires its own specific method of inquiry; each method of inquiry demonstrates how differently the dialecticians can operate and how wide dialectic can reach. For each actual method of inquiry shows how wide is the spectrum of objects that the dialectical enterprise is —but, also and more importantly, can be—concerned with.

Before turning to the manifold character of the dialectical enterprise, I shall address two opposite objections to which my interpretation lies open.

(a) Against my interpretation, one could argue that if what I refer to as different methods differ in their objects, their procedure, and their goal, it is not clear what keeps them together and, more importantly, why it is useful to keep them together. In other words, why not suppose that each individual employment of the method is unique, isolated, and detached from the others? To respond to this objection, it suffices to show that, in the case of the *Sophist* and *Statesman*, the parallels between the different employments of the method are expressly drawn by the *xenos* himself. He is the one connecting the various methods by both underlining their continuity and marking their difference. He is the one encouraging his young interlocutor to proceed as before by continuing to bisect, as happens in the Outer Part of the *Sophist*, or to introduce a difference by operating on a similar level, as happens with Division by limbs in the *Statesman* (where, as I explained above, Young Socrates asks how they should proceed *now*).⁵⁷

(b) Alternatively, and conversely, one could object to my reading that instead of different methods, we should assume one single but extremely flexible method— one which, due to its flexibility, could embrace a variety of objects, procedures, and goals. On my view, if we assume that a single method operates across the late dialogues, we will miss Plato's point of offering dialectical devices that are, at the same time, similar enough to generate a pattern that can be learned and different enough to give this one pattern the versatility that such divergent goals require. In this sense, the dialogues do not confront us with numerically distinct applications of the same "recipe", as it were, to solve different problems, but with strategies that can be applied in a similar fashion to questions and problems that are profoundly distinct from one another.

---

56 For a convincing account of what she calls "Plato's objects-based epistemology", see Moss (2021), pp. 50–85. Her whole book constitutes a defense of the view that the cognitive kinds *epistêmê* and *doxa* are defined by their objects.
57 Even in the case of the taxonomy embedded in D7, which is not properly signposted as something new or at least different from the customary bisection carried out so far in the dialogue, it is noteworthy that Theaetetus himself underlies the double bisection (at *Sph.* 266d5–6).

The methods cannot but be different, since the issues they are meant to tackle are different.[58] Different paths need to be travelled in the investigation of pleasure and in the hunt for the sophist, which, in turn, has to be differentiated from the inquiry into the "greatest kinds". In this sense, it is surely right that the main speaker leads his interlocutor along multiple distinct paths with the goal of stimulating him to find other paths in order to solve new problems he may be confronted with. Along these lines, we are not to think that the family members which we find displayed in the dialogues are or can be the only ones. We are also not to think that beyond the methods depicted in the dialogues there are a few similar others that Plato wants us to discover autonomously. The various methods we find in the dialogues are rather to be thought of as *paradeigmata*, both in the sense of examples and of models, which do not and cannot exhaust the methods of Collection and Division. For the paradigmatic methods we do have are members of a family, one that is always expanding. It is such expansion that accounts for further developments that are still to come, thus encouraging the interlocutors and us readers to be open to new methods adequate to deal with future challenges.

6 Becoming More Inventive: The Dialectical Exercise of Training in Difference and Repetition

In the *Statesman*, the interlocutors openly admit that the inquiry at hand serves the chief purpose of "becoming more dialectical on any given topic" (*Stm.* 285d5–6); in the aftermath, we apprehend that becoming "more dialectical" means (if one takes the *kai* at 287a3 epexegetically), or at least connects closely with, becoming "more inventive in the disclosure of beings by means of *logos*" (*Stm.* 287a3–4). In other words, improving our dialectical skills means increasing our own inventiveness as far as the explanation of beings is concerned. Delcominette articulates this point perfectly when he says that such "inventiveness", as he aptly calls it, develops by means of a variety of unfortunately often neglected procedures, with which the dialectical practice is studded.[59] The inventiveness of the dialectical enterprise can best be appreciated if we keep in mind that the dialogues fulfill a pedagogical function, both for *all* the interlocutors involved and for us readers. Just as the main speaker leads his interlocutor through different paradig-

---

58 For a similar view, see Delcominette (2006), p. 97.
59 Delcomminette (2000), p. 80.

matic methods in order to train him to find new ones, so Plato leads us readers through different methods in order to extend our own dialectical inventiveness.

Considering especially the Outer Part of the *Sophist* and the first stretch of the Division of the *Statesman* (until *Stm.* 268d4), it is undeniable that Dichotomy does envisage a certain amount of almost mechanical repetitiveness, which seems difficult to square with the dialectical inventiveness addressed above. In order to appreciate the dialectical character of repetition and to recognize that the method applied to catch the sophist and the statesman is dialectical not in spite of but precisely because of its repetitiveness, it is crucial to bear in mind the lesson we apprehended from the *Parmenides:* to see the truth clearly, the task of rehearsal must foresee variations. As old Parmenides explains to a still very young Socrates,

> And, in a word, concerning whatever you might ever hypothesize as being or as not being or as having any other property, you must examine the consequences for the thing you hypothesize in relation to itself and in relation to each one of the others, whichever you select, and in relation to several of them and to all of them in the same way; and, in turn, you must examine the others, both in relation to themselves and in relation to whatever other thing you select on each occasion, whether what you hypothesize you hypothesize as being or as not being. All this you must do if, after completing your training (τελέως γυμνασάμενος), you are to achieve a full view of the truth (*Parm.* 136b7–c5, trans. Gill and Ryan).

In this sense, there is never blind and mechanical repetition: in order to be pedagogically fruitful, repetition must embed difference.

As the end of the *Parmenides* passage shows, the task of repetition is intimately connected with the necessity of training. It is first and foremost a question of "having trained till perfection" (τελέως γυμνασάμενος at *Parm.* 136c5). Whereas the very task Parmenides presents can appear to be and is definitively seen by the young Socrates as almost impossible (ἀμήχανος πραγματεία; cf. *Parm.* 136c6), training is often presented in the dialogues and by the dialogues as the only practicable route. As such, each (late) dialogue and not only the second part of the *Parmenides* represents a dialectical exercise. In order to become "more dialectical" (*Stm.* 285d6 and 287a3), the interlocutors need to train. As we apprehend from the beginning of the *Sophist*, such training should start from subjects that the interlocutor sees as more ready-at-hand and should then become increasingly challenging (*Sph.* 218c6–e6). Visual aids, which include but are not limited to *paradeigmata* (at the same time examples and models), play a vital role in improving dialectical skills. Such inducement applies not only to the younger interlocutor, who can best understand something closer to his own experience, but also to the more experienced teacher, who shows his skill as a teacher precisely in the choice and in the discussion of the topics and of the abovementioned visual aids. In the *Statesman*, not only the *paradeigma* of the weaver but even the search for the statesman is said to be carried out not for its own sake but for the sake of becoming "more dialectical on any given topic" (*Stm.* 285d4–7). Exercising through an array of methods in order to internalize them is the main purpose of the inquiry.[60] To make us able to tackle in various ways very different issues, including ones that we cannot conceive yet, Plato offers us not a one single recipe in order to divide and collect but rather a family of methods. It is crucial to stress that the family of the methods of Collection and Division is just one family among the many other families of dialectical strategies that we encounter throughout his work. How exactly this one family crosses and overlaps with other families, such as the "family of hypothetical methods", is a question that requires a further investigation.[61]

Bibliography

Benitez, Eugenio (1989): *Forms in Plato's* Philebus. Assen and Maastricht: Van Gorcum.
Brann, Eva, Kalkavage Peter, and Salem, Eric (1996): *Plato. Sophist. The Professor of Wisdom.* Indianapolis: Hackett.
Brown, Lesley (2010): "Definition and Division in Plato's *Sophist*". In: Charles, David (Ed.): *Definition in Greek Philosophy.* Oxford: Oxford University Press, pp. 151–171.
Burnyeat, Myles (1978): "The Philosophical Sense of Theaetetus' Mathematics". In: *Isis* 69. No. 4, pp. 489–513.
Cavini, Walter (1995): "Naming and Argument: Diaeretic Logic in Plato's *Statesman*". In: Rowe, Christopher (Ed.): *Reading the Statesman. Proceedings of the Third Platonicum.* Sankt Augustin: Academia Verlag, pp. 123–138.
Cherniss, Harold (1962): *Aristotle's Criticism of Plato and the Academy.* Baltimore: The Johns Hopkins University Press.
Cornford, Francis Macdonald (1935): *Plato's Theory of Knowledge. The* Theaetetus *and the* Sophist *of Plato translated with Running Commentary.* London: Routledge & Kegan Paul.
Delcomminette, Sylvain (2000): *L'inventivité dialectique dans le* Politique *de Platon.* Brussels: Ousia.
Delcomminette, Sylvain (2006): *Le* Philèbe *de Platon. Introduction à l'agathologie platonicienne.* Leiden and Boston: Brill.
Deslauriers, Marguerite (1990): "Plato and Aristotle on Division and Definition". In: *Ancient Philosophy* 10. No. 2, pp. 203–219.
Dixsaut, Monique (2001): *Métamorphoses de la dialectique dans les Dialogues de Platon.* Paris: Vrin.
Dorter, Kenneth (2013): "The Method of Division in the *Sophist:* Plato's Second *Deuteros Plous*". In: Bossi, Beatrice and Robinson, Thomas M. (Eds.): *Plato's* Sophist Revisited. Berlin and Boston: de Gruyter, pp. 87–99.
Ferber, Rafael (1989): *Platons Idee des Guten. Zweite, durchgesehene und erweiterte Auflage.* Sankt Augustin: Academia Verlag.

---

60 For a similar view, see Trevaskis (1967), p. 129, Philip (1966), pp. 350–351, and Henry (2011), p. 253.
61 I borrow the label from Rodriguez (2022), who speaks of "Plato's family of hypothetical methods".

Fossheim, Hallvard (2010): "Method in the *Philebus*". In: Dillon, John (Ed.): *Plato's* Philebus: *Selected Paper from the Eighth Symposium Platonicum*. Sankt Augustin: Academia Verlag, pp. 31–35.
Franklin, Lee (2011): "Dichotomy and Platonic Diairesis". In: *History of Philosophy Quarterly* 28, pp. 1–20.
Frede, Dorothea (1993): *Plato. Philebus*. Indianapolis: Hackett.
Gill, Mary Louise (2012): *Philosophos: Plato's Missing Dialogue*. Oxford: Oxford University Press.
Gill, Mary Louise (2019): "The Varieties of Platonic Division". Unpublished Manuscript.
Gill, Mary Louise and Ryan, Paul (1996): *Plato: Parmenides*. Indianapolis: Hackett.
González, Francisco J. (2000): "The Eleatic Stranger: His Master's Voice?" In: Press, Gerald Alan (Ed.): *Who Speaks for Plato? Studies in Platonic Anatomy*. Lanham: Rowman Littlefield, pp. 161–181.
Hackforth, Reginald (1952): *Plato's* Phaedrus. Cambridge: Cambridge University Press.
Hackforth, Reginald (1972): *Plato's Examination of Pleasure: A Translation of the* Philebus *with an Introduction and Commentary*. Cambridge: Cambridge University Press.
Hampton, Cynthia (1990): *Pleasure, Knowledge, and Being. An Analysis of Plato's* Philebus. Albany: State University of New York Press.
Hayase, Atsushi (2016): "Dialectic in the *Phaedrus*". In: *Phronesis* 61, pp. 111–141.
Helmig, Christoph (2004): "What is the Systematic Place of Abstraction and Concept Formation in Plato's Philosophy? Ancient and Modern Readings of *Phaedrus* 249 b–c". In: Van Riel, Gerd and Macé, Caroline (Eds.): *Platonic Ideas and Concept Formation in Ancient and Medieval Thought*. Leuven: Leuven University Press, pp. 83–97.
Henry, Devin (2011): "A Sharp Eye for Kinds: Plato on Collection and Division". In: *Oxford Studies in Ancient Philosophy* 41, pp. 229–255.
Ionescu, Cristina (2014): "Dialectical Method and Myth in Plato's *Statesman*". In: *Ancient Philosophy* 34, pp. 29–46.
Kahn, Charles H. (1996): *Plato and the Socratic Dialogue: The Philosophical Use of a Literary Form*. Cambridge: Cambridge University Press.
Koller, Hermann (1961): "Die dihäretische Methode". In: *Glotta* 39, pp. 6–24.
Krohs, Ulrich (1998): "Platons Dialektik im *Sophistes* vor dem Hintergrund des *Parmenides*". In: *Zeitschrift für philosophische Forschung* 52. No. 2, pp. 237–256.
Lane, Melissa (1998): *Method and Politics in Plato's "Statesman"*. Cambridge: Cambridge University Press.
Larsen, Kristian (2020): "Differentiating Philosopher from Statesman According to Work and Worth". In: *Polis. The Journal for Ancient Greek and Roman Political Thought* 37, pp. 550–566.
Liu, Xin (2021): "On Diairesis, Parallel Division, and Chiasmus". *Plato Journal: Journal of the International Plato Society*. Forthcoming.
Moss, Jessica (2021): *Plato's Epistemology: Being and Seeming*. Oxford: Oxford University Press.
Muniz, Fernando and Rudebusch, George (2018): "Dividing Plato's Kinds". In: *Phronesis* 63, pp. 392–407.
Napolitano, Linda M. (2011): "Teodoro, Teeteto, Socrate il Giovane. I matematici deuteragonisti nel Teeteto, Sofista e Politico". In: Lisi, Francisco, Migliori, Maurizio, and Monserrat-Molas, Josep (Eds.): *Formal Structures in Plato's Dialogues. Theatetus, Sophist and Statesman*. Sankt Augustin: Academia Verlag, pp. 72–83.
Negrepontis, Stylianos (2012): "Plato's Theory of Knowledge of Forms by Division and Collection in the *Sophistes* is a Philosophic Analogue of Periodic Anthyphairesis (and Modern Continued Fractions)". Unpublished Manuscript.

Notomi, Noburu (1999): *The Unity of Plato's "Sophist": Between the Sophist and the Philosopher.* Cambridge: Cambridge University Press.

Pellegrin, Pierre (1991): "Le *Sophiste* ou de la division. Aristote-Platon-Aristote". In: Aubenque, Pierre (Ed.): *Études sur le* Sophiste *de Platon.* Naples: Bibliopolis, pp. 389–416.

Philip, James A. (1966): "Platonic dihairesis". In: *Transactions and Proceedings of the American Philological Association* 97, pp. 335–358.

Press, Gerald Alan (2000): "Introduction". In: Press, Gerald Alan (Ed.): *Who Speaks for Plato? Studies in Platonic Anatomy.* Lanham: Rowman Littlefield, pp. 1–11.

Robin, Leon (1985): *Platon IV 3: Phèdre.* Paris: Belles Lettres.

Rodriguez, Evan (2022): "A Long Lost Relative in the *Parmenides?* Plato's Family of Hypothetical Methods". In: *Apeiron* 55. No 1, pp. 141–166.

Rowe, Christopher (1986): *Plato. Phaedrus.* Warminster: Aris & Phillips.

Runciman, Walter Garrison (1962): *Plato's Later Epistemology.* Cambridge: Cambridge University Press.

Ryan, Paul (2012): *Plato's* Phaedrus: *A Commentary for Greek Readers.* Norman: University of Oklahoma Press.

Smith, Colin C. (2019): "Dialectical Methods and the *Stoicheia* Paradigm in Plato's Trilogy and *Philebus*". In: *Plato Journal: Journal of the International Plato Society* 19, pp. 9–29.

Taylor, Alfred Edward (1957): *Plato, the man and his work.* New York: Meridian Books.

Trabattoni, Franco (2003): "Il dialogo come portavoce dell'opinione di Platone. Il caso del *Parmenide*". In: *Quaderni di acme,* pp.151–178.

Trevaskis, John R. (1967): "Division and Its Relation to Dialectic and Ontology in Plato". In: *Phronesis* 12, pp. 118–129.

Wedin, Michael V. (1987): "Collection and Division in the *Phaedrus* and *Statesman*". In: *Revue de Philosophie Ancienne* 5, pp. 207–233.

Wittgenstein, Ludwig (1953): *Philosophische Untersuchungen.* Oxford: Basil Blackwell.

Lucas Angioni
Sophistical Demonstrations: A Class of Arguments Entangled with False Peirastic and Pseudographemata

1 Introduction

There is a class of sophistical arguments that are entangled both with peirastic arguments and with pseudographemata. They have recently received attention in the literature, especially because of difficulties they pose both to Aristotle's classification of sophistical fallacies in *Sophistical Refutations* 8 and his demarcation between dialectic and scientific expertise.[1] Sophistical arguments of this kind are valid (and even sound) deductions of their conclusions. What makes them sophistical is a false appearance both of scientific expertise and of success in the task proper to peirastic, namely, exposing false claims of scientific expertise. Now, that combination sounds surprising: for, since peirastic success in exposing false claims of expertise does not depend on having any sort of expert knowledge on the discipline at issue, it is unclear how a sophistical argument can emulate peirastic exposure of false pretenders and scientific expertise at the same time.

In order to tackle these difficulties, I will concentrate on two important passages in which Aristotle employs expressions that, although similar to each other ("κατὰ τὸ πρᾶγμα", "κατὰ τὴν τέχνην"), convey different messages. Several issues depend on these two related questions: first, what exactly is encoded in Aristotle's use of the expression "κατὰ τὸ πρᾶγμα" in the relevant passages; second, how is the notion of "appropriateness to its object" related to the idea of being in accordance with an expertise ("κατὰ τὴν τέχνην"). I will discuss the sort of "appropriateness to its object" which sophistical arguments of this kind aim at but fail to attain, and whether this kind of appropriateness belongs to peirastic arguments too. I will also discuss how these sophistical arguments can be understood as *refutations* while simultaneously emulating scientific demonstrations. This will lead me to another intricate issue, namely, what similarities pseudographemata have with the kind of sophistical arguments that aim at a false appearance of scientific demonstration. My final conclusion is that a consistent picture about all those kinds of argument emerges from careful consideration of the relevant passages.

---
[1] See Angioni (2012), Fait (2013), Hasper (2013a), Swanson (2013), and Fait 2016.

2 Setting the Questions

The kind of sophistical argument that concerns me is presented by Aristotle at 169b20–29, which I split into two texts in order to make discussion easier to follow. The first text is this:

T1:
[169b20] "[i] Λέγω δὲ σοφιστικὸν ἔλεγχον καὶ
[21] συλλογισμὸν οὐ μόνον τὸν φαινόμενον συλλογισμὸν ἢ ἔλεγ–
[22] χον μὴ ὄντα δέ, ἀλλὰ καὶ τὸν ὄντα μὲν φαινόμενον δὲ
[23] οἰκεῖον τοῦ πράγματος. [ii] εἰσὶ δ' οὗτοι οἱ μὴ κατὰ τὸ πρᾶγμα
[24] ἐλέγχοντες καὶ δεικνύντες ἀγνοοῦντας, [iii] ὅπερ ἦν τῆς πειραστι–
[25] κῆς".

[i] By a sophistical refutation or syllogism I mean not only what appears to be a syllogism or refutation without being one, but also what is a syllogism, but merely appears appropriate to its object. [ii] These are arguments that fail to refute in accordance with its object and to unmask ignorant people—[iii] precisely what the task of critical examination was (169b20–25, trans. Hasper, modified).

Five relevant points must be highlighted from T1:

(1) The kind of sophistical refutation hereby introduced is different from the kinds presented earlier in *SE:* they are valid (or even sound) deductions. They do not have a false appearance of being a syllogism or a deduction: actually, they are really successful in deducing the conclusion from the premises.[2]

(2) However, they aim at something else, i.e., something beyond mere validity or sound deduction of the conclusion: they aim at being *appropriate* to its object. However, they fail in this aim, for they only produce a false appearance of appropriateness to their objects. I will discuss in more detail what this exactly means. For the time being, I only highlight that the appearance of *appropriateness to its object* is the more specific aim of this kind of sophistical argument.

(3) Now, taking (1) and (2) together, one can say that this kind of sophistical argument fails in being appropriate to its object without failing in being a valid (or sound) deduction of the intended conclusion. There is no need to take Aristotle as saying that this is a universal analysis of what false appropriateness to its object amounts to. It is possible for an argument to fail in being appropriate to its object and at the same time fail in being a valid deduction of the conclusion. However, I emphasize that Aristotle has spotted something important here, namely, a kind of

---

[2] See Fait (2007), p. 135, Acerbi (2008), pp. 536–537, Hasper (2013a), p. 307, Swanson (2013), p. 204, King (2013), p. 193, and Fait (2013), p. 240.

sophistical argument which fails to be appropriate to its object without failing to be a valid (or sound) deduction of the intended conclusion.

(4) This kind of sophistical argument aims at exposing ignorant people. They are supposed to work as refutations directed towards a given piece of knowledge that someone claims to have, but (presumably) does not have. This feature is what authorizes us to refer to them as sophistical *refutations*.

(5) These sophistical refutations are compared to peirastic arguments, namely, dialectical arguments engaged in a specific sort of critical examination of someone who claims to have expert knowledge.³

Point (1) is not controversial (at least when compared to the other points). All the other points offer many difficulties, and I will discuss all of them in particular, but I start with a more general, pressing question. Aristotle ascribes *two* aims and *two* failures to the kind of sophistical argument at stake: they aim at being appropriate to its object, but produce only a false semblance of appropriateness [(2), (3)]; they aim at exposing ignorant people (4), but are not successful at this task. As it is at this juncture—in step [iii] of T1—that the comparison with peirastic (5) is made, one might be tempted to attribute *both* aims to peirastic too, as if, besides being directed towards exposing false claims of expert knowledge, peirastic were also directed towards appropriateness to its object (in the way in which this notion is understood in T1). But is that the most plausible reading of the text?

The next passage adds some important data to the picture:

T2:
[169b25] "[iv] ἔστι δ' ἡ πειραστικὴ μέρος τῆς διαλεκτικῆς· [v] αὕτη δὲ
[26] δύναται συλλογίζεσθαι ψεῦδος δι' ἄγνοιαν τοῦ διδόντος τὸν
[27] λόγον. [vi] οἱ δὲ σοφιστικοὶ ἔλεγχοι, ἂν καὶ συλλογίζωνται τὴν
[28] ἀντίφασιν, οὐ ποιοῦσι δῆλον εἰ ἀγνοεῖ· [vii] καὶ γὰρ τὸν εἰδότα
[29] ἐμποδίζουσι τούτοις τοῖς λόγοις".

[iv] Peirastic is, however, a part of dialectic; [v] and it is able to deduce something false due to the ignorance of the one who concedes the argument. [vi] But *these* sophistical refutations, even if they deduce a contradictory, do not make clear whether someone is ignorant; [vii] for, indeed, they embarrass even someone who has knowledge with these arguments (169b25–29, trans. Hasper, modified).

Step [v] says that peirastic arguments do succeed in their target of exposing ignorant people, and this is contrasted with what steps [vi] and [vii] add: that the kind

---

3 I will employ both "peirastic" and "critical examination" to refer to πειραστική—for stylistic reasons.

of sophistical refutation at stake does not succeed in the same target.⁴ There is a lot to be discussed about those steps, specially about the discomfort which these sophistical refutations cause in the experts even without making clear that they are ignorant. But let me first focus on the pressing issue I raised: what is the relation, if any, between peirastic arguments and refutation in accordance with its object? False appearance of appropriateness is the prominent characteristic by which this kind of sophistical refutation has been introduced in T1. But how are *peirastic* arguments related to this sort of appropriateness? Is Aristotle suggesting or assuming that peirastic arguments also aim at this sort of appropriateness?

The answer to this question also depends, among other things, on how we read step [iii] in T1 and, more precisely, on how we take the pronoun "ὅπερ" in 169b24. Is the pronoun "ὅπερ" in 169b24 referring to *both* tasks in which sophistical arguments fail? Or is it referring only to the last one, namely, the task of exposing false claims of knowledge?

My answer is that the pronoun "ὅπερ" in 169b24 refers only to the latter task at which sophistical arguments fail, namely, that of exposing ignorant people. My argument for that is not philological. One could argue that, as the pronoun is in the singular form, it is more likely that it refers back to only one of those tasks, not to both of them. But one might object that there are several instances of a singular pronoun referring to a plurality of things (trivially, because that plurality is taken as *one* group). My argument is not relying on anything like that. It is rather relying on how I understand the notion of *refuting in accordance with its object*, which, in this context, refers back to the notion of an argument being appropriate to its object. I argue that the sophistical arguments at stake aim at the kind of *appropriateness* Aristotle is talking about in T1, but peirastic arguments have nothing to do with it.

There are three important issues I need to discuss in order to substantiate this claim. First, there seems to be *prima facie* evidence for taking "κατὰ τὸ πρᾶγμα ἐλέγχοντες" in 169b23–24 as ranging over peirastic too, for the same expression "κατὰ τὸ πρᾶγμα" is employed in 171b6 to describe the same task of peirastic, namely, that of spotting the ignorant person who claims to know.⁵ However, I

---

4 The definite article in the expression "οἱ δὲ σοφιστικοὶ ἔλεγχοι" (169b27) has a *determinative* force in this context: Aristotle refers to the particular kind of sophistical refutation at stake (Swanson 2013, p. 209), namely, those that are deductions but have only a false appearance of being appropriate to their objects.
5 Some are tempted to embrace this *prima facie* evidence: Dorion (1995), p. 287, Bolton (1994), p. 124, Fait (2007), pp. 150–151, Swanson (2013), p. 206, and Bolton (2013), p. 273. But there are exceptions: see Fait (2013), p. 240, and, especially, Fait (2016), p. 38 ("the formulation [sc. In 171b6] sounds illogical—*by definition* common items are *not* in accordance with the object").

will argue that the same expression "κατὰ τὸ πρᾶγμα" is used differently in each of these contexts (without any significant change of meaning: the differences come from the context, more precisely, from the specific argumentative performances the expression is applied to). Second, something that might seem obvious needs to be discussed: whether *refuting in accordance with its object* (κατὰ τὸ πρᾶγμα ἐλέγχοντες)—as employed in 169b23–24—is equivalent to the notion of being appropriate to its object (οἰκεῖον τοῦ πράγματος, 169b23). The third important issue concerns what the notion of refuting in accordance with its object in T1 belongs to if not to peirastic. The idea behind this issue is simple. Sophistical arguments emulate successful arguments. Thus, for each aim that sophistical refutations fail to reach, there seems to be a kind of skill or expertise which is normally successful in attaining it. So, if sophistical refutations fail to refute in accordance with its object, what is the skill or expertise that is successful in doing it?

All these questions primarily depend on the notion of an argument (or premises in an argument) being appropriate to its object. For this reason, my next step is to discuss how this notion should be understood in T1–T2.

3 Being Appropriate to Its Object (Pragma)

There are many ways in which an argument can be said to be appropriate to its object. How this notion is to be cashed out depends on two things, namely, how we understand the *object* (πρᾶγμα), and the exact nature of the appropriateness. Πρᾶγμα can be taken in many contexts as equivalent to the subject-matter underlying a scientific discipline or expertise, or even as the subject-matter or topic an argument is concerned with in a general way (e.g., when one says that a discussion is concerned with the next elections). But this does not work for T1.[6] In a nutshell, I argue that πρᾶγμα in T1 must be taken in the following way (satisfying jointly the two conditions):

[a] πρᾶγμα is the proposition targeted by a given argument as the end-product (i.e., the conclusion) the argument is meant to deliver;[7]

---

[6] See Swanson (2013), pp. 205–206. However, her proposal is different from mine: she takes πρᾶγμα as the subject of the conclusion targeted by an argument.

[7] It might seem redundant to say "targeted as the end-product", for the target can be taken as identical to the end-product. However, it makes sense to say that (e.g.) a dialectical argument *targets* the opponent's thesis, but the *end-product* to be delivered in the argument is the contradictory of that thesis. Thus, "target" is a more general term, while "targeted as end-product" has a more restrictive application.

[b] and envisaged exactly under the aspect which the argument is supposed to deliver.

In order to clarify condition [b], I stress that one and the same proposition can be envisaged under different aspects, and these aspects are related to different questions:

[b.1] one question is *on what conditions (if any) one might be justified in taking that proposition to be the case*;

[b.2] another question is *whether it is the case*;

[b.3] another question is *what are the explanatory factors that make it what it is*.

There are many other aspects under which a given proposition can be envisaged, but the contrast between these three aspects is enough for my purposes.[8] Question [b.1] conveys the aspect under which (e.g.) a dialectic debate envisages a given proposition: once the opponent has chosen his *position*, the dialectical questioner looks for ἔνδοξα from which the contradictory of that position can be deduced. In this case, the contradictory of the opponent's position, being the targeted conclusion of the dialectic argument, must be taken *as following from premises conceded by the opponent* (whether they are true or not).[9] On the other hand, forensic rhetoric deals with questions [b.2] when it seeks for premises from which it follows the conclusion (e.g.) that the defendant did commit murderer. And some stages of scientific inquiry are also focused on establishing *that* a given state of affairs holds.[10] As for question [b.3], it conveys the aspect under which scientific demonstration envisages a given targeted proposition. Thus, in the last case, πρᾶγμα will stand for an explanandum that, on normal circumstances, is already known to be the case.[11]

I argue that, in 169b23, πρᾶγμα must be taken according to aspect [b.3]: the πρᾶγμα is a given predication taken as an explanandum in a given scientific discipline, which is tantamount to being taken as a target conclusion for which the

---

[8] "Πρᾶγμα" can be taken in many ways; the way I propose here is commonly found in Aristotle's theory of argumentation. More details can be found in Angioni (2019a), pp. 152–154.

[9] I am not suggesting that dialectic's primary concern is justification. I am only highlighting the fact that premises conceded by the opponent as *endoxa* work as possible reasons for justifying one in abandoning the original position and embracing its contradictory. This is enough for my contrast with other aspects under which a targeted conclusion can be taken. See Hasper (2013c), pp. 36–37, for the structure of a dialectical argument aiming at refuting the opponent's thesis.

[10] There are more aspects in scientific inquiry. The notion of establishing *that something is* (ὅτι ἐστι) in *Posterior Analytics* II.2 is much more demanding: it involves grasping that a given phenomenon is a regular kind produced by causes that can be tracked within a given discipline. But I can avoid these details. See Charles (2000), pp. 70–71, Karbowski (2019), and Karbowski (2016), pp. 121–126.

[11] See Angioni (2016), pp. 142–143.

correlated argument should identify the explanatory factors—more precisely, the *appropriate, primary* explanatory factors.

Thus, I argue that "being appropriate to its object" in T1 stands for the sort of explanatory appropriateness that characterizes scientific demonstrations.

Given all these considerations, let me introduce the label "sophistical demonstrations" to refer to the kind of sophistical arguments at stake. In short, as Aristotle suggests in 171b29, these arguments pretend to be appropriate demonstrations, but they end up having only a (false) appearance of being an appropriate demonstration.

With the notion of appropriateness to its object (as employed in T1) clarified, I should return to what I have called the three important issues. However, those issues can be better tackled if a further, pressing question is previously tackled: how does a sophistical demonstration attempt to expose ignorant people (i.e., to expose false claims of expert knowledge)? A first glance at T1 suggests that sophistical demonstrations *imitate* peirastic arguments in most respects (if not in all).[12] In accordance with this suggestion, one would say that sophistical demonstrations share with peirastic arguments not only the *aim* of detecting false claims of knowledge, but also the *argumentative procedures* by which ignorance is detected and exposed. However, I argue that this *prima facie* plausible suggestion is wrong. Sophistical demonstrations do not follow the same pattern as peirastic in attempting to expose false claims of knowledge. In order to show this, I will dwell on how sophistical demonstrations attempt to emulate scientific demonstrations—and scientific *refutations*.

4 Scientific Refutations: Detecting the Lack of Expert Knowledge

Ignorance can be taken in several ways, even in the restricted domain of Aristotle's account of refutations. Actually, in the context of *SE* 8–11, "ignorance" stands for lack of expert knowledge in a given domain (see also *Posterior Analytics* 77b19).[13] But even this narrower kind of ignorance is multiple: it can be just the lack of knowing that a given proposition is true (and its contradictory is false)—and this kind of mistake can be produced by arguments too (see *Posterior Analytics*

---

12 See Swanson (2013), p. 206.
13 The occurrences of "ἄγνοια" in 169b26 and of the verb "ἀγνοεῖν" in 169b24, 28, 171b5, 172a27, have nothing to do with *ignoratio elenchii* (which is what is meant in several other passages such as 168a18, 168b15). On this notion, see Swanson (2017).

79b24). But ignorance can also be understood as the lack of knowledge about an explanation: a false expert chooses a wrong explanation and thereby fails to know the right one.

When ignorance basically hinges on the principles belonging to a given scientific expertise, it is incumbent on the expert scientist herself to detect it and correct it (cf. 170a23–31, 36–38). Due to ignorance of this kind, one might be led, e. g., to believe that the diagonal of a square is commensurable with its sides (cf. 170a24–26). A geometer will be able both to spot this kind of mistake and to refute it with the right sort of argument. There will be a *scientific* refutation, which cannot be conducted by any dialectician *as* dialectician (cf. 170a23–39, *aPo* 77b9–14).

However, there are two important features of scientific refutations.[14] First, a scientific refutation does not need to be different from a scientific *demonstration* itself. Normally, a refutation involves a particular procedure against a thesis to be rejected. The person conducting the refutation sets the contradictory of the thesis as the conclusion to be deduced and then looks for premises from which that conclusion will follow. This pattern of argumentation is very well known from the *Topics*.[15] However, Aristotle suggests that there is another route open for the scientific expert. A geometer can refute the false claim that the diagonal of a square is commensurable with its sides *just by demonstrating that (or why) it is incommensurable* (170a26).

This is the passage:

T3:
[170a23] ἔλεγχοι δ' εἰσὶ καὶ ἀλη-
[24] θεῖς· ὅσα γὰρ ἔστιν ἀποδεῖξαι, ἔστι καὶ ἐλέγξαι τὸν θέμε-
[25] νον τὴν ἀντίφασιν τοῦ ἀληθοῦς· οἷον εἰ σύμμετρον τὴν διά-
[26] μετρον ἔθηκεν, ἐλέγξειεν ἄν τις τῇ ἀποδείξει ὅτι ἀσύμμετρος.

And these are also correct refutations, since for everything that can be demonstrated, it is also possible to refute the proponent of the contradictory of the truth. For example, if someone has claimed that the diagonal of a square is commensurable with its sides, one can refute him through the demonstration that it is incommensurable (trans. Hasper, modified).

---

[14] There is no need to apply these features to every scientific refutation. I am only claiming that they exist and should be taken into account. Besides, I am not implying that refutations of this kind are the only ones available to a scientific expert. The same person can have both scientific expertise and dialectical skills, and in that case she will be able to produce other kinds of refutations, besides the *scientific* ones. I will avoid the classical problem of how first principles are "discussed" (*Topics* I.2, 101a36 ff.). Nothing important for this paper hinges on that issue.
[15] See Hasper (2013c), pp. 36–37.

Aristotle's last sentence in T3 is very compressed, but I take him to be saying that *the scientific demonstration itself*, by establishing and explaining the true conclusion from its appropriate principles, produces the additional effect of *refuting* the opposite claim, e.g., that the diagonal of a square is commensurable with its sides.[16] Thus, the *genuine, right refutations* (170a23–24) which the scientific expert produces can be extensionally equivalent to the demonstrations themselves. The expert does not need to put a different set of propositions to work as a refutation: it is her demonstration itself that simultaneously works as a refutation. I emphasize that I am not claiming that this kind of refutation by direct demonstration of the contradictory is the *only* kind of refutation available to the expert. I am only insisting that this kind of refutation has been acknowledged as an available option for the expert and, furthermore, plays an important role in understanding the specific nature of sophistical demonstrations.

A second feature of a scientific refutation is that it contains several layers of *being right* or of *hitting the point*. Scientific refutations of this kind (which are extensionally equivalent to scientific demonstrations) will aim at and will have true conclusions. However, I want to stress that there is another important factor in scientific refutations, besides the opposition between true and false conclusions. In Aristotle's example in T3, what needs to be refuted is a predication that is itself false, namely, that the diagonal of a square is commensurable with its sides. But there is no need to take this example as displaying the *only possible* picture, or even the *central* one. There are cases in which what goes wrong in scientific attempts is not the truth value of any predication involved in an argument, but the explanatory claim itself. The case about planets not twinkling and being near the Earth, in *APo* I.13 (78a30–b4), is very well known. The syllogism of the fact (ὅτι) depicted in those lines (78a30–b4) is correct as long as it is taken as merely establishing the fact in the conclusion. But think of someone who mistakenly believes that non-twinkling is the prior explanatory factor explaining why planets are near the Earth. His explanatory claim is wrong, precisely *as* an explanation, but it does not involve any false predication at the level of the ingredients of the syllogism. The relevant point I wish to highlight is this: a wrong explanatory attempt sometimes does not involve any false predication, but consists in a wrong identification of which predication plays the explanatory role for the targeted explanandum. From this perspective, the error which scientific refutations should identify and refute is not located at the level of the truth value of any sen-

---

16 This is not incompatible with the picture suggested in *Posterior Analytics* 72b1–3, according to which the expert will be able to formulate, and reject as wrong, an argument opposed to the demonstration.

tence within the bad argument, but in the *explanatory role* of the premises selected as explanatory factors.[17]

An additional step is needed for these two features of scientific refutations to cohere. The basic idea behind the first feature is that the scientist who demonstrates *p thereby* also refutes the *opposite* of *p*. Here, some caution is required. Indeed, according to Aristotle's example, *p* is the true conclusion of the demonstration. Now, one might be tempted to take "ἀντίφασις" in 170a25 in the strict sense of *contradictory* sentence, i.e., the contradictory predication (as in the square of opposition). However, several occurrences of "ἀντίφασις" do not have exactly the force of contradictory predication.[18] I argue that some occurrences of "ἀντίφασις", including the one in 170a25, should be taken in a broader way. What is relevant in this broader way—for my purposes—is not merely that contraries are covered too (as contrasted with contradictories according to the square of opposition). What is most relevant for my purposes is that a broader use of "ἀντίφασις" also covers *opposites at a supra-predicational level*, namely, opposite arguments—for instance, explanatory claims that are opposite to (and incompatible with) one another. Actually, in 175a36, Aristotle says that the refutation itself is an ἀντίφασις, and this can smoothly be taken as saying that the refutation is a speech act which *opposes* a given thesis or a rival argument. Furthermore, the adjective "ἀληθής" has just been applied to "ἔλεγχος" in 170a23, and the resulting expression refers to a genuine, successful refutation: a "true" refutation is that refutation which is fully successful in delivering its promised product.

Thus, although the example given by Aristotle in 170a25–26 might suggest that "ἀντίφασις τοῦ ἀληθοῦς" in 170a25 strictly refers to a proposition that is *the contradictory of the true one* (the "true one" being the conclusion of the scientific demonstration), the previous context, in which "*alethes*" has been applied to "ἔλεγχος", together with Aristotle's sentence in 175a36, suggests a different, more inclusive option: "ἀντίφασις τοῦ ἀληθοῦς" in 170a25 can be taken as also referring to the argument, or attempted refutation, by which one *opposes* the genuine, scientific refutation.

---

17 The argument I am suggesting from 78a30–b4 is a wrong explanatory attempt that takes convertibles terms in the *reverse order*, but other kinds of explanatory mistakes do not necessarily involve a *reverse* order; for instance: attempting to explain 2R through the specific feature of ἰσοσκελές triangles *as* ἰσοσκελές (74a16–17, 85b5–7); taking the parallels *as perpendiculars* (74a13–16).
18 See a list in Smith (1989), p. 253. The same author (Smith 1997) has preferred to translate "ἀντίφασις" as "negation". Besides, a strict contradiction (according to the square of opposition) sometimes is expressed by other terms, such as "ἀντικείμενον" (e.g., 61a31; 61b18, 32; 62a11; 62b26; 63b15; cf. Ackrill 1963, p. 128).

Consequently, the key sentence which expresses the basic idea behind the first feature of scientific refutations must be rephrased. Instead of the former phrasing:
—"the scientist who demonstrates *p thereby* also refutes the *opposite* of *p*";
let us make room for this formulation:
—"the scientist who demonstrates *why p thereby* also refutes the *opposite demonstration*".[19]

Take "*p*" as referring to the conclusion of the scientific demonstration. Whereas the previous version could only accommodate the case in which the target of the scientific refutation is the false proposition that is the contradictory of the scientific conclusion, the rephrased version can accommodate *both* this case and the further case in which the rival argument is a bad explanation, in which nothing is wrong in the truth value of the predications themselves.

Suppose someone is attempting to explain why elephants have long gestations. Paying attention to the fact that elephants are long-lived, this person can jump to the explanatory claim that elephants have long gestations *because* they are long-lived. Now, the scientific expert will refute this explanatory attempt, but *not* by denying the truth of the predication selected as an explanatory premise: she will, instead, deny its explanatory power for the targeted explanandum. The genuine expert will admit that elephants, which have long gestations, also are long-lived; but she will deny that their longevity is the appropriate factor to explain why they have long gestations.[20] The refutation of a wrong explanatory attempt of this kind aims at the explanatory power, but not the truth value, of the predication that elephants are long-lived.

This example helps to clarify once more the first feature of scientific refutations. The scientific expert does not need to produce a new, specific argument—different from her genuine demonstration—in order to refute the false claim of scientific knowledge. It is her genuine demonstration itself that has the force of refuting the wrong explanation. By demonstrating that, e.g., "elephants have long gestations *because* of the size of their bodies", the expert thereby refutes the rival explanation, namely, that "elephants have long gestations *because* they are long-lived". The idea is that the scientific explanation has (in the context of my example) the intended perlocutionary effect of contradicting the rival explanation, without any need of articulating explicitly the denial of the rival explanatory factor.

---

19 The "ὅτι" employed in 170a26 can smoothly be taken as *why* or as *that*.
20 The example is adapted from Aristotle's discussion in *Gen. Anim.* 777a32–b16.

5 Some Issues Answered

From all these considerations, I am able to answer two important issues I have raised in § 2. The second issue was whether *refuting in accordance with its object* (κατὰ τὸ πρᾶγμα ἐλέγχοντες, 169b23–24) is equivalent to the notion of being appropriate to its object (οἰκεῖον τοῦ πράγματος, 169b23). My answer is positive, but with some qualifications. Appropriateness to the object (as encoded in the expression "οἰκεῖον τοῦ πράγματος", 169b23) has been introduced as a shorthand for the sort of explanatory appropriateness that is the most important mark of scientific expertise. The πρᾶγμα, in this context, is the explanandum envisaged exactly under the aspect on which it must be delivered in a successful scientific demonstration. Thus, the notion of being appropriate to its object is, in this context, the mark of a scientific demonstration that is successful in delivering its product, namely, one that has identified the primary, appropriate principles for the explanandum in question.

Now, since it is open for an expert to employ the same demonstration *as a refutation of the opposite explanatory claim*, I argue that this is what Aristotle has in mind with the notion of *refuting in accordance with its object* (κατὰ τὸ πρᾶγμα ἐλέγχοντες, 169b23–24). The refutation at stake, in this case, is exactly the refutation performed by the expert when producing the direct demonstration that is appropriate to its object (οἰκεῖον τοῦ πράγματος). Consequently, such a scientific refutation is exactly what the sophist emulates: the sophistical demonstration aims at producing an appearance of *refuting in accordance with its object*. In other words, the sophist "imitates" the expert in this way: he produces an argument that [i] seems to be an appropriate demonstration of its πρᾶγμα and thereby [ii] appears to be a refutation of the opposite demonstration of the same πρᾶγμα by the expert.[21] This is how the sophistical demonstration also produces a counterfeit of peirastic: the sophist purposely expects his false demonstration ("φαινομένη ἀπόδειξις", 171b29) to have the perlocutionary effect of appearing to *debunk* the explanation offered by the expert.

Given all these considerations, my answer to the third important issue almost goes without saying. The third issue was this: if the notion of refuting in accordance with the object as employed in T1 does not belong to peirastic, to what does it properly belong? I still need to discuss the antecedent in the question, but the answer is straightforward: refuting in accordance with its object belongs

---

[21] This is why the sophistical demonstration (or, at least, its most dangerous version) can be a sound deduction: the explanatory deception does not necessarily involve a false predication. This is why I disagree with (helpful) suggestions in Fait (2013), pp. 242 and 261.

properly to the expert.[22] Actually, the refutations that are in accordance with their objects, as this notion is employed in T1, are equivalent to (or significantly overlap with) the *right* or "true" refutations introduced in 170a23—which fall beyond the competence of the dialectician and the critical examiner (cf. 170a34–39). These are the refutations that depend strictly on the specific expert knowledge concerning each scientific discipline (cf. 170a36–38).

However, I still need to substantiate my claims with a discussion of the first important issue I have mentioned. Indeed, even if one agrees with all points I have made so far, one might still object that there seems to be *prima facie* evidence for taking "κατὰ τὸ πρᾶγμα ἐλέγχοντες" in 169b23–24 as ranging over peirastic too, given that the same expression "κατὰ τὸ πρᾶγμα" appears in 171b6 to describe an essential task of peirastic. My next step, then, will be a sketchy outline of peirastic arguments (§ 6). Then I will turn to "κατὰ τὸ πρᾶγμα" in 171b6 (§ 7).

6 Peirastic Arguments in Outline

Since my focus is T2—where a specific kind of procedure is ascribed to peirastic in such a way that is relevant to understand the contrast with sophistical demonstrations—I will only explain in outline what peirastic arguments are and how my view does not conflict with the textual evidence most appealed to in the literature.[23] Thus, my sketchy outline of peirastic is not meant to embrace all sorts of arguments that can be classified under that heading.

Step [v] of T2 unfolds the comparison between peirastic arguments and sophistical demonstrations. Both kinds of arguments have the purpose of exposing false claims of *expert* knowledge—for "ignorance" in 169b26 corresponds to the lack of expert knowledge in someone who claims to have it (171b4–6, 172a21–27). Peirastic succeeds in its aim because "it is able to deduce something false due to the ignorance of the one who concedes the argument". But how does this argument pattern work?

Aristotle is talking about a kind of peirastic procedure in which the resulting argument is valid, but not sound: its conclusion is false, and the falsity of the conclusion depends exactly on a false premise that was accepted by the answerer.[24]

---

22 For a similar view (although different on many details), see Swanson (2013), p. 205.
23 For general discussion of peirastic, see Bolton (1990), pp. 218–219, Fait (2007), pp. 160–161, Bolton (2013), Hasper (2013a), and Swanson (2013).
24 My argument does not depend on the stronger claim (which I would *never* endorse) that *every* occurrence of "συλλογίζεσθαι ψεῦδος" must be taken in terms of valid deduction with false conclusion (and at least one false premise). Indeed, "συλλογίζεσθαι ψεῦδος" can be taken in a different

Now, the answerer, in that kind of argument, is precisely the false expert submitted to critical examination. Therefore, peirastic succeeds in exposing the false expert by showing a false conclusion hanging exactly on a false premise advanced by the false expert. However, this characterization of peirastic arguments can potentially clash with two pieces of evidence. On the one hand, it seems to clash with the fact that peirastic must fit the general description of dialectical arguments, since peirastic is a part of dialectic. On the other hand, if Aristotle implies that the critical examiner must be able to *tell that* a given conclusion is false, this picture will hardly be compatible with the characterization of peirastic as an ability that does not depend on any definite knowledge about the subject in question (172a21–27).[25]

Let me dwell on the first potential problem. I will settle some elements in this section, but the full solution to the problem will come in the next section (together with the solution to my first important issue, how peirastic arguments can be said to be "κατὰ τὸ πρᾶγμα").

As long as peirastic is part of dialectic, a peirastic refutation can be depicted as an argument in which, given a proposition claimed as a piece of expert knowledge, the dialectician aims at deducing its contradictory from premises accepted by the answerer. As an example, take someone claiming to have expert knowledge about animals in general. Suppose this person presents the false proposition that no aquatic animal breathes (either due to the pressure of dialectical tricks or out of blatant ignorance of the subject).[26] It would be correct to depict the peirastic refutation of this person as an argument directed towards the conclusion which turns out to be the contradictory of that false proposition. Thus, the argument can be simplified as a *Darapti* or proto-*Darapti* (just for didactic purposes, without implying that the syllogistics of the *Prior Analytics* were available to Aristotle when he wrote the *SE*[27]):

"Whales breathe;
Whales are aquatic animals;
Therefore, some aquatic animals breathe".

---

way (i.e., as pointing to a sound deduction that delivers some *false* result—for instance, if the sound deduction is intended as the appropriate explanation of the conclusion, but the explanation is explanatorily wrong). However, in 169b26, "συλλογίζεσθαι ψεῦδος" describes what happens with some peirastic arguments, and the expression in this restricted domain is better taken in terms of deducing validly a false conclusion (from at least one false premise).

25 For discussion, see Bolton (1994), pp. 121ff.
26 It is immaterial for my point whether this person is really an expert who happens to have an infelicitous moment due to dialectical inability, or a real charlatan—but I believe Aristotle has recognised the existence of both.
27 On this general issue, see Smith (1994).

The conclusion, "some aquatic animals breathe", is the contradictory of the original thesis, "no aquatic animal breathes", so that the argument can be taken as a refutation of the thesis on the basis of premises conceded by the answerer, matching Aristotle's description of the general pattern for dialectical arguments.[28] But how does this argument square with Aristotle's description of peirastic exposure of ignorance in T2? Indeed, the *Darapti* example is sound: the conclusion is true, besides being a consequence of the premises—but T2 seems to require a valid argument with a false conclusion.[29]

In order to depict peirastic exposure of ignorance in T2 as a *successful deduction of something false from a concession of the false expert to be exposed*, I propose this other argument:

"*No aquatic animal breathes* (= thesis);
Whales are aquatic animals;
Therefore, *no whale breathes*".

In this *Celarent* (again, offered just for didactic purposes), we have precisely:
[i] a *false* conclusion successfully deduced from the premises,
[ii] a false premise that is exactly something the adversary to be exposed has said,
[iii] the falsity of the conclusion *depending* exactly on the falsity of the premise conceded by the adversary.

All three features are important, but it is [iii] that clearly makes the peirastic argument fully successful in its specific purpose. Its being an exposure of a false claim of expert knowledge depends on [iii], and this decisive fact is what is encapsulated in Aristotle's phrase "due to the ignorance of the one who conceded the argument" (169b26–27).

However, this interpretive suggestion leads me to the second potential problem. For, it seems to clash with some features normally ascribed to peirastic. Peir-

---

28 In this case, the conclusion is the contradictory according to the square of opposition; however, I am not thereby suggesting that *all* occurrences of "ἀντίφασις" in the *SE* must be taken in this strict sense.
29 One option—against my proposal—consists in taking "συλλογίζεσθαι ψεῦδος" in 169b26 with the force of "deducing something which, although valid and even sound, is *wrong* in another level, for instance, explanatorily wrong". Actually, this force is associated with some occurrences of "ψευδὴς συλλογισμός" (e. g., 170a31); however, in the context of 169b26, I do not believe this is the case, for "συλλογίζεσθαι ψεῦδος" describes at the same time both [i] the peirastic argument and [ii] its success in exposing ignorance. And this success has nothing to do with detecting *explanatory* failures.

astic arguments are presented by Aristotle as able to detect the false expert (cf. 172a21–27):

[a] not on the basis of expert knowledge (which the critical examiner is not required to have),

[b] nor on the basis of specific principles of the discipline (which the critical examiner is not required to know),

[c] but from *consequences* or *things-that-accompany-others* (ἑπόμενα), which are such that, although knowing them does not entail having the expert knowledge at stake, not knowing them entails lacking the expert knowledge at stake.

But the *Celarent* example seems to imply that the critical examiner must be able to *tell that* the conclusion is false, which seems to clash with [a] and [b].

In order to tackle this second problem, let me focus on feature [c]. The "ἑπόμενα" (172a25) are usually taken as coinciding (or significantly overlapping) with those items which Aristotle calls "κοινά" a few lines ahead (172a29, 32).[30] The κοινά are "such that they are not any given nature or genus, but are like the denials" (172a37–38). "Denial" is employed by Aristotle to refer to sentences (or, sometimes, parts of sentences) which are formed by attaching the negation "not" to a previous proposition. Aristotle believes that denials have a specific semantic *as denials:* one does not need to focus on the content of the proposition to understand that a denial normally means that the underlying affirmation must be rejected as false. There are several difficult details about the semantics of the negation, which I need not address. I only need to pin down the idea that a denial, whatever its propositional content, is understood as rejecting the truth of the underlying affirmation. And another basic notion is connected to this, namely, that an affirmation and the correlated denial cannot be both true at the same time, in the same aspect, etc. Understanding that an affirmation and its correlated denial cannot be both true at the same time does not depend on knowing which one is true (if any). Thus, all this—together with Aristotle's remark that κοινά are not "any given nature or genus"—suggests that κοινά are formal principles which underlie our reasoning.[31]

A few lines ahead, Aristotle presents peirastic as an ability to conduct an examination "from these items" (ἐκ τούτων, 172a39), and the pronoun refers to things that are such as the denials, i.e., what he has previously labelled *common items*. I do not agree with taking "ἐκ τούτων" (in 172a39) as requiring that those items—the

---

30 See Fait (2007), p. 160, and Fait (2016), p. 32.
31 See Fait (2007), p. 160. For a different view of "κοινά", see Bolton (1990), pp. 218–219, Bolton (2013), p. 277, and Reeve (1998), p. 232. For discussion, see Devereux (1990).

κοινά—be explicit premises in the peirastic argument.³² All Aristotle needs is to make the κοινά (such as the denials) the principles on the basis of which the argument runs.³³ This is exactly what we have in my interpretation of T2.

Remember my example concerning aquatic animals. The only thing the critical examiner is required to have is the ability to detect the potential conflict between (at least) three propositions, namely, that no aquatic animal breathes (the thesis to be refuted), whales are aquatic animals and whales breathe. Suppose the last proposition (that whales breathe) was inadvertently dropped by the same false expert who had previously said that no aquatic animal breathes. This will be enough for peirastic refutation, for the person being examined would have to admit the conflict between that proposition (that whales breathe) and the conclusion of the *Celarent*. There is no need to take the proposition that whales breathe as part of the background knowledge settled in the critical examiner's convictions. The critical examiner may ignore whether it is true or false. Even if the critical examiner happens to believe that whales actually breathe (this being common knowledge available to laymen etc.), that belief is not required for the peirastic argument to work. What is relevant for the critical examiner is, first, that the answerer has conceded that proposition, and, secondly, that *that proposition is incompatible with another one previously conceded by the answerer.* Thus, the exact role of the critical examiner qua critical examiner is just to *spot the conflict between the three propositions:* no matter how the question should be settled, the point for a critical examiner is that the three propositions cannot all be true together, and their incompatibility is what allows her to attack the thesis of the false expert.³⁴

Therefore, the knowledge required for the critical examiner is a sort of second-order knowledge about the incompatibility between propositions conceded by the adversary. This knowledge is basically about how denials (and other κοινά) are supposed to work. Importantly, knowing *that* those three propositions are in conflict (and cannot all be true simultaneously) does not entail having expert knowledge about aquatic animals. But ignoring that they are in conflict entails

---

32 The assumption that κοινά *must* be premises is usually taken for granted. See (e.g.) Swanson (2013), p. 209.
33 In general, "ἐκ + genitive" can be taken in several ways (see *Physics* 190a21–31, 190b4–5; *Metaphysics* 994a22–b3, 1023a26ff., 1033a5–23, 1092a21–35; *Gen. Anim.* 724a20–30), and one way that usually passes unnoticed in the literature is that which identifies a dominant or prominent explanatory factor on which something primarily depends (see *Metaphysics* 1043b7 and 10).
34 For a helpful example to understand how peirastic examination works, see Fait (2013), p. 242. Although the issue is controversial, this description of peirastic arguments is also applicable to (at least one minimal version of) the Socratic elenchus as practised in the aporetic dialogues. See Devereux (1990), p. 282, Dorion (2004), p. 259, and Dorion (2013), p. 327.

lacking expert knowledge—for expert knowledge is not merely an amassment of propositions known to be true; rather, expert knowledge, as primarily dependent on identifying and systematizing the appropriate explanatory factors, also involves knowing an intricate web of relations between predications, namely, relations of consequence, convertibility, and compatibility.[35]

Thus, there is no need of taking T2 as saying or implying that the critical examiner will be able to *tell that* the conclusion is false and, consequently, will need to have a previous knowledge about a proposition belonging to the expertise in question. When employing "ψεῦδος" in 169b26, Aristotle is not describing a given conclusion from the point of view of the peirastic examiner: he is describing the conclusion from point of view of the expert (i.e., as false), no matter whether the critical examiner is aware of its falsity or not. Understood in this way, Aristotle's description of peirastic in T2 is perfectly squared with the three distinctive features ([a], [b] and [c]) highlighted above.

7 Peirastic "in Accordance with Its Object"

All these considerations allow me to tackle the first important issue I mentioned in § 2. Aristotle's use of the expression "κατὰ τὸ πρᾶγμα" in 171b6 seems to be a plausible reason for taking "κατὰ τὸ πρᾶγμα ἐλέγχοντες" in T1 (169b23–24) as describing peirastic together with sophistical demonstrations. The relevant passage runs as follows:

T4:
[171b4] ἡ γὰρ πειραστική ἐστι διαλεκτι–
[5] κή τις καὶ θεωρεῖ οὐ τὸν εἰδότα ἀλλὰ τὸν ἀγνοοῦντα καὶ
[6] προσποιούμενον. ὁ μὲν οὖν κατὰ τὸ πρᾶγμα θεωρῶν τὰ κοινὰ
[7] διαλεκτικός, ὁ δὲ τοῦτο φαινομένως ποιῶν σοφιστικός.

> For critical examination is a kind of dialectic and considers not the person with knowledge, but the ignorant person who pretends to have knowledge. Someone who considers common items in accordance with the object is a dialectician, whereas someone who does so in appearance is a sophist (trans. Hasper, modified).

---

35 This is important to explain how I disagree with Bolton (1994), p. 123. Fait (2013, p. 242) has a view similar to mine in most respects: "scientific knowledge would ensure to the real knower a full command of those *koina* used to reveal a fake".

The issue is whether T4 uses the expression "κατὰ τὸ πρᾶγμα" in the *same way* as T1. If the answer is positive, I will be wrong in denying that peirastic shares with sophistical demonstrations the same target of appropriateness to its object.

No one can deny that T4 ascribes to dialectic (and, presumably, to peirastic too) the task of "considering the common items in accordance with the object" (κατὰ τὸ πρᾶγμα θεωρῶν τὰ κοινὰ). However, consideration of the common items "in accordance with the object" (κατὰ τὸ πρᾶγμα) can hardly be taken to introduce the same notion of *explanatory* appropriateness which sophistical demonstrations emulate. This is what I will explore now.

Retrieve what I said about πρᾶγμα:

[a] πρᾶγμα is the proposition targeted by a given argument as the end-product (i.e., the conclusion) the argument is meant to deliver;[36]

[b] and envisaged exactly under the aspect which the argument is supposed to deliver.

Now, in a default dialectical exchange, once the answerer has chosen a *thesis*, the questioner should seek premises from which the contradictory of that thesis will follow—and the πρᾶγμα of a dialectical questioner will be exactly the contradictory of the opponent's thesis. The dialectical questioner must consider, and observe, the *common* items (e.g., denials) as tools [i] from which conflicts and incompatibilities between propositions can be identified and [ii] from which a specific argument leading to the final conclusion can be built. Now, this fits the description employed in T4 perfectly well: the consideration (and observance) of the common items must be in accordance with the πρᾶγμα in the sense that it must render a successful way of attaining the conclusion, exactly under the aspect which the argument is supposed to deliver. The πρᾶγμα in a dialectical argument is the contradictory of the opponent's thesis, precisely as derived from premises accepted by the opponent. Thus, a dialectician fails when his employment of the common items fails in delivering the promised πρᾶγμα. A blatant failure happens when the conclusion does not follow from the premises. But more relevant cases for my purposes is when the conclusion is not really incompatible with the thesis and when the premises employed in the argument have the false appearance of being conceded by the answerer.[37]

---

36 See n. 7 for the expression "targeted as the end-product".
37 The last case fits Aristotle's description of sophistical arguments in 100b24–25. Replace whales with otters in my examples. Since "aquatic" involves some vagueness, the answerer can easily be led into confusion, so that, on further reflexion, she could have rejected premise (2) on the ground that otters do not fully live in water.

Now, take "κατὰ τὸ πρᾶγμα" in 171b6 as saying something about peirastic too.[38] The critical examiner conducts a cross-examination until a conflict between propositions is spotted. Then, the critical examiner puts a particular proposition on her focus.[39] The πρᾶγμα of the peirastic argument, then, will be the contradictory of that particular proposition, envisaged exactly under the relevant aspect, namely, as something to be deduced from premises accepted by the false expert.

This allows me to return to the first potential problem I have discussed in the previous section: how does Aristotle's description of peirastic in T2 fit into the general pattern of dialectical arguments. T2 says that peirastic arguments deduce something false from premises that the opponent accepts. This can be depicted in my *Celarent* example. However, it is unclear how such an argument will meet the general feature of dialectical arguments, namely, attaining the contradictory of the opponent's thesis through premises that the opponent accepts. It seems that the suggested *Celarent* does not fit the bill.

However, the desired result can be easily attained if we combine the *Celarent* and the *Darapti*, as follows:

(1) No aquatic animal breathes; [*thesis*]
(2) Whales are aquatic animals;
(3) Therefore, *no whale breathes*;
(4) but whales breathe;
(5) Whales are aquatic animals [= (2) iterated];
(6) Therefore, some aquatic animals breathe.

Lines (1)–(3) deliver the suggested *Celarent* (which depicts what was said in T2), while lines (4)–(6) deliver the suggested *Darapti* (which fits into the general pattern of dialectic). (Actually, the relation between the chosen terms is exactly what Aristotle has in *Prior Analytics* II.8, 59b20–24, especially 23–24, when examining the *conversions* between syllogisms; see also 61b10–15).[40] Thus, I suggest that the full length peirastic argument should be better depicted as the whole argument from (1) to (6). Importantly, while T2 has focused on (1)–(3) in explaining how peirastic arguments are successful in the task of exposing false claims of expertise, the

---

38 If one takes the paragraph division in Hasper's translation (Hasper 2013b, p. 27) as indicating a change of subject, one might argue that the second sentence, in which "κατὰ τὸ πρᾶγμα" occurs, is not any more saying something about peirastic in particular.
39 See a useful picture in Fait (2016), pp. 37–38.
40 There is no room here to examine the convertibility between direct syllogisms and reductions ad impossibile (*APr.* II.11–13). But the two syllogisms in (1)–(3) and (4)–(6) track exactly that convertibility.

expansion of the argument in lines (4)–(6) explains how the peirastic argument conforms to the general pattern of dialectical arguments.

Thus, I can answer the first important issue I have raised in § 2. The expression "κατὰ τὸ πρᾶγμα" does not deliver the same message in 169b23–24 (T1) and 171b6 (T4). Whereas in 169b23–24 it introduces the specific result that sophistical demonstrations aim to deliver (namely, *explanatory* appropriateness), in 171b6 it refers to the result that dialectical arguments in general are supposed to deliver (namely, deduction of the contradictory from premises conceded by the opponent). The same expression is used differently in each of these contexts, without any significant change of meaning. Its different forces come from its being applied to different kinds of argumentative performances in each context.

Furthermore, as I have shown in § 5, the notion of *refuting in accordance with its object* (κατὰ τὸ πρᾶγμα ἐλέγχοντες)—in the specific context of 169b23–24—overlaps significantly with the notion of being explanatorily appropriate to its object (οἰκεῖον τοῦ πράγματος): whereas the latter introduces the ultimate desideratum to be met in a scientific demonstration, the former can be taken to introduce the collateral (sometimes perlocutionary) effect which a scientific demonstration delivers, as refuting an opposite claim. Therefore, the notion of refuting in accordance with its object as employed in T1 properly belongs to scientific demonstrations, and, in appearance, to sophistical demonstrations—but has nothing to do with peirastic.

8 Taking Stock

Let me emphasize the results of this sketchy exploration of peirastic arguments. Aristotle says that sophistical demonstrations share a purpose with peirastic arguments: both kinds of argument aim at exposing false claims of expert knowledge. However, we should be careful about the next move. One might be tempted to take Aristotle's words as implying that the purposes of sophistical demonstrations and peirastic arguments are completely identical. But this will not be true. Actually, their purposes significantly overlap without being identical, for peirastic arguments do not aim at having (or appearing to have) explanatory appropriateness. The critical examiner will be successful in her aim as long as she spots conflicts and inconsistencies that depend only on the logical behavior of the *common* items. And her success does not depend on having definite knowledge about the object in question (aquatic animals, in my example), let alone on attaining explanatory appropriateness.

Furthermore, one might be tempted to take Aristotle's words in T2 as implying that sophistical demonstrations and peirastic arguments share *not only their tar-*

gets *but also the means* to attain them, that is, the same patterns of argument, the same strategies etc. However, as I have argued, this suggestion is wrong. On the one hand, peirastic arguments fulfill their aim by detecting inconsistencies and conflicts between the propositions presented by someone who claims to be an expert—as Aristotle says in 169b25–27, by deducing (what the real expert will know to be) a falsity from a false premise accepted by the opponent. On the other hand, sophistical demonstrations attempt to fulfill their aim by pursuing a *different* route: the sophist formulates a direct demonstration that, presenting itself as an "alternative" to the expert's demonstration, has the effect of suggesting that the expert has been superseded and, thereby, refuted.

9 Back to Explanatory Appropriateness

It makes sense to say that explanatory appropriateness has levels, or degrees. However, talking in terms of "levels", between a minimum and a maximum degree, might be misleading, if it happens to suggest that the knots located within this space are always produced by reference to one and the same criterion. Actually, the criterion changes. One key question to be raised is the following: to *what* exactly is an argument (or a premise) said to be appropriate? The criterion in reference to which an argument (or a premise) is said to be appropriate or inappropriate changes: it can be either the discipline (in general) or the explanandum in question.

Thus, the minimum degree of explanatory appropriateness consists in the employment of explanatory premises that belong to the discipline in question. What we call "explanatory appropriateness" in this case is what Aristotle sometimes marks with the expression "κατὰ τὴν τέχνην" (170a33–34, 171b12)—an expression which, importantly, does not necessarily converge with other similar expressions such as "κατὰ τὸ πρᾶγμα" (169b23, 171b19, 21) and "οἰκεῖον τοῦ πράγματος" (169b23).[41] Actually, an important part of my argument is that the expression "κατὰ τὴν τέχνην" (in 170a33–34, 171b12) is used to introduce a requirement that is broader, and weaker, than what is introduced with the expressions "κατὰ τὸ πρᾶγμα" and "οἰκεῖον τοῦ πράγματος" (in 169b23). There is no need to discuss here the traditional view of kind-crossing (μετάβασις εἰς ἄλλο γένος)—according to which no one tackling problems in a given discipline would be allowed to appeal

---

[41] However, the minimum degree of explanatory appropriateness can also be encoded by different expressions: "οἰκεῖα τῇ ἐπιστήμῃ" (101a14), "τὰ περί τινας ἐπιστήμας οἰκεῖα" (101a6).

to premises from a different discipline.⁴² All I need is to stress that some sophistical arguments fail in attaining this minimum degree of explanatory appropriateness. This happens, for instance, when a sophist appeals to Zeno's paradox about the impossibility of crossing a given space in attempting to explain why one should not walk after dinner (172a8–9). The premises doing the important job in such an argument are alien to the medical expertise.

Another level of explanatory appropriateness consists in the employment of premises that are pertinent, *in general*, to the explanandum at stake. In italicizing "in general", I indicate that those premises need not be the most decisive ones in the explanatory story, nor the ultimate principle that brings the finishing touch to the explanatory story or that encapsulate what is the most important explanatory factor. In order to count as explanatorily appropriate to a given explanandum at this level, it is enough for a premise to be employed in any relevant way in the explanatory story. Thus, e.g., the definition of straight line is explanatorily appropriate for explaining why triangles have 2R. However, the same definition can also be employed in the explanatory story about other explananda that are significantly different from the attribute 2R. Thus, no one would say that the triangle's being composed of lines is a principle appropriate to the attribute 2R in the sense that nothing else could be explained on the basis of the same principle.

Finally, there is the maximum degree of explanatory appropriateness. At this level, being pertinent *in general* to the explanandum in question (as described in the previous paragraph) is not any more sufficient for being called "appropriate to its object". The maximum degree is the most stringent: a premise (or set of premises) is appropriate to its object in the highest level when it is the most decisive principle in the explanatory story in the sense that it brings the finishing touch to the explanatory story or encapsulates what is the most important, primary explanatory factor for *that* explanandum—in such a way that there is nothing else that could also be primarily explained by the same factor.⁴³ Thus, for instance, having angles equivalent to angles around a point (*Metaphysics* 1051a24–25) is a principle of this kind for explaining why triangles have 2R. One can say that the triangle's having angles that are equal to angles around a point is *the* principle appropriate to the attribute 2R in the sense that nothing else can be primarily explained on the basis of the same principle.

---

42 For discussion, see Steinkrüger (2018) and Judson (2019).
43 In the *Posterior Analytics*, this ultimate level of explanatory appropriateness is introduced in 71b23 and 72a5–7 and becomes the central point in I.9. I claim that "οἰκεῖον τοῦ πράγματος" in 169b23 refers to this ultimate level.

10 Sophistical Demonstrations

Sophistical demonstrations can conform to the first level of explanatory appropriateness while failing to attain the intermediate and the ultimate level. I emphasize this point. It is not the case that all sophistical demonstrations fail to attain the first level of explanatory appropriateness—as if all of them were arguments like the one which relies on Zeno's paradoxes in producing a false semblance of medical expertise (172a8–9). Just a few lines before the argument of false medical expertise, other arguments are mentioned (172a6), such as Bryson's quadrature of the circle (henceforth, just "Bryson's argument"). Given that Bryson's argument has been previously characterized as *not being in accordance with its object* (171b16–22, more on this below), one might believe that this is good reason to conclude that Bryson's argument, as every sophistical refutation which is not in accordance with its object, is an argument that fails in attaining the minimum level of explanatory appropriateness: instead of employing premises appropriate to the discipline in question, it employs "common premises" (cf. 172a9), which are not confined to any specific domain of expertise.

However, there is reason to believe that Bryson's argument, exactly *qua* a sophistical demonstration, cannot be charged with failure in the lowest level of explanatory appropriateness.[44] Bryson's argument is very controversial, the information we have about it is very scarce and indirect, and there is no room to offer here a full discussion of the sources. However, at least on one reasonable interpretation, Bryson's argument can be understood as a sophistical demonstration that satisfies the lowest level of explanatory appropriateness, but fails at the higher levels of it.

A sketchy outline of Bryson's argument will be enough for my purposes.[45] Bryson's argument stems from the proposition that:

> For a given thing (of type F), if there is both a given thing (of type G) that is greater than it and a given thing (of type G) than is smaller than it, then there is a given thing (of type G) that is equal to it.[46]

---

[44] See Hasper (2013a), p. 314. For a different view, see Dorion (1995), p. 288 and Swanson (2013), pp. 222–223 (however, Swanson does not conflatethe two levels of appropriateness).

[45] It is enough for my point to follow in general lines the interpretation of Mueller (1982), pp. 161–164 (as in Angioni (2012), pp. 205–208). See also Hasper (2013a), pp. 314–320. For discussion, see Dorion (1995), p. 288, Fait (2007), p. 154, Acerbi (2008), pp. 516–517, Swanson (2013), pp. 221–226, Fait (2013), p. 260, Bolton (2013), pp. 282–284, and Fait (2016), pp. 30–31.

[46] The principle is clear, although its linguistic formulation inevitably sounds twisted. For a formal version, see Hasper (2013a), p. 315, n. 37.

This proposition sounds like a common principle, applicable to a variety of things. However, there is nothing intrinsically wrong with its being common, for the same procedure pertinent to common axioms can be employed here: the proposition will get a "cashed out version" restricted to the particular domain the argument is concerned with. This is, actually, what Aristotle attests as being the experts' practice (see *Posterior Analytics* 76a38–b2 and *Metaphysics* 1005a23–27). Thus, cashing out the common proposition (and using "C" for circle-type and "S" for square-type), Bryson's argument can be reasonably construed thus:[47]

> For every given thing (of type C), there being something (of type S) greater than it and something (of type S) smaller than it entails there being something (of type S) equal to it;
> There is a circle such that a square S1 is greater than it and a square S2 is smaller than it.
> Therefore, there is a circle such that a square S3 is equal to it.

On this interpretation, Bryson's argument has premises that (cashed out in the proper way) belong to the discipline in question, satisfying the first level of explanatory appropriateness. However, Bryson's argument produces only a *false* appearance of being explanatorily appropriate to its specific object. Retrieve what I said about πρᾶγμα. The πρᾶγμα in any argument is not merely the proposition targeted as conclusion, but that proposition considered under the relevant aspect on which it is "promised" as the end-product of an argument with specific purposes. In scientific demonstrations, the relevant aspect in question is the explanation through the appropriate principles. That is why a mere sound deduction of the conclusion from premises that lack explanatory appropriateness and primariness does not count as a scientific demonstration, even if those premises are *in general* appropriate to the domain in question.[48]

In a geometrical demonstration, the πρᾶγμα is not an indefinite proposition (such as "there is *a circle* which etc.") to be merely deduced from principles that are pertinent to the discipline in question "in general". Rather, the πρᾶγμα is a proposition involving (e.g.) a definite particular circle, which has been *settled* (and drawn) as a starting point in a diagram that will show what a geometer must *do* in order to construe a square with the required feature. The formulation of the πρᾶγμα must be something like this: "this circle AB is equal to the square CDEF". Now, suppose the conclusion to be demonstrated has been formulated in a loose way: "there is *a circle* which is equal to *a square*". What geometers normal-

---

[47] My formulation is as near as possible to a syllogism in *Barbara* to follow Aristotle's suggestions in *APo* I.9. For the relation between syllogistic formulation and mathematical demonstrations, I follow Mendell (1989), but nothing important for this paper hinges on this issue.
[48] Angioni (2016), pp. 149–150, and Hasper (2013a), p. 319.

ly want with this loose formula—if they happen to employ it—actually amounts to the determinate proposition about the circle in the diagram ("this circle AB is equal to the square CDEF"), and the demonstration must show how the diagram allows the generalization of its results to *every* circle in virtue of its being a circle. Arguments like Bryson's, I suggest, attempt to take advantage of the inaccuracy of the loose, indefinite formulation of the demonstrandum.[49] Bryson's conclusion matches the loose formulation of the demonstrandum ("there is a circle equal to a square"), but the aspect under which the πρᾶγμα has been promised as the end-product of the geometrical demonstration has been completely neglected. Bryson's argument does not show how a particular square can be construed out of points and lines related to a particular circle and, therefore, does not explain how it is that *every* circle, *by being what it is,* is such that turns out to be equal to a particular square. Therefore, Bryson's argument turns out to be deceptive about the *pragma*, as Aristotle says in this important passage from *SE* 11:

T5:
[171b16] ἀλλ' ὡς Βρύσων ἐτετραγώνιζε τὸν κύκλον, εἰ καὶ τε-
[17] τραγωνίζεται ὁ κύκλος, ἀλλ' ὅτι οὐ κατὰ τὸ πρᾶγμα, διὰ
[18] τοῦτο σοφιστικός. [...]
[19]   [...], καὶ ὁ κατὰ τὸ πρᾶγμα φαινόμε-
[20] νος συλλογισμός, κἂν ᾖ συλλογισμός, ἐριστικὸς λόγος· φαι-
[21] νόμενος γάρ ἐστι κατὰ τὸ πρᾶγμα, ὥστ' ἀπατητικὸς καὶ
[22] ἄδικος (171b16–22).

But the way Bryson tried to square the circle is sophistical—even if the circle is squared—for the very reason that it is not in accordance with the object. Hence [...] a syllogism that appears to be in accordance with its object is an eristic argument, even if it does constitute a syllogism; for it appears to be in accordance with its object and is thus deceptive and dishonest (trans. Hasper, modified).

Bryson's argument does not rely on the same sort of trick as the false medical expert relying on Zeno's paradoxes. Bryson's argument is not appropriate to its object because—even if it deduces a proposition that is extensionally the same as the targeted conclusion—it does not deliver that proposition under the aspect on which it

---

[49] Proclus (*In Prim. Eucl.*, 374.18–21 and 375.8–12) attests (about Euclid's Proposition I.30) that there have been several sophistical moves like this to give "trouble" to ancient geometers. Besides, in 235.15–236.8, Proclus praises Euclid's formulation of Proposition I.4 because it prevents a completely wrong interpretation of the given. Both sophistical moves suggested in these texts are different from Bryson's (because they end up falsifying the Proposition), but what I want to stress is *just* the sophistical attitude of taking advantage of a loose formulation to give trouble to the experts (either by falsifying the Proposition, as in Proclus' examples, or by introducing some deceptive trick in the explanatory story, as in Bryson's argument).

was promised as the final product of a geometrical demonstration. But Bryson's argument has not relied on any premise that would count as blatantly alien to geometry (in the same way as Zeno's paradox is blatantly alien to medical expertise).

11 Sophistical Demonstrations and Pseudographemata

Aristotle has introduced Bryson's argument as an important example of sophistical refutations that fail to be in accordance with its object (i. e., sophistical demonstrations). In the passage immediately before, other kinds of argument give additional complexity to the picture. Aristotle mentions *pseudographemata* as fallacies that somehow are in accordance with a given expertise. But it is very easy to be misled by the passage in question, which runs thus:

> **T6:**
> [171b11] καὶ ὅσοι μὴ ὄντες κατὰ τὴν ἑκάστου μέθοδον παρα–
> [12] λογισμοὶ δοκοῦσιν εἶναι κατὰ τὴν τέχνην. τὰ γὰρ ψευδο–
> [13] γραφήματα οὐκ ἐριστικά (κατὰ γὰρ τὰ ὑπὸ τὴν τέχνην οἱ
> [14] παραλογισμοί), οὐδέ γ' εἴ τί ἐστι ψευδογράφημα περὶ ἀλη–
> [15] θές, οἷον τὸ Ἱπποκράτους ἢ ὁ τετραγωνισμὸς ὁ διὰ τῶν μη–
> [16] νίσκων.
>
> [Sophistical are also the syllogisms] that, [i] not being in accordance with the procedure appropriate to their respective objects [μὴ ὄντες κατὰ τὴν ἑκάστου μέθοδον], [ii] seem to be 'paralogisms' in accordance with the expertise [παραλογισμοὶ δοκοῦσιν εἶναι κατὰ τὴν τέχνην]. [iii] For pseudographemata are not eristic (for these paralogisms [*sc.* the pseudographemata] are in conformity with what falls under the expertise), nor [*sc.* is it eristic] if it is a pseudographema about something true, as that of Hippocrates, or the squaring of the circle by way of lunules (171b11–16, my translation).

I parse the text differently from several translations.[50] The most delicate issue concerns what is included within the range of the negation "μὴ" in 171b11. It is agreed that the negation ranges over the participle "ὄντες", but not over the verb "δοκοῦσιν". However, while some translators take "παραλογισμοί" as a predicate of the participle "ὄντες" (thus, within the range of the negation), I take it as a predicate attached to the expression "δοκοῦσιν εἶναι κατὰ τὴν τέχνην". This interpretation is related to how I understand the difference between, on the one hand, the notion of being appropriate to their respective objects (κατὰ τὴν ἑκάστου μέθοδον) and, on

---

50 For contrast, see Dorion (1995), p. 147, Fait (2007), p. 45, and Hasper (2013b), p. 27.

the other hand, the notion of being in accordance with the expertise (κατὰ τὴν τέχνην).

First, not being in accordance with the procedure appropriate to their respective objects (μὴ ὄντες κατὰ τὴν ἑκάστου μέθοδον) is, in this context, tantamount to not being appropriate to its object, which was encoded by Aristotle in 169b23 with two expressions, "οἰκεῖον τοῦ πράγματος" (with the negation from the previous line implied) and "μὴ κατὰ τὸ πρᾶγμα". Thus, T6 relies on Aristotle's previous characterization of sophistical demonstrations in *SE* 8 (T1). The word "πρᾶγμα" does not occur in 171b11, but I take the pronoun "ἕκαστον" (in this context) with the same force: it introduces exactly what "πρᾶγμα" has introduced before, namely, the proposition desired as the final result (the conclusion) in a given argument, exactly under the aspect on which it is promised in that argument. Thus, in this restricted context, concerned with scientific expertise and sophistical demonstrations, the pronoun "ἕκαστον" has the force of introducing the explanandum in a given argument within a specific discipline.[51] Furthermore, "μέθοδος" (in this context) stands for the principled search for premises that conform to the specific kind of final result the argument is meant to deliver.[52] Besides, the use of the genitive as complement of "μέθοδος" should be taken here in the strictest sense (as is commonly found in Aristotle): the expression "τὴν ἑκάστου μέθοδον" is not introducing the weaker idea of a premise-search generally suited to the domain (or discipline) to which the πρᾶγμα belongs; rather, it introduces the stronger idea of a premise-search specifically suited to the πρᾶγμα in question, namely, to the proposition taken as explanandum, exactly under the aspect on which it is so considered within the expertise in question. Thus, a μέθοδος appropriate to its respective object is (in this context) exactly that premise-search which finds out, and selects, the premises strictly capable of delivering the appropriate explanation.

Thus, the notion Aristotle retrieves in step [i] of T6 is the kind of sophistical refutation that aims at producing an appearance of appropriate explanation: what I have labelled a sophistical demonstration. Step [ii] adds a new, important point: sophistical demonstrations *seem* to be, and *look like*, a kind of paralogism (or fallacy, if one prefers that terminology) that, even failing to deliver the final product it is supposed to deliver (i.e., truth and appropriate explanation), is at least in conformity with several requirements of the expertise in question.

---

51 There are several occurrences of "ἕκαστον" in Aristotle's theory of scientific explanation with the same force: see, e.g., *Posterior Analytics* 71b9, 75b38, 76a4, and 76a27.

52 Actually, this is what "μέθοδος" basically means in some occurrences in *Topics* (e.g., 162b8), *Sophistici Elenchi* (e.g., 172b8) and *Prior Analytics* (e.g., 53a2); the more abstract idea of *study*, even if more frequent, is derivative. For a different view, see Smith (1997), pp. 41–42.

These paralogisms are the pseudographemata, which, however, are not strictly sophistical (171b12–13).

A pressing issue consists in discerning, even if in general terms, what a pseudographema is.[53] But before exploring this question, I emphasize that the expression "δοκοῦσιν εἶναι", in 171b12, should be taken with a specific force that is commonly found in many occurrences: the expression ties two items that have both some similarities *and some differences*. In certain contexts, when one says that A *seems* to be B, one is not saying (nor suggesting) that A and B are similar in all the relevant features that can be tracked. Nor is one saying (or suggesting) that A and B are identical. All one is saying is that A and B have *some* relevant similarities, even when their differences are also prominent and obvious.

This is why a sketchy outline of pseudographemata will be enough for my purposes.[54] No matter what exactly pseudographemata are, all I need is to identify the relevant features that account for sophistical demonstrations *looking like* them.

Aristotle describes the kind of paralogisms in question at the beginning of the *Topics*. His employment of the participle "ψευδογραφῶν" in 101a10 suggests that he is talking about the same thing labelled as *pseudographema*. These are the two most important parts of Aristotle's remarks:

T7:
[101a5] Ἔτι δὲ παρὰ τοὺς εἰρημένους ἅπαντας συλλογισμοὺς
[6] οἱ ἐκ τῶν περί τινας ἐπιστήμας οἰκείων γινόμενοι παραλογι–
[7] σμοί, καθάπερ ἐπὶ τῆς γεωμετρίας καὶ τῶν ταύτῃ συγ–
[8] γενῶν συμβέβηκεν ἔχειν.

Next, apart from all the deductions that have been mentioned, there are the paralogisms [or fallacies] based on what is appropriate to specific sciences, as we find in the case of geometry and its kindred sciences (trans. Smith, modified).

A few lines ahead:

T8:
[101a13]   ἀλλ' ἐκ τῶν
[14] οἰκείων μὲν τῇ ἐπιστήμῃ λημμάτων οὐκ ἀληθῶν δὲ τὸν συλ–
[15] λογισμὸν ποιεῖται. τῷ γὰρ ἢ τὰ ἡμικύκλια περιγράφειν
[16] μὴ ὡς δεῖ ἢ γραμμάς τινας ἄγειν μὴ ὡς ἂν ἀχθείησαν
[17] τὸν παραλογισμὸν ποιεῖται.

---

53 For discussion, see Acerbi (2008), Smith (1997), pp. 49–50, and Hasper (2013a), pp. 289–291.
54 For discussion, see Acerbi (2008), pp. 544–545, and Hasper (2013a), pp. 290 ff.

> Instead, he makes his argument from premises which are appropriate to the science but not true [or not right]: for he produces the paralogism by drawing semicircles not as he should, or by extending certain lines not as they would be extended (trans. Smith, modified).

Arguments of this sort do not fit any of the three descriptions provided before in the same chapter (100a27–30, 100b23–101a4): they are not demonstrative arguments, nor dialectical, nor sophistical. They are mistakes in a particular way: [a] although they start from premises or principles that are appropriate to the expertise in question, at least at large, [b] they deliver something wrong, due to a bad application of those premises or principles. Besides, I suggest that they do not spring from the purpose of producing a false semblance of being knowledgeable and, for this reason, do not count as sophistic or eristic—actually, this explains Aristotle's remarks in 171b12–13 very well.[55] These arguments can be just mistakes that someone significantly acquainted with the expert domain in question may produce.[56]

One crucial point in my interpretation concerns the premises not being true, despite their being appropriate in general to the domain in question. First, being appropriate in general to a *given discipline* is not equivalent to being appropriate to a *given object* (πρᾶγμα): the latter notion of appropriateness usually involves the former, but not vice-versa, and that is why some arguments, such as Bryson's (and the pseudographemata), can have premises that are appropriate in general to the domain in question without being appropriate to its πρᾶγμα.[57]

Second, I suggest that "ἀληθές" in 101a14 should not be taken as exclusively focusing on the truth value of a given proposition in the pseudographema. I am not denying that "ἀληθές" can be taken in this way in some (or even most) cases. I am arguing in favor of an additional option: "ἀληθές" in 101a14 can be taken as saying something about the explanatory power of premises employed as principles in the

---

[55] Sophistical arguments can be described by many objective features, but the essential trait of all of them is the *purpose* (προαίρεσις) of producing a false appearance of σοφία (see *Metaphysics* 1004b18–25; *SE* 165a31, 172b10–11; *Rhetorics* 1355b16–21). See Striker (1996), p. 9, Dorion (1995), p. 212, Fait (2007), pp. 104–105, and, indirectly, Fait (2013), p. 164 ("the only reason why they can be deemed sophistical is the deceptive use they are put to"). For discussion of the ethical dimension of eristic arguments, see King (2013), pp. 190–191.

[56] Besides, if what makes an argument sophistical is, *ultimately*, its purpose (see previous note), arguments that are extensionally the same can be classified differently when employed mistakenly by someone in *bona fide* and when employed purposely with the aim of producing a false appearance of σοφία. However, see the helpful points made by Fait (2016, pp. 25 and 36–37) to avoid overestimating purpose as a criterion for sophistry.

[57] Even if one does not agree that Bryson's argument counts as a good example, the important point is that Aristotle acknowledges the situation (also in *Topics* 101a13–17, as I argue).

attempted demonstration. In this case, instead of saying that a given proposition used in the pseudographema is *not true* (and is, therefore, *false*), "οὐκ ἀληθῶν" in 101a14 can be taken as saying that the proposition, although being appropriate to geometry, is not the *right one* to deliver the desired final result.[58]

Thus, the crucial point is *not* that someone draws a semi-circle in such a wrong way that it is not a semi-circle after all (because, e.g., someone has moved the position of the original center when rotating the *radius*, or has not put the center in the right position—as suggested in Alexander's interpretation of *Topics* 101a15–17, *in Top.* 23.22–25.9, see Acerbi 2008); it is *not* that someone ignores a basic rule (such as Euclid's Postulate I.1) and draws a line which, supposed to be straight, turns out to be curve. Those mistakes would be too obtuse, and it would be hard to understand how they could still be identified as appropriate to geometry. On the other hand, take "ἀληθές" in 101a14 in the way I suggest: then, "οὐκ ἀληθές" applies to a proposition that is itself true (or a step that is itself well-constructed), but was wrongly employed in the diagram—*wrongly*, because it is not suited to deliver the desired result.[59] Therefore, the pseudographema turns out to be a mistake which—although consisting of propositions that are, in general, appropriate to geometry and, in most cases, true—employs at least one premise that is not "the right one" to deliver the desired result. For this reason, a pseudographema [i] is not appropriate to its object (cf. μὴ κατὰ τὴν ἑκάστου μέθοδον in 171b11), although [ii] it is in accordance with geometry in general (cf. κατὰ τὴν τέχνην in 171b12). Besides, a pseudographema is not sophistical because it does not have the original purpose of producing an appearance of expertise.

12 Conclusion

If this suggestion is correct, the result emerging from T6 is similar to the results emerging from T1–T2. Aristotle introduces a kind of sophistical argument that, differently from most kinds previously studied in the *SE*, is a successful deduction of the conclusion. However, the other features of this kind of sophistical argument need to be disentangled from intricate comparisons both with peirastic arguments

---

58 Aristotle employs "ἀληθής" (and "ψευδής") in this way in many cases. See Angioni (2019a), p. 164; for contrast, see my point (Angioni 2017, p. 224) about Aristotle's uses of the interesting phrase "true but not enlightening" in *Eudemian Ethics* 1216b32–3, 1220a16–18, and 1249b5–6.
59 Similarly, the attempt to explain long gestation in elephants by their longevity is wrong [i] not because the premise "elephants are long-lived" is *false* (actually, it is not false), but [ii] because that premise, even being true, does not have the explanatory power appropriate to the object in question.

and with pseudographemata. Like the pseudographemata, these sophistical arguments have premises that are in general appropriate to the scientific discipline in question, and even appropriate to the targeted conclusion *in a general way*, but fail to identifythe principles specifically appropriate to their πρᾶγμα. However, unlike pseudographemata, these sophistical arguments are purposely deceitful: they are essentially directed towards producing a false appearance of being knowledgeable. They are what they are by primarily aiming at a false appearance of explanatory appropriateness to their πρᾶγμα. I have labelled them sophistical demonstrations because being explanatorily appropriate to its πρᾶγμα is the mark of demonstrations.

On the other hand, sophistical demonstrations emulate the aim of peirastic arguments, namely, to expose false claims of scientific expertise. But they explore a route completely different from peirastic. Critical examiners are supposed to detect conflicts and incompatibilities between propositions conceded by the false expert without relying on definite knowledge belonging to the scientific discipline involved in the examination. Once such a conflict is spotted, the critical examiner produces a refutation which includes a false conclusion deduced from a false premise conceded by the false expert (169b25–27). Sophists, instead, produce direct "demonstrations" that appear to be explanatorily appropriate to their πρᾶγμα and, on the basis of that appearance, are expected to provoke the collateral, perlocutionary effect of discrediting the explanatory story offered by the real expert.

Inasmuch as the real experts still remain the experts they are when they are falsely debunked by the sophist (and, importantly, because the things they know are not changed by the false debunking effect), Aristotle can say that sophistical demonstrations are actually not successful in exposing false claims of expertise, as we read in T2:

> T2 [vi] But these sophistical refutations, even if they deduce a contradictory, do not make clear whether someone is ignorant (169b27–28, trans. Hasper, modified).

The gist of this sentence is that sophistical demonstrations do not really debunk the experts. They seem to debunk them, but do not really succeed. This is how Aristotle sees the point, and how the experts themselves can see the point.[60] However, as for the experts themselves, Aristotle sounds more alarmed in the next step of the same passage:

> T2 [vii] they disconcert [or discomfit, or entangle: ἐμποδίζουσι] even someone who has knowledge with these arguments (169b28–29, trans. Hasper, modified).

---

60 See Bolton (1994), p. 123, Fait (2013), p. 241, and Hasper (2013a), p. 288.

It might seem that step [vii] is in conflict with step [vi], for Aristotle recognizes that experts are indeed disconcerted with these sophistical arguments, even if the arguments are not really successful in their debunking purpose. Why experts would be disconcerted, after all, if these arguments do not disclose any ignorance on their part? What happens (besides other things) is that, if a discussion is conducted among non-experts—i.e., conducted in front of an audience of laymen—the poorness of a poor discussion will pass unnoticed, as we see in this passage from the *Posterior Analytics:*

T9:
[77b12] ὥστ' οὐκ ἂν εἴη ἐν
[13] ἀγεωμετρήτοις περὶ γεωμετρίας διαλεκτέον· λήσει γὰρ ὁ
[14] φαύλως διαλεγόμενος

Consequently, one should not discuss geometry among non-geometers: for the one who discusses in a poor way will pass unnoticed (my translation).

Thus, the debunking, even not being real on Aristotle's eyes (and for the experts themselves), may seem to be effective to an uneducated audience.[61] For an uneducated audience might be unable to distinguish between the verbal result of the argument and the real things the arguments are supposed to encode—and will surely be unable to identify which argument is really appropriate to its object and which argument only seems to be so (cf. *Ethica Eudemia* 1217a7–10).

Moreover, sophists can be more successful in their deceitful purpose when they attain a given kind of false appearance of explanatory appropriateness—namely, the kind that Aristotle sometimes associates with the expression "κατὰ συμβεβηκὸς" applied to demonstrative attempts:

T10:
[168b6] ἀλλὰ παρὰ τοῦτο καὶ οἱ τεχνῖται καὶ ὅλως οἱ ἐπιστήμονες
[7] ὑπὸ τῶν ἀνεπιστημόνων ἐλέγχονται· κατὰ συμβεβηκὸς γὰρ
[8] ποιοῦνται τοὺς συλλογισμοὺς πρὸς τοὺς εἰδότας· (168b5–8).

Nevertheless, both specialists and experts generally are refuted on the basis of this by non-experts, for they [*sc.* the non-experts] produce syllogisms on the basis of a concomitant factor against those who have expert knowledge (trans. Hasper, modified).

---

61 For the importance of the audience to sophistry as a public performance, see Swanson (2013), p. 214, and Fait (2013), pp. 241–243. See also interesting remarks about 169b27–29 in Merry (2016), p. 91, and Gentzler (1995), p. 24.

Syllogisms on the basis of a concomitant factor, in this context, refer exactly to those explanatory attempts that rely on premises that are themselves true, and appropriate to the discipline in question, but not primarily appropriate to the explanandum.[62] Now, if the experts themselves are not reasonably skilled in dialectical discussions—if they are not properly educated—a dangerous effect can arise:

T11:
[168b8] οἱ δ' οὐ δυνά-
[9] μενοι διαιρεῖν ἢ ἐρωτώμενοι διδόασιν ἢ οὐ δόντες οἴονται δε-
[10] δωκέναι. (168b8–10).

And those who are not able to make the relevant distinctions, either concede when questioned, or, without having conceded, are thought to have conceded (trans. Hasper, modified).

I take the verb "οἴονται" (168b9) as *passive:* experts who are not able to find the relevant distinctions and, consequently, do not appropriately respond to the debunking attempt, are *thought* (by the audience) to have accepted the results of the sophistical arguments and, thereby, are *thought* to be defeated.[63]

It was against this kind of sophistical argument that Aristotle has advanced his characterization of scientifically knowing an explanandum within a discipline (ἐπίστασθαι ἕκαστον ἁπλῶς , 71b9)—the key notion which occupies him in the first chapters of the *Posterior Analytics*. Indeed, this is how he introduces that notion as a definiendum: "We think we have knowledge of something *simpliciter* (and not in the sophistical way, on the basis of a concomitant factor)" (71b9–10).[64] The contrast with a specific sophistical way of knowing (or appearing to know) was central to his project. What he had in mind was exactly the kind of sophistical argument that produces a false appearance of explanatory appropriateness to its object. But this is a story for another time.[65]

---

[62] Here, I am relying on previous accounts in Angioni (2012), pp. 208–220, and Angioni (2016), pp. 156–163. For a different construal of the argument referred to in 168a 40–b10, see Fait (2016), pp. 42–44.

[63] See *Eudemian Ethics* 1216b40–1217a10: Aristotle was really worried with the phenomenon of *false debunking*. See Angioni (2017), pp. 229–234. On T11, see Hasper (2013a), p. 288.

[64] I am not concerned here with Aristotle's definiens, but only with the way in which he introduces the definiendum. Another important reference to the sophistical way of emulating, and running against, scientific demonstrations is found in 74a28–29—on which see Hasper (2006) and Angioni (2019b), pp. 169–173.

[65] A previous version of this paper was presented at the conference "Ancient Dialectic and its Reception" at the University of Patras in June 2021. I am thankful to Melina Mouzala for organizing the conference, and for helpful comments. I also profited from remarks made by Inna Kupreeva, Paulo Ferreira and others. Joe Karbowski, Adam Crager and Sean Kelsey have read a previous draft and helped me to improve style and content. I profited from exchanges and conversations about

Bibliography

Acerbi, Fabio (2008): "Euclid's *Pseudaria*". In: *Archive of the History of Exact Sciences* 62, pp. 511–551.
Ackrill, John (1963): *Aristotle: Categories and De Interpretatione*. Oxford: Oxford University Press.
Angioni, Lucas (2012): "Três Tipos de Argumento Sofístico". In: *Dissertatio* 36, pp. 187–220.
Angioni, Lucas (2016): "Aristotle's Definition of Scientific Knowledge (*aPo* 71b 9–12)". In: *Logical Analysis and History of Philosophy* 19, pp. 140–166.
Angioni, Lucas (2017): "Explanation and Method in *Eudemian Ethics* I.6". In: *Archai* 20, pp. 191–229.
Angioni, Lucas (2019a): "What Really Characterizes Explananda: *Prior Analytics* I.30". In: *Eirene: Studia Graeca et Latina* 55, pp. 147–177.
Angioni, Lucas (2019b): "Aristotle's Contrast between *episteme* and *doxa* in its context (*Posterior Analytics* I.33)". In: *Manuscrito* 42. No. 4, pp. 157–210.
Bolton, Robert (1990): "The Epistemological Basis of Aristotelian Dialectic". In: Devereux, Daniel and Pellegrin, Pierre (Eds.): *Biologie, Logique et Métaphysique chez Aristote*. Paris: Éditions du CNRS, pp. 184–236.
Bolton, Robert (1994): "The Problem of Dialectical Reasoning (Συλλογισμός) in Aristotle". In: *Ancient Philosophy* 14, pp. 99–132.
Bolton, Robert (2013): "Dialectic, Peirastic and Scientific Method in Aristotle's *Sophistical Refutations*". In: *Logical Analysis and History of Philosophy* 15, pp. 267–285.
Charles, David (2000): *Aristotle on Meaning and Essence*. Oxford: Oxford University Press.
Devereux, Daniel (1990): "Comments on Robert Bolton's 'The Epistemological Basis of Aristotelian Dialectic'". In: Devereux, Daniel and Pellegrin, Pierre (Eds.): *Biologie, Logique et Métaphysique chez Aristote*. Paris: Éditions du CNRS, pp. 263–286.
Dorion, Louis-André (1995): *Les réfutations sofistiques*. Paris and Laval: Vrin.
Dorion, Louis-André (2012): "Aristotle's definition of *elenchos* in the light of Plato's *Sophist*". In: Fink, Jakob (Ed.): *The Development of Dialectic from Plato to Aristotle*. Cambridge: Cambridge University Press, pp. 251–269.
Dorion, Louis-André (2013): "Aristotle and the Socratic *elenchos*". In: *Logical Analysis and History of Philosophy* 15, pp. 323–342.
Fait, Paolo (2007): *Le confutazione sofistiche*. Rome and Bari: Laterza.
Fait, Paolo (2013): "The 'false validating premiss' in Aristotle's doctrine of fallacies". In: *Logical Analysis and History of Philosophy* 15, pp. 238–266.
Fait, Paolo (2016): "Aristotle on the demarcation of dialectical and sophistical arguments". In: *Antiquorum Philosophia* 10, pp. 25–46.
Friedlein, Gottfried (1873): *Procli Diadochi in Primum Euclidis Elementorum Librum Commentarii*. Leipzig: Teubner.
Gentzler, Jyl (1995): "The Sophistic Cross-Examination of Callicles in the *Gorgias*". In: *Ancient Philosophy* 15, pp. 17–43.
Hasper, Pieter Sjoerd (2006): "Sources of delusion in *Analytica Posteriora* I 5". In: *Phronesis* 51, pp. 252–284.

---

the subject over the years with Paolo Fait, Pieter Sjoerd Hasper, Carrie Swanson, Laura Castelli, Fernando Mendonça and Paulo Ferreira.

Hasper, Pieter Sjoerd (2013a): "Between Science and Dialetic: Aristotle's Account of Good and Bad Peirastic Arguments in the Sophistical Refutations". In: *Logical Analysis and History of Philosophy* 15, pp. 286–322.
Hasper, Pieter Sjoerd (2013b): "Aristotle's *Sophistical Refutations:* A Translation". In: *Logical Analysis and History of Philosophy* 15, pp. 13–54.
Hasper, Pieter Sjoerd (2013c): "The Ingredients of Aristotle's Theory of Fallacy". In: *Argumentation* 27, pp. 31–47.
Judson, Lindsay (2019): "Aristotle and Crossing the Boundaries between the sciences". In: *Archiv für Geschichte der Philosophie* 101. No. 2, pp. 177–204.
Karbowski, Joseph (2016): "Justification 'by Argument' in Aristotle's Natural Sciences". In: *Oxford Studies in Ancient Philosophy* 51, p. 119–160.
Karbowski, Joseph (2019): "Syllogisms and Existence in Aristotle's *Posterior Analytics*". In: *Manuscrito* 42. No. 4, pp. 211–242.
King, Colin Guthrie (2013): "False ἔνδοξα and fallacious argumentation". In: *Logical Analysis and History of Philosophy* 15, pp. 185–199.
Mendell, Henry (1998): "Making Sense of Aristotelian Demonstration". In: *Oxford Studies in Ancient Philosophy* 16, pp. 161–225.
Merry, David (2016): "The Philosopher and the Dialectician in Aristotle's *Topics*". In: *History and Philosophy of Logic* 37, pp. 78–100.
Mueller, Ian (1982): "Aristotle and the Quadrature of the Circle". In: Kretzmann, Norman (Ed.): *Infinity and Continuity in Ancient and Medieval Thought*. Cornell: Cornell University Press, pp. 146–164.
Reeve, Christopher David C. (1998): "Dialectic and Philosophy in Aristotle". In: Jyl, Gentzler (Ed.): *Method in Ancient Philosophy*. Oxford: Oxford University Press, pp. 227–252.
Smith, Robin (1989): *Aristotle: Prior Analytics*. Indianapolis: Hackett.
Smith, Robin (1994): "Dialectic and the Syllogism". In: *Ancient Philosophy* 14, pp. 133–151.
Smith, Robin (1997): *Aristotle: Topics, Books I and VIII*. Oxford: Oxford University Press.
Striker, Gisela (1996): "Methods of Sophistry". In: *Essays on Hellenistic Epistemology and Ethics*. Cambridge: Cambridge University Press, pp. 3–21.
Steinkruger, Philipp (2018): "Aristotle on Kind-Crossing". In: *Oxford Studies in Ancient Philosophy* 54, pp. 107–158.
Swanson, Carrie (2013): "Aristotle's Expansion of the Taxonomy of Fallacy in *De Sophisticis Elenchis* 8". In: *Logical Analysis and History of Philosophy* 15, pp. 200–237.
Swanson, Carrie (2017): "Aristotle on Ignorance of the Definition of Refutation". In: *Apeiron* 50, pp. 153–196.

Part II: **Reception, Interpretation, Development and Influence of Ancient Greek Dialectic in Late Antiquity and Byzantium**

Gweltaz Guyomarc'h
The Services of Dialectic: Dialectic as an Instrument for Metaphysics in Alexander of Aphrodisias

It is commonplace to think that ancient commentators aimed to systematize Aristotle. In order to "complete" and "unify"[1] Aristotelian doctrine, the commentators would thus have forced the corpus into consistency by introducing claims or arguments taken from a branch of the corpus in the exegesis of a passage found in another branch. The intended application of the *Analytics*' epistemological rules to the sciences—and especially to first philosophy—allegedly exemplifies the issue.[2] In so doing, the commentators would have ignored all that was exploratory, problematic, zetetic and inchoate in Aristotelian thought.[3] Alexander of Aphrodisias' attempt to make metaphysics into a demonstrative science would be a prime example of such tendency.[4] More generally, his exegetical method is held to rest on a "systematic presupposition" and to aim for a "unified" or even "dogmatic Aristotelianism".[5]

To put it plainly—I think this is painting too unilateral and simple a picture, and I would like to contribute, here, following others,[6] to enrich and detail it. To do so, I will look into Alexander's usage of dialectical method in metaphysics, with particular interest for his exegesis of book *Beta* of the *Metaphysics* and his use

---

Note: Thanks are due to Melina Mouzala for inviting me to contribute to this volume, allowing me to return to some issues I had already explored—namely during a 2017 seminar in Lille, organized by the Universities of Lille, Liège and Bruxelles, whose participants deserve warm thanks, notably Thomas Bénatouïl, Sylvain Delcomminette and Marc-Antoigne Gavray. I hope time will have allowed me to clarify my thoughts and to give them a more intelligible form. I thank Jeanne Allard once more for her patient translation labor.

1 Aubenque (1926b), p. 5.
2 Aubenque (1961), p. 3.
3 An English account can be found in Aubenque (1962a). See also Donini (1994).
4 Bonelli (2001).
5 Donini (1994), pp. 5035 and 5042. See also, for instance, Moraux (2001), p. 252, Cerami (2016), p. 164 ("The ultimate goal of this agenda is to establish an all-embracing philosophical system capable of responding in the best possible way to the philosophical issues debated by his contemporaries"), and Frede (2017) ("In general, Alexander goes on the assumption that Aristotelian philosophy is a unified whole, providing systematically connected answers to virtually all the questions of philosophy recognized in his own time").
6 In particular Kupreeva (2017), with whom I am in complete agreement.

https://doi.org/10.1515/9783110744149-013

of the aporetic method. Alexander's aporetic method in the *Quaestiones*[7] as well as the one he puts to use in his commentary on *Metaphysics Beta* has led to the same diagnosis. In both cases, no "honest perplexity"[8] is displayed, and the *Beta* aporiae are not treated like *genuine puzzles* but rather as simple exposition devices.[9] In contrast to this view, I would like to show two things: first, that aporia retains an authentically exploratory function for Alexander; and, second, that Alexander's use of aporia in metaphysics does not originate in systematization, but rather in the fulfillment of dialectic's status as an *organon* within Aristotelian tradition. Inna Kupreeva has already given strong arguments in favor of the first part of this claim. She has shown that some aporiae in *Beta* are not read by Alexander in a circular way—i.e., that Alexander's exegesis of these aporiae does not already presuppose the Aristotelian solutions—and how, on the contrary, some aporiae remain truly open.[10] One could well join in on her efforts, and show that some aporiae become much more problematic for Alexander, due precisely to his exegesis of Aristotle, than they were for Aristotle himself.[11] But I would like to pursue another path in this paper and examine the role of dialectic in metaphysics. I will claim that dialectic allows Alexander to retain the exploratory aspect of aporiae within a scientific investigation. If we show that the heuristic role proper to dialectic is an integral part of science, we will be better able to support the idea that Alexander retains the exploratory aspect of aporiae.

1 Dialectic in *Metaphysics Beta*

I will start by showing that, for Alexander, *Metaphysics* B uses dialectic—without being a *dialectical book*. Of course, when Alexander writes his commentary on *Beta*, he does so with the entire *Metaphysics* at his disposition. The issue is then not to know whether he has the solutions brought (or not) to the aporiae in the rest of the treatise in mind—because he doubtlessly has. And indeed, in his outline of the *Metaphysics*, Alexander explicitly introduces Γ as the book where the solutions to the aporiae of B start:

---

7 Fazzo (2002), pp. 17–18. For a more nuanced view, see Rashed (2007), pp. 3–4.
8 Cf. Madigan in Madigan and Dooley (1992), p. 79.
9 The expression is in Aubenque (1961).
10 Kupreeva (2017), especially pp. 241–247, which study the case of the eighth aporia, and show that Alexander's interpretation is not circular and that his references to Aristotelian hylomorphism do not hinder the dialectical exploration of the difficulty.
11 On this pursuit, see Lavaud and Guyomarc'h (2021), pp. 111 ff.

> ... and further, as it is useful and necessary for the discovery of the objects proposed to wisdom, he raised certain aporiae concerning being, the principles, and related matters. After the aporiae he begins the present book *Gamma*, in which he finally tells and establishes his own positions and solves the points of aporia (*In Met.* 237.13–238.3, trans. Madigan, modified).[12]

This is not an isolated claim. The rest of the commentary on book Γ mentions the aporiae from Β reprised in Γ, that is, the first, second, third and fourth aporiae.[13] In going from Β to Γ, we thus proceed from the presentation of problems to their solutions.[14] Alexander is, however, peculiar in the fact that he reads *Beta* as the true beginning of the *Metaphysics*, taking *Alpha* and *Alpha elatton* to be "preambles", following Aristotle's own remark (995b4):

> For Aristotle this is the starting point of the proposed treatise; for here he begins to speak of matters which have a necessary bearing on the issues proposed. The matters discussed in *Alpha* would be preliminary to this treatise and contribute to putting it on the right footing. This is why some have thought that the present book is the first book of the treatise *Metaphysics* (172, 18–22, trans. Madigan).

How can a book relying on "dialectic" and putting forward a number of "logical" arguments (as Alexander himself admits)[15] be both the start of the treatise and the beginning of the science discussed in it? Could it be because *Beta* is merely a summary? One could think so, based on the passage from the proemium to the commentary on *Gamma* cited above. But throughout his commentary, Alexander refrains from mentioning the solutions to the aporiae found in the rest of the treatise[16]—with one exception when he announces book Λ.[17] This amounts to say that the alleged summary does not perform its function. In addition, when he says that solving the aporiae is "the chief task (κεφάλαιον) of the proposed science" (180.32–33), Alexander could suggest that this is not the *sole* task of such sci-

---

12 ...Καὶ ἐπὶ τούτοις ὡς χρήσιμον καὶ ἀναγκαῖον πρὸς τὴν εὕρεσιν τῶν τῇ σοφίᾳ προκειμένων ἀπορήσας τινὰς ἀπορίας περί τε τοῦ ὄντος καὶ τῶν ἀρχῶν καὶ τῶν τούτοις παρακειμένων, μετὰ τὰς ἀπορίας ἄρχεται τοῦ προκειμένου τοῦ Γ βιβλίου, λοιπὸν ἐν τούτῳ λέγων τε καὶ κατασκευάζων τὰ αὐτῷ δοκοῦντα καὶ λύων τὰ ἠπορημένα.
13 Aporia 1: is there one science for all causes? Aporia 2: is it the same science which studies the principles of substance and the principles of demonstration? Aporia 3: is there one science of all substances? Aporia 4: is it the same science which studies substance and its essential properties? Cf. Aristotle, *Met.* Γ 2, 1004a32 and 34 (but the first occurrence is problematic in the manuscripts). In Alexander: *In Met.* 246.13–24; 250.3–5; 251.7–9; 257.12–16; 264.23–27; 264.31–34.
14 On Alexander's outline of the first five books of the *Metaphysics*, see Guyomarc'h (2015), pp. 85–93. Concerning *Beta* specifically, see Lavaud and Guyomarc'h (2021), pp. 113–119.
15 For instance, at *In Met.* 206.12–13 or 218.17, but especially at 173.21–174.4.
16 As Arthur Madigan himself notes; see Madigan (1992), p. 79.
17 *In Met.* 178.19–21.

ence. Incidentally, this is shown in the commentary on book *Gamma*—a book which (despite being explicitly depicted by Alexander at 238.2–3 as containing solutions to the aporiae) does not deal exclusively with these aporiae. Finally, we must make note (although this is merely an *a silentio* argument) that Alexander never draws on the expressions Aristotle uses which point to the comprehensive aim of book *Beta*. Indeed, Aristotle uses such expressions at least four times in B 1: "all the difficulties, "all the contending arguments", etc.[18] None of them receives a specific commentary from Alexander.

If the *Metaphysics* begins in *Beta*, it is first because of the book's subject matter. For Alexander, the two previous books are in an ambiguous position, being halfway between physics and metaphysics. This is the case in *Alpha* because the theory of the four causes has already been discussed in the *Physics* (it must only obtain "confirmation"[19] in A) and involves a discussion of the claims of ancient physicists.[20] It is the case in *Alpha elatton* because the book could also be used as an introduction to theoretical philosophy in general, or even to the *Physics* in particular.[21] By contrast, in *Beta*, the ambiguity disappears, according to Alexander: "here he begins to speak of matters which have a necessary bearing[22] on the issues proposed" (172.18–19). This is because the objects of the aporiae cannot be studied by other sciences. For they are, foremost and generally, principles,[23] or, more precisely, for instance, the possibility of a "first cause" being "thoroughly immaterial".[24] But it is also because the aporiae are opportunities for first philosophy to reflect on its own nature, on its unity and on the instruments it uses— i.e., on notions like the same and the other, the like and the unlike, etc., the study of which is, in a certain way, "appropriate for the first philosopher".[25]

But the other feature setting *Beta* apart from the previous books is its method —aporia. Such method is different from the inquiry (ἱστορία[26]) used in the previous books. The task here is not to review previous doctrines and list their premises,

---

[18] See especially 995a25 and 34; 995b4.
[19] *In Met.* 23.2–3: "τὴν τῶν αἰτίων τῶν εἰρημένων βεβαίωσιν".
[20] As shown in a transition passage in the commentary on A 8 (*In Met.* 70.12–71.4), Aristotle comes nearer to the matters appropriate for the current treatise ("ὡς οἰκειοτέρας τῆς προκειμένης πραγματείας") only through the study of Pythagoreans and Plato.
[21] See the commentary to α 3, 995a17–19 at *In Met.* 169.20 ff., but the hypothesis is announced as early as 137.15 ff.
[22] On a similar usage of this verb, see, for instance, *In Met.* 170.3.
[23] See, for instance, the beginning of the development on the eighth aporia at *In Met.* 210.25–26.
[24] At *In Met.* 171.9–11 and 178.19–21 (with a reference to *Met.* Λ).
[25] Cf. *In Met.* 177.8–9. On this issue, see also Moraux (2001), pp. 467–468. On book *Beta* as the opportunity for metaphysics to reflect upon itself, see Guyomarc'h (2021), p. 117.
[26] See, for instance, the use of this term at *In Met.* 9.6 or 41.17.

nor to map their genealogy. The task is not even to refute these doctrines for their own sake. For, in *Beta*, names rather come up as branches in an aporia, and as tags for views determined by the requisites of a problem posed by Aristotle himself. History of philosophy now takes place within the framework of diaporia, i.e., the development of the two branches of an aporia. Many of these problems, of course, originate with Aristotle's predecessors, and the difficulties that come forth are in fact "all the points on which some have held different views" (995a25–26). But, according to Alexander, it is for the sake of diaporia ("in order to explore these aporiae" ("ὡς δὲ καὶ περὶ τούτων διαπορήσων", 172.8–9) that Aristotle reports these views and dedicates time to these issues.

Certainly, the goal pursued through *Beta*'s aporetic method remains the same as the one pursued through the first two books' inquiry: to discover truth in matters relevant to the treatise.[27] Also certain is the fact that the terminology relative to aporia and even diaporia in Alexander is not limited to his commentary on book *Beta*. The main reason for this is that Aristotle himself uses διαπορέω as early as the first book of the *Metaphysics*.[28] Aporiae are also involved in Alexander's strategy to defend the location of *Alpha elatton:* since the end of Chapter A 10 announces a return to the difficulties which could be raised about causes ("ἀπορήσειεν", 993a25), Alexander—while acknowledging that the more obvious reference is *Beta* —justifies the location of *Alpha elatton* by pointing out that this book also carries out an inquiry and brings forth an aporia concerning causes.[29] Still—aporia is the name Alexander uses to summarize book *Beta* when he refers to it in other places,[30] and it even becomes much like the book's title ("ἐν τοῖς ἠπορημένοις", 251.7).

2 *Aporia* and Problem

If *Beta* is indeed the true beginning of the *Metaphysics* and if the book uses dialectic, it means that Alexander considers the use of dialectic in *Beta* to be an integral part of first philosophy. This use of dialectic most markedly focuses on the aporetic method. Aporia has at least as many senses in Alexander as it does in Aristotle: Alexander reprises nearly all Aristotelian senses of the term[31] (except for the eco-

---

27 Compare, for instance, *In Met.* 78.2–4 and 174.1 or 180.31–33.
28 At *Met.* A 2, 982b15 and A 9, 991a9.
29 *In Met.* 136.15–17.
30 See especially *In Met.* 136.11–14; 138.4–6; 237.15–16; 246.14–15; 264.31. Aristotle had already paved the way for this title—see *Met.* Γ 2, 1004a34.
31 See Madigan (1992), p. 87, n. 3, Kupreeva (2017), p. 229, and Guyomarc'h (2021), p. 125. For its senses in Aristotle, see Motte and Rutten (2001), particularly pp. 152 and 367.

nomic sense) and adds an exegetical sense to the list (for instance, to refer to a difficulty encountered in understanding a passage of Aristotle). But its broad extension does not prevent the term from having a technical sense, in which it is a tool within a method:

> These remarks about the need to begin with exploring aporiae would also show the usefulness of dialectic for philosophy and for the discovery of truth. For it is characteristic of dialectic to explore aporiae, *i.e.*, to argue on both sides [of a case]. So, what was said in the *Topics* (1.2), that dialectic is useful for philosophical inquiries, is true (*In Met.* 173.27–174.4, trans. Madigan, modified).[32]

The relation with dialectic is indeed only instrumental (χρήσιμον), because book *Beta* is still not a logical book:[33]

> As Aristotle goes on he will show in what respect the inquiry into and the consideration of these things is also appropriate to the primary philosopher. For this treatise is not logical as some have thought due to the fact that many such things are objects of inquiry in it (177.8–10, trans. Madigan, modified).[34]

This passage is found in the introduction of the 4th aporia. "These things" refer to the predicates which are often called "dialectical": same, other, like, unlike, contrary, etc. The rest of the commentary will show that such study belongs within the science of being *qua* being because these predicates are species of the one and the many.[35] We do not know who are the "some" (τισιν) who have read the *Metaphysics* as a logical treatise—we might imagine they are other Peripatetics, or perhaps Stoics, like those who had taken the *Categories* to be a bad treatise on grammar and rhetoric.[36] Whatever the case may be, the reference shows that, due to such a reading, Alexander needs to distinguish metaphysics from logic in general and from dialectic in particular. One cannot examine the aporiae in *Beta* without

---

[32] Διὰ δὲ τῶν προειρημένων περὶ τοῦ δεῖν διαπορεῖν πρῶτον εἴη ἂν αὐτῷ δεικνύμενον ἅμα καὶ τὸ χρήσιμον τῆς διαλεκτικῆς πρὸς φιλοσοφίαν καὶ τὴν τῆς ἀληθείας εὕρεσιν· τῆς γὰρ διαλεκτικῆς τὸ διαπορεῖν καὶ ἐπιχειρεῖν εἰς ἑκάτερα. ἀληθὲς ἄρα τὸ ἐν τοῖς Τοπικοῖς εἰρημένον τὸ χρήσιμον εἶναι τὴν διαλεκτικὴν πρὸς τὰς κατὰ φιλοσοφίαν ζητήσεις.
[33] Alexander often uses λογικῶς or λογική and διαλεκτικῶς or διαλεχτική interchangeably. See, for instance, *In An. Pr.* 1.3 ff.; 3.7, etc.; *In Top.* 4.5 and 30.12–13 (which accounts for the synonymy), etc.; see also Bonelli (2001), p. 147, n. 40, Guyomarc'h (2014), p. 89, n. 14, and Kupreeva (2017) pp. 239–240. For examples of the synonymy in *Beta*, see, among other cases, *In Met.* 218.17.
[34] Προελθὼν δὲ αὐτὸς δείξει κατὰ τί καὶ ἡ περὶ τούτων ζήτησίς τε καὶ θεωρία οἰκεία τῷ πρώτῳ φιλοσόφῳ. Οὐ γὰρ λογικὴ ἡ πραγματεία, ὥς τισιν ἔδοξε διὰ τὸ πολλὰ τοιαῦτα ζητεῖσθαι ἐν αὐτῇ.
[35] *In Met.* 247.2–8.
[36] Simplicius, *In Cat.* 18.28–19.7. Cf. Moraux (1984), pp. 587–591.

using logical and endoxic arguments,[37] but the annexation of metaphysics to logic does not follow from it.

This could explain why Alexander so carefully avoids the πρόβλημα terminology when commenting *Beta*—in contrast with Syrianus, whose commentary on *Beta* nonetheless clearly depends on Alexander's. In Syrianus, ἀπορία and πρόβλημα are correlated, and even interchangeable.[38] A reminder of the senses of both terms in Alexander will allow us, by contrast, to better understand why aporia, but not the notion of problem, can be used in *Beta*.

The term ἀπορία no doubt has a broader usage than "problem" does: there is no problem which is not a difficulty, but not every ἀπορία is a πρόβλημα.[39] But especially in the *Commentary on the Topics*, Alexander lays out several criteria to specify what is a dialectical problem and how they differ from scientific problems. Such specification is first informed by the kind of questions used: in the four interrogation types distinguished in *A.Po* II.1, a dialectical problem appears under the form ὅτι ἔστι or εἰ ἔστιν, but never as δι' ὅ τι ἐστί or τί ἐστιν, which are questions belonging to science—a claim which Alexander says he takes from the lost treatise Περὶ προβλημάτων.[40] However, this criterion does not seem to apply to *Beta* and does not allow us to say whether the aporiae in *Beta* are dialectical problems in a strict sense or not: a large part of the aporiae in *Beta* are in the form ὅτι ἔστι or εἰ ἔστιν.[41] But the remark is interesting beyond this difference in formulation, since it assigns exclusively to science the inquiries on cause and essence—a point other texts confirm. For instance, in his commentary on *Topics* I.10, Alexander mentions that the question "Is it the case that form and matter are the elements of beings?"[42] cannot be dialectical. The case is significant since it echoes the terminology used in *Metaphysics* A and B.[43] The reason he offers refers to *Top.* I.1 and the distinction between the different kinds of syllogisms based on the nature of their premises: the question is about principles and thus requires ἄμεσοι καὶ πρῶται premises. In this sense, it is scientific rather than dialectical or eristical. The passage does concern premises—rather than problems—but I.10 cov-

---

[37] See, for instance, *In Met.* 236.28: "μὴ λογικαῖς ἐπιχειρήσεσι χρήσασθαι" and, on this passage, Kupreeva (2017), p. 240.
[38] Luna (2004), p. 54.
[39] *In Top.* 68.19–21, stressing the διὸ οὐδέ.
[40] *In Top.* 63.9–19. Cf. Castelli (2013), pp. 78–79.
[41] The 13th aporia (not in B1, but in B6, and commented at *In Met.* 233.1–235.6) appears to be an exception. But we could say that its formulation is only a more complex (and perhaps a second-order) version of an εἰ ἔστιν question.
[42] *In Top.* 70.5–6: ἆρά γε στοιχεῖα τῶν ὄντων εἶδος καὶ ὕλη;
[43] For instance, 986a2; 987b19; 992b19; 998b9.

ers both premises and problems (as was already the case in I.4), and it is not always easy to distinguish between them.[44] In his commentary to I.4, Alexander distinguishes them using their answers: a premise is a "request for an answer" (ἀποκρίσεως αἴτησις), while a problem is a "request for proving one part of the contradictory pair" (δείξεως τοῦ ἑτέρου μορίου τῆς ἀντιφάσεως αἴτησις).[45]

The kind of δεῖξις is precisely what is involved in the difference between scientific problem and dialectical problem, as the commentary on I.11 shows. At the end of the chapter, Aristotle excludes some problems from those to be covered in dialectic: first come the problems which do not cause any real perplexity—for, according to the above-mentioned criterion, a true problem must be difficult. But the problems "whose demonstration is near at hand, [and] those whose demonstration is too remote"[46] are also excluded. Alexander illustrates the former, i.e., problems "easy and well known" (ῥᾴδια καὶ εὔγνωστα, 84.15), with a reference to the Stoic inquiry into befitting actions (καθήκοντα), for instance: "Whether when listening to a philosopher we should have our legs crossed". Turning to the latter possibility, i.e., when the demonstration would be too elaborate, Alexander introduces a distinction (not found in Aristotle's text) between dialectical problems and scientific problems:

> For this reason, all those problems in mathematics and the sciences which admit of more general argumentation[47] may be regarded as dialectical problems, but all those which differ from these because they involve more theoretical study than is in keeping with a training, would be excluded from the (class of) dialectical problems. Thus "Whether or not in every triangle the three internal angles are equal to two right angles" is not a dialectical problem: a more powerful and more accurate method is needed for establishing problems of this kind; it is the part of the geometrician to prove how it is with this. For the same reason all of the following problems in philosophy are not dialectical either, as e.g., "Whether or not there is one matter for all things", "Whether or not matter is one", "Whether or not atoms are the principles of all there is", "Whether or not everything that moves another thing does this because it is itself moved", "Whether or not motion is eternal": questions like these require fuller and more accurate attention (*In Top.* 85.7–19, trans. Van Ophuijsen, modified).[48]

---

44 Brunschwig (1967), pp. 120–121, n. 6.
45 *In Top.* 40.28–29.
46 *Top.* I.11, 105a7–8, "οὐδὲ δὴ ὧν σύνεγγυς ἡ ἀπόδειξις, οὐδ᾽ ὧν λίαν πόρρω" (trans. Van Ophuijsen).
47 Unlike Johannes M. Van Ophuijsen, I think that ἐπιχείρησις no longer means "argumentative attack" in Alexander—as it did in Aristotle, and as Jacques Brunschwig also translates it. Alexander frequently uses ἐπιχείρησις and its cognate verb ἐπιχειρεῖν to speak of properly dialectical argumentation. In the commentary to *Beta*, see for instance *In Met.* 174.2, 176.35, 236.26, etc.
48 Διὸ ὅσα μὲν τῶν κατὰ τὰς ἐπιστήμας τε καὶ τὰ μαθήματα κοινοτέρας ἐπιχειρήσεις δέχεται, διαλεκτικὰ ἂν εἴη προβλήματα· ὅσα δὲ μὴ τοιαῦτα τῷ πλείω θεωρίαν ἔχειν ἢ <κατὰ γυμναστικήν,> ἐκπίπτοι ἂν τῶν διαλεκτικῶν προβλημάτων. Οὐ γὰρ διαλεκτικὸν τὸ 'πότερον πᾶν τρίγωνον τὰς

The possibility of finding some scientific problems among dialectical problems originates in Aristotle's claim at the beginning of the chapter: dialectical problems concern "either choice and avoidance or truth and knowledge" (πρὸς αἵρεσιν καὶ φυγὴν ἢ πρὸς ἀλήθειαν καὶ γνῶσιν, 104b2). Dialectical problems are general because dialectic can discuss any question, which should mean that it can also discuss scientific questions. The claim in this passage echoes the way in which Alexander commented the universal ambitions of dialectic evoked in the first sentence of the treatise, which is cited just before our passage at 85.6–7. The treatise provides a method to make syllogisms on any topics or problems: περὶ παντὸς τοῦ προτεθέντος [προβλήματος].⁴⁹ The first sentence of our passage thus has a concessive structure: some problems may be general and may require an argumentation general in scope—*but* dialectic's ability to discuss any kind of problem does not make such problems dialectical. Some problems encountered in the sciences can be covered by dialectic—but this does not mean that all dialectical problems are scientific, and conversely. Here, Alexander is being perfectly consistent with his commentary of the first sentence of the treatise, at 6.21–7.2: according to him, when Aristotle says that dialectic is "about everything" (περὶ παντός), we should not read "everything" "without qualification" (οὐχ ἁπλῶς, 6.25). And he already refers to I.11 during his commentary on I.1, which shows clearly enough that he intends to distinguish dialectic and science.

Examples of these scientific problems requiring more elaborate study are often from mathematics and physics. The question whether matter is unified or not reminds the *De mixtione*, for instance, as the question concerning movers could refer to the *Refutation of Galen*. Also to be noted is the difference between "mathematics" and "philosophy" in accordance with other passages of the Alexandrinian corpus.⁵⁰ But mathematics and philosophy are both subject to a "precision" requirement that demands longer demonstrations. Conversely, as Alexander already established in his commentary, reasonings established "through what is approved are divorced from scientific precision of speech" (κεχώρισται τῆς ἐπιστημονικῆς ἀκριβολογίας, 26.12–13).

ἐντὸς τρεῖς γωνίας δυσὶν ὀρθαῖς ἴσας ἔχει ἢ οὔ;'· κρείττονος γὰρ καὶ ἀκριβεστέρας μεθόδου δεῖ πρὸς τὴν τῶν τοιούτων κατασκευὴν προβλημάτων· τοῦ γὰρ γεωμέτρου τὸ δεῖξαι τοῦτο ὅπως ἔχει. Διὸ οὐδὲ τῶν κατὰ φιλοσοφίαν ἐστὶ πάντα τὰ τοιαῦτα διαλεκτικὰ προβλήματα, οἷον πότερον μία ὕλη πάντων ἢ οὔ, καὶ πότερον ἡ ὕλη ἥνωται ἢ οὔ, καὶ πότερον αἱ ἄτομοι ἀρχαὶ τῶν ὄντων ἢ οὔ, καὶ πότερον πᾶν τὸ κινοῦν κινούμενον κινεῖ ἢ οὔ, καὶ πότερον ἀίδιός ἐστιν ἡ κίνησις ἢ οὔ· πλείονος γὰρ καὶ ἀκριβεστέρας τὰ τοιαῦτα ἐπιστάσεως δεῖται.
49 The word is omitted at 5.20 but appears at 7.1. Cf. Brunschwig (1967), p. 114, n. 4.
50 For instance, *In A.Pr.* 3.20–24. Alexander's commentary on *Metaphysics* Γ never places mathematics among theoretical sciences.

In both his commentary on *Metaphysics* B and his commentary on the *Topics*, Alexander's intention seems to be to establish the scientific *use* of dialectic while maintaining (or even consolidating) the boundary between science and dialectic. For nothing in Aristotle's *Top.* I.1 calls for a commentary aiming to restrict dialectic's scope—on the contrary. Universality is also named as a common trait of dialectic and philosophy in *Metaphysics* Γ 2, 1004b19–20 (since dialecticians, like philosophers, discuss all things)—and Alexander cannot not know about this passage. But the insistence to separate science and dialectic makes sense considering the annexation of metaphysics by dialectics performed by "some". It explains why Alexander is so careful never to speak of "problems" in his commentary on *Metaphysics* B. The expression can be interpreted specifically as "scientific problems"— but it still has a dialectical ring to it, and thus could be confusing. As we have seen, however, Alexander admits that *Beta* uses dialectic. For him, the issue at stake is then to establish such use, and to do so in a controlled manner, i.e., so as to guard metaphysics against a possible invasion.

3 The Importation of the Dialectical Method

The claim that Aristotle uses dialectic in scientific discussions would today be seen as quite on the nose—but Alexander may precisely be the one responsible for its being commonplace. Against a looser usage of "dialectical" to refer to a number of passages from the Aristotelian corpus, and following some contemporary interpreters,[51] I must remind that dialectic in fact consists in a codified procedure. It is not obvious that dialectic could be used in a non-dialogical context, outside "dialectical meetings",[52] and such claim calls for justification. What I would like to show now is that it is precisely the conditions of such non-dialogical use of dialectic that Alexander tries to identify. Further, I would like to show that, for him, such use is the main purpose of dialectic. In other words: what was in Aristotle a collateral benefit of dialectic (at least in part due to its opposition to Platonic doctrine) becomes, in Alexander, its central purpose.

The shift operated by Alexander can be seen when he comments on the services which dialectic can provide, while discussing the well-known Chapter I.2 from the *Topics*. As a reminder, the chapter lists three services which can be expected from the current "πραγματεία". Here, πραγματεία arguably refers both to the trea-

---

[51] Cf. the fascinating remarks of Rapp (2017), p. 116 (the issue of "dialectical method in non-dialogical contexts" has been developed further in an early and unfortunately unpublished draft of this paper); see also Primavesi (1996), pp. 52 ff.
[52] See the ἐν δὲ ταῖς διαλεκτικαῖς συνόδοις phrase at *Top.* VIII 5, 159a32.

tise itself and to dialectic as a discipline, since, right after, Aristotle mentions that we have a method[53]—either way, Alexander himself must take the word in both senses, since it is in accordance with his own usage.[54] The three services are the following: intellectual training or intellectual gymnastics; meetings or contacts with others; and what concerns the philosophical sciences.[55] Alexander ponders a possible fourth service (which I will not cover here, since it would lead us elsewhere): the one concerning the "first things" or principles in each science (τὰ πρῶτα τῶν περὶ ἑκάστην ἐπιστήμην, 101a36–37)—but, in the end, he judges it to be an extension of the third service.[56]

The first service—concerning intellectual training—refers directly to dialectical discussions. For Aristotle, the very first advantage of dialectic as a method is thus to be of use in the dispute which is also called dialectical, and the *Topics* are first useful for training in the subject of dialectic itself. In fact, it is as if Aristotle was listing the three services in increasing order of externality. Dialectic thus has some interest for science—as Aristotle points out next—since it teaches us to διαπορεῖν (101a35). But this is a supplemental benefit, not its primary purpose. That is, to claim that dialectic is useful to science does not mean that dialectic is *solely* an instrument in the service of science. Yet Alexander precisely seems to reduce dialectic to its instrumental purpose. For this is how he comments on the first service:

> By training he either (i) means that which occurs in discussions with others—as a form of training they try, receiving certain problems from their interlocutors, to defend these problems by producing argumentations through what is approved—or (ii) he means by it argumentation on either side of a question. This kind of speech was customary among the older philosophers, who set up most of their classes in this way—not on the basis of books as it is now done (since at the time there were not yet any *books of this kind*), but, after a thesis had been posited, they trained their aptitude at finding argumentations by producing arguments about this thesis, establishing and refuting the position through what is approved. There are *books of this kind* written by Aristotle and Theophrastus, containing argumentations on both sides of a question through what is approved (*In Top.* 27.8–18, trans. Van Ophuijsen, modified).[57]

---

53 Cf. Rapp (Unpublished version), n. 5.
54 Guyomarc'h (2015), pp. 62–63.
55 Cf. *Top.* I.2, 101a28–29: πρὸς γυμνασίαν, πρὸς τὰς ἐντεύξεις, πρὸς τὰς κατὰ φιλοσοφίαν ἐπιστήμας.
56 *In Top.* 29.19–20 and 30.9–12: "And so this fourth use of dialectic can be subsumed under its usefulness for philosophy, as an explicit addition that dialectic is in this respect useful for other sciences in the same way that it is for philosophy" (trans. Van Ophuijsen).
57 Λέγει δὲ γυμνασίαν ἤτοι τὴν γινομένην ἐν τῷ διαλέγεσθαι πρός τινας· δεχόμενοι γάρ τινα προβλήματα παρὰ τῶν προσδιαλεγομένων γυμναζόμενοι πειρῶνται τούτοις παρίστασθαι, δι' ἐνδόξων

Alexander gives two interpretations of the reference to γυμνασία, and it is sometimes difficult to tell them apart. In the first interpretation (i) the oral disputes are less codified than in the second interpretation (ii), which further specifies the nature of the discussion, *i.e.*, an argumentation comprising both contradictory positions (εἰς ἑκάτερον μέρος ἐπιχείρησιν). The translation of "τούτοις παρίστασθαι" is problematic in this sentence. It may mean "to help the interlocutors", which is what Van Ophuijsen opts for (τούτοις refers to τῶν προσδιαλεγομένων and Van Ophuijsen thus translates "to assist these"). The sentence then discusses the assistance provided to an interlocutor: when being helped in the examination of their position, they are trained in solving problems. This could refer to the Socratic practice still in use at the Academy.[58] But the verb may also mean "*to defend*", i.e., to provide reasons supporting a claim. There is good evidence, in Alexander, for this use of παρίστασθαι in the middle voice with its object in the dative—whether it concerns the defense of an opinion, a claim[59] or, precisely, a "problem".[60] To defend a problem then means, via synecdoche, to defend one of the claims constituting the problem: at *In Top.* 149.22–25, for instance, the phrase ὁ τῷ προβλήματι παριστάμενος, "the defender of the problem", is used to explicate the mention of an *opponent* in the Aristotelian text.[61] Not only is this meaning better attested —it also seems to me more direct and more consistent with the context, since it averts the need for a third party to intervene in the dialectical dispute to assist one of the two disputers. Consequently—put more clearly—option (i) differs from option (ii) in that one of the interlocutors must defend only one position, while option (ii) refers to the exercise in which one argues on either side of a question (εἰς ἑκάτερον μέρος).[62]

τὰς ἐπιχειρήσεις ποιούμενοι· ἢ γυμνασίαν λέγοι ἂν τὴν εἰς ἑκάτερον μέρος ἐπιχείρησιν. ἦν δὲ σύνηθες τὸ τοιοῦτον εἶδος τῶν λόγων τοῖς ἀρχαίοις, καὶ τὰς συνουσίας τὰς πλείστας τοῦτον ἐποιοῦντο τὸν τρόπον, οὐκ ἐπὶ βιβλίων ὥσπερ νῦν (οὐ γὰρ ἦν πω τότε τοιαῦτα βιβλία), ἀλλὰ θέσεώς τινος τεθείσης εἰς ταύτην γυμνάζοντες αὑτῶν τὸ πρὸς τὰς ἐπιχειρήσεις εὑρετικὸν ἐπεχείρουν, κατασκευάζοντές τε καὶ ἀνασκευάζοντες δι' ἐνδόξων τὸ κείμενον. Καὶ ἔστι δὲ βιβλία τοιαῦτα Ἀριστοτέλους τε καὶ Θεοφράστου γεγραμμένα ἔχοντα τὴν εἰς τὰ ἀντικείμενα δι' ἐνδόξων ἐπιχείρησιν.
58 Van Ophuijsen (2001), p. 150, n. 265; see also Fortenbaugh (2005), p. 186.
59 Cf. for instance *De fato* 177.6 and 204.4, *Quaestiones* 25.21, and *In Top.* 18.33.
60 *In Top.* 149.23, 548.1 and 15, etc.
61 Compare *Top.* II.3, 110a26–27 (ἐὰν γὰρ μὴ λανθάνῃ πολλαχῶς λεγόμενον, *ἐνστήσεται ὅτι*...) and *In Top.* 149.22–23 (ἂν γὰρ μὴ λανθάνῃ, προχείρως *ἐνστήσεται ὁ τῷ προβλήματι παριστάμενος, ὅτι*...). I detail these specifications because Van Ophuijsen had knowingly rejected this translation, cf. Van Ophuijsen (2001), pp. 150–151, n. 265. The possibility of the evocation of Socratic discussions in what follows does not seem sufficient to reject the more direct translation I propose here.
62 This is also how Moraux (1968), p. 301, reads it.

The second option, as we have said, thus refers to a more specific exercise, i.e., a codified dialectical meeting and, perhaps more specifically still, to a Socratic practice: passages in Xenophon and Plato mention Socrates' discussions (συνουσίαι).[63] In the second part of our passage, the thorny issue is the opposition between the Ancients (ἀρχαίοις) and now (νῦν) due to the two mentions of "books of this kind" (τοιαῦτα βιβλία), at l. 14 and 17. At first sight, the text seems to say that there were no such books among the Ancients, but that Aristotle and Theophrastus have written some. There is mostly no doubt that Aristotle and Theophrastus have indeed written "books of this kind", as a consultation of Diogenes Laertius' bibliographical lists will confirm.[64] But, as other readers of this passage have noted, it is rather unlikely that Alexander would take himself to be contemporaneous with Aristotle and Theophrastus.[65] Alexander does often discuss past philosophical views using the present tense, but it has no temporal value. Furthermore, the subject of "ἀλλὰ θέσεώς τινος τεθείσης εἰς ταύτην γυμνάζοντες" are the Ancients. Yet the account Alexander offers does apply the practice of dialectic as Aristotle describes it in the *Topics*. It would thus be surprising for this kind of dialectical discussion not to refer to some practice of Aristotle's.

This leads to two possible solutions. Johannes M. Van Ophuijsen has cleverly suggested to change the parenthetical text at line 14 into τοσαῦτα βιβλία, "not so many books".[66] The distinction between then and now thus refers to a quantitative difference. It is, however, possible to refrain from changing a reading attested in all manuscripts: we can take the "Ancients" to refer to philosophers older than Aristotle or Theophrastus. Such reference is common usage in Aristotle, and Alexander almost systematically reprises it:[67]: to my knowledge, Alexander never speaks of Aristotle as an "Ancient". We then need to think of the history of dialectic in three distinct moments. First, the Ancients—including Socrates—lead discussions (συνουσίας) in a codified format (τὸ τοιοῦτον εἶδος τῶν λόγων). The general method used to lead these oral discussions has then been described by Aristotle in the *Topics*—as well as in other treatises on more specific subjects by him and Theo-

---

[63] For instance, Xenophon *Mem.* 1.2.60; Plato, *Prot.* 318a3...
[64] For Theophrastus, see the mention of the Προβλημάτων συναγωγῆς at DL 5.45.11, or of the Περὶ ψυχῆς θέσις μία at 5.46.12 (Van Ophuijsen 1994, p. 151, n. 173). For Aristotle, see Θέσεις ἐπιχειρηματικαί, Θέσεις ἐρωτικαί, Θέσεις φιλικαί, Θέσεις περὶ ψυχῆς at 5.24, cf. Van Ophuijsen (1994), p. 151 alongside the indications given in *Topics* VIII.
[65] Fortenbaugh (2005), p. 187; see also Sharples (2010), p. 38, n. 4.
[66] Van Ophuijsen (1994), p. 150, n. 64, and (2001), p. 151, n. 268.
[67] The only exception I located is in the *Commentary on the Prior Analytics* (262.28–32 et 263.26), where ἀρχαῖοι may be used to refer to philosophers more recent than Aristotle.

phrastus.⁶⁸ The rest of the passage mentions a "method for finding arguments" (μέθοδον τινα εὑρετικὴν τῶν ἐπιχειρημάτων, 27.19–20). But—third moment— such discussions, held in an educational setting, are "now" led using books (ἐπὶ βιβλίων), meaning that books are the starting points of the discussion and support it. This brings to mind the description of the first part of Epictetus' teachings⁶⁹ or the *lectio* method mentioned in Aulus Gellius.⁷⁰ With these three stages distinguished, the historical transformation Alexander quickly narrates is more of an incremental change. He does not describe an abrupt transformation but rather a gradual change from a civilization of the spoken word to a culture of the written word.⁷¹

We likely cannot infer from this text that the practice of dialectic as an oral discipline had entirely disappeared in Alexander's time—but the emphasis still moves from discussion to book. Our text has indeed long been read along with two well-known passages from Strabo (on the Skepsis cave in *Geography* XIII.1) and Cicero,⁷² both of which seem to testify to the intense practice of dialectic in the Hellenistic Lyceum, a practice which would apparently have disappeared afterward. In addition, a passage of Chrysippus (*via* Plutarch) indicates that the successors of Plato and Aristotle were "serious about dialectic", "up to Polemo and Strato"⁷³ and one sometimes take this to mean that the Peripatos lost interest in dialectic after Strato.⁷⁴ The passage in Alexander does not go this far since it assumes that there such discussions (συνουσίαι) still take place "today", even if they are led "on the basis of books". But, in what immediately follows this passage, rhetoric is mentioned as if it was the best example of such oral argumentation. This change in emphasis could mean the service of dialectical method has lost ground to dialectical practice itself.

Crucially, Alexander directly reintroduces scientific finality to conclude the development on the first service of dialectic:

> And such a training in argumentations is useful for finding what is investigated and what is true, as Aristotle himself will say when he sets out its usefulness for philosophy as a preliminary mental preparation. For just as exercises of the body, performed according to the rules

---

68 See above, n. 64.
69 *Diss.* 1.10.8–9; 1.26.1; 2.14.1…
70 See, e.g., *Attic Nights* 1.26.
71 See also the remarks of Hadot (1995), pp. 163–170 and 231–235.
72 At *Tusc.* 2.3.9, cf. Moraux (1968), pp. 301–303, Aubenque (1962b), p. 256, and Aubenque (1961; reprinted in 2009), pp. 51–52, n. 3.
73 Strato, Fr. 19 Wehrli = 14 Sharples. Cf. Van Ophuijsen (1994), p. 132, n. 4, and on dialectic in the Hellenistic Peripatos more generally, Crivelli (2018).
74 See the discussion of this issue in Van Ophuijsen (1994).

of the art, produce fitness for the body, so exercises of the mind in argumentations, performed according to method, produce the fitness which is peculiar to the mind; and the peculiar fitness of the rational soul is the capacity by which it becomes apt at finding and judging what is true (*In Top.* 27.24–31, trans. van Ophuijsen slightly modified).[75]

This passage of the commentary is independent, having no equivalent in Aristotle. It clearly shows that Alexander interprets the first service of dialectic in light of the third service—while nothing in Aristotle suggests that the first service is a preparation for the other ones, or that it has less value than them. Contrastingly, Alexander establishes here a non-dialogical usage of dialectical method as the main purpose of dialectic. We can then better understand why Alexander's updates in the *Topics* (i.e., when he pauses his strict exegesis and gives examples from his own philosophical environment) feature debates with the Stoics and the Epicureans on "scientific" issues, especially in physics and ethics.[76] Likewise, when he comments I.11, Alexander says in passing that such an exercise (γυμναστική)—i.e., dialectical exercise—is a "preparation for philosophy" (προπαρασκευὴ πρὸς φιλοσοφίαν, 83.29–32). Other texts confirm this—for instance, the passage commenting the very beginning of I.2, in which Alexander claims that dialectic is useful for the discovery of truth and that, in *this* sense, it is not external to philosophy (οὐκ ἔξω φιλοσοφίας ἡ προκειμένη πραγματεία, 27.3–4).

Such a claim could surprise us, given what we have said above of Alexander's effort to distinguish dialectic from the sciences in general and from metaphysics in particular. Yet such an effort is indeed found at the very beginning of the *Commentary on the Topics*, when Alexander distinguishes Aristotelian dialectic from its Stoic and Platonic rivals (in that order).[77] It is crucial for him to show that his rivals are mistaken in their wish to turn dialectic into a science, or at least into a capacity to draw inferences from true premises. For "dialectic does not have its being in syllogizing through what is true but through what is approved".[78] Any other usage of the name 'dialectic' is "improper" (οὐκ οἰκείως, 3.23–24). In fact, the cat-

---

[75] Χρήσιμος δὲ ἡ τοιαύτη κατὰ τοὺς λόγους γυμνασία πρὸς εὕρεσιν τῶν ζητουμένων τε καὶ ἀληθῶν, ὡς καὶ αὐτὸς ἐρεῖ δι' ὧν τὸ πρὸς φιλοσοφίαν αὐτῆς ἐκθήσεται χρήσιμον· προπαρασκευάζει γὰρ τὴν ψυχήν. ὡς γὰρ τὰ τοῦ σώματος γυμνάσια γινόμενα κατὰ τέχνην εὐεξίαν περιποιεῖ τῷ σώματι, οὕτω καὶ τὰ τῆς ψυχῆς ἐν λόγοις γυμνάσια κατὰ μέθοδον γινόμενα τὴν οἰκείαν εὐεξίαν τῇ ψυχῇ περιποιεῖ· οἰκεία δὲ εὐεξία ψυχῆς λογικῆς ἡ δύναμις καθ' ἣν εὑρετική τε τοῦ ἀληθοῦς καὶ κριτικὴ γίνεται.
[76] Castelli (2015), p. 21.
[77] For an enlightening explanation of the non-chronological order, see Ierodiakonou (2018), pp. 116–117.
[78] *In Top.* 3.21–23: ὥστε οὐκ ἐν τῷ δι' ἀληθῶν συλλογίζεσθαι ἡ διαλεκτικὴ τὸ εἶναι ἂν ἔχοι ἀλλ' ἐν τῷ δι' ἐνδόξων, (trans. Van Ophuijsen).

egory of usage and the notion of instrument are precisely what allow us to reconcile the claim that dialectic is not external to philosophy with the claim that this same dialectic is not a science.

Alexander develops on the instrumental status of logic at the beginning of his *Commentary on the Prior Analytics*. At this point, such status is well established: it goes back to at least Andronicus of Rhodes,[79] and perhaps to others before him, as Diogenes Laertius reports,[80] and is also found in Galen.[81] However, it remains a minority view: beyond Aristotelian circles, the three-part structure of the *logos* (logic—physics—ethics) is dominant, especially the Stoic claim according to which logic is a part—and even the first daughter—of philosophy.[82] Alexander does not use *Organon* as the name referring to a set of treatises, but he does attempt to defend its meaning. The problem is posed at the beginning of the *Commentary on the Prior Analytics:* everyone agrees to admit logic as the product (ἔργον) of philosophy, but is it so as a part (μέρος) or as an instrument (ὄργανον)?[83] It would not serve my purpose in this paper to cover in detail the resulting extensive discussion of the issue by Alexander.[84] But I should at least point out the following: if the initial consensus makes logic the ἔργον of philosophy, then making logic an instrument does not mean it is external to philosophy. Logic is as much within the discipline of the philosopher as the hammer and anvil are within the blacksmith's workshop: "A hammer and anvil are not precluded from being an instrument of the smith's art by the fact that they are its product".[85] Not only are the roles of product and instrument not contradictory—but the analogy further implies that, among the blacksmith's productions, we will necessarily find second-order objects (like a hammer) which make possible the production of those artefacts which are the proper end of the blacksmith's craft (like a sword).

Expanding on the analogy, we may say that the metaphysician and the logician are one and the same person, but that they exercise two distinct activities. Likewise, the blacksmith is the best suited person to produce an anvil, but while making it, he is not—strictly speaking—exercising his craft, whose end is not to produce its own tools. This allows us to understand the passages where Alexander

---

[79] See Moraux (1973), pp. 76–79, Barnes (1997b), pp. 33–37, Griffin (2015), pp. 31 and 33–34, and Hatzimichali (2016).
[80] Cf. Griffin (2015), pp. 33–34.
[81] Cf. Chiaradonna (2008).
[82] Cf. Ammonius, *In A.Pr.* 8.20–9.2 (SVF II.49) and, on this view, Barnes (1997a), p. 20.
[83] Alexandre, *In A.Pr.* 1.7–2.34.
[84] On this point, see Guyomarc'h (2017).
[85] *In A.Pr.* 2.20–22: "οὐδὲ γὰρ ἡ σφῦρα καὶ ὁ ἄκμων ὄργανον κωλύεται τῆς χαλκευτικῆς εἶναι, διότι αὐτῆς ἐστιν ἔργα" trans. Barnes et al.

claims that it is the metaphysician's task to develop a theory of demonstration—when he says, e.g., that "the general discussion of what demonstration is and how it is carried on" belongs to the metaphysician[86]—or those passages where he claims that "the division of being into genera, which he carried out in the *Categories* belongs to first philosophy"[87] or where he maintains that it is the metaphysician's task to study dialectical predicates.[88] As I have tried to show elsewhere,[89] the metaphysician and the dialectician both study some of the same objects—for instance, dialectical predicates—but they conduct such study from different perspectives: ἐπιστημονικῶς on the one hand and κατὰ τὸ ἔνδοξον on the other.[90] This partial overlap in objects may be extended to all of logic—including categories or demonstrations. But the distinction between disciplines is established via distinctions in the modalities of discourses (scientific or not), and this very distinction ensures that dialectic can serve as an instrument for metaphysics.

The inclusion in philosophy of logic in general and of dialectic in particular, therefore, does not contradict the external role they have due to their instrumental status. It also explains why Alexander tends to reduce dialectic to its instrumental function and to exhaust it in its scientific purpose. His insistence on dialectic as a preparation for philosophy (προπαρασκευὴ πρὸς φιλοσοφίαν, 83.32) is not a commonplace statement but rather a proper claim on the nature of dialectic. We may even wonder if dialectic can be something beyond the preparation to scientific activity for Alexander: unless I am mistaken, all other references to dialectical meetings in the rest of the *Commentary on the Topics* directly and accurately echo Aristotle's text.

Strictly speaking, the only non-scientific use of dialectic seems to be found in rhetoric—as shown in the commentary to dialectic's second service, which I will cover quickly. In his comment, Alexander points out the need not to use true premises and demonstrations (ἀληθῶν τε καὶ ἀποδεικτικῶν, 28.4) to persuade the multitude, since "they are absolutely not even able to understand any of these things, and do not submit to being instructed about them either" (trans. Van Ophuijsen modified).[91] Here, the service of dialectic is taken to fall under moral rhetoric and exhortative speeches—as one does in order to persuade the crowd that, for instance, pleasure is not the good. But—once again—we must understand the re-

---

86 *In Met.* 266.24–25.
87 *In Met.* 245.33–35.
88 *In Met.* 177.8–10.
89 I defended this claim in Guyomarc'h (2017).
90 *In Met.* 344.14–15.
91 *In Top.* 28.5–6: "οὐδὲ γὰρ τὴν ἀρχὴν συνιέναι τῶν τοιούτων δύνανται· ἀλλ' οὐδὲ μανθάνειν ὑπομένουσι".

sort to "those things which are approved and are held to be so by these people themselves" (28.7–8) only insofar as they provide a starting point: "if one *starts* (ὁρμώμενος) from that which is common and approved, and so examines whether the interlocutor...".[92] As our examples indicate, such starting point would ideally be dropped and eventually replaced as we move to a scientific view of moral issues— a view where, for instance, pleasure is technically defined as "a perceptible process toward a natural state" (γένεσιν εἰς φύσιν αἰσθητήν, 28.12).

4 The Usefulness of Dialectic: The Standpoint of Plausibility

The third service is the one which is of the greatest interest to us:

> The third way that Aristotle sets out in which the study of dialectic is beneficial, is in its use for philosophy and scientific discernment, that is towards the finding and discerning of the truth. By 'sciences which make up philosophy' he means physics, ethics, logic and metaphysics.
>
> (1) For those who can discern what is plausible as contributing to opposite conclusions, and can argue on either side of a question, will find out more easily on which side of the contradiction the truth lies, as if they had listened to both parties in a lawsuit. For just as the judge comes to know what is right through listening to both parties, so in philosophical inquiries at many points, it is not possible to find the truth easily without first having argued on both sides.
>
> (2) What Plato says in the *Parmenides* accords with this: "Accustom and train yourself more, while you are young, in that art which is held to be useless and is called by the many 'idle talking'"—otherwise the truth will escape you".[93]
>
> (3) Further, the person who knows the nature of what is plausible will not be led astray by it as if it were true, but will first distinguish what looks as if it were true from what is not true by comparing them to each other: for no one will be led astray by those who try to make the truth disappear if he is versed in the means by which they do so.
>
> (4) In addition, the person who is apt at finding what looks just like the truth—i.e., what is plausible—is the better prepared for finding out what is actually true.
>
> (5) And further, if the person who speaks soundly and correctly about a subject is the one who argues in such a way that his arguments suffice also to solve the puzzles surrounding it, then it is clearly useful to be well-trained in the puzzles that may be raised with respect to it, for

---

92 *In Top.* 28.16–17 (trans. Van Ophuijsen).
93 *Parm.* 135d3–6: "ἕλκυσον δὲ σαυτὸν καὶ γύμνασαι μᾶλλον διὰ τῆς δοκούσης ἀχρήστου εἶναι καὶ καλουμένης ὑπὸ τῶν πολλῶν ἀδολεσχίας, ἕως ἔτι νέος εἶ· εἰ δὲ μή, σὲ διαφεύξεται ἡ ἀλήθεια".

thus one could at once have a comprehensive view of the solutions to these puzzles (*In Top.* 28.23–29.16, trans. Van Ophuijsen, modified).[94]

On the one hand, this text deserves to be better known—since it is the only undeniable occurrence of "μετὰ τὰ φυσικά" referring to a science (rather than a treatise) in Alexander, and perhaps even in the whole of Antiquity.[95] On the other hand, the presence of logic among "sciences which make up philosophy" is certainly anomalous—but it can be accounted for via the distinction made at *Topics* I.14, 105b19–21 between three kinds of premises—ethical, physical and logical—in a passage well-known because it is sometimes considered to be one of the sources for the Hellenistic tripartition of philosophy.[96]

But the most striking and novel aspect of this passage is the introduction of an element absent from Aristotle's text, i.e., τὸ πιθανόν, what is "plausible" or "persuasive". Alexander uses the concept to build a bridge in Aristotle's text between "πρὸς ἀμφότερα διαπορῆσαι" (to develop an aporia by arguing on both sides) and "ἐν ἑκάστοις κατοψόμεθα τἀληθές τε καὶ τὸ ψεῦδος" (to discern, in each subject matter, the true and the false). Put concisely, Alexander asks the following question: how does diaporia allow us to discern the true and the false better? His answer: in exercising our capacity to discern the plausible, diaporia starts us on the

---

94 Τρίτον τῆς ὠφελείας αὐτῆς ἐκτίθεται τρόπον τὸν πρὸς φιλοσοφίαν καὶ τὴν κατ᾽ ἐπιστήμην γνῶσιν, τοῦτ᾽ ἔστι πρὸς τὴν τοῦ ἀληθοῦς εὕρεσίν τε καὶ γνῶσιν. <κατὰ φιλοσοφίαν> δὲ <ἐπιστήμας> εἶπε τὴν φυσικήν, τὴν ἠθικήν, τὴν λογικήν, τὴν μετὰ τὰ φυσικά. (1) οἱ γὰρ δυνάμενοι τὰ πιθανὰ πρὸς τὰ ἀντικείμενα συντελοῦντα διορᾶν καὶ εἰς ἀμφότερα ἐπιχειρεῖν ῥᾷον ἂν εὑρίσκοιεν ἐν ποτέρῳ αὐτῶν μέρει τῆς ἀντιφάσεως τὸ ἀληθές ἐστιν, ὥσπερ ἀντιδίκων ἀμφοτέρων τῶν μερῶν ἀκηκοότες. ὡς γὰρ ὁ δικαστὴς διὰ τοῦ ἀμφοτέρων ἀκοῦσαι τὸ δίκαιον γνωρίζει, οὕτως καὶ ἐν ταῖς κατὰ φιλοσοφίαν ζητήσεσιν ἐπὶ πολλῶν οὐχ οἷόν τε τὸ ἀληθὲς εὑρεῖν ῥᾳδίως μὴ πρότερον εἰς ἑκάτερον ἐπιχειρήσαντα. (2) συνᾴδει τούτῳ καὶ τὸ ὑπὸ Πλάτωνος εἰρημένον ἐν τῷ Παρμενίδῃ τὸ "ἔθισον σαυτὸν καὶ γύμνασον μᾶλλον διὰ τῆς δοκούσης ἀχρήστου εἶναι καὶ καλουμένης ὑπὸ τῶν πολλῶν ἀδολεσχίας, ἕως ἔτι νέος εἶ· εἰ δὲ μή, διαφεύξεταί σε ἡ ἀλήθεια". (3) ἔτι ὁ εἰδὼς τὴν τοῦ πιθανοῦ φύσιν οὐκ ἂν ὑπ᾽ αὐτοῦ παραχθείη ποτὲ ὡς ἀληθοῦς ὄντος, ἀλλὰ προκρίνοι ἂν τὰ φαινόμενα ἀληθῆ τῶν μὴ ἀληθῶν τῇ παραβολῇ αὐτῶν τῇ πρὸς ἄλληλα· δι᾽ ὧν γὰρ τὸ ἀληθές τινες ἀφανίζειν πειρῶνται, τούτοις τις ἐγγεγυμνασμένος οὐκ ἂν ὑπ᾽ αὐτῶν παράγοιτο. (4) πρὸς δὲ τούτοις ὁ τῶν ὁμοίως φαινομένων τῷ ἀληθεῖ εὑρετικός (τοιαῦτα δὲ τὰ πιθανά) καὶ πρὸς τὴν τοῦ ἀληθοῦς εὕρεσιν ἑτοιμότερος. (5) ἔτι δὲ εἰ δεῖ τὸν περί τινος ὑγιῶς τε καὶ ὀρθῶς λέγοντα τοιούτους τοὺς περὶ αὐτοῦ ποιεῖσθαι λόγους ὡς δύνασθαι δι᾽ αὐτῶν λύεσθαι καὶ τὰ ἀπορούμενα περὶ αὐτοῦ, δῆλον ὡς χρήσιμον τὸ γεγυμνάσθαι ἐν τοῖς πρὸς αὐτὸ ἀπορεῖσθαι δυναμένοις· οὕτω γὰρ συνορᾶν δύναιτ᾽ ἂν καὶ τὰς λύσεις τῶν ἀπορουμένων.

95 Cf. Brisson (1999) and Narcy (2003). Concerning Alexander on this point, also see Guyomarc'h (2015), pp. 65–66.

96 Cf. Hadot (1979), pp. 207–208, who shows convincingly that this text "ne peut faire allusion à une véritable division des parties de la philosophie".

path to truth. The premise for such an answer is a definition of the plausible as what resembles the true. The claim will appear stronger (and less banal) if we recall that Galen has precisely restricted metaphysical claims to πιθανά claims, theoretical philosophy being unable to test its claims and thus to reach truth.[97] This underlines how πιθανόν is already, at the time, an epistemological norm which can be used to characterize a discourse and situate it in the domain of knowledge. Alexander's determination to make πιθανόν discourse a tool to serve in the search for truth is far from insignificant: on the contrary, it is an iconic feature of what has been called an "epistemocentric interpretation of [Aristotelian] logic".[98]

To justify the scientific use of dialectic, Alexander offers five arguments. The analogy with the judge (1) comes from *Metaphysics* B 1, 995b2–4. Drawing on this reference first shows how book *Beta* provides a prime example of such scientific use of dialectic. But it especially shows that, in his view, the non-dialogical use of dialectic is a part of the *Topics* from the outset, confirming that this is indeed his main interest in the matter. The analogy illustrates the comparative effect of diaporia: considering both sides of a question allows one to ascertain which position is strongest. Truth—like Justice—appears once the act of judging and deciding has taken place. Thus, the standpoint of plausibility is adopted during what might be called a trial, i.e., a moment dedicated to the evaluation of the respective plausibility of both positions.

The second argument contains a reference to Plato—a fact so uncommon in Alexandrinian commentaries that it deserves our attention. Such references are indeed rare enough, but are not isolated phenomena either: for instance, on the subject of matter, Alexander cites the 'bastard reasoning' from *Timaeus* 52b several times[99] and refers elsewhere to the *Parmenides*.[100] The passage mentioned here is found at the end of the first part of the dialogue, after the reference to the young Aristotle at 135d (the one who was mixed up with Aristotle of Stagira in ancient interpretations). As we have said before, when Alexander introduces the rival accounts of dialectic at the beginning of his commentary, he starts with the Stoics (1.8–14), but he cannot be unaware of the chronological anteriority of the Platonists. Here, he restores the proper chronological order by placing Aristotelian dialectic under Platonic auspices. Holding the *Parmenides* to be the origin of Aristotle's dialectical method is not an exclusively recent interpretation[101]—it rather

---

97 Cf. Chiaradonna (2014).
98 Cf. Brunschwig (1991), p. 425. For its application to Alexander, see Rashed (2000).
99 For instance, *In Met.* 164.20–21, *Quaestio* I 1, 4.10–11, and *In Phys. Scol.* 21 (Simplicius, *In Phys.* 542.19–22).
100 For instance, *In Met.* 52.6.
101 For instance, Berti (1980).

originates in Antiquity. In his *Commentary on the Parmenides*, Proclus points out that, for some exegetes, the "exercise" mentioned in the *Parmenides* passage refers to the method from the *Topics*:

> However, since some commentators, relying on the word exercise would have it that this exercise is the dialectical method of the Peripatetics (for Aristotle, in stating its usefulness says that it contributes to exercise), although I have said a good deal in refutation of these in the Preface, yet now I would like to say something again briefly... (Proclus, *In Parm.* 981.1–9, trans. Morrow and Dillon).[102]

Reading this, one may be tempted to infer that Alexander and Proclus simply draw from the same sources. For the "commentators" Proclus objects to, the connection between the two passages comes from the γύμνασαι at *Parmenides* 135d4 and the γυμνασία at *Top.* I.2, 101a27 ff. Therefore, they are connecting the *Parmenides* passage with the first service of dialectic in *Top.* I.2, rather than with the third service —as Alexander does. Proclus reminds the reader, with some weariness, that he has already spent time refuting this attempt at reconciling Plato and Aristotle. In the proemium, the main reason he brings forward is precisely that Aristotelian dialectic does not allow us to see the true (κατίδοι τὸ ἀληθές, 653.1), by contrast with the dialectic Plato mentions in the *Parmenides* passage. The relation to truth is indeed exactly what is at stake, for the anonymous commentators as well as for Alexander and Proclus. Alexander and Proclus both maintain that Aristotelian dialectic does not depend on true premises, and thus does not in itself conduce to see truth. Yet, as Alexander points out at the very start of his commentary, dialectic as an instrument "contributes to finding the truth, which is the goal of philosophical study" (πρὸς τὴν εὕρεσιν τῆς ἀληθείας αὐτοῖς συντελοῦσα, ὃ τέλος ἐστὶ τῆς φιλοσόφου θεωρίας, 1.7–8, trans. Van Ophuijsen). Here, the commentary explicates the modalities of this "contribution", i.e., it explains how dialectic keeps truth from "escaping" us.

This explication is the subject matter of the third and fourth arguments, focused on the πιθανόν. In other texts, Alexander also defines the πιθανόν based on its relation of resemblance or proximity with truth. The plausible is "what lies close to what is true" (τὸ παρακείμενον τἀληθεῖ).[103] But "close" is a scalar predicate, i.e., it can vary in degree. It covers a range going from what appears to be true but is really false to what appears to be true and is really true. When Alexand-

---

[102] ἐπειδὴ δέ τινες, τοῦ τῆς γυμνασίας ὀνόματος δραξάμενοι, τὴν παρὰ τοῖς Περιπατητικοῖς ἐπιχειρηματικὴν οἴονται μέθοδον ταύτην εἶναι τὴν γυμνασίαν (καὶ γὰρ ἐκείνης τὸ χρήσιμον λέγων ὁ Ἀριστοτέλης πρὸς τὴν γυμνασίαν αὐτὴν εἶναί φησι συντελοῦσαν), εἴρηται μὲν ἡμῖν ἐκ προοιμίων πολλὰ πρὸς τούτους· νυνὶ δὲ συντόμως τι πάλιν εἰπεῖν...
[103] *In A.Pr.* 8.25.

er speaks of the πιθανόν, he can also refer to what is simply false—for instance, to sophistical claims which are "superficially plausible" (ἐπιπόλαιον τὸ πιθανὸν)[104] or, using it in an evidently pejorative sense, to what is *only* likely.[105] This large scope of usages may be quite common at the time, as shown in a passage from Sextus Empiricus (drawing on Carneades, and which could in fact be Stoic in origin)[106] where three senses of πιθανόν are distinguished: what is true and appears true; what is false but appears true; what appears true and is common to both previous senses. In the rest of the Alexandrinian corpus, the πιθανόν is also found, for instance, along with reasons and arguments, by contrast with the obviousness of facts,[107] or along with induction (an instrument of the dialectician) and contingency, by contrast with necessity.

In the *Commentary on the Topics*, πιθανόν is omnipresent to refer to the prime objective of the dialectician, which he must most aim at, and it is frequently associated with received ideas (ἔνδοξα).[108] It is quite unlikely that a term with such strong connotations as πιθανόν in Hellenistic and post-Hellenistic philosophy would be used innocently, *a fortiori* if it were used to comment Aristotelian texts where it is conspicuously absent.[109] Some Aristotelian passages could authorize using it—for instance, when Aristotle defines the dialectical premise as what is neither universally rejected, nor obvious to everyone (I.10), he circumscribes an intermediary spot which plausibility could fill. But the reference remains distant. Alexander's own use of πιθανόν thus testifies to a deliberate intervention and strategy on his part. In fact, if we search for the Aristotelian source of the term, we must turn to the *Rhetoric*. In *Rhet*. I.1, Aristotle announces that the task of rhetoric is to discern what is plausible (1355b10–11). But while *Rhet*. I.1 persistently draws dialectic and rhetoric closer together, the standpoint of plausibility is never explicitly attributed to dialectic. Yet, as we have seen already in his comment to the first service of dialectic, Alexander clearly reads the *Topics* with the *Rhetoric* in mind. Thus, he says at the beginning of the commentary:

---

104 *In A.Pr.* 14.17.
105 For instance, *De anima* 100.1.
106 Cf. LS 69D; Sextus, *AM* VIII.174–175. Cf. Allen (1994) and Chiaradonna (2014), p. 76.
107 For instance, *De fato*, 196.21 and *Quaest.* 3.12, 101.18.
108 Including at *In Top.* 3.18ff.; 5.5; 29.6; 62.16; 87.7, etc. The association of "ἔνδοξα" with "πιθανά" is found in the commentary to *Metaphysics* Γ, and concerns—aptly—the dialectician (238.26 and 260.26).
109 It is the case in *Metaphysics* B. It is less true in the *Topics* (see, for instance, I.11, 104b14 and VIII.11, 161b35) which could have been Alexander's sources.

> Given the nature of dialectic, Aristotle reasonably calls it a counterpart to rhetoric, since it also deals with what is plausible, which is such because it is approved as well (*In Top.* 3.25–26, trans. Van Ophuijsen, heavily modified).[110]

The alignment of dialectic with rhetoric takes place as early as the proemium of the *Commentary on the Topics*. In what is likely a quite standard wording, Alexander brings forward the notion that the range of rhetoric is less broad than the one of dialectic, insofar as the former is mostly political.[111] As Van Ophuijsen has aptly seen, the alignment goes both ways:[112] rhetoric is described in terms reminding one of dialectical activity,[113] and, conversely, dialectic adopts the standpoint of plausibility as well. The difference between rhetoric and dialectic resides not only in the more or less broad range of their object, but also in their form: one is performed via questions and answers; the other is continuous, ἐν ἐρωτήσει τε καὶ ἀποκρίσει, διεξοδικός. As Van Ophuijsen notes, the difference is, incidentally, never made by Aristotle but was well established in Alexander's time—we will find traces of it in Diogenes Laertius concerning the Stoics, even using similar terminology.[114] Finally, it is likely quite banal, in Alexander's time, to conceive of ἔνδοξα as plausible or probable opinions—in the sense of *veri simile*—since it is already the case in Cicero, whose interest for *disputatio in utramque partem* is known.[115]

Alexander's interpretation is original, because of how it imports πιθανόν in the *Topics* to justify the instrumental function of dialectic as a discipline serving the discovery of truth. The fifth argument paraphrases Aristotle's text quite closely: the exercise of diaporia makes one capable to "discern the true as well as the false in any subject" (ἐν ἑκάστοις κατοψόμεθα τἀληθές τε καὶ τὸ ψεῦδος, *Top.* I.2, 101a35–36, trans. Robin Smith), which Alexander reprises as "thus one could at once have a comprehensive view of the solutions to these puzzles" (οὕτω γὰρ συνορᾶν δύναι' ἂν καὶ τὰς λύσεις τῶν ἀπορουμένων). Dialectical training serves the search for truth because it leads to the development of a kind of intellectual intuition, a capacity to sort and distinguish arguments and positions. The meaning of the fifth argument becomes clearer, however, if we consider it in light of what we have previously said. Negatively, plausibility training serves as a safeguard against error

---

110 Τοιαύτην δὲ οὖσαν αὐτὴν εἰκότως καὶ ἀντίστροφόν φησιν Ἀριστοτέλης εἶναι τῇ ῥητορικῇ, ἐπειδὴ κἀκείνη περὶ τὰ πιθανά, ἃ τῷ ἔνδοξα εἶναι καὶ αὐτά ἐστι τοιαῦτα·
111 *In Top.* 5.5 ff. and 6.18 ff.
112 On all that follows, cf. Van Ophuijsen (1994), pp. 151–154.
113 *In Top.* 27.21–24.
114 DL VII.42. Cf. Van Ophuijsen (1994), pp. 153, n. 189 and 190.
115 Cf., for instance, *De inventione* I.46, Glucker (1995), and Spranzi (2011), pp. 45–46.

and dissimulation (in sophists?); positively, as we have seen, it serves to develop our capacity to approach the truth, *i.e.*, more precisely, to solve difficulties posed by the chosen subject matter.

This is indeed the capacity at work in *Metaphysics* B, whose commentary Alexander concludes by saying:

> Here are the aporiae presented in *Beta*,[116] whose arguments are [drawn] from accepted opinions (ἐνδόξων) and [conducted] on the level of plausibility (κατὰ τὸ πιθανόν). And indeed, it is impossible for people to argue for opposed positions, except by using dialectical (λογικαῖς) arguments; nor, for that matter, could the aporiae be solved, if this were not the case (*In Met.* 236.26–29, trans. Madigan modified).[117]

According to Alexander, what Aristotle displays in book *Beta* is precisely this diaporetic work where all involved sides are contrasted and the degree of likelihood of their views are evaluated. In such work, plausibility takes on the role of norm and evaluation criterion. Perhaps this is the way in which Alexander reconnects with the ambition of comprehensiveness which Aristotle seems to direct at book *Beta*: *Beta* is not comprehensive in the sense that one should lay out all aporiae once and for all, as if the catalogue were thereafter closed, but in the sense that, on a given issue, we must let all plausible opinions be voiced and then follow them to their last development. For such is, following the *Commentary on the Topics*, the task of the dialectician who must "omit nothing" of what is plausible when he examines a question—just like the doctor (and other practitioners of stochastic crafts) who must do all in his power to help his patient.[118] This involves, Alexander says, some measure of luck ("ἀπὸ τύχης", 33.19), in such a way that the goal of stochastic crafts is in part found outside defined procedures. In non-stochastic crafts (for instance, house-building), the function of the craft extinguishes itself in the production of the goal pursued. In medicine and dialectic, the goal pursued is not to heal the patient (*sic*, 33.1–2), nor is it to necessarily lead the interlocutor to contradict himself, but rather to implement all things conducing to such end, whose realization depends on other, external factors.

Concerning the aporiae from *Beta* precisely, this aspect is found in Aristotle's continued efforts to develop the arguments involved in a given issue as much as

---

[116] The best manuscripts, A¹ and O, mention the "second" book (δευτέρῳ), but given the unstable place of book α and the possible confusions between the letters as book names and as numbers, these variations can be explained (see also, for instance, *In Met.* 344.22–25).

[117] Ταῦτα τὰ ἐν τῷ Β ἠπορημένα, ἐξ ἐνδόξων τὰς ἐπιχειρήσεις ἔχοντα καὶ κατὰ τὸ πιθανόν· καὶ γὰρ οὐδὲ οἷόν τε εἰς τὰ ἀντικείμενα ἐπιχειροῦντας μὴ λογικαῖς ἐπιχειρήσεσι χρήσασθαι· οὐδὲ γὰρ ἂν λύεσθαι δύναιντο, εἰ μὴ εἶχεν οὕτως.

[118] *In Top.* 32.18–26 and Kupreeva (2017), pp. 238–239.

possible, thus testing their plausibility. This is what Alexander means when, over the course of his commentary, he calls some arguments simply "logical" or merely "dialectical".[119] Bringing views to their most extant development is also how one might reveal that their consequences oppose facts or turn out to be counter-intuitive. We may thus identify some positions as absurd in as early a stage of an investigation as dialectical exploration.[120]

That diaporia may be used for authentic exploration does not mean it is neutral; quite the opposite, even, since it tests plausibility. In this sense, the alternative —in which book *Beta* is either an open-ended and purely preliminary exploration (with no positive claims involved), or a settled summary, an exposition segment leading to a foregone conclusion—is biased. Diaporetic exploration requires argumentation and evaluation, and therefore involvement and engagement by a philosopher in this dialectic activity.

In other words, on dialectic's path to the discovery of truth, the plausibility of one argumentation will place it closer or further away from truth. Dialectical work is precisely to try and determine *where* we stand on this path. Therein resides the heuristic character of dialectic. Alexander can thus also recognize in *Beta* moments where, according to him, Aristotle initiates a solution. For him, Aristotle will indeed undertake to "solve" the aporiae, and propose "λύσεις" for them.[121] And this resolution work starts as early as book *Beta* itself (see for instance 195.13–14). In *Beta*, Aristotle also starts to refute some positions. It is this engagement on Aristotle's part that explains why book *Beta* can be considered the true starting point of the *Metaphysics*—but a starting point only, since the standpoint of plausibility calls for its own overtaking.

Is Alexander's interpretation then as systematizing as it is said to be? In *Beta*, at least, Alexander has not gone as far as Nicolaus of Damascus—who, according to Averroes, had sprinkled the aporiae of *Beta* all over the treatise,[122] since he cites, as we have seen, only book Λ, and since Γ begins the resolution of some aporiae while not restricting itself to this task. Assuredly, Alexander has retrieved in the *Topics* the materials necessary to nourish a profitable interpretation of *Metaphysics* B— and *vice versa*. But he develops this interpretation so that the heuristic character of the aporetic method and its exploratory function are preserved, reinforcing, to that purpose, the purely instrumental status of dialectic. Thus, it is inaccurate to think that he completes the Aristotelian system by binding together domains of object, like the Stoic system does it—if only because Aristotelian logic is not a science

---

119 For instance, 206.12–13 and 218.17.
120 For instance, 184.7, 191.5, 193.12–16, etc.
121 See especially the very clear text of 136.8–11.
122 Fazzo (2008), pp. 116–117.

and does not have any genus for an object. In this at least, the Exegete par excellence pays more attention to the deep-rooted concern for research which is inchoate and inherent to Aristotelian philosophy than he is commonly said to be.

Bibliography

Allen, James (1994): "Academic probabilism and Stoic epistemology". In: *The Classical Quarterly*, 44. No. 1, pp. 85–113.
Aubenque, Pierre (1961): "Sur la notion aristotélicienne d'aporie". In: Mansion, Suzanne (Ed.): *Aristote et les problèmes de méthode. Communications présentées au Symposium Aristotelicum tenu à Louvain du 24 août au 1er septembre 1960*. Louvain: Publications Universitaires, pp. 3–19.
Aubenque, Pierre (1962a): "Aristotle and the Problem of Metaphysics". In: *Philosophy Today*, 6. No. 2, pp. 75–84.
Aubenque, Pierre (1962b): *Le Problème de l'être chez Aristote. Essai sur la problématique aristotélicienne*. Paris: Presses Universitaires de France.
Aubenque, Pierre (2009): *Problèmes aristotéliciens. Philosophie théorique*. Paris: Vrin.
Barnes, Jonathan (1997a): *Logic and the Imperial Stoa*. Leiden and Boston: Brill.
Barnes, Jonathan (1997b): "Roman Aristotle". In: Griffin, Miriam T. and Barnes, Jonathan (Eds.): *Philosophia togata II. Plato and Aristotle at Rome*. Oxford: Clarendon Press, pp. 1–69.
Berti, Enrico (1980): "Aristote et la méthode dialectique du *Parménide* de Platon". In: *Revue Internationale de Philosophie* 34. No. 133–134, pp. 341–358.
Bonelli, Maddalena (2001): *Alessandro di Afrodisia e la metafisica come scienza dimostrativa*. Napoli: Bibliopolis.
Brisson, Luc (1999): "Un si long anonymat!". In: Narbonne, Jean-Marc and Langlois, Luc (Eds.): *La métaphysique. Son histoire, sa critique, ses enjeux*. Paris and Québec: Vrin and Presses de l'Université Laval, pp. 37–60.
Brunschwig, Jacques (1967): *Aristote. Topiques (Livres I-IV)*. Paris: Les Belles Lettres.
Brunschwig, Jacques (1991): "Sur quelques malentendus concernant la logique d'Aristote". In: Sinaceur, Mohammed A. (Ed.): *Penser avec Aristote*. Toulouse: Erès, pp. 423–427.
Castelli, Laura M. (2013): "Collections of *Topoi* and the Structure of Aristotle's *Topics*: Notes on an Ancient Debate (Aristotle, Theophrastus, Alexander and Themistius)". In: *ANTIQVORVM PHILOSOPHIA an International Journal* 7, pp. 65–92.
Castelli, Laura M. (2015): "Alexander of Aphrodisias: Methodological Issues and Argumentative Strategies between the *Ethical Problems* and the *Commentary on the Topics*". In: Bonelli, Maddalena (Ed.): *Aristotele e Alessandro di Afrodisia (Questioni etiche e Mantissa): metodo e oggetto dell'etica peripatetica*. Napoli: Bibliopolis, pp. 19–42.
Cerami, Cristina (2016): "Alexander of Aphrodisias". In: Falcon, Andrea (Ed.): *Brill's Companion to the Reception of Aristotle in Antiquity*. Leiden: Brill, pp. 160–179.
Chiaradonna, Riccardo (2008): "Scienza e contingenza in Galeno". In: Perfetti, Stefano (Ed.): *Conoscenza e contingenza nella tradizione aristotelica medievale*. Pisa: ETS, pp. 13–30.
Chiaradonna, Riccardo (2014): "Galen on what is persuasive (*pithanon*) and what approximates to truth". In: *Bulletin of the Institute of Classical Studies (Supplement)* 114, pp. 61–88.

Crivelli, Paolo (2018): "Dialectic in the Early Peripatos". In: Ierodiakonou, Katerina and Bénatouïl, Thomas (Eds.): *Dialectic after Plato and Aristotle*. Cambridge: Cambridge University Press, pp. 47–81.
Donini, Pierluigi (1994): "Testi e commenti, manuali e insegnamento: la forma sistematica e i metodi della filosofia in età postellenistica". In: Haase, Wolfgang (Ed.): *Philosophie, Wissenschaften, Technik. Philosophie (Systematische Themen; Indirekte Überlieferungen; Allgemeines; Nachträge)*. Berlin and Boston: de Gruyter, pp. 5027–5100.
Fazzo, Silvia (2002): *Aporia e sistema: la materia, la forma, il divino nelle "Quaestiones" di Alessandro di Afrodisia*, Pisa: Edizioni Ets.
Fazzo, Silvia (2008): "Nicolas, l'auteur du Sommaire de la philosophie d'Aristote : doutes sur son identité, sa datation, son origine". In: *Revue des Études Grecques* 121. No. 1, pp. 99–126.
Fortenbaugh, William W. (2005): *Theophrastus of Eresus: Sources for his life, writings, thought and influence. Commentary. Volume VIII: Sources on rhetoric and poetics*. Leiden and Boston: Brill.
Frede, Dorothea (2017): "Alexander of Aphrodisias". In: Zalta, Edward N. (Ed.): *The Stanford Encyclopedia of Philosophy*. Stanford: Metaphysics Research Lab, Stanford University.
Glucker, John (1995): "*Probabile, Veri Simile* and Related Terms". In: Powell, Jonathan G. F. (Ed.): *Cicero the Philosopher: Twelve Papers*. Oxford: Clarendon Press, pp. 115–144.
Griffin, Michael J. (2015): *Aristotle's Categories in the early Roman Empire*. Oxford: Oxford University Press.
Guyomarc'h, Gweltaz (2015): *L'Unité de la métaphysique selon Alexandre d'Aphrodise*. Paris: Vrin.
Guyomarc'h, Gweltaz (2017): "*Métaphysique* et *Organon* selon Alexandre d'Aphrodise. L'utilité de la logique pour la philosophie première". In: Balansard, Anne and Jaulin, Annick (Eds.): *Alexandre d'Aphrodise et la métaphysique aristotélicienne*. Louvain: Peeters, pp. 83–112.
Hadot, Pierre (1979): "Les divisions des parties de la philosophie dans l'Antiquité". In: *Museum Helveticum* 36. No. 4, pp. 201–223.
Hadot, Pierre (1995): *Qu'est-ce que la philosophie antique ?*. Paris: Gallimard.
Hatzimichali, Myrto (2016): "Andronicus of Rhodes and the Construction of the Aristotelian Corpus". In: Falcon, Andrea (Ed.): *Brill's Companion to the Reception of Aristotle in Antiquity*. Leiden: Brill, pp. 81–100.
Ierodiakonou, Katerina (2018): "Dialectic as a Subpart of Stoic Philosophy". In: Ierodiakonou, Katerina and Bénatouïl, Thomas (Eds.): *Dialectic after Plato and Aristotle*, Cambridge: Cambridge University Press, pp. 114–133.
Kupreeva, Inna (2017): "*Aporia* and Exegesis: Alexander of Aphrodisias". In: Karamanolis, George and Politis, Vasilis (Eds.): *The Aporetic Tradition in Ancient Philosophy*. Cambridge: Cambridge University Press, pp. 228–247.
Lavaud, Laurent, and Guyomarc'h, Gweltaz (2021): *Alexandre d'Aphrodise. Commentaire à la Métaphysique d'Aristote, Livres Petit alpha et Beta*. Paris: Vrin.
Luna, Concetta (2004): "Alessandro di Afrodisia e Siriano sul Libro B della *Metafisica*. Tecnica e Struttura del Commento". In: *Documenti E Studi Sulla Tradizione Filosofica Medievale* 15, pp. 39–79.
Madigan, Arthur, and Dooley, William E. (Trans.) (1992): *Alexander of Aphrodisias: On Aristotle's Metaphysics 2 and 3*. Ithaca: Cornell University Press.
Moraux, Paul (1968): "La joute dialectique d'après le huitième livre des *Topiques*". In: Owen, Gwilym E. L. (Ed.): *Aristotle on Dialectic: The Topics. Proceedings of the Third Symposium Aristotelicum*. Oxford: Clarendon Press, pp. 277–311.

Moraux, Paul (1973): *Der Aristotelismus bei den Griechen: von Andronikos bis Alexander von Aphrodisias. Band 1 Die Renaissance des Aristotelismus im I. Jh. V. Chr.* Berlin and New York: de Gruyter.

Moraux, Paul (1984): *Der Aristotelismus bei den Griechen. Band 2: Der Aristotelismus im I. und II. Jh. N.Chr.* Berlin and New York: de Gruyter.

Moraux, Paul (2001): *Der Aristotelismus bei den Griechen. Band 3: Alexander von Aphrodisias* Wulfgang Kullmann, Robert W. Sharples, and Jürgen Wiesner (Eds.). Berlin and New York: de Gruyter.

Motte, André, and Rutten, Christian (Eds.) (2001): *Aporia dans la philosophie grecque des origines à Aristote.* Louvain: Peeters.

Narcy, Michel (2003): "Aristote de Stagire, la *Métaphysique*. Tradition grecque". In: Goulet, Richard (Ed.): *Dictionnaire des philosophes antiques.* Vol. Supplément. Paris: CNRS éditions, pp. 224–258.

Primavesi, Oliver (1996): *Die Aristotelische Topik: ein Interpretationsmodell und seine Erprobung am Beispiel von Topik B.* Munich: C.H. Beck.

Rapp, Christof (2017): "*Aporia* and Dialectical Method in Aristotle". In: Karamanolis, George and Politis, Vasilis (Eds.): *The Aporetic Tradition in Ancient Philosophy.* Cambridge: Cambridge University Press, pp. 112–136.

Rashed, Marwan (2000): "Alexandre d'Aphrodise lecteur du *Protreptique*". In: Hamesse, Jacqueline (Ed.): *Les Prologues médiévaux.* Turnhout: Brepols, pp. 1–37.

Rashed, Marwan (2007): *Essentialisme: Alexandre d'Aphrodise entre logique, physique et cosmologie.* Berlin and New York: de Gruyter.

Sharples, Robert W. (2010): *Peripatetic Philosophy, 200 BC to AD 200: An Introduction and Collection of Sources in Translation.* Cambridge: Cambridge University Press.

Spranzi, Marta (2011): *The Art of Dialectic Between Dialogue and Rhetoric: The Aristotelian Tradition.* Amsterdam and Philadelphia: John Benjamins.

Van Ophuijsen, Johannes M. (1994): "Where Have the *Topics* Gone?". In: Fortenbaugh, William W. and Mirhady, David C. (Eds.): *Peripatetic Rhetoric after Aristotle.* New Brunswick and London: Transaction Publishers, pp. 131–173.

Van Ophuijsen, Johannes M. (Trans.) (2001): *Alexander of Aphrodisias: On Aristotle's Topics 1.* Ithaca and New York: Cornell University Press.

Silvia Fazzo
Aporiai with Multiple Solutions in Alexander of Aphrodisias

1 Introduction: Aporiai in the Narrow Sense in Alexander of Aphrodisias

Alexander's so-called *Quaestiones*, in Greek ἀπορίαι καὶ λύσεις,[1] are of special interest for scholars thinking of Aristotelianism as an exegetical tradition.

In Alexander, an aporia is a peculiarly philosophical kind of problem.[2] Typically, more than one solution can be offered for a single aporia, i.e., for a single exegetical and/or theoretical problem.

Their collection, Alexander's *Aporiai kai lyseis*, has been handed down to us[3] by a *codex vetustissimus*, Ven. Gr. 258 (9th century) and a number of (probably) *descripti*. Their first and only critical edition was promoted by the Berliner Academy in 1892 within one of two volumes of *Alexandri Opera Minora* (1889, 1892). A new critical edition that will take into greater account the crucial role of Ven. Gr. 258 and new marginalia discovered in the meanwhile is thus necessary. However, dealing with their manuscripts reveals a peculiar and telling obstacle as far as the history of dialectics is concerned: in manuscripts and manuscript catalogues belong-

---

[1] Reference edition by Bruns (1892). Translation by Sharples (1992–1994). For a more comprehensive inquiry, see Fazzo (2002), with special reference to pp. 25–29.

[2] Aporiai or *Quaestiones*? The so-called "minor works" (Opera Minora) by Alexander, which bear in Greek the title of *Aporiai and lyseis*, are better known, if ever, as *Quaestiones*, which is the title of their first print in the 16th century: "Alexandri Aphrodisiensis quaestiones naturales, de anima, morales" (Venetiis, Zanetti 1536). Since this title appears in catalogues of manuscripts as well, this implies a frequent confusion between Alexander's *Quaestiones* and pseudo-Alexander's.

[3] These give the general title to a collection that is far from homogeneous, including pieces of different nature as well. See Sharples (1992) pp. 4f.: "A classification of the types of texts found in these collections was undertaken by Bruns (1892: v-xiv) in the preface to his edition. In addition to (i) "problems" in the strict sense with their solutions, these minor works include (ii) expositions (exegeseis) of particularly problematic Aristotelian texts, (iii) short expositions of Aristotelian doctrine on a particular topic, and (iv) straightforward and sometimes tedious paraphrases of passages in Aristotle's writings; both (iii) and (iv) alike seem to be described as epidromai or 'summaries'. There are also (v) collections, one might almost say batteries, of arguments for a particular Aristotelian position (these are characteristic of the *Mantissa* rather than of the *Quaestiones*; they also occur in the *Ethical Problems*); and (vi) what appear to be fragments of more elaborate literary works apparently never finished".

ing to the late tradition, aporiai are often mixed up and confused with spurious *Problemata* of pseudo-[Alexander].

This happens because both sets of texts are liable to be called *Quaestiones* when a Latin title occurs in catalogues. In fact, as I wish to recall first of all, Alexander's aporiai should not be mixed up and confused with *Problemata*.

Once this is made clear, I will attempt a closer understanding of these aporiai within the Aristotelian tradition. My proposed reading will be relevant for the sake of the one controversial issue where the value of multiple solutions is concerned.

The interpretation of Alexander's aporiai is thus related to a special kind of dialectical discussion, namely, aporetic discussion: what is the narrow sense of "aporetic" and of "aporia"? In this frame, for the purposes of this conference on ancient dialectics, I will propose focusing on a main issue typical of some of these texts: the use of multiple solutions for a difficult philosophical problem. This may be considered proper of Alexander's aporiai in the narrow sense.

In this regard, Alexander's aporiai show a common trend which seems to be proper of exegetical traditions as such—probably including legal and religious books, as commented in the very first centuries of the Christian era: in such cases as well, multiple solutions may be often suggested for a single problem.

In several sections of the collection of Alexander's aporiai, more than one solution is thus offered for a single aporia, i.e., for a single exegetical and theoretical problem: see *Quaestiones* 1.7, 1.8, 1.10, 1.16, 2.3, 2.25, 4.5, 4.16.

This makes Alexander's aporiai paradigmatic in a sense, because after Alexander philosophical exegesis became the proper way of making philosophy based on Aristotle's texts. Multiple solutions often arose on the part of different philosophers whose names became famous in the Middle Ages and Renaissance.

Nonetheless, in Alexander's aporiai, the different solutions are not attributed to other authors. Does this mean that Alexander produced all of them? Or rather that a canon was built in Alexander's time as a given standard system on Aristotle's name?

Up to now, it seems that the way aporiai were produced and intended to work still requires investigation.[4] The issue is relevant for that part of the history of ancient dialectics which is concerned with the use of philosophical questions. For the present purpose, I will initially frame my argument rather more widely.

---

4 See also Kupreeva (2018), p. 228, n. 1, "*Much still remains to be done*" (my italics), and Fazzo (2002).

2 Aporiai vs. *Problemata:* The Difference in Scope

Asking questions as a part of ancient dialectics displays a wide range of meanings and uses. It could not be otherwise since dialogue, διαλέγεσθαι, is the standard means human beings employ to share and improve their views. In this frame, asking questions is recognized in classical philosophy as the most effective means of διαλέγεσθαι.

The primacy in philosophy of asking questions, dialectics, applies both to Plato and Aristotle, but in different ways. Indeed, in Plato and Aristotle the very act of raising questions can be seen to have two kinds of scope:
—progress in knowledge on the respondent's part: a pedagogical scope for questioning; —progress in knowledge on the speaker's part, i.e., on the questioning subject, on the question's part as such: a zetetic, or strictly aporetical scope.

As for pedagogical questions aiming at philosophical progress on the respondent's part, these show, in turn, a comprehensive range of uses: this mood is remarkably broad in meaning and timeframe, spanning at least from Socratic maieutic (late 5th century BCE) to the school problemata in handbooks of Late antiquity. They thus differ crucially in many regards: in school handbooks and teaching, the master is *generally* expected to know the answer, whereas in Socratic dialogues he does not strictly need to. Their common point is that the master, in asking the questions, compels the pupil, the listener, or the reader to find the true path in himself.

The other kind of scope is a properly aporetic or zetetic aim. This includes from the outset Aristotle's aporiai, the ones to which *Metaphysics* book *Beta* is devoted. But it also reaches and includes—as I posit—Alexander's "Questions" in a narrow sense, i.e., Alexander's ἀπορίαι καὶ λύσεις, our subject for ongoing research.

How do Alexander's ἀπορίαι καὶ λύσεις work? No consensus has been reached so far.

Here I owe a preliminary remark to Laks and Most, the editors of Theophrastus' Metaphysics.[5] In itself—they say—raising an aporia might well say nothing about the underlying reason for it. I believe this view is correct because of a number of cases in philosophical literature where the most disputable issue is indeed whether or not the author knows the answer already, when putting forth the question. This is precisely what one would need to know in order to assess whether an aporia is entirely zetetic or not.

---

5 See the introduction by Laks and Most (1993), p. **xviii**.

The issue is further complicated by the fact that hermeneutical currents promptly deny the relevance of the original author's point of view, both in general and in this very regard. Some, e.g., Scholastic and neo-Scholastic thinkers, might appreciate the final result, regardless of the author's consciousness; others might better like questions in themselves and discard answers as such. The latter might even seek for the ultimate and most original questions regardless—once more—of the author's consciousness of what is the *very* question to which he was about to respond.

3 A Crucial Aporia in Aristotle's *Metaphysics* as a Case Study

The mood of strictly aporetic, i.e., zetetic, aporiai might include Aristotle's companions, Theophrastus and Eudemus, especially the former's so-called *Metaphysics*. Theophrastus' book, whether complete or fragmentary (another ongoing discussion) was originally intended as an aporetic περὶ ἀρχῶν, a text devoted to an aporetical re-examination of Aristotle's theory of principles, especially as expounded in *Metaphysics* book *Lambda*.

Theophrastus' aporetical book is thus a good parallel case study. This is evident in that it is precisely with this reference that the general principle stated above has been formulated: in itself, raising an aporia does not say anything about its underlying reason. In this case as well, scholars might disagree about Theophrastus' general scope: did Aristotle's friend know the answer from the outset when wondering how the prime unmoved mover moves the celestial spheres? If Reale's reading is remarkably positive,[6] Laks and Most's is not: they do not find a reason to attribute any preconceived view to Theophrastus, and insist that any answer is a matter of interpretation.

The same doubt, to tell the truth, affects Aristotle's book *Beta*. Some (most remarkably Aubenque) take this book as a paradigm of the aporetic, i.e., truly philosophical, path. Others (especially Menn [forthcoming]) use the book as a preliminary agenda, pointing for development to the other books of Aristotle's *Metaphysics*.[7]

This latter view *might* imply that Aristotle already knows the answer when raising aporiai in this book.

---

[6] Reale (1984).
[7] Aubenque (1962); see also Menn (forthcoming).

Such a doubt affects aporetical texts in general, but is especially relevant for Alexander's aporiai with multiple solutions.

For, after all, if Alexander knows the final answer before openly raising doubts, this implies that offered solutions have no independent value in themselves, save one: they are put forward for the sake of the contrast in order to enhance the value of the preferred solution—which is often the final one. If so, the multiple solutions would be pedagogical in character, since a number of them aim to increase the value of a single one.

Admittedly, the whole issue is a matter of interpretation and will remain disputable. Regardless, I wish to bring as an argument a kind of thought experiment, picking out the most controversial question as far as Aristotle's theory of principles is concerned.

Let us examine the interpretation of these two phrases from the beginning of Aristotle's *Metaphysics* Book *Lambda*, where he states:

> (a) The object of desire and the object of thought move in this way; they move without being moved. κινεῖ δὲ ὧδε τὸ ὀρεκτὸν καὶ τὸ νοητόν· κινεῖ οὐ κινούμενα (*Met. Lambda* 7, 1072a26).

Further on, he says:

> (b) "It produces motion by being loved, and it moves the other things by the body which is moved. κινεῖ δὲ ὡς ἐρώμενον, κινουμένῳ δὲ τἆλλα κινεῖ (*Met. Lambda* 7, 1072b3f.[8]).

The two phrases describe the way the prime unmoved mover moves the universe. How can an unmoved mover move while being unmoved? This is the problem.

We recalled already that Theophrastus was highly puzzled by this theory.[9] This is his main reason for writing the book we now call Theophrastus' *Metaphysics*.

Half a millennium later, a final interpretation of *Lambda* 7 was established based on possible interpretations of (a) and (b): the prime unmoved mover not only moves being unmoved, so as the object of love and thought only can move —a possible interpretation of (a)—and not only move as if it were beloved, namely, while being unmoved—a possible interpretation of (b) (interpretations supported even in our times by Enrico Berti[10])—but is actually loved by the celestial sphere, which is moved by the desire to imitate its perfection and immobility.

---

8 The present discussion of 1072b4 (κινουμένῳ as opposed to Ross (1924)' κινούμενα) is based on the critical edition by Fazzo (2012). The *Revised Oxford Translation* by Barnes (1984), which is the source of the quotations, has been modified accordingly.
9 This is the main subject of Theophrastus' so called "Metaphysics", edited by Laks and Most (1994). Theophrastus' perplexity prompted the discussion by Berti (2002).
10 Broadie (1993), Berti (2000) and Laks (2000).

This interpretation, which underlies the standard reading of the books up to today, is in no way perfect. The theory lacks support in Aristotle's corpus. Aristotle never talks of the highest principle of the physical universe as an object of thought or as a final cause. This continues to cause lively debate, as can be seen in Broadie and Berti and Laks vs. Berti.

We do not know exactly how many tentative solutions to the question "How can an unmoved mover move while being unmoved?" were proposed before the final one was established.

Let us suppose a first one was given and then another one.

A striking sequence of facts can be detected as follows (see Fazzo 2014): once the question was raised initially, more than one answer was formulated to compete; subsequently, the comparatively better one, even if not entirely satisfactory, was regarded, first as being better than others are; then as the best one; hence later on as the most suitable one, as the only good one, and ultimately as the *true* one.

Dealing with multiple solutions is in no way singled out as belonging exclusively to philosophical *aporiai*. It probably also applies to exegetical traditions as such. This is why it helps to consider Aristotelianism, the main path of Middle Eastern and European thought, as an *exegetical tradition*.[11]

4 Alexander's *Quaestio* 2.3 as a Case Study in the Exegetical Tradition: The Historical and Theoretical Relevance of the Last Proposed Solution

We now return to the *aporiai* by Alexander. Let us discuss in particular one of those with more than one solution, *Quaestio* 2.3. Here is the transmitted title: "What the power is that comes to be, from the movement of the divine body, in the body adjacent to it which is mortal and subject to coming-to-be".[12] The aporia is thus: how does heavenly movement affect the sublunary world? More solutions are offered within the text. The first solution is the mild one established in Aristotle's *Meteorologica* and *De Generatione et Corruptione* II.10: the power of stars is added from outside to existing sublunary bodies. The second solution is quite re-

---

11 Fazzo (2004).
12 English trans. by Sharples (1992). See the critical edition in Fazzo (2002).

markable: sublunary bodies split into two groups, simple bodies and ensouled bodies, which are such because of the "divine power coming from the celestial bodies".

> As many bodies as come to be from the natural mixture and blending of these simple bodies (...), it is on account of this power that these no longer possess in themselves a principle only of motion in accordance with inclination, but have acquired in addition also a certain psychic [motion] which possesses its origin and coming-to-be from the divine power.

This second solution is connected with the theory of pseudo-Aristotle's *De mundo* 6.[13] But it is also remarkable that this theory is referred to with some distance and incertitude (a detail which shows, by the way, that Alexander does not regard *De mundo* as authentic).

Finally, a third, even more rewarding and remarkable solution, while including the former assessment about the relevance of divine, heavenly power for living and ensouled beings as such, endorses heavenly bodies with the power to confer basic qualities—hot, cold, dry, wet, to a completely uniform prime matter:

> Or rather: one could say that the power from the divine bodies is the cause of the <difference> between the simple bodies and of their coming-to-be, itself coming to be their form and nature. For matter, which is in itself without quality or shape is and comes to be body in actuality, is given form and shaped, by the power which comes to be in it from the divine bodies.

This is a very strong claim.

Hence the disputed question among contemporary scholars whether or not the last one is the preferred one, i.e., the most advanced one. Donini denies this, positing that the final solution is so far from Aristotle's thought that Alexander could not possibly have held such a view.[14]

However, it would be a stretch to classify and label multiple solutions according to their proximity to Aristotle's own wording. After all, did the school not establish the text of Aristotle's corpus for centuries to come? It seems that exegesis and textual edition have been dealt with in a joint venture, so as to answer both the old and new aporiai concurrently.[15]

Moreover, one might wonder whether or not it makes sense for an interpreter to put forward an explanation which he regards as worse than previous ones. The broader issue can be seen in our *Quaestio* 2.3.

---

13 Detailed analysis of the whole *Quaestio* 2.3 and comparison with *De mundo* in Fazzo (2002), pp. 175–212.
14 Donini (1996).
15 I will discuss this topic further in my future book "The Emergence of Aristotle's Metaphysics".

Further hints corroborate the last solution to *Quaestio* 2.3 as being the last added, and the preferred one at the end of the tradition of the problem. Strikingly, a single manuscript (Ven 194) independent from the *vetustissimus* (Ven 258[16]) transmits the same text in a different form, without this solution, so it seems that this was indeed added at the end of a process from within the school.

Prime matter does not exist in Aristotle.[17] But the idea was highly suggestive for the history of alchemy. The Arab alchemist Jabir was one of the latest readers of Alexander's commentary *On De Generatione et Corruptione* and uses this theory as the basic creed of alchemical practice.[18] This encourages the view that a theory of prime matter was a final solution in *Quaestio* 2.3 and perhaps in Alexander's commentary *On De Generatione et Corruptione* 2 as well.

Thus, Alexander's theory of prime matter was Aristotelian in sources and frame, albeit not expressed nor, in all likelihood, conceived by Aristotle. Based on Alexander's exegesis, the theory was reworked and commented upon, most remarkably by Arab, then western scholars from within the Aristotelian tradition. At the end of the story, working along this path alchemy evolved into chemistry and prepared a theoretical foundation for subsequent inquiries into the very structure of matter.

---

16 Fazzo (2002), pp. 36–41. On the new sets of *marginalia*, see Fazzo (1999).
17 Seminal discussions in King (1956), de Haas (1996), and Fazzo (2002); further developments will take into account Kupreeva (2004). I would like once again to thank Richard Sorabji for encouraging the discussion when my 2002 book was in preparation.
18 A telling parallel can be gathered from the description of Jabir's theory in Holmyard (1957), see especially Chapter 5, "Islamic Alchemy", in part. § "Jabir Ibn Hayyan", with reference to Kraus 1942–1943 "Holmyard, 1957 (Chapter 5, "Islamic Alchemy", in part. § "Jabir Ibn Hayyan", with reference to Kraus 1942–1943): "On the constitution of matter, Jabir held the Aristotelian conception of the four elements: fire, air, water, earth: but developed it on different lines. He postulated first the existence of four elementary qualities or 'natures', namely hotness, coldness, dryness, and moistness. When these natures united with substance they formed compounds of the first degree, namely hot, cold, dry, moist. Union of two of these gave rise to fire (hot + dry + substance); air (hot + moist + substance); water (cold + moist + substance) and earth (cold + dry + substance). In metals, two of the "natures" are external and two internal, a point to which further reference is made later (p. 77). Thus, in his "Seventy Books" Jabir says that lead is cold and dry externally and hot and moist internally; gold, on the other hand, is hot and moist externally and cold and dry internally. He believed that, under the influence of the planets, metals were formed in the earth by the union of sulphur (which would provide the hot and dry "natures") and mercury (providing the cold and moist). This theory, which appears to have been unknown to the ancients, represents one of Jabir's principal contributions to alchemical thought; it may have been wholly original, though perhaps Jabir found the germs of it in Apollonius of Tyana. It was generally accepted by later generations of alchemists and chemists, and survived until the rise of the phlogiston theory of combustion in the concluding years of the 17th century.

Overall, the power of dialectics is thus best shown when a tradition remains in dialogue with the past and dares to offer new solutions for emerging problems and queries.

Bibliography

Aubenque, Pierre (1962): *Le problème de l'être chez Aristote: Essai sur la problématique aristotélicienne.* Paris: Bibliothèque de philosophie contemporaine.

Barnes, Jonathan (1984): *The Complete Works of Aristotle: The Revised Oxford Translation.* Princeton: Princeton University Press.

Berti, Enrico (2000): *Unmoved Mover(s) as Efficient Cause(s) in* Metaphysics Λ 6. In: Frede, Michael and Charles, David (Eds.): *Aristotle's* Metaphysics Lambda. *Symposium Aristotelicum (Oxford, August 26–30, 1996).* Oxford: Clarendon Press, pp. 181–206.

Berti, Enrico (2002): "Teofrasto e gli Accademici sul moto. Gigantomachia. Convergenze e divergenze fra Platone e Aristotele. Maurizio Migliori (Ed.): Brescia: Morcelliana.

Berti, Enrico (2022): "Ancora sulla causalità del motore immobile". In: *Méthexis* 20, pp. 7–28, http://www.jstor.org/stable/43739180, last accessed on November 7, 2022.

Broadie, Sarah (1993): "Que fait le premier moteur d'Aristote. Sur la théologie du livre *Lambda* de la *Métaphysique*". In: *Revue de la France et de l'étranger* 183, pp. 375–411.

Bruns, Ivo (Ed.) (1892): *CAG Suppl. 2.2 Quaestiones, De fato, De mixtione.* Berlin: Ivo Bruns.

Donini, Pierluigi (1996): "*Theia dunamis* in Alessandro di Afrodisia". In: Romano, Francesco and Cardullo, Loredana (Eds.): *Dunamis nel neoplatonismo: atti del II colloquio internazionale del centro di istema sul neoplatonismo. Symbolon* 16, pp. 12–29. [Reprinted in Donini, Pierluigi (2011): *Commentary and Tradition Aristotelianism, Platonism, and Post-Hellenistic Philosophy.* Mauro Bonazzi (Ed.). Berlin: de Gruyter, pp. 125–139.

Fazzo, Silvia (1999): "Philology and Philosophy on the Margins of Early Printed Editions of the Ancient Greek Commentators on Aristotle, with special reference to copies held in the Biblioteca Nazionale Braidense, Milan". In: Blackwell, Constance and Kusukawa, Sachiko (Eds.): *Philosophy in the Sixteenth and Seventeenth Centuries: Conversations with Aristotle.* New York: Routledge, pp. 48–75.

Fazzo, Silvia (2002): *Aporia e Sistema: la materia, la forma, il divino nelle Quaestiones di Alessandro di Afrodisia.* Pisa: ETS.

Fazzo, Silvia (2004): "Aristotelianism as a commentary tradition". In: *Bulletin of the Institute of Classical Studies* (Supplements) 47, pp. 1–19.

Fazzo, Silvia (2012): *Il libro Lambda della Metafisica di Aristotele.* Naples: Bibliopolis.

Haas, Frans A. J. de (1996): *John Philoponus' New Definition of Prime Matter: Aspects of its Background in Neoplatonism and the Ancient Commentary Tradition.* Leiden: Brill.

Holmyard, Eric John (1957): *Alchemy.* Harmondsworth: Penguin.

King, Hugh R. (1956): "Aristotle without prima materia". In: *Journal of the History of Ideas* 17, pp. 370–389.

Kraus, Paul (1942–1943, 1986): *Jābir ibn Ḥayyān. Contribution à l'histoire des idées scientifiques dans l'Islam,* II. *Jābir et la science grecque.* Cairo and Paris: Maktabat al-Khanji.

Kupreeva, Inna (2004): "Alexander of Aphrodisias on Mixture and Growth". In: *Oxford Studies in Ancient Philosophy* 27, pp. 297–334.

Kupreeva, Inna (2018O): "Aporia and Exegesis: Alexander of Aphrodisias". In George Karamanolis, George and Politis Vasilis (Eds.): *The Aporetic Tradition in Ancient Philosophy*. Cambridge and New York: Cambridge University Press, pp. 228–247.

Laks, André (2000): *Metaphysics Λ 7*. In: Frede, Michael and Charles, David (Eds.): *Aristotle's* Metaphysics Lambda. *Symposium Aristotelicum (Oxford, August 26–30, 1996)*. Oxford: Clarendon Press, pp. 207–243.

Laks, André and Most, Glenn (Eds.) (1993): *Théophraste, Métaphysique. Texte édité, traduit et annoté par André Laks et Glenn W. Most avec la collaboration de Charles Larmore et Enno Rudolph et pour la traduction arabe de Michel Crubellier.* Paris: Les Belles Lettres, Collection des Universités de France.

Menn, Stephen (forthcoming): *The Aim and the Argument of Aristotle's* Metaphysics.Oxford: Oxford University Press.

Reale, Giovanni (1984): *Il concetto di filosofia prima e l'unità della metafisica di Aristotele. Con traduzione integrale e commentario della "Metafisica" di Teofrasto.* Milan: Pubblicazioni dell'Università Cattolica del Sacro Cuore.

Sharples, Robert W. (1992): *Alexander of Aphrodisias Quaestiones 1.1–2.15. Translated by R. W. Sharples* [Ancients Commentators on Aristotle]. London: Duckworth.

Inna Kupreeva
Alexander of Aphrodisias on the Principle of Non-Contradiction: The Argument "from Signification"

In *Metaph.* 4.4, Aristotle sets out to refute the views of those who deny the principle of non-contradiction. The nature and scope of these arguments have been a matter of scholarly controversy. Elizabeth Anscombe accurately described this text as "long, difficult and bad-tempered" but pointed out that in it Aristotle was making a connection between his ontology and the principle of non-contradiction.[1] Some prominent logicians questioned the status of PNC as an indemonstrable first axiom.[2] Some leading Aristotle scholars saw his arguments in support of PNC in *Metaphysics* as a sign of his departure from a demonstrative model of science developed in the *Posterior Analytics*.[3] Clearly this difficult text has been used to support several very weighty claims, and even this alone would make understanding it important for students of Aristotle. Alexander's interpretation has received some critical attention in recent scholarship,[4] but many questions still remain.

In this paper, I would like to provide an outline of Alexander's reading of Aristotle's argument which will allow us to see how his position stands in the light of contemporary discussions of Aristotle's argument. After a very brief summary of the main points of Aristotle's argument in Γ 4 in § 1, I discuss (in § 2) Alexander's interpretation of elenctic demonstration (with special attention to his distinction between the *elenchos* proper and the more general argument from signification), and in § 3, I try to show that Alexander develops his own version of unrestricted

Acknowledgements: I am very grateful to Prof. Melina Mouzala for inviting me to present at the Patras conference and to all the conference participants for their excellent contributions and discussion. I owe special thanks to Prof. Mouzala for her help, encouragement and patience during my work on this paper. Earlier drafts were presented to CNRS seminar in Sorbonne on Alexander's Metaphysics commentary and at a Philosophy Department research seminar in the University of Leiden. I am very grateful to the audiences and organisers, particularly to Cristina Cerami, Annick Jaulin and Gweltaz Guyomarc'h and to Frans de Haas, Bert van den Berg and Maria van der Schaar for stimulating discussions and feedback. The paper further benefited from comments and criticisms by Stephen Menn and George Medvedev. Any remaining errors are mine.

1 Anscombe (1961), p. 39.
2 For a good analytical summary, see Berti (2014).
3 Irwin (1977) and Irwin (1988), pp. 179–198, Bolton (1994).
4 Flannery (2003) and Mignucci (2003).

essentialist interpretation of Aristotle's argument which has some philosophical merits.

1 Outline of Aristotle's Argument

The discussion of the principle of non-contradiction is answering the question raised in the second aporia of *Metaph.* 3.2: "Is it the task of a single science to investigate both the ultimate principles of being and the basic principles of reasoning (e.g., the principle of non-contradiction)? Or is it the task of fundamentally different sciences?" (996b26–997a15). In *Metaph.* 4.3, the question is reformulated to ask whether it belongs to the same or different sciences to treat of what is called in theoretical disciplines "axioms" and of being.[5] Aristotle's answer supports the first disjunct: it is the task of the single science, for these principles belong to all the things that are rather than to some particular kind separate from others (1005a22–23).

In the space of a dozen lines in *Metaph.* 4.3, we have several formulations of the principle:[6]

> PNC-I *([A] belonging and not belonging to [B]):* It is impossible for the same thing both to belong and not to belong to the same thing at the same time and in the same respect—and let us assume we have drawn all the further distinctions that might be drawn to meet logical complaints (1005b19–20)

> PNC-II *(contrary attributes belonging to the same thing):* It is impossible that contrary attributes should belong at the same time to the same subject (the usual qualifications must be presupposed in this premiss too) (1005b26–27)

> PNC-III *([A] being and not being [B]):* It is impossible for anything at the same time to be and not to be (F) (1006a3–4)

> PNC-IV *(veridical or semantic):* It is impossible that it should at the same time be true to say of the same thing both that it is man and that it is not man. (1006b33–34)

In *Metaph.* 4.3, Aristotle establishes that the PNC is the "firmest" principle about which it is impossible to be mistaken. Aristotle's goal, as he explains, is not to demonstrate (i.e., scientifically prove) the PNC: this is impossible as it would require

---

5 For a recent study of Aristotle's overall strategy in *Metaph.* 4.3–6, see Crubellier (2008).
6 For a recent detailed study of these formulations and on relation of PNC to other logical principles making up the Law of Contradiction in Aristotle and later tradition, see Cavini (2007 and 2008).

a derivation of PNC from some prior principle. Rather, his goal is to show that *given that it is true*, it is necessary to believe it and rejecting it is implausible.[7]

Aristotle's argument in *Metaph.* 4.4 can be divided into three parts. The first part (Aristotle, 1005b35–1006a28, Alexander's commentary *in Metaph.* 271, 24–275, 20) opens with a methodological preamble, in which Aristotle tells us that there are some people who deny the PNC in full earnest, and some others still who ask for its demonstration. This latter query is, for Aristotle, a sign of poor education (1005b35–1006a11).

Despite this disparaging remark, Aristotle still goes on to say that it is indeed possible to demonstrate the PNC, not by an unqualified demonstration, which would require deriving the PNC from a more fundamental principle, but by a qualified one, which he calls "elenctic", or "refutative". This kind of demonstration will be effective against those people who subvert the proof of the "firmest principle" in *Metaph.* 4.3 by claiming that they are happy to deny its assumption, i.e., the truth of PNC, and so also on rational grounds deny it as the principle governing their beliefs. Aristotle promises to show how they cannot avoid committing to PNC on pain of being unable to engage in any rational discourse. (1006a11–28).

In the second part (Aristotle 1006a28–1007b18, Alexander 275, 23–290, 21), after the introduction of the concept of elenctic demonstration and the general explanation of the way it works, Aristotle sets out the details of the argument from signification. It is the longest single argument in this chapter. Its main line is punctuated by a number of departures building up towards additional arguments that result in several important points and distinctions which Aristotle uses to respond to possible difficulties or counterexamples and stave off any misreadings of the proposed argument. I'll look at its main steps in the course of my discussion of Alexander's interpretation.

The third part contains six shorter arguments in which Aristotle goes on to show further implausible consequences of the denial of the principle of non-contradiction.[8]

My main focus in this paper is on the elenctic demonstration and the argument from signification which correspond to the first and second parts of the overall argument of Gamma 4.

---

[7] This argument has been at the centre of controversy; see Lukasiewicz (1993), pp. 22–48, Barnes (1969), and Wedin (2004).

[8] Argument from monism (1007b18–1008a2, Alexander 290, 24–292, 21), argument from the law of excluded middle (1008a2–7/292, 24–293, 32), arguments against partial and total denial of PNC (1008a7–34/293,35–297,6), argument from bivalence (1008a34–b2/297,7–25), argument from truth and pragmatic (1008b2–31, 297, 28–300, 22), argument from the more and less (1008b31–1009a5/300, 24–301, 25).

2 Alexander on Elenctic Demonstration

Aristotle's argument in *Metaph.* 4.4 starts with an explanation of the goals and epistemic status of his proofs in defense of the PNC. These proofs are addressed to those opponents who would not accept the PNC and would ask for its demonstration. Aristotle indicates that although the principle does not need a demonstration, there must be a rational way of establishing the truth of the principle for its opponents, by showing that some of their assumptions depend on it, and so they cannot deny it without denying also some of those assumptions.

> (T1) (1) But it is possible to demonstrate by way of refutation also about this, that it is impossible, *if only the disputant says something.* If he says nothing, it is ridiculous to look for a speech[9] in response to one who has a speech of nothing, in so far as he has not; such a person, in so far as he is such is similar to a plant. (2) I say that an elenctic demonstration differs from a demonstration because someone who is demonstrating would seem to be begging the question, whereas if another were to be responsible in such a way it would be refutation but not a demonstration (1006a11–18)[10]

This "demonstration by refutation" or "elenctic demonstration"[11] must be different from both a standard demonstration and a standard *elenchos*. We already know the reason why it cannot be a standard demonstration: a demonstration would need an even more fundamental principle as its premiss, and this Aristotle will not concede. It cannot be an *elenchos* proper, because an *elenchos* is a reduction to a contradiction,[12] but Aristotle's opponents will not be taking a contradiction in their claims as a sign of defeat, as it is their position that it is not problematic.

The logical structure of the elenctic demonstration of the impossibility of the denial of PNC mirrors the dialectical *reductio ad absurdum* in the following way.

---

[9] 'Speech' translates λόγος, the word that is rendered differently in different translations. In this paper I largely follow Kirwan's translation of Metaphysics Gamma and Madigan's translation of Alexander's commentary. In both cases light modifications will be obvious (needed in part for some consistency of vocabulary). Kirwan translates λόγος as 'statement', while Madigan prefers 'speech'. There are many other possibilities. I will keep the translation 'speech' for both Aristotle and Alexander, understanding by it a meaningful speech which can be used by the parties in a dialectical argument rather than any utterance.

[10] (1) ἔστι δ' ἀποδεῖξαι ἐλεγκτικῶς καὶ περὶ τούτου ὅτι ἀδύνατον, ἂν μόνον τι λέγῃ ὁ ἀμφισβητῶν· ἂν δὲ μηθέν, γελοῖον τὸ ζητεῖν λόγον πρὸς τὸν μηθενὸς ἔχοντα λόγον, ᾗ μὴ ἔχει· ὅμοιος γὰρ φυτῷ ὁ τοιοῦτος ᾗ τοιοῦτος ἤδη. (2) τὸ δ' ἐλεγκτικῶς ἀποδεῖξαι λέγω διαφέρειν καὶ τὸ ἀποδεῖξαι, ὅτι ἀποδεικνύων μὲν ἂν δόξειεν αἰτεῖσθαι τὸ ἐν ἀρχῇ, ἄλλου δὲ τοῦ τοιούτου αἰτίου ὄντος ἔλεγχος ἂν εἴη καὶ οὐκ ἀπόδειξις.

[11] *Metaph.* 4.4: 1006a15. See a fine discussion of this passage in Crubellier (2008).

[12] *An. Pr.* 2.20, 66b11, 8. See also *Soph. El.* 9, 170b1.10, 171a2.4.

While in the *reductio*, the respondent aims to hold on to his professed beliefs and avoid the trap that will lead him to a contradiction, in our special case the respondent is happily embracing any contradictory statement and the dialectician's task is to force him into making a statement that would prove that he cannot be committed to a contradiction.

Formalizing a standard dialectical discussion will give us something like this:

$$\{\Gamma, \{B \in \Gamma |\ A\&\sim A\}\} \vdash \sim B^{13}$$

Here Γ stands for any beliefs the respondent may have, and B is a particular belief that entails a contradiction. In order to avoid the contradiction, the respondent must reject B (deriving ~ B).

Formalizing the elenctic demonstration will give us the following strange result:

$$\{\Gamma, \{B \in \Gamma, (A\&\sim A) |\ \sim B\}\} \vdash \sim(A\&\sim A)$$

Here, Γ stands for any beliefs the respondent may have, and B is a particular belief that is so important to the respondent, that whatever formula entails its denial must be rejected, even if this leads the respondent to the denial of the contradiction, to which he was committed to begin with.

P. Gottlieb has aptly called this proof "the elenctic demonstration turned upside down".[14] The respondent is forced to deny the contradiction because denying one of his other premisses, B, is still less acceptable to him. It is very hard to imagine what kind of a dilemma could be so powerful as to force the denier of the PNC to drop this commitment of his. As M. Crubellier has pointed out, it is the task of a questioner to discover the kind of premiss that the respondent will not be able to deny.[15]

An important constraint for the questioner is that his proposed premiss cannot involve any reference to the law of contradiction. Aristotle describes this step, i.e., the premiss of the proof supplied by the opponent to make the refutation possible, as satisfying a condition "if only the disputant says something" (T1.1). The

---

13 I am grateful to Stephen Menn for discussing the formalism with me (he is not responsible for the final version).
14 Gottlieb (2009).
15 Crubellier (2008), p. 391.

meaning of "says something" has been controversial. We will take a look at Alexander's understanding of this condition.

Alexander generally seems to be treating Aristotle's argument as dialectical, because its main purpose is the refutation of those who object to the PNC, and because by refuting them and showing the reasons for accepting the PNC, one still contributes to the elucidation of the principles of science, which is one of the main tasks of dialectic outlined by Aristotle in *Topics* 1.2.[16]

Alexander begins his analysis of this proof by describing the difference between the demonstration *simpliciter* and the elenctic demonstration.

> (T2) (1) The one who demonstrates something without qualification assumes certain things as primary and more familiar than the thing to be demonstrated, and attempts in this way to demonstrate the proposition. (2) But it is not possible to assume anything primary and more familiar than this axiom, as has been said [1005b12–25], and so it is not possible to demonstrate it either. (3) Further, one who assumes from himself[17] and posits that everything either is or is not what he says it is will seem to be begging the question, i.e., to be assuming in advance the object of the inquiry, for this is what the inquiry was about. (4) Since, however, the refutation is carried on in reply to someone else, i.e., is derived from what the respondent posits (a refutation is a syllogism that leads by way of questioning to a contradiction), it can also be carried on by way of such premises. (5) For such a syllogism is not carried on by way of premises that are primary, nor will one appear to beg the question [by assuming this] from himself, if the respondent agrees *that everything either is or is not something which it is said to be*. (6) For it is he who is responsible for such a syllogism: the one who thinks that things immediately familiar should be proven, and who is being forced out of shame to grant these things; (7) if the one demonstrating posited these things from himself, he would seem to be begging the question. (8) For as Theophrastus said in his *On Affirmation*, demonstration of this axiom is forced and contrary to nature.[18]

---

[16] I have shown in Kupreeva (2017) that Alexander takes this task very seriously.

[17] 'From himself' (ἀφ' ἑαυτοῦ) means from own decision rather than being proposed by the opposing party in a dialectical debate. This formula is standardly used by Alexander in this discussion for this kind of signposting. Thanks to Stephen Menn for flagging and discussing this.

[18] (T2) (1) ὁ μὲν γὰρ ἁπλῶς ἀποδεικνὺς λαβὼν πρῶτά τινα καὶ γνωριμώτερα ὄντα τοῦ δεικνυμένου, οὕτω τὸ προκείμενον ἀποδεικνύναι πειρᾶται. (2) οὐχ οἷόν τε δὲ πρῶτόν τι καὶ γνωριμώτερον τούτου λαβεῖν, ὡς προείρηται, ὥστε οὐδὲ ἀποδεῖξαι αὐτὸ οἷόν τε. (3) ἔτι ἀφ' ἑαυτοῦ λαμβάνων τε καὶ τιθεὶς πᾶν εἶναι τοῦτο ὃ λέγει ἢ μὴ εἶναι δόξει τὸ ἐν ἀρχῇ αἰτεῖσθαι καὶ τὸ ζητούμενον προλαμβάνειν, ἐπεὶ περὶ τούτου ἡ ζήτησις ἦν. (4) ὁ μέντοι ἔλεγχος ἐπεὶ πρὸς ἄλλον γίνεται καὶ ἐξ ὧν ὁ προσδιαλεγόμενος τίθησιν (ἔστι γὰρ ἔλεγχος συλλογισμὸς δι' ἐρωτήσεως εἰς ἀντίφασιν ἄγων), δύναται γίνεσθαι καὶ διὰ τῶν τοιούτων. (5) οὐκέτι γὰρ οὔτε διὰ πρώτων ὁ τοιοῦτος συλλογισμός, οὔτε τὸ ἐν ἀρχῇ λαμβάνειν τις παρ' αὑτοῦ δόξει, ἂν ὁ προσδιαλεγόμενος συγχωρήσῃ πᾶν ἢ εἶναί τι ὃ λέγεται ἢ μὴ εἶναι. (6) ἐκεῖνος γὰρ αἴτιος τοῦ τοιούτου ὁ ἀξιῶν δείκνυσθαί τε τὰ αὐτόθεν γνώριμα καὶ διδόναι ταῦτα ἀναγκαζόμενός τε καὶ δυσωπούμενος, ἃ εἰ ὁ ἀποδεικνὺς ἐτίθει ἀφ' ἑαυτοῦ, ἐδόκει ἂν τὸ ἐν ἀρχῇ αἰτεῖσθαι· (7) ὡς γὰρ εἶπε Θεόφραστος ἐν τῷ Περὶ καταφάσεως, βίαιος καὶ παρὰ φύσιν ἡ τούτου τοῦ ἀξιώματος ἀπόδειξις. (273, 4–19)

In (T2.1–2), Alexander follows Aristotle in seeing the PNC as the most primary and best known, indemonstrable principle. He further explains (T2.3) that this principle cannot be taken for granted by a questioner in a dialectical debate against the opponent of PNC, for this will be begging the question. But (T2.4) a valid refutation can have its premisss posited by the respondent. Alexander points out that the proof will be question-begging if the questioner "posits that everything is or is not what he says" (T2.3), but not if the same assumption is made by the respondent (T2.5). This expression "if the respondent agrees that everything either is or is not something which it is said to be" (T2.5) seems to be Alexander's paraphrase of Aristotle's "if only the disputant says something" (T1.1 above). So, Alexander's interpretation of this condition is not yet explicitly connected with the argument from signification, but takes it to be the most general description of the respondent's role as a source of premisses in this dialectical exchange. The nature of the premisses is not yet discussed.[19] The report about Theophrastus' appraisal of this demonstration as "being against nature" at (T2.8) suggests that Alexander's interpretation at this point might have as its source Theophrastus' work *On affirmation and negation*, which he seems to know well.[20]

In a standard dialectical discussion, with both interlocutors recognising the decisive role of the PNC in the argument, the questioner asks general questions, presupposing "yes" or "no" answers, selecting these questions so as to challenge some known stock beliefs of the respondent. In our case of elenctic demonstration, the respondent denies the PNC. Therefore, the questioner can't expect the respondent to be impressed if a contradiction is discovered in his views: it was licensed by the respondent to begin with, so having him reiterate it in his answers will not make a refutation. The only way to produce a refutation is somehow to force the respondent to give up on his denial of PNC without resorting to the PNC. In doing so, the questioner has to ask a question so that the respondent in his answer must voluntarily give up on his denial of PNC without a standard *reductio ad absurdum*. This is

---

[19] So, Kirwan's analysis of the phrase at 1006a12–13 as presupposing not only the argument from signification but in fact, the definitive formula as a part of the respondent's speech is a bit farfetched from the point of view offered by Alexander's interpretation. K. Flannery sees it as a flaw of Alexander's interpretation of Aristotle, attributing it to Alexander's failure to appreciate the difference between the semantic and the dialectical argument (Flannery (2003), pp. 121–123). I am going to show that Alexander does indeed treat the elenctic refutation as a dialectical argument, but distinguishes it from the argument from signification which follows after the elenchos.
[20] Fr. 68.3c and 85 A FHS&G and see next note.

why, probably, Theophrastus called this a "forced" proof, going against the nature of the proof.[21]

Aristotle gives us some more details about the nature of the premiss that can be elicited from the respondent:

> (T3) The starting point, in reply to all such arguments, is *not** to insist that [the respondent] say that something is the case or is not the case, but to signify something for himself and for another: for this is necessary if he were to say something that is meaningful. (1006a18–21)
>
> ἀρχὴ δὲ πρὸς ἅπαντα τὰ τοιαῦτα <u>οὐ</u> τὸ ἀξιοῦν ἢ εἶναί τι λέγειν ἢ μὴ εἶναι (τοῦτο μὲν γὰρ τάχ' ἄν τις ὑπολάβοι τὸ ἐξ ἀρχῆς αἰτεῖν), ἀλλὰ σημαίνειν γέ τι καὶ αὐτῷ καὶ ἄλλῳ· τοῦτο γὰρ ἀνάγκη, εἴπερ λέγοι τι.

The only request that comes from the questioner is that the respondent should "signify something for himself and for another", i.e., commit to some meaningful speech. The option of not assigning any meaning to his own statements is ruled out by Aristotle in strong terms: someone who does not "say something" is not entitled to a search for the *logos*, as long as he has not got any *logos* himself, and this type of interlocutor is similar to a plant. (T 1.1, 1006a13–16) Thus the respondent's choices are limited.

Alexander's manuscript reading of Aristotle's text at 1006a18–20 **(T3)** is different from our received text in the crucial bit of missing the negative particle οὐ where it is particularly important.[22] So Alexander would have to comment on the text which *prima facie* says that the starting point in such arguments is precisely to insist that the respondent say that something is the case or not.[23] Alexander spends some time developing a tortuous interpretation of this reading that would exempt the questioner from a charge of *petitio principii*. But Alexander is famously good at developing tortuous explanations to save the right meanings, and in this case, he gets the faulty text to say the same thing as the correct one. He explains that Aristotle means: "i.e., he [the respondent] should be asked whether it does not seem to him that everything that is said [to be something] is that or not, e.g., whether 'man' is or is not man, and likewise horse, dog, etc.".[24] Where the problematic text would have a yes-or-no question concerned with something being

---

[21] Cf. Bocheński (1947), p. 46; Graeser (1973), pp. 66–67; Repici (1977), pp. 62–64; Huby (2006), pp. 44–45.

[22] Our Ross' text has "is not" (on the basis of EJ, with Alexander being the only source to report the reading without negation). See Kotwick (2012), pp. 130–134.

[23] Flannery thinks Alexander is attracted to this reading; I don't see any evidence for this in the arguments.

[24] 273, 24–26: τουτέστιν ἐρωτητέον αὐτὸν εἰ μὴ δοκεῖ αὐτῷ πᾶν ἢ εἶναι τοῦτο ὃ λέγεται ἢ μὴ εἶναι, οἷον ἄνθρωπον ἢ εἶναι ἄνθρωπον ἢ μὴ εἶναι, ὁμοίως ἵππον, κύνα, τὰ ἄλλα.

the case, Alexander's interpretation renders it as the question concerned with the *use of language*, asking whether the objects to which the names are applied are or are not such as their names say they are.[25]

Alexander is still not too happy with this construal of the text[26] and informs us that some manuscripts have the reading which seems to him to make better sense —this is the reading of the rest of our tradition of this text so far.[27] On the basis of this superior reading he develops an interpretation of what "saying something" could mean as a step in a dialectical argument.

(T4) Alexander *in Metaph.* 274, 1–13

(1) And this way [i.e., with negation] the meaning is more perspicuous. (2) He thinks, then, that one should ask the respondent whether, when speaking, he signifies anything, to himself and to another, by the words he speaks and by the names he gives to things, and whether there is something to which he applies the name, i.e., which he wishes to signify when he speaks and produces the name. (3) For one who says that he signifies nothing by what he says and the answers he makes could not be saying anything to himself or to another, (4) nor will his thoughts which he uses [speaking] to himself in place of names and speech, be signifying anything; (5) for if these were signifying anything, it would be possible to put names on the things signified by the thoughts, and thus the spoken sound would also be signifying. (6) In this way the respondent would be agreeing that he was not using speech; for speech is the significant spoken sound (7) But he would likewise destroy the speech (logos) if he said that it no more signifies than does not signify. For if he did not signify whenever he said something, he would not be in conversation: for this one again is like a plant.[28]

---

25 273,22–26: Τοῦ ἐλεγκτικοῦ συλλογισμοῦ καὶ τοῦ πρὸς ἄλλον γινομένου περὶ τῆς τοῦ προκειμένου ἀξιώματος δείξεως ἀρχήν φησιν εἶναι τὸ ἀξιοῦν τὸν προσδιαλεγόμενον ἢ εἶναί τι λέγειν ἢ μὴ εἶναι, τουτέστιν ἐρωτητέον αὐτὸν εἰ μὴ δοκεῖ αὐτῷ πᾶν ἢ εἶναι τοῦτο ὃ λέγεται ἢ μὴ εἶναι, οἷον ἄνθρωπον ἢ εἶναι ἄνθρωπον ἢ μὴ εἶναι, ὁμοίως ἵππον, κύνα, τὰ ἄλλα.
26 My interpretation here and in what follows differs from Flannery (2003). I think that Alexander realises the impossibility of the position described by the transmitted text and when interpreting it he uses the explanation he finds further down in Aristotle's argument from signification to get the right sense. He is not very happy with the reading he has, *pace* Flannery.
27 This reading is adopted by modern editions of the *Metaphysics*, see n. 22 above.
28 (1) καὶ ἔστι γνωριμώτερον τὸ λεγόμενον οὕτως. (2) ἀξιοῖ δὴ ἐρωτᾶν τὸν προσδιαλεγόμενον, εἰ λέγων τι σημαίνει δι' ὧν λέγει τε καὶ ὀνομάζει καὶ ἑαυτῷ καὶ ἄλλῳ, καὶ ἔστι τι καθ'οὗ φέρων τὸ ὄνομα καὶ ὃ σημαίνειν βουλόμενος λέγει τε καὶ προφέρεται αὐτό. (3) ὁ γὰρ λέγων μηδὲν σημαίνειν δι' ὧν λέγει τε καὶ ἀποκρίνεται, οὗτος οὔτ' ἂν πρὸς αὐτὸν λέγοι τι οὔτε πρὸς ἄλλον. (4) οὐδὲ γὰρ τὰ νοήματα ἔσται σημαίνοντά τι, οἷς ἀντὶ τῶν ὀνομάτων καὶ τοῦ λόγου χρῆται πρὸς αὐτόν. (5) εἰ γὰρ ἦν τι ταῦτα σημαίνοντα, ἦν ἂν τοῖς σημαινομένοις ὑπ' αὐτῶν καὶ ὀνόματα τίθεσθαι, καὶ οὕτως ἂν ἦν καὶ ἡ φωνὴ σημαίνουσα. (6) οὕτω δὲ αὐτὸς ἂν ὁμολογοίη μὴ χρῆσθαι λόγῳ· ὁ γὰρ λόγος φωνὴ σημαντική. (7) ὁμοίως δὲ ἀναιρεῖ λόγον κἂν λέγῃ μηδὲν μᾶλλον σημαίνειν αὐτὸν ἢ μὴ σημαίνειν. εἰ οὖν μὴ σημαίνοι ὅταν λέγοι τι, οὔτ' ἂν διαλέγεσθαι εἴη· πάλιν γὰρ οὗτος ὅμοιος γίνεται φυτῷ.

According to his explanation, the question asked by the dialectical questioner is whether the respondent signifies anything, to himself and to another, by the words he speaks and the names he gives to things. This is further paraphrased as asking whether there is something to which he applies the name and which he wants to signify when speaking and producing the name. We can now see that this question is exactly the same as the one that Alexander teases out of the incorrect MS reading, so there is no reason to think that he is somehow swayed by the wrong text or otherwise biased.[29] If the respondent answers this question by "no, I don't signify anything at all with what I say" or "I signify no more this than that", he will destroy the *logos*.

In his summary of the elenctic demonstration, Alexander gives us two different versions of the argument:

> (T5) Alexander in *Metaph.* 274, 13–20
>
> **(1)** [First version] But if he grants and agrees that he signifies something by what he says, it will be possible for us to assume this and carry out a refutation of his proposition. He uses "demonstration" for "refutation". **(a)** For the one who grants this is by that very fact treating something as definite. **(b)** For he grants that what is signified by each speech is something definite. **(c)** For what is no more this than that is indefinite. **(2)** [Second version] **(a)** Or the statement "there will, then, be something definite" [1006a24–25] is this very thing that is granted, that the respondent signifies something. **(b)** This being granted, then, "there will be demonstration" [1006a24], that is, refutation, of which refutation the cause will be not the refuter but the one being refuted, for he will be refuted on the basis of what he grants.[30]

Both versions presuppose that the respondent grants that he "signifies something" with his speech. What is different is the explanation of how the *elenchos* is achieved in the two cases. According to **(T5.1)**, the respondent grants that he signifies something by what he says. It is then possible for the questioner ("us" in the sentence) to proceed with the refutation from this assumption. "We" will establish that the respondent by granting us that he signifies something treats this something as definite. At **(T5.1b)** the proof moves from "granting that he signifies some-

---

[29] The only difference between the two readings, on his presentation, is that the incorrect reading without οὐ formulates the question about meaning as a dilemma: something that the correct reading purports to be avoiding.

[30] (1) ἂν δὲ διδῷ καὶ συγχωρῇ σημαίνειν δι' ὧν λέγει, τοῦτο λαβόντας δυνατὸν ἔσται ἔλεγχον τοῦ προκειμένου ποιήσασθαι· ἀντὶ γὰρ τοῦ ἐλέγχου τῇ ἀποδείξει ἐχρήσατο. (a) ὁ γὰρ τοῦτο δοὺς ἤδη τι ὁρίζει (b) δίδωσι γὰρ ὡρισμένον τι εἶναι τὸ ὑφ' ἑκάστου λόγου σημαινόμενον (c) ἀόριστον γὰρ τὸ οὐδὲν μᾶλλον τοῦτο ἢ τοῦτο. (2) (a) ἢ τὸ ἔσται δή τι ὡρισμένον αὐτὸ τοῦτο δεδομένον τὸ σημαίνειν τι (b) τούτου δὴ δοθέντος ἔσται ἀπόδειξις τουτέστιν ἔλεγχος, οὗ ἐλέγχου αἴτιος οὐχ ὁ ἐλέγχων ἀλλ' ὁ ἐλεγχόμενος· ἐκ γὰρ ὧν δίδωσιν ἐλέγχεται.

thing by what he says" and that thing being definite to a universal claim that every *logos* signifies something that is definite. It is not clear whether we get an assent of the respondent to this move: this could be a part of this interpretation, or this generalisation could be derived by the questioner as a consequence of the single assumption granted at the beginning. The final step **(T5.1.c)** "but what is no more this than that is indefinite" entails that the denial of the claim that any x is no more this than that must now be given up by the respondent because it does not satisfy the condition on every *logos* that follows from the assumption he has earlier granted. This seems to be the rough idea of the first explanation. Alexander may be not totally happy with it, because it is not clear whether it is dialectically effective: what if the respondent only grants the signification claim but denies a more general claim and any further derivations?

On the second version **(T5.2)**, the very thing that is granted, namely that the respondent signifies something, is understood to be "something definite" mentioned by Aristotle.[31] Once this response has been obtained, there will be a "demonstration, i.e., a refutation" of the denial of the PNC, for which the respondent and not the questioner will be responsible.

Alexander seems to make use of both interpretations in his commentary. The first interpretation based on the idea that the respondent must signify something definite is at the foundation of Alexander's interpretation of the argument from signification, as we shall see in § 3. It aims to give universal, or "global", refutation of PNC-denial. This interpretation has its difficulties. We already mentioned the need for justification of the generalised premiss (T5.1.b): it may be dependent on the argument from signification, but in this case the elenctic refutation as a whole will depend on this argument, which has not yet been fully explained at this point. Alexander may think that taken in a narrow exegetical sense, this understanding does not do justice to Aristotle's idea in this text.

The second interpretation, on the other hand, seems to be in better agreement with Aristotle's claim that the respondent, the denier of PNC, is responsible for this proof. While on the first version, the conclusion that something definite has been granted is *derived* from the initial concession by the respondent that he signifies something and then considered in a generalised form that is applicable to any similar objection, on the second version this conclusion that there will be something definite is seen as a part of what is being granted, so that the "demonstration" is taken to be produced by the respondent himself rather than derived by the ques-

---

31 Unless this is a paraphrase, we seem to have a different reading at 1006a24–25: where the MSS have ἤδη γὰρ τι ἔσται ὡρισμένον, Alexander reports ἔσται δή τι ὡρισμένον (274, 18).

tioner from his concession about signifying.[32] This argument lacks a generalised premiss which the first version has, so this refutation will have a "local" rather than "global" scope. But its merit is that it raises no questions about the provenance of the generalised premiss, and it agrees better with the terms of the discussion described by Aristotle so far. The fact that in Alexander's interpretation this version is given as the last one may indicate that Alexander himself prefers it to the first one.[33]

> (T6) Alexander *in Metaph.* 274, 20–32
>
> **(1)** And the one responsible for the refutation will be the one refuted, not the one refuting; for he is refuted out of what he himself grants. **(2)** For if [the one refuting] tried to argue from himself[34] that to be the case and not to be the case cannot be both true of the same thing at the same time, and assumed that speech signifies something and treated this as definite, he would appear to beg the question; **(3)** but since it is the one who supports that thesis who is forced to agree to this position, he is responsible for the refutation, because he does away with speech. **(4)** For one who says that everything is not more so than not so (this was to do away with [the notion of] contradiction) does away with speech. But in doing away with speech he makes use of speech. This is [the meaning of] "he abides by the speech" [1006a27]. **(5)** Aristotle proves by what he says that **(i)** one who says that speech signifies nothing says that speech has been done away with. **(ii)** Or if there is speech, there is also something posited as definite, and it is not true in every case that nothing is this rather than that. **(6)** And he was wary of appearing to be producing a demonstration in a strict sense because he set it out as indemonstrable and the principle of all the axioms.[35]

Alexander here explains how the refutation depends on the respondent. Alexander takes the structure of the refutation to be summarized in Aristotle's statement

---

32 Alexander appears to be drawing this distinction between the two arguments at 276, 3–6 (T10.1 below, discussed in section 3.2 below, pp. 305–307).
33 It may also indicate (although we have no evidence for this) that Alexander finds the first version in an earlier commentary which he is using while writing his own (e.g. by Aspasius, or his teacher Aristotle, sometimes identified as Aristotle of Mytilene, or some other earlier commentator on the *Metaphysics*)
34 "From himself" (ἀφ' ἑαυτοῦ): see n. 17 above.
35 **(1)** οὗ ἐλέγχου αἴτιος οὐχ ὁ ἐλέγχων ἀλλ' ὁ ἐλεγχόμενος· ἐκ γὰρ ὧν δίδωσιν ἐλέγχεται. **(2)** εἰ μὲν γὰρ ἀφ' αὑτοῦ δεικνὺς τὸ μὴ δύνασθαι ἐπὶ τοῦ αὐτοῦ ἅμα τὸ εἶναί τε καὶ μὴ εἶναι ἀληθεύεσθαι ἐλάμβανε τὸ τὸν λόγον σημαίνειν τι καὶ ὥριζε τοῦτο, ἐδόκει ἂν τὸ ἐν ἀρχῇ λαμβάνειν· **(3)** ἐπεὶ δὲ ὁ τῇ θέσει παριστάμενος ἐκείνῃ αὐτός ἐστιν ὁ ἀναγκασθεὶς συγχωρῆσαι τοῦτο, ἐκεῖνος τοῦ ἐλέγχου αἴτιος, ἀναιρῶν λόγον. **(4)** ὁ γὰρ λέγων πᾶν οὐδὲν μᾶλλον οὕτως οὐχ οὕτως ἔχειν (τοῦτο γὰρ ἦν τὸ ἀναιρεῖν τὴν ἀντίφασιν) ἀναιρεῖ λόγον. ἀναιρῶν δὲ λόγον χρῆται λόγῳ· τοῦτο γὰρ τὸ ὑπομένει λόγον. ἔδειξε δὲ δι' ὧν εἶπεν ὅτι ἀνῃρῆσθαι λέγει λόγον ὁ μηδὲν σημαίνειν αὐτὸν λέγων. **(5)** ἢ εἰ ἔστι λόγος, ἔστι τι καὶ ὡρισμένον τιθέμενον, καὶ οὐκ ἐπὶ παντὸς ἀληθὲς τὸ οὐδὲν μᾶλλον τοῦτο ἢ τοῦτο. **(6)** ἐφυλάξατο δὲ τὸ δοκεῖν κυρίως αὐτοῦ ἀπόδειξιν πεποιῆσθαι, ὅτι ἔκειτο αὐτὸ εἶναι ἀναπόδεικτόν τε καὶ ἁπάντων τῶν ἀξιωμάτων ἀρχή.

at 1006a26: 'for even doing away with the speech he abides by the speech' (ἀναιρῶν γὰρ λόγον ὑπομένει λόγον).³⁶ **(T6.2)** may be directed against the proof at **(T5.1)**, where what is granted by the respondent is clearly separated from the demonstration conducted by "us" on behalf of the questioner. It is clear from these texts and from what follows that Alexander has no quarrel with the nature of the moves made by "us" in that demonstration. However, because it is not clear how the acceptance of these further moves is secured in our elenchos, he probably sees these details as obscuring the general form of elenctic argument at this stage. What is important at this point is to show that the PNC-denier himself accepts the key premiss that leads to the refutation, on pain of being excluded from the rational conversation, rather than granting a neutral point from which a refutation can be derived by his questioner defending the PNC.

In **(T6.4)**, Alexander explains why denying the PNC puts one outside rational discourse. The PNC-denier holds that everything is no more thus than not thus, no more F than ~F, where F is any predicate. This does away with contradiction, not by reducing it to a further contradiction, but by denying its very concept, so that there is no work for the PNC to do. This position lacks consistency in a way more fundamental than could be acceptable for a respondent with inconsistent opinions: in this latter case the discovery of inconsistency by registering a contradiction does not need to annul the respondent's participation in the dialogue at least up to this point, whereas the PNC-denier who admits to not signifying anything locks himself out of rational discourse. Yet somebody who in this way destroys meaningful speech must still use meaningful speech to do so.

In **(T6.5)** Alexander summarizes Aristotle's refutation in the form of a dilemma which faces the PNC-denier: "Accept that you are not making any statements and take your exit, or agree that you signify something with your statements, and then you've lost". The first horn **(T6.5i)** corresponds to the "exit" option: if the respondent does not signify anything, he confirms that he makes no statement, and the discussion is over. The second horn **(T6.5ii)** corresponds to the "staying" option: the denier agrees that he signifies something and stays in the discussion, sticking to his denial of PNC, only to be forced to give up on it.³⁷ The following pas-

---

36 The expression translated here as 'abides by the speech' (ὑπομένει λόγον) is rare and occurs in rhetorical and polemical contexts in the meaning of countenancing or tolerating somebody's speech (e.g, Dionysius of Halicarnassus, *Antiquit. Rom.* VIII. 71.1.3, X.40.2.1, X.41,2,3) Alexander seems to take *logos* here as speech or account in general rather than the adversarial speech of the opponent more specifically (so too Asclepius *in Metaph.* 260, 33–34)

37 This passage has been analyzed differently by Madigan, who saw (i) and (ii) as two different interpretations of the refutation (Madigan 1993, p. 56 and n. 341), and Flannery, who took (T6.5ii)

sage confirms that this is Alexander's understanding of the overall structure of the argument and gives us a new version which more fully illustrates how the respondent, the PNC-denier, is in charge of his own refutation:

> **(T7) (1)** This person is responsible for his own refutation: not the one asking questions (if he who assumes this had assumed it from himself,[38] he would have appeared, as Aristotle said [1006a16–17, 20–21] to beg the question), but rather the respondent, who is forced either to do away with speech or to give answers that conflict with his own thesis. **(2)** And he adds the explanation. *"While, then, he completely does away with speech"* by his thesis[39] ((i) For one who says that in every case nothing is any more this than the opposite of this does away with speech. (ii) For he will say, as he says about other things, that speech too is no more than it is not. (iii) But this is [the stance] of one who does away with speech, because speech is significant vocal sound, and one who says that speech no more signifies than it does not signify would also do away with speech; (iv) but this is what one who says that in every case a contradictory pair can be true at the same time would say.) **(3)** *While he indeed does away with speech by his thesis, he agrees to it by his answers*; for when he says that speech signifies, he posits that there is speech, positing by his answer what he destroyed by his thesis (275, 8–20).[40]

In (T7.2 (i)–(iv)), we have the position of the PNC-denier presented as a reasoning concluding that speech is neither signifying nor not signifying, the claim that does away with speech (T7.2iii). So, the step whereby the respondent admits to signifying something supports rational discourse and destroys his own reasoning that nullified it.

---

to be a third version of the argument (Flannery 2003, p. 129). Neither construal seems to me to be borne out by the text.

38 As before: i.e. as his own premiss (see nn.17, 34 above)

39 Alexander paraphrases and explains Aristotle's claim at 1006a26: ἀναιρῶν γὰρ λόγον ὑπομένει λόγον.

40 T7 (1) οὗτος δὲ τοῦ ἐλέγχου αἴτιος, οὐχ ὁ ἐρωτῶν (εἰ γὰρ αὐτὸς ἀφ' ἑαυτοῦ ἔλαβε τοῦτο ὁ λαμβάνων τοῦτο, ὡς εἶπεν, ἔδοξεν αὐτὸ τὸ ἐν ἀρχῇ λαμβάνειν), ἀλλ' ὁ ἀποκρινόμενος, ἀναγκαζόμενος ἢ ἀναιρεῖν λόγον ἢ μαχόμενα ἀποκριθῆναι τῇ θέσει τῇ αὐτοῦ. (2) καὶ προστίθησι τὴν αἰτίαν. ἀναιρῶν γὰρ ὅλως λόγον διὰ τῆς θέσεως ((i) ὁ γὰρ λέγων ἐπὶ παντὸς οὐδὲν μᾶλλον τοῦτο ἢ τὸ ἀντικείμενον αὐτοῦ λόγον ἀναιρεῖ (ii) ἐρεῖ γὰρ ὁμοίως τοῖς ἄλλοις καὶ τὸν λόγον μηδὲν μᾶλλον εἶναι ἢ μὴ εἶναι (iii) τοῦτο δὲ ἀναιροῦντος λόγον, ὅτι ὁ μὲν λόγος φωνὴ σημαντική, ὁ δὲ μηδὲν μᾶλλον σημαίνειν ἢ μὴ σημαίνειν λέγων αὐτὸν ἀναιροῖ ἂν καὶ οὗτος λόγον·(iv) λέγοι δ' ἂν τοῦτο ὁ λέγων ἅμα τὴν ἀντίφασιν ἀληθεύειν ἐπὶ παντὸς δύνασθαι) —(3) ἀναιρῶν δὴ λόγον διὰ τῆς θέσεως, δι' ὧν ἀποκρίνεται συγχωρεῖ· λέγων γὰρ τὸν λόγον σημαίνειν, τίθησιν εἶναι λόγον, ὃ διὰ τῆς θέσεως ἀνῄρει, τοῦτο διὰ τοῦ ἀποκρίνεσθαι τιθείς.

3 Argument from Signification

This section of Chapter 4 (1006a28–1007b18) is devoted to the explanation of the way in which signification depends on the PNC. This has to do with the notion of signifying and Aristotle explains the way it works using many examples.

3.1 Signifying Something: An Instance of Elenctic Reasoning?

Aristotle begins by explaining what he means by "signifying something" and why signifying nothing makes any discussion impossible.

> **(T8) (1)** First of all, it is clear that this itself is true, that the name signifies "being an F" or "not being an F", so that it is not the case that everything is so-and-so and not so-and-so. **(2)** Further, if "man" signifies one thing, let that be "biped animal". What I mean by "signifying one [thing]" is this: if that thing is a man, then if anything is a man, that thing will be to be a man. **(3)** But it makes no difference even if someone were to assert that it signified more than one thing, provided that these were definite; for a different name could be assigned to each formula: I mean, for instance, if one did not say that "man" signifies one, but many, of which one would have one formula "biped animal", and there were several others, but limited in number, he could assign a peculiar name for each formula. **(4)** But if, instead of so assigning, he were to assert that it signified infinitely many things, it is obvious that there would be no speech. **(5)** For not to signify one thing is to signify nothing, and if names do not signify, discussion with others is eliminated; and, in truth, even with oneself, since it is not possible even to think of anything for someone who is not thinking on one thing, and if it is possible, one name could be assigned to the object [of thought] (1006a28–b11).[41]

---

41 (1) πρῶτον μὲν οὖν δῆλον ὡς τοῦτό γ' αὐτὸ ἀληθές, ὅτι σημαίνει τὸ ὄνομα τὸ εἶναι ἢ μὴ εἶναι τοδί, ὥστ' οὐκ ἂν πᾶν οὕτως καὶ οὐχ οὕτως ἔχοι· (2) ἔτι εἰ τὸ ἄνθρωπος σημαίνει ἕν, ἔστω τοῦτο τὸ ζῷον δίπουν. λέγω δὲ τὸ ἓν σημαίνειν τοῦτο· εἰ τοῦτ' ἔστιν ἄνθρωπος, ἂν ᾖ τι ἄνθρωπος, τοῦτ' ἔσται τὸ ἀνθρώπῳ εἶναι (3) (διαφέρει δ' οὐθὲν οὐδ' εἰ πλείω τις φαίη σημαίνειν μόνον δὲ ὡρισμένα, τεθείη γὰρ ἂν ἐφ' ἑκάστῳ λόγῳ ἕτερον ὄνομα· λέγω δ' οἷον, εἰ μὴ φαίη τὸ ἄνθρωπος ἓν σημαίνειν, πολλὰ δέ, ὧν ἑνὸς μὲν εἷς λόγος τὸ ζῷον δίπουν, εἶεν δὲ καὶ ἕτεροι πλείους, ὡρισμένοι δὲ τὸν ἀριθμόν· τεθείη γὰρ ἂν ἴδιον ὄνομα καθ' ἕκαστον τὸν λόγον. (4) εἰ δὲ μὴ [τεθείη], ἀλλ' ἄπειρα σημαίνειν φαίη, φανερὸν ὅτι οὐκ ἂν εἴη λόγος·(5) τὸ γὰρ μὴ ἓν σημαίνειν οὐθὲν σημαίνειν ἐστίν, μὴ σημαινόντων δὲ τῶν ὀνομάτων ἀνῄρηται τὸ διαλέγεσθαι πρὸς ἀλλήλους, κατὰ δὲ τὴν ἀλήθειαν καὶ πρὸς αὐτόν· οὐθὲν γὰρ ἐνδέχεται νοεῖν μὴ νοοῦντα ἕν, εἰ δ' ἐνδέχεται, τεθείη ἂν ὄνομα τούτῳ τῷ πράγματι ἕν).

Aristotle's first sentence **(T8.1)** is difficult: both what is taken to be true and the inference from it allow for several different interpretations.[42] As a further difficulty, Alexander's commentary here reports a different text: where the received text of Aristotle has "or" (ἤ) Alexander's lemma at 1006a28 has "and" (καί).[43] Whether or not this different reading influenced his interpretation of the passage is hard to say, but his interpretation differs from all others that have been offered.[44]

> **(T9) (1)** First, he says, given that things said signify something, it is true that the one who says that something "is" signifies something, and likewise the one who says that something "is not", but not that he signifies and does not signify. **(2)** And he referred to "being" and "not being" as the names. **(3)** If it is true of them that they signify and not true that they do not signify, then it is no longer the case that both the affirmation and the negation are true of everything. **(4)** For since "not signifying" is the negation of "signifying", it is not true of that which has been conceded to signify something. **(5)** For even if it does not signify, signifying would not be conceded, and so there would not be any speech. **(6)** And when he says "first of all, it is clear that this itself is true, that the name "being" signifies", having said this, he has added "a this", which indicates that the signifying [expression] signifies some definite nature. **(7)** For someone who signifies something for himself or for another always signifies a this or a that; namely, he signifies some of the things that are and something that differs from other things that are not signified by the same [expression]. For it is not the case that all [expressions] signify the same [things], nor are all things signified by one expression (275, 23–31).[45]

Alexander seems to take Aristotle's argument in **(T8.1)** as a concluding round of the elenctic demonstration. In what has been seen as a problematic move, he

---

[42] Kirwan, in his commentary (ad loc.), lists three interpretations different from Alexander's: (1) 'The name chosen signifies e.g. "(to be) man" or "not (to be) man" but never both ['or' exclusive]; so it is impossible to be both man and not man'....(2) 'One who says that x is e.g. a man signifies that x is, or is not, something in particular; so it is not everything whatever (whatever you like)'...(3) 'The name chosen signifies being or not being something; and that is the starting point from which we proceed to prove PNC'. (Kirwan 1971, p. 93)

[43] πρῶτον μὲν οὖν δῆλον ὡς τοῦτο αὐτὸ ἀληθές, ὅτι σημαίνει τὸ ὄνομα τὸ εἶναι καὶ μὴ εἶναι [τοδί] (275, 21–22). This reading does not seem to be attested in other sources; see Kotwick (2016), p. 290.

[44] Kirwan attributes two different interpretations to Alexander: '(4) 'The name chosen signifies to be or not to be something and does not also not signify that; so at least one predicate, "signify" does not share its contradictory with any of its subjects'. (5) 'the name "to be" or "not to be" signifies this particular thing' (Kirwan 1971, p.93) He considers both to be unsuccessful.

[45] Τὸ πρῶτον, φησί, δοθέντος τοῦ τὰ λεγόμενα σημαίνειν τι, ἀληθές ἐστι τὸ σημαίνειν τι τὸν εἶναί τι λέγοντα καὶ τὸν μὴ εἶναί τι ὁμοίως, ἀλλ' οὐχὶ καὶ σημαίνειν καὶ μὴ σημαίνειν. ὀνόματα δὲ εἶπε καὶ τὸ εἶναι καὶ τὸ μὴ εἶναι. Εἰ δ'ἀληθὲς ἐπ'αὐτῶν τὸ σημαίνειν, οὐκ ἀληθὲς δὲ τὸ μὴ σημαίνειν, οὐκέτι ἂν ἐπὶ παντὸς καὶ ἡ κατάφασις καὶ ἡ ἀπόφασις ἀληθὴς ἂν εἴη· τοῦ γὰρ σημαίνειν τὸ μηδὲν σημαίνειν ἀπόφασις ὂν οὐκ ἔστιν ἀληθὲς ἐπὶ τῶν σημαίνειν τι συγχωρηθέντων. Εἰ γὰρ καὶ μὴ σημαίνοι, οὐκέτ' ἂν εἴη τὸ σημαίνειν συγχωρούμενον, οὕτω δὲ οὐκέτ' ἂν οὐδὲ ὁ λόγος εἴη.

takes τὸ εἶναι and τὸ μὴ εἶναι in Aristotle's text to be names **(T9.2)**.[46] Alexander may be relying on *De interpetatione* 3, 16b19–22, where Aristotle says that verbs pronounced by themselves are names, so there is nothing to forbid μὴ εἶναι to be a naming and signifying expression.[47] It is perhaps important to see that Aristotle in the *De int.* 3 passage says that verbs spoken by themselves are names because they 'signify something' (σημαίνει τι) insofar as 'the speaker arrests his thought and the hearer pauses',[48] but they do not yet signify whether something is the case or not.[49] This latter clause defines the force of his claim that follows, that taken separately, "being" or "not being" do not signify "the thing" (τὸ πρᾶγμα). "Signifying the thing" involves signifying whether or not something is the case, whereas "bare" substantivated infinitives or participial constructions cannot do this latter job.[50] Aristotle here does not retract what he said in lines 16b19–21 about the verbs spoken by themselves being names. In Ammonius' commentary, we find the explanation of this point according to which the 'name' here stands for a signifying expression (54, 24 Busse) which may be going back to Alexander's lost commentary.[51]

Alexander explains in **(T9.3)** that each of these names does signify, and therefore the affirmative statement which says that "being" signifies is true, as is the similar affirmative statement about "not being". Both the negative statements which deny signifying to each of these two names, respectively, are false. Thus **(T9.4)** for each of these names we have a contradictory pair of affirmation and negation such that when one of the pair is true, the other is false. This would provide us with a position that differs from the total denial of PNC. We have to understand that these truth values underlying this position have also been granted by the opponent of the PNC. Avoiding the *petitio principii*, Alexander points out that, on the other hand, the case when the affirmation of signifying of either of these two names is taken to be false **(T9.5)** and the respective negation true will amount to the respondent's refusal to grant signifying in the elenctic demonstra-

---

[46] Kirwan has this worry about Alexander's reading, pointing out a possible inconsistency with *De int.* 16b22–25, and he seems to be seconded by Whitaker, who does not name Alexander (Whitaker 1996, p.190, n. 11).
[47] See further and n. 51 below.
[48] 16b20–21: ἴστησι γὰρ ὁ λέγων τὴν διάνοιαν, καὶ ὁ ἀκούσας ἠρέμησεν (Ackrill trans.)
[49] 16b21–22: ἀλλ' εἰ ἔστιν ἢ μή, οὔπω σημαίνει.
[50] οὐ γὰρ τὸ εἶναι ἢ μὴ εἶναι σημεῖόν ἐστι τοῦ πράγματος, οὐδ' ἐὰν τὸ ὂν εἴπῃς ψιλόν (1006a22–23)
[51] Alexander reiterates this point shortly in our text arguing that all the names according to Aristotle are by convention: τό τε γὰρ εἶναι καὶ τὸ μὴ εἶναι ὀνόματα αὐτοῦ. οὐδὲν γὰρ κωλύει καὶ τοῦτο γίνεσθαι ὄνομα, ὄντων κατὰ συνθήκην τῶν ὀνομάτων (280, 30–31). For Alexander's view on the nature of names, see also his *Quaest.* 3.11.

tion, thus closing the rational discussion in line with Aristotle's dialectical framework discussed in the previous section.

So, this argument is taken by Alexander as the application of the argument "from signifying" to the case of our denier of the PNC. Perhaps it has to be read as supporting his interpretation of the nature of elenctic refutation in **(T5.2)** above.

Modern interpretations of this passage differ on how to take the expressions τὸ (μὴ) εἶναι τοδί in **(T8.1)**. Anscombe translates τὸ εἶναι τοδί or τὸ μὴ εἶναι τοδί, "to be (or not to be) a this" and argues that the expression τὸ εἶναι τοδί means the formula of essence, taking it to be in the category of substance.[52] Whitaker more recently has argued that τοδί need not be a substance and stand for a particular (τόδε τι), as well as quality, quantity, etc.[53] On his reading, the requirement of definiteness does not seem to entail any essentialist interpretation.

Now, despite an altogether different construal of the text, Alexander's interpretation of this passage seems to have something in common with both these approaches, but as I will try to show below it is distinct. Alexander in **(T9.6)** does read its account of signifying along essentialist lines, agreeing in this with Anscombe. He says that the names "being" and "not being" signify "a this", which in turn refers to *some definite nature* that differs from all other things. The meaning of this "definite nature" is not specified any further, and moreover, Alexander does not tell us what the nature signified by "being" and "not-being" is on his reading of the argument, postponing this explanation till the next step in Aristotle's argument. We will see in § 3.3 that his essentialist interpretation of the expression does not entail a substantialist interpretation of the category of τοδί: on this, he agrees with Whitaker.

3.2 Signifying One Thing vs. Many

In **(T8.2)**, Aristotle introduces a further requirement for a signifying expression: it should signify "one thing" (ἕν). In order to explain what he means by "signifying one thing" Aristotle introduces a stipulative definitional formula of a thing being signified in this way. When our respondent agrees that he signifies something by "man", for instance, he has to accept that the name "man" in all its occurrences stands for such a formula, e.g., "animal biped". In the case where a name is used homonymously **(T8.3)**, a different formula and a corresponding name can

---

52 Anscombe (1961), p. 39.
53 Whitaker (1996), p. 191.

be assigned to each different use to avoid saying that such-and-such a name signifies indefinitely many things. In this latter case, i.e., when a name signifies indefinitely many things **(T8.4)** there will be no rational discourse, and even thought itself will be eliminated **(T8.5)**.

The exact force of this unity requirement has been a matter of controversy, and the example of "man" (ἄνθρωπος) does not make things clearer: are we to understand that this requirement can be met only by substances?[54] In order to see what Alexander's position is, it will be good to study his reading of Aristotle in some more detail.

Alexander provides us with his take on Aristotle's argument in the following summary. It is worth citing in full despite its length because Alexander himself introduces it as his synopsis of Aristotle's argument from signification.

(T10) Alexander in *Metaph.* 276, 3–28 (ad 1006a31–b11)

(1) Having assumed that the names signify and each of them signifies something definite, and having proved on this basis that someone who has granted this will no longer be able to state that this is no more than its opposite, **(2)** now using this he proves that the contradictory pair (ἀντίφασις) can in no case be jointly true (συναληθεύειν). **(3)** But he states and proves what he set out to prove by means of many arguments. The reason for this is that he takes instead those [premises] through which he proves what he wants to prove and establishes *them* as the ones that must be assumed; this is why the text (λέξις) is less clear. **(4)** But for someone who has grasped it, the sense of the reasoning is as follows. **(5)** If each name signifies, it signifies some one thing. For that which signifies, signifies something, and what is something is one. For even if homonyms are said of several things, one who uses them and signifies something by them is not signifying all these things at the same time. **(6) (i)** The names that are different and do not signify the same thing as the name that was posited as signifying one thing clearly will not be said of that thing. **(ii)** For otherwise that [thing] would no longer have *the one nature*, if the names that are signifying different things were true of it. **(7) (a)** "Man" and "not-man" are different names, and not the names *of one thing*, **(b)** for they will be neither predicated of the same (ἐπὶ ταὐτοῦ), **(c)** nor will that which is signified by "not-man" be true of that which is signified by "man", **(d)** given that the names themselves differ from one another, and the different names express different [things], and *each name expresses one* [thing], *i.e., one nature* **(e)** For that which is signified by "man" will no longer be one, as has been said, if "not-man", which signifies another object, were also true of it. (The same argument as applies in the case of "man" and "not-man" applies in all other cases as well.) **(8)** But if the things revealed by these names do not belong together, then neither

---

54 Some commentators took it this way raising worries about a potential *petitio principii* committed by a questioner on this scenario and about potential confusion of sense and reference by Aristotle himself. This is an influential reading defended by Anscombe (1961). Other scholars have resisted this approach arguing for the unrestricted scope of signifying one thing in Aristotle's argument. More on this point below in § 3.3.

would the affirmation saying that so-and-so is a man have as simultaneously true together the negation saying that so-and-so is not a man.⁵⁵

Alexander first **(T10.1)** outlines the general structure of Aristotle's argument.⁵⁶ As already pointed out in the previous section,⁵⁷ he seems to take Aristotle's explanation at **(T8.1)** as an instance of elenctic demonstration against someone who denies the PNC but concedes to signifying something with his speech. The denial of this concession will exclude the respondent from rational discourse. So, on pain of that, the questioner can elicit from the respondent a concession of signifying and the rejection of non-signifying, which together can give us a position different from the denial of PNC. As Alexander is aware, this position is forced upon the opponent by the elenctic demonstration and it does not amount to establishing the PNC. It is not even an instance of PNC. The logical form of the outcome of this elenchos would be:

Sa & ~~Sa,

---

55 (1) Λαβὼν ὅτι σημαίνει τὰ ὀνόματα καὶ ἕκαστον αὐτῶν τοδί, καὶ δείξας δι' αὐτοῦ ὅτι μηκέτι δυνήσεται λέγειν ὁ τοῦτο συγχωρήσας τὸ οὐδὲν μᾶλλον τοῦτο ἢ τὸ ἀντικείμενον αὐτοῦ, (2) νῦν τούτῳ προσχρώμενος δείκνυσιν ὅτι ἐπὶ μηδενὸς οἷόν τε τὴν ἀντίφασιν συναληθεύειν. (3) διὰ πλειόνων δὲ λέγει τε καὶ δείκνυσι τὸ προκείμενον. αἴτιον δὲ τούτου ὅτι μεταλαμβάνει τὰ δι' ὧν δείκνυσιν ὃ βούλεται, καὶ κατασκευάζει αὐτὰ ὡς δεόντως λαμβανόμενα· διὸ καὶ ἀσαφεστέρα ἡ λέξις. (4) ἐκλαβόντι δὲ ὁ νοῦς τῆς ἐπιχειρήσεως τοιοῦτος. (5) εἰ ἕκαστον ὄνομα σημαίνει, καὶ ἕν τι σημαίνει· τὸ γὰρ σημαῖνον τὶ σημαίνει, τὸ δὲ τὶ καὶ ἕν· καὶ γὰρ εἰ κατὰ πλειόνων τὰ ὁμώνυμα, ἀλλ' οὐχ ἅμα γε πάντα σημαίνει ὁ χρώμενος αὐτοῖς καὶ σημαίνων τι δι' αὐτῶν. (6) (i) τὰ διαφέροντα δὴ ὀνόματα καὶ μὴ ταὐτὸν σημαίνοντα τῷ κειμένῳ ἓν σημαίνειν δῆλον ὡς οὐ ῥηθήσεται κατ' ἐκείνου· (ii) εἴη γὰρ ἂν οὐκέτι μία φύσις ἐκείνου, εἰ τὰ διαφερόντων πραγμάτων σημαντικὰ ὀνόματα ἀληθεύοιτο κατὰ τοῦ αὐτοῦ. (7) διαφέροντα δὲ ὀνόματα τό τε ἄνθρωπος καὶ τὸ οὐκ ἄνθρωπος, καὶ οὐκ ἔστιν ἑνὸς ὀνόματα πράγματος· οὐ γὰρ ἐπὶ ταὐτοῦ κατηγορηθήσεται, οὐδὲ ἔσται ἀληθευόμενον κατὰ τοῦ σημαινομένου ὑπὸ τοῦ ἀνθρώπου τὸ σημαινόμενον ὑπὸ τοῦ οὐκ ἀνθρώπου, εἴ γε διαφέρει τε αὐτὰ τὰ ὀνόματα ἀλλήλων, καὶ τὰ διαφέροντα διαφερόντων ἐστὶ δηλωτικά, καὶ ἑνὸς ἕκαστον καὶ φύσεως μιᾶς· οὐ γὰρ ἂν ἔτι εἴη τὸ ὑπ' ἀνθρώπου σημαινόμενον ἕν, ὡς ἐρρέθη, εἰ ἀληθεύοιτο ἐπ' αὐτοῦ καὶ τὸ οὐκ ἄνθρωπος, ἄλλου πράγματος ὂν σημαντικόν. ὁ δὲ αὐτὸς λόγος ὅς ἐπ' ἀνθρώπου καὶ οὐκ ἀνθρώπου, καὶ ἐπὶ τῶν ἄλλων πάντων. (8) εἰ δὲ μὴ τὰ ὑπὸ τῶν ὀνομάτων τούτων δηλούμενα συνυπάρχει, οὐκ ἂν τῇ καταφάσει τῇ λεγούσῃ εἶναι ἄνθρωπον τόδε, ἅμα συναληθεύοιτο ἡ ἀπόφασις ἡ λέγουσα μὴ εἶναι ἄνθρωπον τόδε. (276, 17–27)

56 This part is skipped by Mignucci in his discussion and therefore he does not sufficiently appreciate that Alexander distinguishes between the elenctic refutation in a dialectical exchange as set out by Aristotle in 1006a18–31 and the argument from signification in 1006a31–b34 (which may be seen as directed against the strong form of the denial of PNC). See Mignucci (2003), p. 112.

57 See pp. 302–303 above.

where a is a respondent and S stands for 'signifies something with his speech'. The resulting statement is not logically equivalent to PNC, even though it provides a sort of case against the denial of PNC by showing the statement that such a denial is not true as *conceded* by the opponent of PNC.

At the next step, according to Alexander, Aristotle wants to prove that in no case can both parts of a contradictory pair be true together **(T10.2)**. Aristotle conducts this next proof "making use" (προσχρώμενος) of what has been established in the elenctic refutation. Alexander tells us that this second proof is conducted διὰ πλειόνων, probably referring to a complex structure of the argument from signification. In the sentence that follows in **(T10.3)** he says that the reason for this more complex structure of the argument is that Aristotle additionally adopts, μεταλαμβάνειν, those premisses 'through which he will prove what he intends to prove' and establishes them as necessary assumptions, which gives an impression of unclarity. Μεταλαμβάνειν can also mean 'take instead' and in this meaning it may be a reflection on the use Aristotle makes of the elenctic demonstration, to which Alexander refers in the previous sentence **(T10.2)**. While in the elenctic demonstration we have the main premiss provided by the respondent, and the conclusion has the force within this very narrow, "local" scope, the task of the argument from signification is to show that it is possible to have this kind of proof for a general case, starting from the premiss that each name signifies something. This kind of argument requires a more detailed discussion of the nature of signifying. Establishing these new assumptions 'as necessary' means showing that they are true and that the conclusion follows from them. This involves further additional steps in the argument.

In **(T10.4)** Alexander signals the opening of a "synoptic" argument which will provide the proof of the PNC based on the agreed assumption about signification, in its universal version, i.e., where the impossibility of contradiction depends already not on one "forced" quasi-counterexample, but on some universal feature of signification. The overall schema of the argument seems to be as follows:

If each name signifies, it signifies some one thing (T10.5)

The names that are different and do not signify the same thing will not be said of that thing. (T10.6)

'Man' and 'not-man'—and all the other names of this form (F and not-F)—are different names, and not the names of one thing (T10.7)

But if the things revealed by these names do not belong together, then neither would the affirmation saying that so-and-so is a man (x is F) have as simultaneously true together the negation saying that so-and-so is not a man (x is not F) **(T10.8)**

Each of these claims is established by means of clarifying arguments. So the proof 'by means of several arguments' refers to both several inferences involved and several clarifications which establish the key concepts and show how inferences work.

In **(T10.5)** Alexander follows Aristotle as he explains what "signifying one thing" means by drawing a contrast between univocal and multivocal (homonymous) expressions. This contrast is helpful because it gives us an idea of what is ruled out by the requirement of the unity of the signified object. Homonyms share the same name but have different definitional formulae associated with these names, e.g., the name "bank" can signify a financial institution or a landscape feature (as in 'the left bank of the Forth') or a piece of garden furniture. This multivocity can be disambiguated if we assign a different name to each definitional formula (e.g., $bank_1$, $bank_2$, $bank_3$). If the same name, as a linguistic item, signifies more than one definitional formula, this will still not create a multivocity as long as the formulae are definite in number **(T8.3)** Aristotle in this whole argument seems to have in mind the opponents who will be looking for different methods of mitigating their concession of signifying one thing by making its scope more "inclusive", so as to include, ultimately, some possibility for deriving a contradiction. One of such moves towards a more "inclusive" treatment of the one is to say that each name signifies something one, but also many other things, as in the case of homonyms. Aristotle posits a limit to such an inclusion by demanding that the number of different things signified by one name be finite.

Alexander devotes a small digression to the question: "Why if each name were to signify more [things] that are definite, the rational discourse is not destroyed, even though the signified thing is not one, and if it were to signify unlimited [things], it is destroyed?" (278, 21–23). His discussion has a structure characteristic of some short treatises preserved in Alexander's school collections, *Mantissa* and *Quaestiones*, where we have the main question based on Aristotle's claim and different types of answer that could be given to this question.[58] Alexander gives five different arguments to explain and support Aristotle's position.[59]

> **(T10.5 A)** Either [1], first, everything definite and comprehensible is in a way one, so that even [a name] that signifies more things [than one], but a limited number, in a way signifies one thing. It is possible to circumscribe them, i.e., to separate them from the things that are not being signified, and to say that these are the things that are being signified, whereas with those that are unlimited this is impossible.

---

[58] This type of a question belongs to *vera problemata* in Bruns' classification, where we have the statement of the puzzle and the arguments for proposed solutions (see Bruns 1892, pp. V-VII and Sharples 1992, p. 4).

[59] Madigan notes arguments (i) and (ii) as two different explanations, but it its clear that the three short arguments that follow respond to the same question and all five form a single small treatise suitable as a basis for classroom discussion.

Or [2] [Aristotle says] "to signify more [things], but a definite number" because it is possible for all those things that are taken as signified to be distinguished from each other and made definite by distinctive names, in this way [the names] would be signifying something.[60] But if each of the names signified an unlimited number of things, none of the things signified by a name could have a distinctive name or sign. But in this way even the names given to them could not be signifying anything, as they could no more be predicated of the things to which they are given than of those to which they were not given.

Further, [3] if each of the names signifies an unlimited number of things, then each of the names would signify the same things. But if all names signify the same things, discussion and signification are destroyed. For no name signifies one thing if it is posited that all names signify the same things as each other. For to say that each name signifies all things is equivalent to saying that each name signifies nothing.

It is also possible for someone who proceeds methodically to show in the following way that the things expressed by each name are not unlimited in number, though the proof is more dialectical. [4] If "man" signifies an unlimited number of things, then either [i] "man" also signifies "not-man" (for this too is included among the unlimited number of things), so that one who says "man" would also have signified "not-man"; or, if this is absurd, [ii] there will be some thing over and above the infinite number of things; the addition of not-man to the infinite number of things that "man" signified makes the things even more numerous; and thus the infinite number will be less than some higher number.

Further, [5] either [i] the name "infinite" signifies non-infinite as well, and thus each name no more signifies an infinite number of things than it signifies a non-infinite number of things; or [ii] if it does not signify this as well, then, first, (a) it will not be possible [in that case] for the negation also to be true of that of which the affirmation is true; and further, (b) that which is signified by the name "non-infinite" will be outside the infinite number of things; the infinite number of things will be more numerous with the addition than they were by themselves. (278, 24–279, 14).

Alexander elucidates Aristotle's point that homonymy cannot be used by the opponents of PNC as a legitimate case to argue that a thing can be an F and not an F because a bank can be a financial institution and not a financial institution (insofar as we mean a riverbank). Aristotle resolves this problem explaining that any homonymy could be disambiguated and presented as a finite series of unambiguous (non-homonymous) names, each with its own signified object.

Alexander considers the question why such a series should be finite or limited. In the first argument the main reason he gives is that there should be a definite number of different signified things for each name, so that it would be possible to separate what is signified by this name from what is not signified by it. The con-

---

[60] The text at 278, 27–30, is uncertain. I read: ἢ τὸ 'πλείω σημαίνειν, ὡρισμένα δὲ' τῷ δύνασθαι πάντα τὰ σημαινόμενα ληφθέντα διακριθῆναί τε ἀπ'ἀλλήλων καὶ ἰδίοις ὀνόμασι ὁρισθῆναι, τούτῳ ἂν εἴη σημαίνοντά τι.

trast seems to be not with the case when we have an infinite number of signified objects, i.e., definitional formulae, corresponding to one ambiguous name.[61] The claim is not that even if this original ambiguity is resolved in a distinct way by each of these formulae, we cannot say the same with regard to the original ambiguous term on the whole. Rather the claim seems to be that in the case of infinite number of significata, it will be impossible to assign a distinct name to each of them, and the original term will be permanently ambiguous.[62]

This point seems to be spelled out more clearly in the second argument, where Alexander says that if each name signified an unlimited number of things, it would be impossible to have a distinct name for any of those things. The limit on the number of "things" signified by a name is imposed by the definitional formula whose negation serves as a definitional formula for those things that are not signified by this name.

What is being resisted in both this and the previous argument is an attempt to introduce the concept of signifying bypassing the question of how this signification is fixed. As soon as we have an Aristotelian answer to this question, namely, that it is fixed by means of a definitional formula, it becomes impossible for the name so defined to signify an infinite number of things. In the argument [3] Alexander points out that signifying an infinite number of things means that all names will signify the same things, and this will destroy the signification.

These three short arguments are distinguished by Alexander from the following two, which are described as "more dialectical": this striking description immediately precedes argument [4], but the nature of the proof in both arguments seems very similar and most likely the description is intended for both. While the arguments [1]-[3] are directly showing how taking the object of signification to be un-

---

[61] In this sense, Alexander agrees with Kirwan and Dancy that the unlimited in question is not an infinite series but lack of definition in the case of each signified object, i.e., each definitional formula. Neither Aristotle nor Alexander say much about what provides such a formula with unity and limitedness, but it is clear from counterexamples that a formula must answer stronger criteria than just any syntactically well-formed combination of names. See Kirwan (1993), p. 94, and Dancy (1975), pp. 83–87.

[62] The case where we could have an infinite number of finite well-formed definitional formulae does not seem to be considered by either Aristotle or Alexander. The reason may be once again that Aristotle's concern is to rule out the case where accepting an indefinite number of names as a potentially infinite series (in accordance with Aristotle's own definition of the infinite in Physics 3.6, as that of which some part is always beyond), leaves open a possibility that at one of the future assignments of meaning to an ambiguous term will bring about an ambiguous or, worse, contradictory definitional formula. The actually infinite multiplicity of good formulae is much less problematic in this respect (this agrees with the argument about Aristotle's actual infinities developed recently by Jacob Rosen in Rosen 2021).

limited leads to the impossibility of signifying and rational discourse, in the two last arguments, [4] and [5], the discourse is destroyed by the absurd consequences of the assumption that a name can signify things without a limit. In the argument [4], the first absurd consequence [i] is that "man" signifies the same as "not-man", and that is described as absurd. But of course, it is absurd for Alexander and his students, whereas for Aristotle's opponents this is exactly a desideratum as a position to take in the elenctic demonstration. So, this particular reduction to absurdity by Alexander cannot be very strong because it begs the question. The second consequence [ii] is based on a familiar kind of reduction to what is greater than infinity.

Argument [5] treats "infinite" as a name and shows that it either [i] does not properly signify "infinite" in that it also signifies the "finite" or [ii] if the opponent denies that it signifies the finite, this will (a) amount to his denial of his rejection of PNC and (b) lead to the thing "greater than the infinity", because if what is finite is outside the scope of what is infinite, the two together would make something that is greater than the infinite.

The names that are different and do not signify the same thing as the name that has been posited as signifying one thing will not be spoken of that thing **(T10.6)**: e.g., bank$_2$ will not be spoken of a financial institution, nor bank$_1$ of a piece of garden furniture. Alexander explains the reason for this: the nature of that thing would not be one, or single, if the names that were signifying different things were true of it. It is clear from our text that **(T10.6)** continues the discussion of the possible counterexample of homonyms started in **(T10.5)**.[63]

In order to understand these arguments, it is good to keep it in mind that by "things signified" Alexander, following Aristotle, means not the extensions of the predicates or formulae used to explain the signification but the objects as specified by stipulative definitional formulae. "Circumscribing" definite names is thus done by these definitive formulae—and also, as we can see from some of these arguments, by their negations, which too, in their turn, have the function of circumscribing and will thus "signify something one".

Two problems have been raised in connection with this view. The first was originally a question for Aristotle, which we can now forward to Alexander: are the things signified by names only substances? The second was raised recently by Mario Mignucci and had to do with the coherence of this argument and the account of signification that underlies it.

---

[63] Mario Mignucci in his reconstruction of Alexander's argument omits the discussion of homonyms and takes **(T10.6)** as a separate self-standing premiss (γ) which he considers to be false or at least in need of a supplement. But his reconstruction omits some crucial parts of the context. More on this in § 3.4. See Mignucci (2003), pp. 112–116, with a reconstruction of premisses on p. 112.

Mario Mignucci argued that the fact that two different names may have no single nature does not forbid us to predicate them of the same subject: e.g., the names "man" and "white" can be both truly predicated of Callias without any contradiction. The suggestion he makes is that Alexander should distinguish between the predicate which expresses the signified relatively to the subject ("identifying predication"), and the predicate which does not express the signified relatively to the subject ("non-identifying predication"). In the former case, the predicate will express the nature of the subject and in the latter case not: e.g., "man" expresses the nature of the subject (Callias) and "white" does not express the nature of the subject.[64] In this way, presumably, it will be possible to derive the PNC for both the identifying and non-identifying predication by reasoning about signification.

In order to see how this argument works, it will be useful to take into account a further distinction drawn by Aristotle, between "signifying one thing" and "signifying about one thing". Alexander presupposes it in this argument.

3.3 Signifying One Thing vs. Signifying About One Thing

For his conclusion in this argument, Aristotle still needs a further distinction between signifying *one thing* and signifying *about one thing*. This distinction will allow him to establish that what is signified by a name cannot be signified by its negation, a key step in deriving the PNC "from signification".

This is a complex argument, and before reading it with Alexander, it will be good to have a general outline of its structure. In doing so, I will not discuss in detail the problems of each step in order to keep the overall framework clear.

> **(T11) (1)** Let the name, then, as was said originally, signify something and signify one thing. **(2)** "To be man" cannot signify the same as "not to be man", if "man" signifies not just *about one thing*, but *signifies even that which is one*. For we do not consider that as signifying one thing, namely that which signifies about one thing, since in this way also the "musical", the "white", and the "man" would signify one thing, so that all will be one: for they would be synonyms. **(3)** And it ["to be man"] will not be to be and not to be the same thing, except homonymously, as if, for instance, what we were calling "man", others would be calling "not-man". But the question is not whether something can be the same and not the same in respect of name, but in respect of the thing. **(4) (a)** And if "man" and "not man" do not signify different things, **(b)** it is clear that there will be no difference between "not to be man" and "to be man", **(c)** so that "to be man" will be "not to be man"; **(d)** for they will be one. For being one means this: as in the case of garment and cloak, if their definition is one.**(e)** And if

---
64 Mignucci (2003), pp. 115–116.

["man" and "not man"] are one, **(f)** then "to be man" and "not to be man" will signify one thing. **(g)** But it has been shown that they signify different things. **(5) [Concluding argument] (a)** Therefore it is necessary, in order to make a true statement that something is a man, to say that it is a biped animal (for this was what "man" signifies). **(b)** But if this is necessary, then it cannot fail to be the biped animal (for "necessary" means "impossible not to"). **(c)** Therefore it is impossible to say that it is true that the same thing simultaneously is and is not a man (1006b11–b34).⁶⁵

In (T11.1) Aristotle restates the assumptions of his argument. (T11.2–4) provides us with the claim that "to be man" and "not to be man" signify different things (T11.2); the expression of the form "to be man" will not signify being and not being the same thing, except by homonymy (T11.3); (T11.4) proves that "man" and "not-man" signify different things.

The concluding argument (T11.5) is: (a) if it is true to say of something that it is a man, then it must be a biped animal (the signified of the name "man"); (b) if this is necessary, then it is impossible for it not to be a biped animal. (c) Then it is impossible to say that it is true that the same thing simultaneously is and is not a man. I'll consider the concluding argument (T11.5) in the next section. Now, let us concentrate on the distinction between the two types of signifying.

The distinction between signifying one thing and signifying about one thing is drawn as an auxiliary step in the proof of the claim that "to be man" cannot signify the same as "not to be man" (T11.2). Without this distinction, the names "musical", "white", and "man", when signifying *about one thing*, will all be *signifying one thing*.

The *one thing* these three different names are signifying *about* is taken to be an individual, e.g., Socrates, of whom all these three predicates are true. This is

---

65 (T11)(1) ἔστω δή, ὥσπερ ἐλέχθη κατ' ἀρχάς, σημαῖνόν τι τὸ ὄνομα καὶ σημαῖνον ἕν· (2) οὐ δὴ ἐνδέχεται τὸ ἀνθρώπῳ εἶναι σημαίνειν ὅπερ ἀνθρώπῳ μὴ εἶναι, εἰ τὸ ἄνθρωπος σημαίνει μὴ μόνον καθ' ἑνὸς ἀλλὰ καὶ ἕν (οὐ γὰρ τοῦτο ἀξιοῦμεν τὸ ἓν σημαίνειν, τὸ καθ' ἑνός, ἐπεὶ οὕτω γε κἂν τὸ μουσικὸν καὶ τὸ λευκὸν καὶ τὸ ἄνθρωπος ἓν ἐσήμαινεν, ὥστε ἓν ἅπαντα ἔσται· συνώνυμα γάρ). (3) καὶ οὐκ ἔσται εἶναι καὶ μὴ εἶναι τὸ αὐτὸ ἀλλ' ἢ καθ' ὁμωνυμίαν, ὥσπερ ἂν εἰ ὃν ἡμεῖς ἄνθρωπον καλοῦμεν, ἄλλοι μὴ ἄνθρωπον καλοῖεν· τὸ δ' ἀπορούμενον οὐ τοῦτό ἐστιν, εἰ ἐνδέχεται τὸ αὐτὸ ἅμα εἶναι καὶ μὴ εἶναι ἄνθρωπον τὸ ὄνομα, ἀλλὰ τὸ πρᾶγμα. (4)(a) εἰ δὲ μὴ σημαίνει ἕτερον τὸ ἄνθρωπος καὶ τὸ μὴ ἄνθρωπος, (b) δῆλον ὅτι καὶ τὸ μὴ εἶναι ἀνθρώπῳ τοῦ εἶναι ἀνθρώπῳ, (c) ὥστ' ἔσται τὸ ἀνθρώπῳ εἶναι μὴ ἀνθρώπῳ εἶναι· (d) ἓν γὰρ ἔσται. τοῦτο γὰρ σημαίνει τὸ εἶναι ἕν, τὸ ὡς λώπιον καὶ ἱμάτιον, εἰ ὁ λόγος εἷς· (e) εἰ δὲ ἔσται ἕν, (f) ἓν σημανεῖ τὸ ἀνθρώπῳ εἶναι καὶ μὴ ἀνθρώπῳ. (g) ἀλλ' ἐδέδεικτο ὅτι ἕτερον σημαίνει. (5) (a) ἀνάγκη τοίνυν, εἴ τί ἐστιν ἀληθὲς εἰπεῖν ὅτι ἄνθρωπος, ζῷον εἶναι δίπουν (τοῦτο γὰρ ἦν ὃ ἐσήμαινε τὸ ἄνθρωπος)· (b) εἰ δ' ἀνάγκη τοῦτο, οὐκ ἐνδέχεται μὴ εἶναι <τότε> τὸ αὐτὸ ζῷον δίπουν (τοῦτο γὰρ σημαίνει τὸ ἀνάγκη εἶναι, τὸ ἀδύνατον εἶναι μὴ εἶναι [ἄνθρωπον])· (c) οὐκ ἄρα ἐνδέχεται ἅμα ἀληθὲς εἶναι εἰπεῖν τὸ αὐτὸ ἄνθρωπον εἶναι καὶ μὴ εἶναι ἄνθρωπον.

suggested by Aristotle's own use of an example of an individual,[66] and Alexander understands this example in the same way.[67] This understanding of the subject of predication as an individual has led some scholars to identify an individual with an individual substance of the *Categories* and thus to an interpretation of the scope of the argument "from signification" as restricted to substances only. On such a view, of our three names, "man", "white", and "educated", only "man" signifies *one thing* when thus signifying *about one thing*, whereas the other two expressions signify something *about one thing*, but do not signify *one thing* in the sense required by Aristotle.

The argument to this effect, connecting this passage with Aristotle's discussion of substance in the *Categories* and *Metaphysics* Z, has been given by Elizabeth Anscombe.[68] In relation to our text, the reasoning behind such a reading would go as follows: "man" is the most important of all the names predicated of Socrates, because it expresses what Socrates is as an individual, and it must hold of Socrates as long as Socrates exists. The expressions "white" and "musical", although true of Socrates, do not express what he is and must not hold of Socrates always in the same way as "man", but sometimes their negations can also be true of Socrates, such as "not-white" and "not-musical". Even if we assume that "white" can signify as "something one", the individual white colour that is "in" Socrates as "in a subject", in the sense of the *Categories* 2, there exists an insoluble ambiguity about whether in "signifying one thing that is white" the "one thing" is to be understood as (a) a white colour present in Socrates as in a subject—if so, our signification would be *per se*, having as its object a *per se* existent which is not said of anything further (*Cat.* 2, 1a25–29), or (b) a thing which has white colour, as in the reading of our example mentioned above, e.g., "white man", where the signified of a name "man" does not possess an attribute "white", but can have it accidentally—this meaning of a white thing (τὸ λευκόν) will not satisfy the criteria for Aristotle's "one thing" because it belongs to an entity of which the predicate "not-white" can also truly hold.[69] This ambiguity cannot be clarified in the case of such predicates as "white" and "educated" because the presence of a subject in which the *per se* existents they signify (i.e., the individual properties of being white and educated "owned" by a given individual substance) must always be presupposed in their signification. Anscombe thought we could get support for this interpretation—or evidence of Aristotle's confusion on this point—by clarifying the scope of the modal

---

66 In the argument "from substances", 1007b5–15.
67 See T12 below.
68 Anscombe (1961); a similar view is expressed by Lear (1980), pp. 106–110 and Furth (1986), pp. 376–381.
69 Anscombe (1961), pp. 41–43.

operator of necessity in the concluding argument where Aristotle derived an instance of the PNC for "man" (1006b28–34). But she admitted that *prima facie* the concluding derivation permits a construal that accounts for the unrestricted scope of the whole argument.[70]

Several more recent works argue for the unrestricted scope of PNC in the argument from signification but do not address the role of the distinction between "signifying about one" and "signifying one" in their interpretations.[71]

What about Alexander? He sees the distinction between *signifying one* and *signifying about one thing* as implicit in Aristotle's account of signification which he uses in his "synoptic argument" (in **(T10)**). Explaining the concept of "signifying one thing", he says:

> **(T12) (1)** What the signified object is, he defines and makes known by means of showing that it is not the case that if something is said about something, this already signifies one thing, as he will also say further down. **(2)** For it is not the case that if "white" is said *about* Socrates, then Socrates and "to be Socrates" are also signified by the white. **(3)** As for all those things that are predicated *so as to signify nature and essence, as is the case with those predicated in "what it is"*, they are taken as signifying one thing, and in this way something one is signified by each of the names. **(4)** And he explained in what sense he said "signifying one thing" lest someone should think that he means one numerically, nor that a numerically one thing should be called one, as he will say, but *a certain one nature*. **(5)** For if "man" signifies a biped animal, then "to be man" will consist in being a biped animal (276, 37–277, 9).[72]

In (T12.1), the formula "something is said about something" from Aristotle's definition of premiss[73] refers to any well-formed statement, such as "Socrates is white" (T12.2) which is contrasted with predication in the genus "what is it?" (T12.3) Alexander points out in (T12.2) that in the ordinary predication, the name that ex-

---

70 Anscombe (1961), p. 44.
71 The "unrestricted" reading is supported, on different grounds, by Dancy (1975), Kirwan (1993), Whitaker (1996), and Wedin (1999). Kirwan and Wedin argue for it on the basis of the logical structure of the concluding argument (1006b28–34). Whitaker's unrestricted interpretation seems to be based on the view of Aristotle's signification according to which words pick out things (Whittaker 1996, p. 195), but the role of essence in the process of "picking" does not seem to be very important.
72 (1) τί δέ ἐστι τὸ σημαινόμενον, ὁρίζεται καὶ γνώριμον ποιεῖ ὑπὲρ τοῦ δεῖξαι ὅτι μή, εἴ τι κατά τινος λέγεται, ἤδη τοῦτο καὶ ἓν σημαίνει, ὡς καὶ προϊὼν ἐρεῖ. (2) οὐ γὰρ εἰ τὸ λευκὸν κατὰ τοῦ Σωκράτους λέγοιτο, ὁ Σωκράτης καὶ τὸ εἶναι Σωκράτει ὑπὸ τοῦ λευκοῦ σημανθήσεται. (3) ὅσα δὴ οὕτω κατηγορεῖται ὡς φύσιν τε καὶ οὐσίαν σημαίνειν, ὡς ἔχει τὰ ἐν τῷ τί ἐστι κατηγορούμενα, ταῦτα ὡς ἓν σημαίνοντα λαμβάνεται, καὶ οὕτως ὑφ' ἑκάστου τῶν ὀνομάτων ἕν τι σημαίνεται. (4) ἐξηγήσατο δὲ πῶς εἶπε τὸ ἓν σημαίνειν, ἵνα μὴ ἓν κατὰ τὸν ἀριθμόν τις αὐτὸν ἡγῆται λέγειν, μηδὲ ἓν τὸ καθ' ἑνὸς τῷ ἀριθμῷ λέγεσθαι, ὡς ἐρεῖ, ἀλλὰ μίαν τινὰ φύσιν. (5) εἰ γὰρ ὁ ἄνθρωπος ζῷον δίπουν σημαίνει, καὶ τὸ ἀνθρώπῳ εἶναι ἐν τούτῳ ἔσται ἐν τῷ ζῷῳ δίποδι εἶναι.
73 *An. Pr.* 1.1, 24a16–17.

presses the predicate does not necessarily signify the subject of the statement. By contrast, in the case of the predication "in what-it-is", the predicate expression will signify the subject of the statement (e.g., "Socrates is a man"). (T12.3) This distinction is introduced in order to make it clear that "signifying one thing" should not be taken as referring to the thing which is one numerically (T12.4). "Signifying one thing" means signifying *nature* and *essence* (*ousia*) rather than something that is numerically one (T12.3–4).

The terminology of "*ousia*", "nature", and "things predicated in what it is" in (T12.3) might suggest that Alexander takes "signifying one thing" to be restricted to the category of substance, along the lines of Anscombe's reading above. Therefore, it is important to show that this is not the case.

As we can see in (T12), Alexander relates the distinction between the two kinds of signifying to the difference between two types of predication. Signifying *about one thing* is what a name is doing as a standard predicate of one thing about another, *ti kata tinos* (T12.1) Signifying *one thing*, which means signifying *nature and substance (ousia)*, characterizes the predicates in "what it is" (ὡς ἔχει τὰ ἐν τῷ τί ἐστι κατηγορούμενα) (**T12.3**).

One way of understanding the latter kind of predicates would be indeed to take them to be substances, the first of the ten highest genera of being: in his list of ten categories in the *Topics* 1.9 Aristotle uses the expression τί ἐστι instead of οὐσία for what we translate as "substance" (103b20), a variation duly noted by Alexander in his commentary who, however, remarks also that τί ἐστι is used in many ways.[74]

The weakness of this interpretation of the predicates 'in what it is' (from **T12.3**) is that it reduces the difference between the two kinds of signifying to the categorial difference between the predicates in the category of substance and the predicates in the non-substantial categories. But there is nothing in either Aristotle or Alexander to suggest such a reduction. In *Metaphysics* Γ, Aristotle does not exclude substances from "signifying about one thing": in his example, "man", as well as "white" and "educated", can signify "about one thing", e.g., individual Socrates. "Man" is then taken as an example of "signifying one thing", but this is not enough to suggest that such signifying is reserved for the class of substance-predicates.[75] If this example entailed that the names "white" and "educat-

---

[74] Alexander *in Top.* 65, 17–19: "Instead of 'substance' Aristotle adopts the phrase 'what-it-is': for substance is what is in the strict sense, and the 'what it is' and the definition are strictly of substance, even if 'what it is' is used in more than one way" (trans. Van Ophuijsen).

[75] Marco Zingano is one of very few authors who discusses this problem and correctly rejects this version of "essentialism" (Anscombe-style; see Zingano 2008, p. 409 n. 5). But Zingano treats in the

ed" do not signify, respectively, one thing each, Aristotle probably would have indicated this, as this would be an important exception to his argument from signification. It is more promising therefore not to stick with substantialist interpretation of signifying one thing, but to see if we could make any sense of "signifying one thing" as a property of any predicate, not alternative to, but distinct from "signifying about one thing". What exactly might be this property?

We know that "man", according to the *Categories*, is "said of" "this man, e. g., Socrates" as "man" and "Socrates" share a definitional formula ("biped animal").[76] This would be a special way in which "man" is said of Socrates, but "white" and "educated" are not. On the other hand, "white" and "educated" could conceivably be predicated in a similar way, as "being said of something", of an individual taken to be "this white" and "this educated thing". We have indirect evidence that Aristotle sees it this way in the *Categories* 2, where Aristotle distinguishes the class of things that are in a subject *and* are said of the subject.[77] And there is more direct evidence in *Topics* 1.9, where Aristotle uses the expression τί ἐστι in a sense different from the above mentioned[78] but also technical, as signifying any one of the ten highest kinds, namely, the one to which the predicate expressed in the predication belongs. [79]

> (T12 A) Aristotle, *Topics* 1.9, 103b27–(1) It is clear from this that he who signifies the *"what-it-is"* signifies sometimes substance, sometimes quantity, sometimes quality, and sometimes some of the other categories. (2) For when about a given man he says that the given one is a man or an animal, he says what it is and signifies substance. (3) When about a given white colour he says that the given is white or a colour, he says what it is and signifies quality. (4) Similarly, if about a given length of a cubit he says that the given is a cubit long in respect of magnitude, he says what it is and signifies quantity. (5) And similarly with the rest. (a) For each of such [things], if it is said either about itself or about the genus of it, signifies what it is. (b) But

---

same way a different kind of essentialism, the one that seems to be closer to Alexander's version that I am exploring here.
76 *Categ.* 2, 1a20–22.
77 *Categ.* 2, 1a29–b3. This example is much discussed in connection with the problem of the ontological status of non-substantial individuals, especially with the problem of recurrence. For the current discussion, it is important that Aristotle has room for the case where X is legitimately 'said of' Y, where Y is a non-substantial individual, no matter how clear or felicitous this particular example is, and no matter what we think of the recurrence of Y. What is important is that there are non-substantial universals and they are said of their corresponding individuals.
78 *Top.* 103b20 and n.64 above.
79 This text, and the exact ontological status of τὸ τί ἐστι has been much discussed. See, in particular, Michael Frede's paper on the nature of the categories in Aristotle (Frede 1981). My task here, though not unrelated to the problems discussed by Frede, is more narrowly circumscribed: to find some support for non-substantial predication in 'what-it-is' (proposed by Alexander) in the texts of Aristotle.

when it is said about another thing, it does not signify what it is, but quantity or quality or a certain one of the other categories.[80]

Aristotle distinguishes two types of categorial predication: one when the subject and the predicate are in the same category (say, "man" is predicated of this man, e.g., Socrates, or "colour" of this white colour) and another when the subject and predicate are in different categories (e.g., "Socrates is white").

When predication is in the same category its predicate signifies the "what-it-is" and the category in question. Thus, if we say "the white of this page is white" we signify the "what-it-is" of the white of this page and its quality: both are expressed by "[colour] white". If we say "the white of this page is a colour", we signify the "what-it-is" of the white of this page and its quality: both are expressed by "colour" (the genus of white and a kind in the highest genus of quality).

When subject and predicate are in different categories, the predicate does not signify the "what-it-is" but signifies only its own category. Thus, if we say "Socrates is white", "white" signifies a quality (of whiteness or colour white), but does not signify any "what-it-is".

Alexander, in his commentary on this *Topics* passage, explains that the cases of predication in the same category, when the predicate signifies this category and "what-it-is", include: predication of a thing of itself, predication of a genus of a thing, and predication of the definitional formula of a thing:

(T12B) Alexander *in Top.* 67, 3–(1) That the *what-it-is* is given in each category, and that what is predicated in the *what-it-is* can be obtained for each category, this, as I have said, Aristotle shows by induction. **(2)** For each of the things under each category, whether the thing is itself said of itself, as e.g., that the man is man or the white is white, and so with the other categories, <or whether> it gives the genus appropriate to it, "signifies *what-it-is*". **(3)** Similarly, when someone gives the definition of each thing, he too states the *what-this-thing-is:* for it is equivalent to pronounce the man man or biped land animal, for in both expressions it is said about itself, and both have the same extension, which is why he has taken the name instead of a definition. **(4)** But when the thing predicated is not of the same genus as the subject but is said of another thing (for this is the sense of "but when one thing is said of another thing"), then the genus of the problem will be from the same category as

---

80 (1) δῆλον δ' ἐξ αὐτῶν ὅτι ὁ τὸ τί ἐστι σημαίνων ὁτὲ μὲν οὐσίαν σημαίνει, ὁτὲ δὲ ποσόν, ὁτὲ δὲ ποιόν, ὁτὲ δὲ τῶν ἄλλων τινὰ κατηγοριῶν. (2) ὅταν μὲν γὰρ ἐκκειμένου ἀνθρώπου φῇ τὸ ἐκκείμενον ἄνθρωπον εἶναι ἢ ζῷον, τί ἐστι λέγει καὶ οὐσίαν σημαίνει· (3) ὅταν δὲ χρώματος λευκοῦ ἐκκειμένου φῇ τὸ ἐκκείμενον λευκὸν εἶναι ἢ χρῶμα, τί ἐστι λέγει καὶ ποιὸν σημαίνει. (4) ὁμοίως δὲ καὶ ἐὰν πηχυαίου μεγέθους ἐκκειμένου φῇ τὸ ἐκκείμενον πηχυαῖον εἶναι μέγεθος, τί ἐστι λέγει καὶ ποσὸν σημαίνει. (5) ὁμοίως δὲ καὶ ἐπὶ τῶν ἄλλων· ἕκαστον γὰρ τῶν τοιούτων, ἐάν τε αὐτὸ περὶ αὑτοῦ λέγηται ἐάν τε τὸ γένος περὶ τούτου, τί ἐστι σημαίνει· ὅταν δὲ περὶ ἑτέρου, οὐ τί ἐστι σημαίνει ἀλλὰ ποσὸν ἢ ποιὸν ἢ τινα τῶν ἄλλων κατηγοριῶν.

the thing predicated. (5) It then no longer signifies the what-it-is, for a thing from one category is not predicated of a thing in another category as what-it-is; one thing is stated of another in those in which the thing predicated is an accident (trans. Van Ophuijsen, modified).[81]

Alexander takes Aristotle's examples in (T12 A2–5) to be parts of an inductive proof showing that each category has its own frame for signifying "what-it-is". He spells out the results of Aristotle's argument by stating the types of predication in each category which correspond to signifying "what-it-is". In ordinary predication, where subject and predicate are in different categories, there is no predication in "what-it-is".

Moreover, every category also has its own qualified οὐσία. In the *Metaphysics* Δ, explaining the senses of οὐσία, Aristotle mentions the meaning which he describes as the substance of each thing (οὐσία ἑκάστου): "also the what-it-was-to-be, whose formula is a definition, and this is said to be the substance of each thing".[82] Aristotle devotes just a line and a half to this meaning of οὐσία as οὐσία ἑκάστου, but Alexander's commentary elaborates on it and allows us to see that Alexander takes for granted the essences that correspond to non-substantial categories:

(T12C) Alexander *in Metaph.* 374, 37—375, 9 (1) In addition to the above-mentioned meanings, Aristotle says, they call substance (οὐσία) the 'essence whose formula is definition'. For when asked what essence is we say that it is definition (and he uses a more general expression 'formula' (λόγος) for definition (ὁρισμός)).(2) He says in fact that substance (οὐσία) is also the definition of each thing, namely of that of which it is the definition, not without qualification. In this way there could be the substance (οὐσία) of a quality, a quantity, and likewise all the rest. (3) This would be different from the enmattered form which he mentioned shortly before [1017b14–16] as the cause for each of the substances to be the one it is, because that one is in things put together naturally and is a natural form and is substance (οὐσία) in a strict sense, whereas that which is discussed now is the form according to which each thing has

---

81 (1) ὅτι καθ' ἑκάστην κατηγορίαν τὸ τί ἐστιν ἀποδίδοται, καὶ ὅτι τὸ ἐν τῷ τί ἐστι κατηγορούμενον καθ' ἑκάστην λαμβάνεται, τῇ ἐπαγωγῇ, ὡς εἶπον, δείκνυσιν. (2) ἕκαστον γὰρ τῶν ὑφ' ἑκάστην κατηγορίαν, ἄν τε αὐτὸ περὶ ἑαυτοῦ λέγηται, οἷον ὅτι ὁ ἄνθρωπος ἄνθρωπός ἐστιν ἢ τὸ λευκὸν λευκόν ἐστι (καὶ ἐπὶ τῶν ἄλλων ὁμοίως, <ἄν τε> τὸ οἰκεῖον γένος αὐτοῦ ἀποδιδῷ, τί ἐστι σημαίνει. (3) ὁμοίως δὲ κἂν τὸν ὁρισμόν τις τὸν ἑκάστου ἀποδιδῷ, τὸ τί ἐστιν αὐτὸ λέγει· ἴσον γάρ ἐστι τὸν ἄνθρωπον ἄνθρωπον εἰπεῖν ἢ ζῷον πεζὸν δίπουν· ἐν ἀμφοτέροις γὰρ αὐτὸ περὶ ἑαυτοῦ λέγεται, καὶ ἐπ' ἴσης ἀμφότερα· διὸ ἀντὶ τοῦ ὁρισμοῦ τὸ ὄνομα ἔλαβεν. (4) ὅταν δὲ τὸ κατηγορούμενον μὴ τοῦ αὐτοῦ γένους ᾖ τῷ ὑποκειμένῳ, ἀλλὰ περὶ ἄλλου λέγηται (τοῦτο γάρ ἐστι τὸ ὅταν δὲ περὶ ἑτέρου), ἐξ ἧς ἂν κατηγορίας ᾖ τὸ κατηγορούμενον, ἐξ ἐκείνης καὶ τὸ γένος τοῦ προβλήματος ἔσται. (5) καὶ οὐκέτι τὸ τί ἐστι τὸ ὑποκείμενον σημαίνει· οὐ γὰρ ἐν τῷ τί ἐστι κατηγορεῖται τὸ ἐξ ἄλλης κατηγορίας τοῦ ἐξ ἄλλης. περὶ ἑτέρου δὲ ἕτερον λέγεται, ἐν αἷς συμβεβηκός ἐστι τὸ κατηγορούμενον.
82 *Metaph.* 5.8: 1017b21–22: ἔτι τὸ τί ἦν εἶναι, οὗ ὁ λόγος ὁρισμός, καὶ τοῦτο οὐσία λέγεται ἑκάστου.

its essence (τὸ τί ἦν εἶναι) even if it is not in substances: for essence (τὸ τί ἦν εἶναι) is not only of substances. (4) Therefore such forms are not substances without qualification, but they are substances of those things whose essence they express. This is why it is said that each thing has some substance. (5) Someone can also understand this so that the form is said in many ways: in one way, as the cause of being for that in which it is, as the soul, about this way he said before, and another as the being itself[83] whose cause was perceiving soul, which is not the same as the soul, but soul is its cause. (6) And he would be mentioning this now, not the external form and shape, which is somehow a limit of each thing and defines it.[84]

In (T12C2) Alexander explains the οὐσία ἑκάστου as a qualified definition distinguishing it from an unqualified definition, which he seems to posit as a default case, and which allows us to speak of essences and definitions not only of substances, but also of qualities, quantities, and other categories. The difference of unqualified essence is that it corresponds to the enmattered form which has a causal role in relation to the substance in which it is present (T12C3).

The default case Alexander describes here is in agreement with his reading of *Topics* 1.9 passage (T12 A2–5, T12B) according to which 'signifying what-it-is' characterises predication in the same category. From the three kinds of such signification considered there, the relevant one here is 'when someone gives the definition of a thing' (T12B3).

This seems to be the sense of *ousia* that Alexander is using in his unrestricted interpretation of 'signifying one thing' in *Metaphysics* Γ 4 (T12 above).

It is remarkable that the distinction between the two types of essence is followed in T12C3–5 by the distinction between the two senses of 'form'. Standard enmattered form (Form-1) corresponds to the unqualified essence, which is a formal cause of a hylomorphic compound. Another sense of form (Form-2) picks out the

---

83 'The being itself': reading αὐτοῦ τοῦ εἶναι at 375, 12, following, with Dooley, Bonitz's conjecture from Sepulveda's translation.
84 (1) Πρὸς τοῖς προειρημένοις οὐσίαν φησὶ λέγεσθαι καὶ τὸ τί ἦν εἶναι, οὗ τί ἦν εἶναι λόγος ἐστὶν ὁ ὁρισμός· τί γάρ ἐστι τὸ τί ἦν εἶναι ἀπαιτούμενοι, λέγομεν ὅτι ὁ ὁρισμός, λόγον κοινότερον εἰπὼν τὸν ὁρισμόν. (2) οὐσίαν δή φησι καὶ τὸν ὁρισμὸν τὸν ἑκάστου ἐκείνου εἶναι οὗ ὁρισμός ἐστιν, οὐχ ἁπλῶς· οὕτω γὰρ καὶ ποιοῦ οὐσία ἂν εἴη καὶ ποσοῦ καὶ τῶν ἄλλων ὁμοίως. (3) διαφέροι δ' ἂν τοῦ ἐνύλου εἴδους, περὶ οὗ πρὸ ὀλίγου εἶπεν ὡς αἰτίου ὄντος ἑκάστη τῶν οὐσιῶν τοῦ εἶναι ταύτην ἥτις ἐστίν, ὅτι ἐκείνη μὲν ἐν τοῖς φύσει συνεστῶσιν οὖσα καὶ φυσικὸν εἶδος οὖσα καὶ κυρίως ἦν οὐσία, τὸ δὲ νῦν λεγόμενον εἶδος, καθ' ὃ τὸ τί ἦν εἶναι ἑκάστῳ, καὶ ἐν μὴ οὐσίαις ἐστίν· οὐ γὰρ μόνον τὸ τί ἦν εἶναι ἐπὶ τῶν οὐσιῶν. (4) διὸ οὐδὲ ἁπλῶς τὰ τοιαῦτα εἴδη οὐσίαι, ἀλλ' ἐκείνων οὐσίαι ὧν τὸ τί ἦν εἶναι δηλοῦσι, διὸ καὶ λέγεται ἑκάστου οὐσία τις εἶναι. (5) δύναταί τις καὶ τοῦ εἴδους ὡς πλεοναχῶς λεγομένου ἀκούειν, καὶ ἑνὸς μὲν ὄντος ὡς αἰτίου τοῦ εἶναι τῷ ἐν ᾧ ἐστιν, ὡς ἡ ψυχή, περὶ οὗ εἶπε πρώτου, ἄλλου δὲ ὡς αὐτοῦ εἶναι, οὗ ἦν αἰτία ἡ ψυχὴ ἡ αἰσθητική, ὃ οὐκ ἔστι ταὐτὸν τῇ ψυχῇ, ἐκείνη μέντοι αἰτία τούτου. (6) καὶ εἴη ἂν τούτου νῦν μνημονεύων, ἀλλ' οὐχὶ τῆς ἔξωθεν μορφῆς καὶ τοῦ σχήματος, ὃ πέρας πώς ἐστιν ἑκάστου καὶ ὁρίζει αὐτό.

characteristics of a hylomorphic composite that are caused by the Form-1. Form-2 is not to be taken in the sense of shape contrasted with body, rather it seems to be an ensemble of all the properties that are causally dependent on Form-1. With this distinction Alexander signals a very important exegetical presupposition in his interpretation of Aristotle's hylomorphism.[85] The important point for the current argument is that Alexander sees the 'qualified' essence, which is the definition of a property in any category, including all non-substantial categories, as fully underwritten by his version of hylomorphic theory of substance.

So, Alexander's interpretation of "signifying one thing" seems to be essentialist in so far as "one thing", the definitional formula of a thing, is understood as the nature and essence of the thing, but it is not substantialist, because a thing in question can belong to any of the ten categories. Next, we will see how this distinction works and whether it really can support the concluding argument.

3.4 Negated Names and Two Types of Negation

The distinction between "signifying one thing" and "signifying about one thing" is important in disambiguating certain expressions with negation, of the form "not-man". This is how Alexander explains the use of "signifying one" commenting on Aristotle **(T11)**:

> (T13) Alexander *in Metaph.* 279, 32–280, **20 (1)** [Aristotle] teaches us how one should understand "signifying one". For "man" is one [thing] because it signifies *some one nature*, not because it is said of one thing. **(2)** But as it signifies *one nature* it would not signify also simultaneously some other nature. **(3)** He shows that not everything said *about one* thing is one. For "man", "the white", and "the educated" can be truly predicated about one thing, but it is not the case that because of that they are one. **(4)** And he arrived at proving this because it seems that "discussing" can be predicated about one and the same thing as "man" [signifies], e.g., about Socrates; yet "discussing" is not "man". **(5)** For in this way, of *that about which* (καθ' οὗ) "man" is predicated, the "not-man" would also be predicated accidentally, but it is not the case that thereby already that which is signified by "man" and "not man" are the same. **(6)** For "signifying one thing" is not defined in the same way as "being predicated about one thing". **(7)** Aristotle proved by the additional argument that [names] predicated about one thing do not by the same token also signify one thing. **(8)** For several [names] different from each other are predicated about one thing: those [things] that are ac-

---

85 See Marwan Rashed's discussion of this distinction in relation to the ontological project of the middle books of Metaphysics in Rashed 2007, 232–234. There remains a question of Alexander's motivation for setting this discussion in parallel with the discussion of two types of essence, but discussing it would lead to a major digression in this already very long paper. I plan to discuss this question elsewhere.

cidental to something are similarly predicated about that thing to which they are accidental, which is one, but it is not the case that because of that all [of them] signify one and the same thing. **(9)** Thus, it is not the case that since "educated", "white", and "man" are predicated of one thing, therefore even that which is signified by them is the same. **(10)** For had the signified been the same, so that there would be no difference between being predicated of one thing and signifying one thing, they would have been synonyms, i.e., polyonyms, for he now uses "synonyms" instead of "polyonyms" (279, 32–280, 20).[86]

The concept of "one nature" which Alexander uses in **(T13.1)** is already familiar. The nature of X for Alexander is the "what-it-is" expressed by the definitional formula of X, where X can belong to any of the ten categories. There is an important point that Alexander spells out: as the name signifies one nature it cannot simultaneously signify another nature **(T13.2)**. There is exactly one definitional formula for each name which expresses that which a name signifies. When several different names are said *about* one and the same thing, they do not signify one nature. Alexander uses Aristotle's example of "man", "educated", and "white" predicated of Socrates to illustrate the point **(T13.3)**.

Aristotle's aim in drawing this distinction, according to Alexander, is to disambiguate the negated name "not-man". If we use it to mean anything that is *different* from "man" without any further qualification,[87] so that either "white" or "educated" or "discussing" could be substituted with "not-man", then the following piece of reasoning could get us a purported denial of PNC: "Socrates is a man. But Socrates is also white. To be white is not the same as to be a man. So, Socrates is also not the

---

[86] (1) πῶς τὸ ἓν σημαίνειν ἀκούειν χρή, διδάσκει ἡμᾶς· ὁ γὰρ ἄνθρωπος ἓν τῷ φύσιν τινὰ μίαν σημαίνειν, οὐ τῷ καθ' ἑνός λέγεσθαι. (2) μίαν δὲ φύσιν σημαίνων οὐκ ἂν ἅμα καὶ ἄλλην τινὰ σημαίνοι. (3) ὅτι δὲ τὸ καθ' ἑνός λεγόμενον οὐ πᾶν ἕν, δείκνυσι· καθ' ἑνός γὰρ καὶ τὸ ἄνθρωπος καὶ τὸ λευκὸν καὶ τὸ μουσικὸν δύναται κατηγορεῖσθαι ἀληθῶς, ἀλλ' οὐ διὰ τοῦτο καὶ ἓν ταῦτα. (4) ἦλθε δὲ ἐπὶ τὸ δεικνύναι τοῦτο διὰ τὸ δοκεῖν δύνασθαι καθ' ἑνός καὶ τοῦ αὐτοῦ καθ' οὗ ὁ ἄνθρωπος, οἷον κατὰ Σωκράτους, κατὰ τούτου καὶ τὸ διαλέγεσθαι· τὸ δὲ διαλέγεσθαι οὐκ ἄνθρωπος. (5) εἴη γὰρ ἂν καθ' οὗ ὁ ἄνθρωπος, καὶ οὐκ ἄνθρωπος κατηγορούμενος κατὰ συμβεβηκός· ἀλλ' οὐ διὰ τοῦτο ἤδη καὶ ταὐτόν ἐστι τό τε ὑπὸ τοῦ ἀνθρώπου καὶ τοῦ οὐκ ἀνθρώπου σημαινόμενον. (6) οὐ γὰρ τούτῳ ὥρισται τὸ ἓν σημαίνειν τῷ καθ' ἑνός κατηγορεῖσθαι. (7) ὅτι γὰρ μὴ τὰ κατὰ τοῦ αὐτοῦ κατηγορούμενα ἤδη καὶ ἓν σημαίνει, δι' ὧνπερ παρέθετο ἔδειξε. (8) καθ' ἑνός γὰρ πλείω κατηγορεῖται διαφέροντα ἀλλήλων· τὰ γὰρ συμβεβηκότα τινί, ἕτερα ὄντα καὶ τούτου ᾧ συμβέβηκε καὶ ἀλλήλων, ὁμοίως κατηγορεῖται κατ' ἐκείνου ἑνός ὄντος, ᾧ συμβέβηκεν, οὐ μὴν διὰ τοῦτο πάντα τὸ αὐτὸ καὶ ἓν σημαίνει. (9) οὐ γὰρ ἐπεὶ καθ' ἑνός τὸ μουσικὸν καὶ τὸ λευκὸν καὶ ὁ ἄνθρωπος κατηγορεῖται, ἤδη καὶ ταὐτόν ἐστι τὸ σημαινόμενον ὑπ' αὐτῶν. εἰ γὰρ ταὐτόν εἴη τὸ σημαινόμενον ὡς μὴ διαφέρειν τὸ καθ' ἑνός κατηγορεῖσθαι τοῦ ἓν σημαίνειν, ἔσται συνώνυμα, τουτέστι πολυώνυμα, καθ' ὧν πλείω κατηγορεῖται·τῷ γὰρ συνώνυμα νῦν ἀντὶ τοῦ πολυώνυμα χρῆται.

[87] For instance, if we were to take in this way Plato's suggested solution to the Parmenidean paradox of Not-Being in the Sophist, taking "Not-Being" as "Different from Being" (*Soph.* 255d5–e7)

same as a man. To be not the same as a man is to be a not-man. Hence Socrates is a man and not-man. Hence Socrates is a man and Socrates is not a man".

But Alexander explains that in this case "not-man" would be predicated of Socrates accidentally, just as "white" was. (T13.4–5) So, it should inherit its accidental type of predication from the "white" for which it was substituted: this kind of "not-man" is predicated of Socrates accidentally and does not constitute a real denial of PNC. Being predicated about one thing does not amount to signifying one thing. Had the three terms "man", "white", and "educated", signified one thing, they would have been synonyms, Aristotle says, or polyonyms, as Alexander corrects.[88] This would mean that each of these expressions is just another way of signifying the essence of a thing (Callias or Socrates).

The important corollary of this discussion is that in order to get a genuine denial of PNC, the negated name "not-man" should inherit its type of predication not from an accident, but from the definitional formula itself, so that the function of this kind of negated expression should be not just to "point away in every direction indiscriminately", as is the function of negated names construed as indefinite names in *De interpretatione*.[89] Rather, its role should be a more targeted destruction of the definitional formula itself. It is a different meaning of negation. Alexander explains it in this way, elaborating on Aristotle's text:

> (T14) Alexander in *Metaph*. 281, 1–(1) The object of our investigation was not whether it is possible to impose to the same thing both the name "man" and "not-man", so that their difference would be merely verbal and no longer also according to that which is signified, (2) but whether, when the objects signified by each differ, and "man" signifies such a form whose formula is "animal footed biped", while "not-man" signifies *the destruction of such a form*, it is possible for these to be predicated simultaneously and about the same thing so that it would be the

---

[88] Aristotle uses the concept just once in our extant corpus (*Hist. An.* 289a1–3: Καλεῖται δ' ᾗ μὲν λαμβάνει, στόμα, εἰς ὃ δὲ δέχεται, κοιλία· τὸ δὲ λοιπὸν πολυώνυμόν ἐστιν). This concept goes back to Speusippus' division of names into tautonyms (further divided into homonyms and synonyms) and heteronyms (further divided heteronyms proper, polyonyms and paronyms) of which we know through Boethus' report preserved to us by Simplicius (Simplic. *in Categ.* 38,19–24 Kalbfleisch = Boethus fr. 10 Rashed, see also Barnes 1971). It seems that this classification was used by Peripatetics as a part of their own school legacy: at least Alexander's correction of Aristotle's usage here suggests this much. Cf. also Alexander, in Metaph. 247, 22–29: οὕτως ἔχει πρὸς ἄλληλα καὶ τὸ ἀμερὲς καὶ τὸ ἐλάχιστον, καὶ σπέρμα καὶ καρπός, καὶ ἀνάβασις καὶ κατάβασις, καὶ πάνθ' ὅσα κυρίως ἑτερώνυμα καλεῖται. εἰπὼν δὲ τὸ ὂν καὶ τὸ ἓν ταὐτὰ εἶναι καὶ μίαν τινὰ φύσιν, οὐ μέντοι ὡς ἑνὶ λόγῳ δηλούμενα, ἐπήνεγκεν ὅτι διαφέρει δὲ οὐδὲν οὐδ' ἂν καὶ κατὰ τὸν λόγον αὐτὰ ταὐτὰ ὑπολάβωμεν εἶναι, μὴ μόνον κατὰ τὴν ὑποκειμένην φύσιν, ὡς εἶναι τῶν πολυωνύμων τὸ ὑποκείμενον αὐτοῖς, ὧν πλείω μὲν ὀνόματα, καθ' ἕκαστον δὲ τῶν ὀνομάτων ὁ αὐτὸς λόγος, ὡς φασγάνου καὶ μαχαίρας, καὶ λωπίου καὶ ἱματίου.

[89] I borrow a fine description by Whitaker (Whitaker 1996, p. 64).

same thing to be such a substance and not to be. **(3)** For *the contradictory opposition* is of this kind, while the opposition which is accidental is not contradictory.[90]

Alexander spells out Aristotle's distinction between the accidental and non-accidental predication of a negated name "not-man" of Socrates in terms of signifying. If we predicate "not-man" of Socrates accidentally, meaning that Socrates is white, say, and being white is not the same as being man, the opposition the "not-man" and "man" will be accidental and not contradictory.

Let us use a rather ugly contraption Not-Man$_{ac}$ to stand in for the predicate "not-man" used as such an umbrella term for all the things that are not the same in definition as Man (which will be a predicate form of the name "man"). The pair:

Man (Socrates) and Not-Man$_{ac}$ (Socrates)

is not a contradiction because we can substitute a predicate (White, Educated, Bald, In the Market, 70 years old) instead of our Not-Man$_{ac}$. Note that we cannot substitute 'Not-(a)-Biped-Animal' for Not-Man$_{ac}$, because the 'biped animal' is not an accident of 'man'.

But a contradiction will arise if we take 'not-man' to stand in for the predicate whose only meaning is that the definitional formula associated with 'man' is not applicable to a thing of which 'not-man' is predicated. Let us call it Not-Man$_{def}$, a "definitional" Not-Man: also not an inch-perfect label, because this predicate Not-Man$_{def}$ does not define anything. It only denies the application of a definition (a definitional formula or a corresponding predicate) to a subject. And it can give us a contradictory opposition. The pair:

Man (Socrates) and Not-Man$_{def}$ (Socrates)

will make a contradiction when said of the same subject. The predicate Not-Man$_{def}$ denies the application of the definition of the signifying expression 'man' to Socra-

---

90 (1) οὐκ ἦν δὲ τοῦτο ἡμῖν τὸ ζητούμενον, πότερον δύναται τῷ αὐτῷ πράγματι καὶ ἄνθρωπος ὄνομα καὶ μὴ ἄνθρωπος τεθῆναι, ὡς εἶναι τὴν διαφορὰν αὐτῶν κατὰ τὴν λέξιν μόνην, μηκέτι δὲ καὶ κατὰ τὸ σημαινόμενον, (2) ἀλλ' εἰ ὄντων τῶν σημαινομένων ὑφ' ἑκατέρου διαφερόντων, καὶ τοῦ μὲν ἀνθρώπου τὸ τοιοῦτον σημαίνοντος εἶδος οὗ λόγος ζῷον πεζὸν δίπουν, τοῦ δὲ οὐκ ἀνθρώπου τὴν ἀναίρεσιν τοῦ τοιούτου εἴδους, οἷόν τε ἅμα ταῦτα καὶ κατὰ τοῦ αὐτοῦ κατηγορεῖσθαι οὕτως ὡς εἶναι ταὐτὸν τὸ εἶναί τε τὴν τοιαύτην οὐσίαν καὶ μὴ εἶναι·(3) τοῦ τοιούτου γὰρ δηλωτικὴ ἡ ἀντιφατικὴ ἀντίθεσις, οὐκ ἀντιφατικὴ δὲ ἡ ἀντίθεσις ἡ κατὰ συμβεβηκός.

tes. In order to introduce this kind of opposition Aristotle needs this distinction between 'signifying one thing' and 'signifying about one thing'.

Alexander seems to think that this distinction, in order to work properly, has to be supported by the whole system of categories, in particular by the predication of "what-it-is", and more specifically still, of the definitional formulae, within each category. Aristotle's example of "man" illustrates the way it works with substances, but we could do the same with the predicate in any other category, for instance, we could take "white" and formulate in the same way as above:

White (Socrates) and Not-White$_{ac}$ (Socrates)

is not a contradictory pair if Not-White$_{ac}$ stands for Man (Socrates). But

White (Socrates) and Not-White$_{def}$ (Socrates)

is a contradictory pair if the definition of white is, say: "the colour which dissolves the eye-stream"[91] and the definitional Not-White denies the definitional formula of white. The analysis then could be sketched out as follows. White (Socrates) will say that Socrates is of a colour that satisfies the definition of white and Not-White$_{def}$ (Socrates) will say that Socrates is of a colour that does not satisfy the definition of white (assume Socrates has just one colour: a multicolour version can be worked out as well, using Aristotle's tools for disambiguation we have just seen). Here the definitional formula and the subject to which it is applied also must be in the same category, the way we had it with substance, but in this case it will be a non-substantial category of 'qualified' (ποιόν). Socrates *qua* qualified cannot receive contradictory properties, and this will be true of all categories, and supplying any further qualifications that are needed to rule out any easy counter-examples.

This analysis of Alexander's interpretation of Aristotle might answer the worry raised by Mario Mignucci who accepted that Alexander's interpretation of "signifying one thing" does not restrict it to substances alone, but thought that he did not distinguish sufficiently clearly between understanding a negated name of the form not-X (e.g., "not-man") as a complement of X (the understanding that rules out their joint predication of the same subject, say Y) and understanding not-X as anything that is different from X (the understanding that does not rule out their joint predication of X, as when both "man" and "white" are predicated of Soc-

---

[91] *Top.* 3.5, 119a30; cf. *Metaph.* 11.7, 1057b8.

rates in an ordinary way).[92] Assuming that each name "signifies one thing", i.e., signifies the corresponding proper nature, still, when both "white" and "man" are predicated of Socrates, "man" does signify one thing, but not the "white", because only "man" expresses the nature of Socrates. Mignucci suggested that in this perspective it is necessary to distinguish between "signifying one thing" and "signifying one thing in respect of the subject".[93] For instance, in the statement "Socrates is white" and "Socrates is a man", both "white" and "man" do signify one thing, i.e., signify, respectively, their own proper nature (corresponding definitional formula to each), but with respect to Socrates only "man" signifies one thing, but "white" does not. Mignucci thinks that this further distinction is necessary, but Alexander misses it.[94]

But it we take Alexander's "broad" essentialist interpretation of "signifying one thing" to be linked to the way in which a definitional formula is predicated within each of the ten categories, this further distinction suggested by Mignucci may be unnecessary. For Alexander follows the distinction drawn by Aristotle in the *Topics* 1.9 between the predication in the same category (i.e., when the subject and the predicate belong to the same category) and not in the same category (i.e., when the subject and the predicate belong to different categories). According to this distinction, the definitional formula—which expresses the one nature or the one thing signified by a name—belongs to the former type of predication, where the subject and predicate belong to the same category. This is the only kind of predication that could give us a contradictory pair if name and its negation were predicated on the same subject. And every name that has a definitional formula has its own category and its own subject of predication with which it is synonymous under that category. So, even if "white" is an accident of Socrates, still the pair "White" and "Not-White$_{def}$" will be contradictory and subject to the same rules in this respect as the pair "Man" and "Not-Man$_{def}$".[95]

Aristotle's concluding argument **(T11.5)** above is:

> **(a)** Therefore, it is necessary, in order to make a true statement that something is a man, to say that it is a biped animal (for this was what "man" signifies). (1006b28–30)

---

92 Mignucci (2003), p. 114.
93 Mignucci (2003), p. 115.
94 Mignucci (2003), pp. 115–116.
95 This analysis based on Alexander's interpretation might also provide an answer to a much earlier worry raised by Maier with respect to Aristotle's argument from signification which he saw as working only for definitional formulae and unable to account for all the statements where predicates are accidental (Maier 1896, pp. 55–56).

(b) But if this is necessary, then it cannot fail to be the biped animal (for "necessary" means "impossible not to") (1006b30–33)

(c) Therefore it is impossible to say that it is true that the same thing simultaneously is and is not a man. (1006b33–34).

The crucial role is given to the definitional formula that expresses what is signified by a name used as a predicate term in a statement. It is impossible for it to hold and not to hold of the subject of its predication. Alexander summarizes this argument in his commentary three times: twice as a part of a larger argument (at the beginning, at 276, 17–27 and in the end, 282, 3–19), and once in some more detail to its text (282, 19–36). All three seem to be mutually consistent. Let us consider the one that comes from Alexander's synopsis we have been discussing, our **(T10.7)** above:

(a) "Man" and "not-man" are different names, and not the names *of one thing* (ἑνὸς πράγματος),

(b) for they will be neither predicated of the same (ἐπὶ ταὐτοῦ),

(c) nor will that which is signified by "not-man" be true of that which is signified by "man",

(d) given that the names themselves differ from one another, and the different names express different [things], and *each name expresses one* [thing], *i. e., one nature* (καὶ ἑνὸς ἕκαστον καὶ φύσεως μιᾶς)

(e) For that which is signified by "man" will no longer be one, as has been said, if "not-man", which signifies another object, were also true of it. (The same argument as applies in the case of "man" and "not-man" applies in all other cases as well.)

(f) But if the things expressed by these names do not belong together, then neither would the affirmation saying that this (*tode*) is a man have as simultaneously true together the negation saying that this (*tode*) is not a man.

We know that the difference between the names "man" and "not-man" is accounted for by a difference of a respective signified thing which is expressed by a definitional formula. We have now seen that Alexander is taking both "man" and "not-man" in the strictly definitional sense, so that "not-man" should be taken as our Not-Man$_{def}$, not as Not-Man$_{ac}$ above. The name "Not-Man$_{def}$" signifies that the definitional formula of "man" does not apply. The name and its definitional negation will never be predicated of the same thing (b) because the predication in question is not the accidental predication, but a definitional predication of two different names.

That which is signified by a "not-man" will not be true of what is signified by "man" (c) because two genuinely different names correspond to two different definitional formulae which cannot express the same nature, but must express two different natures signified respectively by each name (d). Had it been possible for "not-man" to signify what is signified by "man", this signified object and nature would no longer be one. (e) Since these things (or natures) do not belong together (συνυπάρχει), it follows that the affirmation saying that this is a man will not be true together with the negation saying that this is not a man (f).

Thus, the refutation of those who deny the PNC seems to be based on the impossibility for the definitional formula and its negation to hold truly of the same subject in a statement. The denial of PNC allows for these two formulae, one of which negates the other, to "belong together" as being signified by the same name, and this destroys signifying, since the first postulate stated by Aristotle was that the name should signify something one, e.g., the name "man" should signify always "animal biped" (if this has been stipulated), and cannot signify both "animal biped" and "not [an] animal biped". The latter case amounts effectively to the opponent's withdrawal of his concession to signifying something and once again to the destruction of rational discourse or to being excluded from it.

4 Conclusion

This is a very incomplete survey of Alexander's commentary on Aristotle's argument from signification in *Metaphysics* Γ 4, not just because it does not cover the whole argument, but also because Alexander's commentary needs to be much better contextualized with his other work on the *Organon* and other relevant parts of the corpus. There is still much to do on all these fronts. However, some of the results we have discovered are interesting for Aristotelian scholarship. In the elenctic demonstration, Alexander seems to distinguish two different (but related) arguments: the elenchos proper, where the concession of "something definite" is secured "locally", from this interlocutor, and the more general argument based on a generalized concession by an imaginary opponent, where the defender of PNC can work out certain general rules applicable in all such cases. Alexander's interpretation of signification develops an unrestricted essentialist, but not substantialist, interpretation of the scope of the argument. Its particular interest for Aristotle scholars is its very important connection with the account of predication that we find in *Topics* and in the *Categories*. Alexander's approach allows us, I think, to see what work the categories do in Aristotle's first philosophy.[96]

---

[96] To this extent, Alexander's interpretation of the argument from signification may be seen as belonging to the project that was described by John Ellis as his defence of Aristotle's *Categories* (Ellis 1994).

Bibliography

Anscombe, Gertrude Elizabeth Margaret (1961): "Aristotle". In: Anscombe, Gertrude Elizabeth Margaret and Geach, Peter Thomas: *Three Philosophers: Aristotle, Aquinas, Frege*. Ithaca and New York: Cornell University Press, pp. 1–63.
Barnes, Jonathan (1969): "The Law of Contradiction". In: *The Philosophical Quarterly* 19. No. 77, pp. 302–309. [Revised version in Barnes, Jonathan (2012): *Logical Matters. Essays in Ancient Philosophy* II. Oxford: Oxford University Press, pp. 353–363.]
Barnes, Jonathan (1971): "Speusippus and Aristotle on Homonymy". In: *Classical Quarterly* 21, pp. 65–80. [Revised version in Barnes, Jonathan (2012): *Logical Matters. Essays in Ancient Philosophy* II, Oxford: Oxford University Press, pp. 284–311.]
Berti, Enrico (2014): "Objections to Aristotle's Defence of the Principle of Non-Contradiction". In: Ficara, Elena (Ed.): *Contradictions: Logic, History, Actuality*. Berlin and Boston: de Gruyter, pp. 97–108.
Bocheński, Józef Maria (1947): *La logique de Théophraste*, Fribourg: Librairie de l'Université.
Bolton, Robert (1994): 'The Problem of Dialectical Reasoning (Συλλογισμός) in Aristotle', *Ancient Philosophy* 14, pp 99–132.
Castagnoli, Luca (2010): *Ancient Self-Refutation*, Cambridge: Cambridge University Press.
Cavini, Walter (2007): "*Principia Contradictionis:* Sui principi aristotelici della contraddizione (§§ 1–3)". In *Antiquorum Philosophia* 1, pp. 123–170.
Cavini, Walter (2008): "*Principia Contradictionis:* Sui principi aristotelici della contraddizione (§ 4)". In: *Antiquorum Philosophia* 2, pp. 159–187.
Crubellier, Michel (2008): "La tactique argumentative de Métaphysique Gamma 3–6". In: Hecquet-Devienne, Myriam and Stevens, Annick (Eds.): *Aristote, Métaphysique Gamme: Édition, Traduction, Études*. Louvain: Peeters, pp. 379–402.
Dancy, Russell (1975): *Sense and Contradiction: A Study in Aristotle*. Dordrecht and Boston: D. Reidel.
Dooley, William E. (1993): *Alexander of Aphrodisias on Aristotle Metaphysics 5*, London: Duckworth.
Ellis, John (1994): 'Alexander's Defense of Aristotle's Categories', *Phronesis* 39/1, pp. 68–89.
Flannery, Kevin (2003): "Logic and Ontology in Alexander of Aphrodisias' Commentary on Metaphysics IV". In: Movia, Giancarlo (Ed.): *Alessandro di Afrodisia e la "Metafisica" di Aristotele*. Milan: Vita e Pensiero, pp. 117–134.
Furth, Montgomery (1986): "A Note on Aristotle's Principle of Non-Contradiction". In *Canadian Journal of Philosophy* 16. No. 3, pp. 371–381.
Gottlieb, Paula (1994): "The Principle of Non-Contradiction and Protagoras: The Strategy of Aristotle's *Metaphysics* IV 4". In: *Proceedings of the Boston Area Colloquium in Ancient Philosophy* 8, pp. 183–209.
Gottlieb, Paula (2019): "Aristotle on "Non-contradiction". In: Zalta, Edward N. (Ed.): *The Stanford Encyclopedia of Philosophy* (Spring 2019 Edition). https://plato.stanford.edu/archives/spr2019/entries/aristotle-noncontradiction, last accessed on November 7, 2022.
Graeser, Andreas (1973): *Die logischen Fragmente des Theophrastus*, Berlin-New York: Walter de Gruyter.
Huby, Pamela, et al. (2006) *Theophrastus of Eresus. Sources for His Life, Writings, Thought and Influence: Commentary, Volume 2: Logic*, Leiden, Brill.
Irwin, Terence (1977): "Aristotle's Discovery of Metaphysics". In: *Review of Metaphysics* 31, pp. 210–229.
Irwin, Terence (1988): *Aristotle's First Principles*. Oxford: Clarendon Press.
Kirwan, Christopher (Ed.) (1971): *Aristotle, Metaphysics*. Books IV, V, and VI. Oxford: Clarendon Press.

Kotwick, Myriam (2016): *Alexander of Aphrodisias and the Text of Aristotle's Metaphysics.* Berkeley: University of California Press.
Kupreeva, Inna (2017): "Aporia and exegesis". In: Karamanolis, George and Politis, Vasilis (Eds.): *The Aporetic Tradition in Ancient Philosophy.* Cambridge: Cambridge University Press, pp. 228–247.
Lear, Jonathan (1980): *Aristotle and Logical Theory.* Cambridge: Cambridge University Press.
Lukasiewicz, Jan (1993): *Über den Satz des Widerspruchs bei Aristoteles.* Hildesheim: G. Olms.
Madigan, Arthur (1993): *Alexander of Aphrodisias on Aristotle Metaphysics 4.* London: Duckworth.
Maier, Heinrich (1896): *Die Syllogistik des Aristoteles, 1. Teil: Die Logische Theorie des Urteils bei Aristoteles.* Tübingen, H. Laupp.
Mignucci, Mario (2003): "Alessandro interprete di Aristotele: luci e ombre del commento a *Metaph.* Γ". In: Movia, Giancarlo (Ed.): *Alessandro di Afrodisia e la "Metafisica" di Aristotele.* Milan: Vita e Pensiero, pp. 93–116.
Politis, Vasilis (2004): *Aristotle and the Metaphysics.* London: Routledge.
Rashed, Marwan (2007): *Essentialisme: Alexandre d'Aphrodise entre logique, physique et cosmologie,* Berlin & New York: Walter De Gruyter
Repici, Luciana (1977): *La logica di Teofrasto: studio critico e raccolta dei frammenti e delle testimonianze,* Bologna: Il Mulino.
Rosen, Jacob (2021) 'Aristotle's Actual Infinities', in: *Oxford Studies in Ancient Philosophy* Volume LIX, pp. 133–185
Sharples, Robert William (1992) *Alexander of Aphrodisias: Quaestiones 1.1–2.15,* London: Duckworth.
Wedin, Michael Vernon (1999): "The Scope of Non-Contradiction: A Note on Aristotle's 'Elenctic' Proof in *Metaphysics* Γ 4". In *Apeiron* 32. No. 3, pp. 231–242.
Wedin, Michael (2004): 'Aristotle on the Firmness of the Principle of Non-Contradiction' in *Phronesis* 49/3, pp.225–265
Whitaker, Christopher (1996): *Aristotle's De interpretatione: Contradiction and Dialectic.* Oxford: Clarendon Press.
Zingano, Marco (2008): "*Sêmainein hen, sêmainein kath' henos* et la preuve de 1006b28–34". In: Hecquet-Devienne, Myriam and Stevens, Annick (Eds.): *Aristote, Métaphysique Gamme: Édition, Traduction, Études.* Louvain: Peeters, pp. 403–421.

Ilaria L.E. Ramelli
Ancient Greek Dialectic and Its Reception in Origen of Alexandria: From Plato to Christ-Logos

1 Introduction: Parmenides, Plato, Origen, and Dialectics

Plato's dialectics (διαλεκτικὴ ἐπιστήμη) may have been influenced by Parmenides, to whom Plato devoted an important dialogue—which seems to have impacted the thought of Clement and Origen.[1] Parmenides is the first philosopher who, in his poem,[2] recommends the use of "critical personal reason" in order to establish a "crisis" (κρίσις) amongst opinions and values, and distinguish opinions from the truth, "what is", "what exists" (τὸ ἐόν). The importance assigned by Parmenides to "critical reason" and to the concept of "crisis" determines the activity and "method" of the philosopher.[3]

This heritage was developed by Plato with his dialectic as well as by the heirs of Plato, who included both Neoplatonists in the "pagan" tradition and Christian (Neo-)Platonists such as Origen of Alexandria. Among Christian philosophers, Origen probably made the most of Plato's dialectics and "zetetic" philosophical strategy, which puts at the center the crisis or discernment of ideas, theories, and arguments. Indeed, Origen appropriated this methodology and placed it at the core of his philosophical theology.

Note: Many thanks to Melina Mouzala for the organization of the conference, the editorship of this important volume, and her kind invitation.

1 Ramelli (forthcoming a).
2 On which see the recent analysis by Mansfeld (2021).
3 Fattal (2018).

2 Plato as the Founder of Dialectics, Socrates, the Sophists, and Hints of Origen's Reception

Plato is the founder of dialectic and probably the inventor of the term. Dialectic is the art that leads to the knowledge of the Forms (especially the Form of the Good) and the first principles through the exercise of philosophy. Indeed, διαλέγεσθαι means "to converse", the way in which philosophy develops in Plato's dialogues. That Plato was the initiator of dialectics is also remarked by Aristotle: "Plato's predecessors did not know dialectics".[4] Plato remembered Socrates' elenchic method, which certainly exerted a strong influence on the development of the former's dialectic. Socrates, at least qua filtered through Plato, depicted dialectical activity as the greatest human good.[5] The dialectician is central to Plato's dialogues and in the *Republic* instruction in dialectic is prescribed as advanced, to be delayed.[6] According to Szlezák, a full account of dialectic is not in the dialogues, but in Plato's unwritten doctrines.[7] Plato did not express everything in his dialogues but, given his suspicious attitude towards writing, also had higher, unwritten doctrines. These were also central to the anonymous 6th-century *Prolegomena to Plato's Philosophy* 13.18–23: beyond his dialogues, of which the standard reading order among Neoplatonists is here recorded (likely going back to Iamblichus),[8] Plato transmitted unwritten doctrines through his lectures about his loftiest teachings, those most easily liable to misunderstanding. The author adduces Aristotle as a witness to Plato's unwritten doctrines on protology and dialectic. Plato, indeed, omits to include in

---

[4] Διαλεκτικῆς οὐ μετεῖχον, *Met.* A6 987b32–33.
[5] *Apol.* 38a.
[6] Broadie (2020) asks why Plato required such an intensive, multi-disciplinary education in mathematics as a propaedeutic to dialectic. Broadie answers that Plato's educational program in mathematics remedies a "vulnerability in reason" that, although inherent to human rationality itself, is compounded by the culture and education of the theoretical city of Callipolis (Broadie 2020, p. 43). Her account of the origin of the vulnerability of reason rests on the interpretation that dialectic in the *Republic* refers to Socrates' elenchic method of testing moral beliefs in the aporetic dialogues. This understanding of dialectic would seem to explain Plato's worry that dialectic threatens to undermine foundational moral beliefs and encourages a zeal for eristic. This led Plato to propose delaying instruction in dialectic (*Resp.* 539b–c).
[7] *Resp.* 533a2; *Phaedr.* 276be; *Seventh Letter* 617–9. On Plato as teaching more than what he wrote, see at least Banner (2018), Chapter 1, with bibliography, Szlezák (2019), and Halfwassen (2021).
[8] Providing the last summary statement of how the Neoplatonists read Plato, what they held to be his most important doctrines, why he wrote dialogues, and how his philosophy was superior to that of all the rival schools. See Müller-Jourdan (2018).

his dialogues details concerning the nature of dialectic, reached through the "long detour" (μακρὰ περίοδος) of oral teaching.⁹

Origen, who made very much of Plato's dialectic, clearly alludes to Plato's unwritten doctrines on protology: "Plato had something more sacred and divine than the teachings he wrote down and transmitted".¹⁰ Similarly, Origen remarks that even Scripture does not contain "some of the more majestic and divine aspects of God's mysteries", being just "the most elementary" and short introduction to "the totality of knowledge".¹¹ Origen's apophatic theology was also inspired by Plato's words on the difficulty in knowing and expressing God in his *Timaeus*.¹² Neither Plato nor the Bible committed to writing the highest teachings. This is one parallel—among many—that Origen saw between Plato and Scripture.

In Plato's *Meno* 75cd, Socrates, discussing virtue with Meno, states that in a debate it is necessary to proceed with "dialectical rigour", which consists in answering truthfully, but also "answering in those terms that the interlocutor explicitly recognizes to know". In his *Sophist*, Plato establishes the diairetic method also thanks to his "parricide" or overcoming of Parmenides' tenet that non-being is not. In 253be, the Stranger in his discussion with Theaetetus argues that dialectic is a "greatest science", μεγίστη ἐπιστήμη, a science for free people, τῶν ἐλευθέρων ἐπιστήμη, which is typical of the philosopher and identifies him or her. This is why the Stranger jokes: "looking for the sophist, we have found the philosopher!" (φιλόσοφον, 253c9). This ideal philosopher can "distinguish according to the genera [κατὰ γένη διαιρεῖσθαι], without taking a species [εἶδος] as identical if it is different, nor as different if it is identical": this is the work of the "dialectical science" (διαλεκτικῆς ἐπιστήμης, 253d). The dialectic person can "individuate one idea through many" and vice versa many ideas, individually separated, and "discern by genera" (διακρίνειν κατὰ γένος, 253d). Plato insists again that dialectic (τὸ διαλεκτικόν) can be achieved only by the person who "philosophizes correctly" (δικαίως φιλοσοφοῦντι).

This substantial identification of dialectics with philosophy is at work in the *Republic* (531–9). Here, dialectic is the supreme science, uniting the first principles of all disciplines under the super-Idea of the Good.¹³ This section is particularly important

---

9 *Phdr.* 273d8–274a3.
10 *Cels.* 6.6, possibly related to the oath of secrecy that he, Plotinus and Herennius took not to divulge Ammonius Saccas' esoteric teaching. Discussed in Ramelli (2009 and 2017a); further work is in progress.
11 *C.Io.* 13.5.27–30.
12 See Ramelli (2014).
13 See Sedley (2015); on Plato's argumentative strategy see Gonzalez (1998) and Irani (2017).

in the present research, both in Plato's conception of dialectic and in several other respects, as well as in view of the reception of Plato's dialectic in Origen.

Socrates describes dialectic as "helpful for the research of the beautiful/noble and the good", χρήσιμον πρὸς τὴν τοῦ καλοῦ τε καὶ ἀγαθοῦ ζήτησιν (531c). This is the description of philosophical exercise and ζήτησις, which from Plato onwards became a marker of philosophical inquiry, and will be central to Origen's philosophico-theological work (see below). The "first-rate dialecticians" (δεινοὶ διαλεκτικοί) are those able "to give and receive reason" of something (δοῦναί τε καὶ ἀποδέξασθαι λόγον, 531e; the dialectician can "give reason of the essence of each being", διαλεκτικὸν καλεῖς τὸν λόγον ἑκάστου λαμβάνοντα τῆς οὐσίας, 534b). For there is no other method to access the essence of each reality (αὐτοῦ γε ἑκάστου περὶ ὃ ἔστιν ἕκαστον, 533b) but "the dialectic method alone" (ἡ διαλεκτικὴ μέθοδος μόνη, 533c), until one reaches the Good itself, the supreme Idea. Indeed, the end of dialectics is to grasp intellectually what the Good Itself is (αὐτὸ ὃ ἔστιν ἀγαθὸν αὐτῇ νοήσει λάβῃ, 532b), to "define through reasoning the Idea of the Good, distinguishing it from all other realities" (διορίσασθαι τῷ λόγῳ, ἀπὸ τῶν ἄλλων πάντων ἀφελών, τὴν τοῦ ἀγαθοῦ ἰδέαν, 534bc). Origen, when reading these lines, identified the Good Itself or the hyper-Idea of the Good with God, which he, like Numenius, called αὐτοαγαθόν.

Dialectic is higher than all disciplines (τοῖς μαθήμασιν ἡ διαλεκτικὴ ἡμῖν ἐπάνω κεῖσθαι, 534e) and must be studied after arithmetic, geometry, and all preparatory disciplines (προπαιδείας, 536D); then, after studying dialectics as the top of all disciplines, Plato's model citizens will spend the next fifteen years in State administration. This stance would be advanced by Origen, who posited all disciplines as inferior to philosophy; theology, in turn, which he calls epoptics, is the highest part of philosophy. In reference to this encyclopedic preparation and the dialectician's mental attitude, Plato states that "the synoptic", the philosopher who has a synoptic or comprehensive vision, is also a dialectician (ὁ συνοπτικὸς διαλεκτικός 537c); conversely, whoever has no synoptic vision and is no philosopher is no dialectician. For Origen, the synoptic is primarily God, who sees and knows everything at once through the exemplarism of the divine Mind—this is why the Origenian Evagrius called God "essential knowledge"—and, derivatively, the Christian philosopher.

Plato insisted that the dialectician-philosopher necessarily had to be φιλόπονος, as he states twice in *Resp.* 535cd: the philosopher-dialectician must have "acumen [δριμύτητα πρὸς τὰ μαθήματα] for the sciences and learn with ease [μὴ χαλεπῶς μανθάνειν]", as Origen was himself and as he wanted his disciples to be. "One must seek a person who possesses good memory [μνήμονα]" and "completely loving of labour", πάντῃ φιλόπονον; φιλοπονία or love of toil—a trait typical of Origen, displayed by him and attributed to him even in the superlative[14]—must not concern, according to Plato, only gymnastics, hunting, and physical work (πάντα διὰ τοῦ σώματος φιλοπονῇ), but also learning (φιλομαθής). Origen was indeed a lover of learning, so as to recognize himself as Plato's φιλομαθής, and sought to inspire such a love in his disciples,[15] hearers, or readers. Origen also had an excellent memory, which helped him in his arguments and in his knowledge of the Bible, which in many cases he knew even by heart. Origen probably had Plato's requirements, analyzed here, in mind.

Plato recommends instructing people beginning in childhood, and doing so in the form of play (τοὺς παῖδας ἐν τοῖς μαθήμασιν παίζοντας τρέφε, 537a). But the instruction in dialectics must be reserved for mature years (359bc), given the utmost importance of this intellectual skill.

In the dialogue, Socrates also remarks that whatever has been said of the education of men, up to dialectics, is also the case for the education of women (540c). Origen will observe this precept in his own school, where he is even reported to have mutilated himself in order to avoid any gossip, since he taught women as well as men.[16] Socrates, the Stoics, Cynics, Pythagoreans, Epicureans, and (many) Platonists also taught women and had women as philosophers and heads of schools: Hypatia, for instance.[17] Porphyry traced this practice back to Pythagoras: "Pythagoras had many disciples, not only men, but also women, among whom Theano" (VP 19). Justin also seems to have had women among his pupils (Mart.Iust. 3–4) as did Origen's admirer Pamphilus. Origen's self-emasculation is reported by Epiphanius AH 64.3 as excessive gossip [φασί], with hypotheses on its precise modality, although scholarship is divided on its historicity. I deem rather improbable that Eusebius (and Pamphilus) invented an embarrassing tradition, which he made efforts to justify[18] by stating that Origen observed Matt. 19:12 "too literally and excessively", in order, as mentioned, to avoid gossip owing to his teaching female students. This saying about Origen's self-mutilation may even have been slander by his opponents aimed at denying his fitness for the priesthood. Still, in C.Cant. 2.1.48 Origen reports Matt. 19:12 without any kind of warning or disagreement, but he strongly approves of those who make themselves eunuchs for the Kingdom. In C.Matt. 15.3, however, Origen quotes Sentences of Sextus 13 and 273 and Philo Det. 176: both passages favored the practice of men's self-mutilation for the sake

---

14 See Ramelli (2011a).
15 As Gregory Thaumaturgus' Thanksgiving Oration makes clear.
16 Eusebius HE 6.8.1–3.
17 See Wider (1986) on female Pythagoreans, Epicureans, and Hypatia as well as Dutsch (2020).
18 HE 6.8.1–3.

of chastity. Although he usually agrees with Sextus and Philo,[19] Origen disagrees here with both for misinterpreting the Bible's spiritual meaning. Matt. 19:12 refers to *spiritual* self-neutering, Origen maintains, and this point was misinterpreted by Sextus and those who nurture "an inordinate love for chastity" (*C.Matt.* 15.1). Origen adduces both medical reasons against self-mutilation and a biblical proof: "whoever is emasculated or has his organ cut off shall not enter into the Lord's assembly".[20] This may well have been what his opponents adduced against his ordination.

In *Cels.* 7.48, also a late work like the Commentary on Matthew, Origen again praises the chastity of many Christians who voluntarily master their desires, "in that the λόγος drives out lust from their minds", as opposed to those involved in the "pagan" Eleusinian mysteries, who rather relied on hemlock.[21] Ambrose remembered Origen's condemnation of a literal exegesis of Matt. 19:12 when he remarked that those who mutilate themselves with a knife act out of weakness, because what makes a man continent is not impotence, but his own will.[22] If Origen was initially "pagan", which is uncertain,[23] his possible mutilation might have been connected with the cult of Serapis, which Egyptian Christians tolerated, even adopting elements of it.[24] Origen's moniker, "Son of Horus", may be an indicator of his connection with the cult of Horus and Serapis. Later, to justify the information about his self-mutilation, Christians may have found the explanation of a literal reading of Matthew about the blessing of "becoming eunuch" and Origen's desire to avoid gossip about the female students in his classroom—unless one assumes that the accusation of being a eunuch arose from a misunderstanding of Origen's ascetic lifestyle as a "spiritual eunuch", without any physical mutilation whatsoever.[25] That Origen understood Matt. 19:12 literally clashes, as Szram observes, with his allegorical exegesis of the Bible; however, we may view it as a very early deed, before he began to deploy allegoresis systematically. If the charge, instead, was forged from scratch in order to deny Origen's suitability for the priesthood and destroy his reputation, this would attest once again to the role of slander against his figure.[26]

---

19 Ramelli (2016), Introduction; Chapter 1.
20 Deut. 23:1.
21 Gordon (2016).
22 *Vid.* 13.75–77.
23 See Ramelli (2022).
24 On the slow process of the Christianization of Egypt, see at least Frankfurter (2017).
25 Szram (2020).
26 See Ramelli (2014 and 2019).

Plato intended to construct his own dialectic in opposition to that of the Sophists, which he attacked as anti-dialectic, and which, through mirrors of appearances, subverted knowledge instead of sustaining it.[27] Indeed, as I point out in this essay, Origen also condemned rhetoric as psychagogy.

3 The Dialektikoi, the Stoics, Aristotle, and Other Socratic Schools: The Value of Dialectics

Aristotle, who was well aware of his teacher Plato's dialectic, deemed Zeno of Elea the founder of dialectic,[28] from whom Parmenides descended philosophically—and I have already highlighted the importance of Parmenides in the history of dialectic and his attitude's (indirect) influence on Origen. It is possible that the beginning of Stoic dialectics, i.e. Zeno's dialectics, was named after the so-called Dialecticians (Διαλεκτικοί), who were called thus "because they were accustomed to construct their arguments in the form of questions and answers".[29] Aristotle criticized Plato's diairetic process (*Part.anim.* 642b–644a) and his dialectics (*Top.* 105b.30–31), but he did use the diairetic process, just as the Aristotelians after him did.[30] Aristotle seems to have subordinated dialectic to syllogistics.[31] This had an enormous impact on Stoic logic and on subsequent thought; Origen himself absorbed part of Aristotelian and Stoic syllogistics, but he was also aware of Platonic dialectic and made the most of it, especially in relation to theology, as I shall argue. Aristotle addressed the dialectical method in his *Topics*. He separates dialectic from scientific "demonstration", since it proceeds from τὰ φαινόμενα, "what appears", or from what is γνωριμώτερον or more known to us. This theory and Aristotle's *Sophistical Refutations* nuances or diminishes Plato's absolute value of dialectic as perfect philosophy. And the Aristotelian tradition followed Aristotle.[32] But Origen, as I shall argue, seems to restore dialectic to the level of the loftiest science, philosophical theology in his case, through the Logos as the highest philosopher-dialectician.

---

27 *Resp.* 454a, 475a–477b, 539b; *Soph.* 230b–235a, 250b–258c, 267e; *Gorg.* 452d–456a.
28 Diogenes Laertius 9. 25.
29 Diogenes Laertius 2.106. So Castagnoli (2010) 157.
30 Diogenes Laertius 5.23; 5.46.
31 Aristotle *Top.* 100a25–105b36 and *Metaph.* 1003a22–37, 1005b14; see also Baltzly (1999).
32 Spranzi (2011).

The Socratic Megaric school, also called the Dialectic School, also disagreed with Plato's concept of dialectic.[33] The Stoics identified the wise person as the only true dialectician—which is reminiscent of Plato's identification of the philosopher with the dialectic person—but they also classified dialectic or logic as a part of philosophy alongside physics and ethics.[34] Origen was acquainted with the Stoic and the Platonic partitions of philosophy. The Epicureans rejected dialectic as superfluous, and they spurned the liberal arts as a preparation for philosophy, let alone philosophy as a preparation for theology or culminating in theology.

4 The Academics and the Pre-Plotinian ("Pagan" and Christian) Platonists: The Division of Dialectics and the Question of the Application of Dialectics to God

While the Academics practiced dialectic but also undercut it through their skeptic stance,[35] in imperial pre-Plotinian Platonism, so-called Middle Platonism, with which Origen was familiar, Alcinous discusses (Platonic) dialectic by dividing it into diairetics (the kind of Platonic dialectic with which Origen was most familiar), horistic, analytics, epagogic, and syllogistic (διαιρετικῶς, ὁριστικῶς, ἀναλυτικῶς, ἐπαγωγικόν, συλλογιστικόν).[36] He details that the substance is the object of dialectics, which he depicts in Plato's terms as consisting of a diairetic process that moves from the universal to the individual, and the opposite process that moves from the individual to the universal. Alcinous refers this description directly to Plato. In *Phaedrus* 266BC, Socrates defines as διαλεκτικοί those who proceed through the processes mentioned later by Alcinous: "division and unification, to be able to speak and to think". In *Sophist* 218dff., Plato offers an example of diairetic procedure; in 264c, we find another example of the same procedure, aimed at defining the sophist. In 267d, Plato observes through the Stranger that none of his philosophical predecessors has applied dialectics: "Our predecessors had an ancient, foolish laziness to divide genera into species". Plato claims the gist of philosophy,

---

33 On which see the chapter by Claudia Marsico in this volume.
34 Diogenes Laertius 7.41–83; see also Ramelli (2013).
35 Cicero *Acad.* 2.91–98.
36 *Did.* 5.156.24–33 (Whittaker, p. 8). On dialectic in Hellenistic and early imperial thought up to Sextus Empiricus and Galen (Chapters 9 and 10), not in Origen but as a background to Origen, see Bénatouïl and Ierodiakonou (2019).

dialectics (here in its diairetic form), for himself as originator, and imperial Platonism, including Origen as a Christian Platonist, acknowledged Plato's pivotal novelty. Origen even developed Platonic dialectic in a theological way, as will be argued.

Alcinous distinguishes in turn five forms of διαίρεσις: of a genus into its species, of the whole into its parts, of a word into its meanings, of the subject into its accidents, and of the accidents into the subjects. The first kind is the only one that allows for the determination of the substance of the being.[37] But in the case of the supreme Godhead, "ineffable [ἄρρητος], graspable only by intellect alone [νῷ μόνῳ ληπτός], since it is neither genus nor species nor difference [ἐπεὶ οὔτε γένος ἐστίν οὔτε εἶδος οὔτε διαφορά], nor has it got any accident [ἀλλ'οὐδὲ συμβέβηκέ τι αὐτῷ]"[38] nor "all" or "parts" (ὅλον, μέρος),[39] then it is not subject to any form of διαίρεσις. It is superior to dialectics. Origen was ready to acknowledge this point, albeit with some adjustments.

The kinds of διαίρεσις listed in the *Didaskalikos* are in fact not found in Plato, although Plato does have διαίρεσις as a form of dialectics—as Origen was well aware—but they are shared by another author roughly contemporary with Alcinous, Galen.[40] The Platonists and the Stoics made ample use of διαίρεσις[41]—what we see at work here in Alcinous, Galen and some Christian Platonists, including Origen.

Clement of Alexandria could be styled a Christian imperial pre-Plotinian Platonist (commonly, a Christian "Middle Platonist"), and, although never cited by Origen in his extant works, was certainly known to him.[42] Clement applies Plato's description of dialectics, and more specifically diairetic dialectics, albeit simplified, to his own depiction of dialectics in *Strom.* 6.10.80.4: "But the gnostic [the perfect Christian] avails himself or herself of dialectics [τῇ διαλεκτικῇ] by means of the division [διαίρεσιν] of the genera into species, and the distinction [διάκρισιν] of beings". Clement refers to *Resp.* 534d and is using Platonic dialectic terminology, such as διαίρεσις. All this, as we shall see, appears again in Origen and in his disciples' report on his teaching. Indeed, Clement associated dialectic primarily with diairetic dialectic: "true dialectic is knowledge that can make divisions among the objects of though" (*Strom.* 1.176.3) and highlights the anagogic value of dialectic as a guide to God, an ascension "to the most mighty substance of all, and even beyond,

---

[37] *Did.* 156.34–157.10 (Whittaker, pp. 8–9).
[38] *Didaskalikos* 10.165.5–6 (Whittaker, p. 23).
[39] 165.5–16 (Whittaker, pp. 23–24).
[40] Ps. Galen, *History of Philosophy* 4 (XIX 237) and Galen, *The Doctrines of Hippocrates and Plato*=- *Plac. Hipp. Plat.* 9.9 (V 796–805).
[41] Galen, *Plac. Hipp. Plat.* 4.13; 7.200.
[42] See Ramelli (2021a).

to the God of the universe" (*Strom.* 1.177.1). The theological function of dialectics, which will be developed by Origen, implies that Christian philosophers should be valiant dialecticians. Jesus himself, as already Clement claimed, was a strong dialectician as he responded to the temptations of the devil (*Strom.* 1.176.3).

While Alcinous listed five forms of διαίρεσις, Clement lists three:[43] what is subject to διαίρεσις (τὸ διαιρούμενον) can be: genera into species (εἰς εἴδη ὡς γένος), the whole into its parts (εἰς μέρη ὡς ὅλον), and something into its accidents (εἰς συμβεβηκότα). The first kind of διαίρεσις, of a genus into species, has a limit in τὸ τί ἦν εἶναι, an Aristotelian expression which in this case is the simplest species (ἁπλούστατον εἶδος), not the single individual but, for instance, the human species: this is ἄτομον or indivisible.[44] It is probable that Origen was familiar with both Alcinous' and Clement's classification of diairetic dialectics.

Interestingly, according to Clement, since God is no substance, genus, etc., God is not subject to διαίρεσις and thus dialectics cannot be applied to the Godhead. This is in full accord with Clement's apophatic tendency[45] and Alcinous' above-mentioned caveat. (Origen probably knew both Alcinous' and Clement's provisos regarding God in dialectics and shared Clement's apophatic tendency, but this did not prevent him from joining dialectics and theology very closely, as I shall demonstrate). Indeed, Clement observes that God "cannot be expressed [ῥητόν]" because God "is neither genus [γένος] nor difference [διαφορά] nor species [εἶδος], neither indivisible [ἄτομον] nor number [ἀριθμός], neither an accident [συμβεβηκός] nor anything that can have accidents [ᾧ συμβέβηκέν τι]. Nobody would be correct in saying that God is a whole [ὅλον], since the whole is ordered on the basis of size [ἐπὶ μεγέθει] and God is the Father of all [τῶν ὅλων πατήρ]. Nor can it be said that there are parts [μέρη] of God, for the One is indivisible [ἀδιαίρετον γὰρ τὸ ἕν], and for this reason it is also unlimited/infinite [διὰ τοῦτο δὲ καὶ ἄπειρον], if this is considered not in the sense of something that cannot be exhausted / infinite in extent or duration [οὐ κατὰ τὸ ἀδιεξίτητον νοούμενον], but in the sense of what is adimensional and unlimited [κατὰ τὸ ἀδιάστατον καὶ μὴ ἔχον πέρας] and therefore without shape and unnamable [ἀσχημάτιστον καὶ ἀνωνόμαστον]".[46] The term ἀδιεξίτητον was used by Aristotle, Alexander of Aphrodisias, and Plotinus.[47] The latter, like Clement, hypothesizes: "if ἄπειρον is taken in the sense of ἀδιεξίτητον" (here in reference to matter), one should know that "there is nothing of the sort among the beings". Clement devoted the

---

[43] *Strom.* 8.19.3 (Havrda, p. 108).
[44] *Strom.* 8.17.4 (Havrda, p. 106).
[45] Hägg (2006) and Ramelli (2009 and 2021b).
[46] *Strom.* 5.12.81.5–82.1: SC 278.158–160.
[47] Aristotle *Phys.* 207B29, Alexander of Aphrodisias *In Arist. Top.* 86.27, and Plotinus *Enn.* 2.4.7.14–15.

so-called Eighth Book of the *Stromateis*—a long section from which I have quoted above—to logic. Here he depends a great deal on Stoic lore, and it is uncertain whether he was acquainted with Aristotle's *Categories* directly or through the channel of Stoic refutations and references.[48]

The long citation just examined also shows that Clement styled God ἄπειρον before Origen. Did God's—and only God's—absolute eternity match a notion of God's infinity in Origen? It is usually assumed that Origen did not deem God infinite—because Greek thought assimilated infinitude to imperfection—while Gregory of Nyssa did. Ekkehard Mühlenberg saw no authentic theological or philosophical antecedents to Gregory of Nyssa's notion of God's infinity.[49] Both Origen and Gregory of Nyssa were acquainted with Philo, who provided starting points for divine infinity.[50] Mark Weedman called attention to another antecedent: Hilary[51]—who, I note, was influenced by Origen. According to Origen, I remark, "God's greatness [μεγαλωσύνη] has no limit [πέρας]" and God's providence runs "from the infinite [ἐξ ἀπείρου] to the infinite [ἐπ'ἄπειρον] and further".[52] In Origen's original and authentic Greek texts, God is expressly declared to be infinite: mindless people "compare the Infinite [ἄπειρον], surpassing in excellence all created nature, with objects that are in no way comparable";[53] God is "from infinities to infinity", ἐξ ἀπείρων ἐπ' ἄπειρον.[54] In Mühlenberg's opinion, Origen deemed God finite because God is Nous (Νοῦς, Intellect).[55] However, Origen conceived of God as both Nous and *beyond Nous:* hence God's infinity. Moreover, Gregory of Nyssa also described God not only as One, but also as "Nous and Logos",[56] which did not prevent him from postulating God's infinity. Ovidiu Sferlea rightly argues that the concept of divine infinity is already in *De hominis opificio* and *De anima*.[57] Although Sferlea, like everybody else, overlooks Origen's influence, it is significant that both works are heavily impacted by Origen. To Origen I also add the important antecedent of Clement, who, as noted above, already deemed God ἄπειρον—[58] and Origen was very probably aware of this. Plotinus, also mentioned above, was a source of

---

48 Havrda (2016).
49 Mühlenberg (1966 and 2015).
50 Geljon (2005).
51 "The Polemical Context of Gregory of Nyssa's Doctrine of Divine Infinity", *JECS* 18 (2010), pp. 81–104.
52 *Sel.Ps.* 144.
53 *Cels.* 3.77 (trans. Chadwick).
54 *Or.* 17.16; see the argument in Ramelli (2018).
55 Mühlenberg (1966), p. 81.
56 *Hom.op.* 5.2.
57 Sferlea (2013); see also Sferlea (2010).
58 *Strom.* 5.12.81.5–82.1: SC 278, 158–160, quoted and analyzed earlier.

inspiration for Gregory of Nyssa's idea of divine infinity and therefore incomprehensibility.

5 The Decline of Dialectis: Ambrose's Testimony, Augustine's Dialogues, and a Comparison with Origen

In the time of Clement and Origen, dialectics was still a vital part of philosophy, whereas by the late 4th century Ambrose attests that the study of dialectics was declining in philosophical schools in his day: "Even in their schools, dialectics is not heard anymore".[59] Clearly, one must take into account Ambrose's rhetorical vehemence here, but his remark seems to contain a nugget of truth. Scholarship remains divided on whether Augustine's early dialogues, written shortly after Ambrose, do indeed express a dialectic approach to philosophical theology. According to Erik Kenyon,[60] Augustine's dialogues *Contra Academicos, De beata vita, De ordine, Soliloquia, De immortalitate animae, De quantitate animae, De musica,* and *De libero arbitrio* (apart from *De magistro*) are real pieces of "zetetic" inquiry and methodological investigations into the act of inquiry itself. The dialogues reflect, indeed, on "the act of inquiry itself: The fact that we can inquire at all tells us various things about ourselves. By reflecting on our own act of inquiry, we are put in a position to improve how we go about inquiring",[61] because reflection on our inquiry yields a discovery of "cognitive norms of thought" (34) operative in "most if not all acts of rational inquiry" (40). These dialogues start with an aporetic debate, reflect on the act of debating "as instances of rational activity... and draws various conclusions about human nature as a way of expanding his interlocutors' stock of self-knowledge" (12); and propose a plausible conclusion. Later, however, Augustine abandoned the genre of dialogue—an intermediate stage might be the inner dialogue in *Soliloquia*[62]—and its philosophical dialectics.[63]

---

59 *De fide* 1.13.74.
60 Kenyon (2018).
61 Kenyon (2018), p. 12.
62 See Ramelli (forthcoming b).
63 Hoffmann (1966) and Voss (1970), continued by the handbook by Rigolio (2019), Clark (2008), Stock (2010 and 2012, pp. 315–347); and Guiu (2021). Stock argues that the tradition of the ancient philosophical dialogue becomes more introspective in Augustine and ultimately becomes an "interior dialogue". Clark goes even further and argues that a threefold abandonment of dialogue takes

Instead, Origen exercised dialectics also by engaging in numerous dialogues throughout his life. These were recorded and published in his mature age—even leaving aside the *Dialogue of Adamantius*, which is probably based on Origen's own arguments (contrary to what is generally assumed) but is literarily reworked and is not authored by him.[64] Part of the literary strategy in this dialogue consists in the figure of the arbiter, who is a pagan philosopher: Eutropius.[65] He functions as an expression of the pure, and expectedly impartial, *logos*, who, at the end of the second book,[66] and later at the end of the whole dialogue, proclaims the victory of Origen-Adamantius. The Philocalists call this dialogue *Dialogue of Origen with Marcionites and other Heretics* (*Philoc.* 24), and Anastasius Sinaita and the *Praedestinatus* also call it a dialogue of Origen.[67] Origen is recorded to have had many public debates against other Christians on doctrinal points on the basis of Scripture and argument. For example, he held one with Heraclides, which is preserved thanks to a Toura papyrus probably based on Pamphilus' and Eusebius' edition, and others that have been lost, with Candidus the Valentinian,[68] with Beryllus in Bostra,[69] with some heretic Arabians,[70] and still others with Jewish rabbis, and other public debates, διαλέξεις ἐπὶ τοῦ κοινοῦ, that are attested by Eusebius.[71] Only from the age of 60 onwards did Origen allow a

---

place in Augustine: "movement away from dialogue as a literary genre; loss of confidence in dialectic as the way to achieve truth; and failure to engage with people who held different views".

64 See the full treatment of all of Origen's dialogues in Ramelli (2012–2013); on arguments on the genesis of the *Dialogue of Adamantius*, see Ramelli 2010, 2012–2013, and 2020).

65 This is not only clear from the dialogue itself but also remarked on in the Prologue, and it is singled out and given special attention in the second Prologue included in ms. F. In the dialogue, Eutropius' role as an arbiter and judge elected by both Adamantius and Megethius is emphasized from the beginning (804b Bakuyzen).

66 At the end of Book II, he proclaims the victory of Adamantius over the two Marcionites, Megethius and Marcus: *Megethius et Marcus et caeteros istiusmodi dogmatis sectatores insipientes simul et ignorantiae uideo errore deceptos ... mihi uidetur illa sola esse recta definitio quam statuit Adamantius de uno Deo creatore et conditore omnium habente uerbum et spiritum per quem et omnipotens est et imperium uniuersorum tenet, cui contrarium potest esse nihil ... quem recta sequitur ecclesia quae dicitur catholica ... cui opto etiam ego congregari et unus effici ex his qui iam recte de deo uel intellegunt uel fatentur.* Adamantius' orthodox faith is confirmed by the philosophical *logos*.

67 Anastasius uses a part of the *Dialogue of Adamantius* (818d–819b van de Sande Bakhuyzen) in his *Quaestiones*, 48, and presents it as a work of Origen. In *Praedestinatus* 21, it is attested that *Marcionitae, cum universalem orientis ecclesiam macularent, ab Origene superati, confutati, et per singulas sunt civitates damnati.*

68 Attested in Origen's *Letter to Friends in Alexandria*, PG 17.625B; Jerome, *Ep.* 33.4: *Dialogus adversus Candidum Valentinianum.* Immediately before this, Jerome also attests Origen's two *dialogi de resurrectione*.

69 In addition to Eusebius *HE* 7.33.3, see Jerome, *De vir. Ill.* 70.

70 Eus. *HE* 6.37.

71 *HE* 6.36; Scherer 1960, 13–14 n. 3.

written record of his speeches, such discussions included. The very same term, διαλέξεις, indicates both Maximus of Tyre's diatribes and Origen's public discussion with Beryllus in Bostra in Eusebius *HE* 7.33.3.

The genre of the dialogue, indeed, was cherished not only by "pagan" philosophers, who emulated the Socratic-Platonic dialogues,[72] but also by Christian philosophers, especially Christian Platonists, such as Bardaisan,[73] Origen, and Gregory of Nyssa; later, philosophical and literary Byzantine dialogues flourished, including Jewish-Christian dialogues.[74] And Origen valued dialectics, founded by Plato,[75] as did Plotinus (his fellow-student at Ammonius Saccas' school), who "restored" Plato's dialectics and deemed it superior to logic.[76]

6 Origen and Dialectics in Scripture and Philosophy: Dialectics vs Psychagogy and the Heritage of Platonic and Stoic Dialectics

Origen, who is virtually never treated in histories of logic or dialectic, not only knows Plato's dialectic, but even attributes dialectic to Scripture, especially in that he reads the latter through the lens of Platonism.[77] Indeed, Origen boldly states that Scripture speaks πάνυ διαλεκτικώτατα, "in a really most dialectical way". This means that the Bible, in his view, is in perfect accord with Plato's diairetic dialectic, already praised by Alcinous[78] and then by Plotinus in a methodological "zetetic" passage,[79] but claimed to belong to rhetoric by Aelius Aristides.[80] Ori-

---

[72] The "pagan" reception is investigated by Stavru and Moore (2018), from Plato to Xenophon, Plutarch, Maximus of Tyre, Themistius, and Proclus.
[73] Ramelli (2009c and 2009–2019).
[74] Ramelli (2007a; 2007b, and 2018c) for Nyssen. On Byzantine dialogues, see Cameron (2016) and Cameron–Gaul (2018).
[75] See, e.g., Gourinat–Lemaire (2016), especially pp. 43–135.
[76] Jean-Baptiste Gourinat, "Logique et dialectique chez Plotin", in Gourinat–Lemaire (2016), pp. 363–382.
[77] See the demonstration in Ramelli (2021c).
[78] "In general, the man [Plato] was supremely competent in, and a connoisseur of, the procedures of definition, division, and analysis, all of which demonstrate particularly well the power of dialectic" (*Did.* 6.159.45–160.3, trans. Dillon).
[79] *Enn.* 1.3.4.2–7 and 12–15: "using the Platonic diairetic method to distinguish the Ideas/Forms... to determine the essential nature of things and the primary kinds [Plato's μέγιστα γένη in *Soph.* 254b–256a, being motion rest sameness and difference, which Plotinus called πρῶτα γένη], interconnecting intellectually the things that come from these kinds, all of this throughout all the intelligible".

gen, after declaring that Scripture speaks "in a really most dialectical way", gives a specific example: Scripture does not say, "before I created [ποιῆσαι] you in the womb, I know you", because it is when the divinity created the human in God's image that God "has created" (πεποίηκε); on the contrary, when God made the human from the earth, God simply "moulded" it (ἔπλασεν). Thus, the human "created" (ποιούμενον) by God is not that which "is formed in the womb", but "what is moulded from the earth is what is founded in the womb",[81] that is, the mortal body. This is the human body after the fall. This, and not the body tout court, is what Origen associates with death and sin.[82] Therefore, what Origen indicates as "dialectics" operating in the Bible, has to do here with the precise conceptual distinction between God's action of creation and God's action of molding, the former applying to the human being in the image of God, namely in its intellect and self-determination, and the latter to the human body. Origen, therefore, seems to allude to diairetic dialectic.

In the prologue to his Commentary on Genesis, Origen warns that the findings of dialectical inquiry are not always certain: "If some profound truth occurs to someone in discussion/debate [disceptatione], this must be said [dicendum], but not with absolute certainty" (cum omni adfirmatione).[83] Origen's humble prudence and awareness of human fallacy comes from the absence of a direct divine revelation (facie ad faciem) as experienced by the apostles; therefore, he concludes, "I am aware of my ignorance", ignorantiam nostri non ignoramus. Now, this is a clear allusion to Socrates: "I don't presume to know what I don't know" in Plato's report in Apol. 21d.[84] Origen often warns that the argument he is going to produce (or has just made) will be, "not a truth of faith, but an object of examination and discussion ... more an object of readers' investigation than an exact definition".[85] This is part of Origen's zetetic, heuristic method, which resonates well with his

---

80 "Dialectic is a constituent part of rhetoric" (τὴν διαλεκτικὴν μέρος τι τῆς ῥητορικῆς, Or. 3.509, ed. Trapp 2017). Note the difference with the Stoic theory. Aristides praised Plato for his natural genius and greatness and a "kingly" scorn for reputation (Or. 3.577, 586, 663), as "the best of the Greeks" (Or. 3.42, 461. 557, 607, 663), but primarily as "the father of oratory" (Or. 2.465). His Platonic Orations (Or. 2 and 3) elicited the refutations of Neoplatonists such as Porphyry, who attacked them in 7 books; Synesius, Dio 3.5, who depicted him as an enemy of philosophy, and Olympiodorus, who, commenting on the Gorgias, called Aristides φλήναφος, "a babbler". According to Trapp (2020), p. 88, Aristides' Platonic orations testify to "the continuing awkwardness of philosophy and of Plato in his day, in relation not just to rhetoric but to classicising paideia more generally".
81 Τὸ πλασσόμενον ἀπὸ τοῦ χοῦ τῆς γῆς, τοῦτο ἐν κοιλίᾳ κτίζεται (H.Ier. 1.10).
82 Ramelli (2018b and 2021c).
83 Pamphilus Apol. 7.
84 See, e.g., Ambury and German (2019).
85 Princ. 2.8.4–5. This is but one example among many.

dialectic frame of mind. In the general prologue to his *First Principles* (Περὶ ἀρχῶν) 6, Origen programmatically declares that it is necessary to apply rational investigation to the issues left unclarified by Scripture and the apostolic tradition. And in Book IV of the same work, he subsumes exegesis of Scripture under philosophy.[86] For the soul must stick to reason and faith together:[87] no reason without faith, since Origen's philosophy is Christian, but no faith without reason, because Origen's Christianity is philosophy. Faith and reason cannot diverge since Christ is the Logos. Neglecting (*neglexisse*) the study of Scripture, the Logos, is more serious than neglecting the Eucharist lest parts of it drop inadvertently.[88]

We are also informed about Origen's attitude to dialectics and his teaching in this respect from Gregory Thaumaturgus, in his *Thanksgiving Oration to Origen*, 8. Although this section (6–9) might have been added later,[89] this is the perspective of Origen's students, who were a first-hand source concerning his courses. According to this testimony, which is in full accord with Origen's own declarations, he taught dialectic (διαλεκτική) to his students. This also functioned, in Origen's intention, as a philosophical, critical attitude against psychagogy. Origen formed in his disciples "the moral form that dialectic alone can straighten" (ὅπερ εἶδος διαλεκτικὴ κατορθοῦν μόνη εἴληχε). Dialectics corrects "the inferior level of our soul" (τὸ ταπεινὸν τῆς ψυχῆς), which is impressed by what is marvelous but is also "irrationally impacted by consternation/terror/passion, and even when we know, we cannot rationally deal with it, like irrational beasts" (ἀλόγως ὑπεπτηκότων δὲ ὑπὸ ἐκπλήξεως, εἰδότων δὲ οὐδ' ὁτιοῦν ἐπιλογίσασθαι, δίκην ἀλόγων ζῴων). Origen means that we are subject to psychagogy, but dialectics can correct this defect.

Origen corrected the liability of the soul to passions and impressions, whereas it should choose reason primarily through dialectic, together with other disciplines (ἀνορθῶν μαθήμασιν ἑτέροις), such as physics, astronomy, and ethics. He expounded "each of the beings" (ἕκαστα τῶν ὄντων), applied διαίρεσις or distinction—a technical term in Plato's dialectic, as seen—by tracing them back to the first principles, and "reassembling by means of reason" (ἐπιπλέκων τῷ λόγῳ). These are the two branches of dialectics described by Plato. In Plato, διαίρεσις is the dialectical process of division of genus into species.[90] The opposite of diairetic dialectic is the συναγωγή or "bringing together", mentioned by Plato *Phaedr.* 266b, parallel to the reassembling of which the apologist[91] speaks in reference to Origen's dialectical ac-

---

86 Ramelli (2022b).
87 *C.Cant.* 2.10.7.
88 *H.Ez.* 13.3.
89 Scholars are divided on this point, which is, however, irrelevant to our research.
90 *Soph.* 267D.
91 Gregory Thaumaturgus or the author of the *Thanksgiving* or *Panegyrical Oration* to Origen.

tivity in his education. Thus, the panegyrist remarks, thanks to his "rational arguments in his teaching [τῆς παρ' αὐτοῦ διδασκαλίας καὶ λόγων], both those he learnt and those he excogitated personally [ὧν τε ἔμαθεν ὧν τε ἐξεύρετο]", Origen "put in our souls a rational admiration instead of an irrational one", ἀντὶ ἀλόγου λογικὸν ταῖς ψυχαῖς ἡμῶν ἐγκατέθετο θαῦμα. The stress is laid on the fact that Origen's teaching, guided by a dialectic method, removed irrational passions from his disciples' souls, replacing them with rational admiration for God and divine providence.

Now, Origen himself, in a work of certain attribution, juxtaposed psychagogy to dialectics, associating the first to rhetoric and the second to rationality, the Logos. He affirmed that the soul ought to be aware whether it is influenced by rhetorical strategies or it relies on reason, whether it judges from likeness or truth, and whether it is moved by the beauty of words: *sed et in intellectu perpensabit semet ipsam huiusmodi animam, ut cognoscat utrum facile eam moveat cuiuscumque verisimilitudinis auditio et subripiatur ei arte vel suavitate sermonum an raro hoc an numquam patiatur.*[92] Origen was well aware of the psychagogic function of rhetoric. The strong and perfect soul is not subject to it, because it judges on the plane of rationality and not at the level of emotions. The level of rationality is the one of dialectic, which Origen taught his disciples, in opposition to liability to passions and rhetorical psychagogy.

Origen likely knew that Plato had contrasted rhetoric with dialectics-philosophy as devoted to truth[93] and had criticized rhetoric as capable of supporting falsehoods and as a "maker of persuasion" in the psychagogic sense (πειθοῦς δημιουργός, *Gorg.* 453a). Actually, Origen may well have been influenced by Plato in contrasting rhetoric with dialectic. Plato warned against rhetorical psychology, as Origen did later. Plato intended to eliminate mere verbal dispute, since philosophy must aim at the search for truth: this is the drift of dialectic.[94] One must tell the truth, including what one really thinks, otherwise one will destroy the Logos (*Gorg.* 495a). Origen would elaborate on this point, also on the basis of the evangelical identification of Christ, the Logos, with Truth.

Alexander of Aphrodisias, an older contemporary of Origen, whom the latter is likely to have known and used in the construction of his philosophical theology, as I argued elsewhere,[95] displays a double attitude towards dialectics-logic, which is

---

92 *C.Cant.* 5.2.14.
93 *Gorg.* 448D; 449B.
94 *Meno* 75CD; *Gorg.* 471E–472C; *Prot.* 337AB; *Phaed.* 101D–102 A; *Theaet.* 167E; 172D; *Phil.* 11C; 17 A; 50B.
95 Ramelli (2014a).

also present in Origen. In his commentary on Aristotle's *Analytica Priora*,[96] Alexander expresses the twofold status of dialectics-logic, mentioning those who considered it "a part" (μέρος) of the philosophical sciences, as one of them (a "treatment", πραγματεία), and those who deemed it a transversal "instrument" (ὄργανον) for all sciences and philosophy, and not as a particular science itself. We shall see that in his preface to his Commentary on the Song of Songs, Origen expressed the same alternative, referring to different views: some philosophers considered it a part of philosophy, others as interwoven with all of its parts.

Origen knew and praised Plato's dialectics, including diairetic dialectic, and employed Aristotle's and especially the Stoics' logic-dialectics (which in turn was influenced by the Megaric school). Origen actively used logic, for instance in *Contra Celsum*, in which Stoic logic is paramount; he was probably familiar with some of Chrysippus' extensive oeuvre at first hand and appears to be "a considerable supplier of material, some of it of great interest to the historian of Stoic logic".[97] Origen's logic has been described as "a mixed Stoic-Aristotelian logic intermediated by Middle Platonism than direct Stoic sources".[98]

The few scholars who have explored Origen's logic have focused on *Contra Celsum*—and this is probably why Origen has been associated primarily with Stoic logic, instead of paying attention to Platonic dialectic as well as a prominent source of Origen's dialectic. In fact, there is much more, as I show here, to Origen's appreciation of dialectic, especially Platonic dialectic. Stoics generally considered dialectic to be a part of logic, along with rhetoric, and their very founder Zeno represented dialectics as a closed fist, and rhetoric as an open hand, as reported by Sextus Empiricus.[99] This similitude probably pointed to the argumentative and more concise and compact nature of dialectics as opposed to the more expanded discourse of rhetoric. Chrysippus seems to have identified logic and dialectic, arguing that logic (λογικόν) is one of the three branches of philosophy along with physics and ethics, studies the Logos (περὶ τὸν λόγον), and "is also called dialectics" (δια-

---

[96] *CAG* 2/1. 1.3–4.29.
[97] See Rist (1981), pp. 64–78 and Hadot (1989), pp. 183–188; see also Somos (2013a), pp. 29–40, (2013b), pp. 409–421, and (2015) on what Origen said about logic, dialectics, and theory of science, and how he structured his argumentation (Somos 2015, pp. 73–92 on *Princ.*; Chapter 9 on *Cels.*): Origen had a professional knowledge of dialectics and logical literature (Somos 2015, pp. 188 and 209) from Platonic, Stoic and Aristotelian logic (Somos 2015, p. 192), and "used, reused, adapted, and reformed Greek philosophical thoughts and methods belonging to logic and created a theory of his own about the relationship between the divine Logos and human logos" (Somos 2015, p. 210).
[98] Somos (2015), p. 202.
[99] *AdvMath.* 2.6–7; SVF 1.75.

λεκτικόν).¹⁰⁰ Other sources on the Stoic tripartition of philosophy, instead, do not identify logic with dialectics but only speak of logic.¹⁰¹

According to Alexander of Aphrodisias, the Stoics maintained that "the sage alone is dialectic", μόνος ὁ σοφὸς διαλεκτικός, and dialectics was the fullest form of philosophy (τῆς τελειοτάτης φιλοσοφίας).¹⁰² Such declarations agreed with Plato's view of dialectics, and Origen received both—but Plato had the priority, so much so that Origen, as we shall see, sometimes criticized Stoic logic, but *not* Platonic dialectic. According to Sextus Empiricus, another semi-contemporary of Origen, the Stoics defined dialectics (διαλεκτική) as the science of true things, false things, and things that are neither true nor false.¹⁰³ Chrysippus defined dialectics as the science of discussing/debating correctly (ἐπιστήμη τοῦ ὀρθῶς διαλέγεσθαι, SVF 2.2.131). Origen, who was a strong debater, as his byname Adamantius was interpreted ("man of stainless steel", "impossible to defeat in argument"), both incarnated and promoted this model of dialectics. Indeed, Origen's surname, Adamantius, was explained by Photius as a reference to his philosophical strength in dialectical argumentation: "They say Origen was also called Adamantius, because, whatever arguments he put together, they seemed interconnected by stainless-steel [ἀδαμαντίνοις] bonds".¹⁰⁴ Already according to Jerome, a proof that Origen was an "immortal genius" is given by his proficiency in dialectics: "He also learnt dialectics in such a way as to have as disciples even scholars in secular literature".¹⁰⁵

Epicurus criticized dialectics,¹⁰⁶ and this was one more element of dissension with Origen, who has many other criticisms.¹⁰⁷ This may also be the reason why

---

100 Aetius *Placita* 1 proem. 2, SVF 2.35.
101 Diogenes Laertius 7.39 = SVF 2.37; Sextus Empiricus *AM* 7.22 = SVF 2.44; *AM* 7.16 = SVF 2.38; Plutarch *Stoic.rep.* 9 = SVF 2.42.
102 In the Preface to his Commentary on Aristotle's *Topics*, SVF 2.2.124. See also Boeri and Salles (2014), pp. 94–252 and Ierodiakonou (2019), Chapter 4.
103 *AdvMath.* 11.187.
104 Origen probably used Ἀδαμάντιος, "man of stainless steel"; Epiphanius confirms that Origen ascribed it to himself (*AH* 2.522.7). Adamantius as Origen's byname is attested by Eusebius, *HE* 6.14.10, the *Dialogue of Adamantius*, Rufinus, and later Christian sources. It was regularly used by Jerome when he was still Origen's follower (*Vir.Ill.* 2;54; *Ep.* 33.3, which refers "Adamantius" to Origen's zeal for the investigation of Scripture). Origen is only attested as "Adamantius" in *Christian* sources, never in "pagan" sources, e.g., Porphyry (*ap.* Eus. *HE* 6.19.4–8; *V.Plot.* 14), Hierocles, and Proclus. Origen's appellative "hardworker" likewise suggests Origen's toil (φιλόπονος/φιλοπονώτατος, Athanasius, *Decr.* 27.1–2; *Qui dixerit verbum* PG 26.649.21 ; Socrates, *HE* 6.13, etc.). Now, this toil was mainly "zetetic" and dialectical.
105 *Illud de immortali eius ingenio non tacens, quod dialecticam quoque ita didicit, ut studiosos quoque saecularium litterarum sectatores haberet* (*Vir. Ill.* 54).
106 Sedley (2019).

Origen, while attacking Celsus in *Contra Celsum*, labelled him an "Epicurean", despite being aware that he was a Platonist. The Christian Platonist Origen felt much more at ease in attacking an Epicurean than a Platonist.

In *Cels.* 2.20, which is significantly included in von Arnim's *Stoicorum Veterum Fragmenta* as 2.957 qua important *testimonium* of Stoic logic, Origen counters Celsus' identification of divine prescience with divine necessity (if God foresees anything, God makes it necessary) using the Stoic ἀργὸς λόγος.[108] He takes the example of Laius, a typical Stoic instance: an oracle told Laius to have no offspring, because he would be killed by his child, while both having and not having children were possible for Laius. Origen explicitly states that the predicament of Laius is similar "to the so-called 'idle argument', ἀργὸς λόγος [ὁ ἀργὸς καλούμενος λόγος], which is a sophism [σόφισμα ὤν]". This sophism endeavors to convince an ill person not to be cured by a doctor; the argument is the following: if it is fated that you recover from an illness, it will be useless for you to go to a physician; if it is fated that you do not recover, it is equally useless to go to a physician. Origen adds the following ἀργὸς λόγος: if it is fated that you will have children, it will be useless for you to have intercourse with a woman; if it is fated that you will have no children, it is equally useless for you to have intercourse with a woman. Origen here objects to the Stoic ἀργὸς λόγος, observing that without intercourse it is impossible to beget children; therefore, the Stoic argument is invalid. Likewise, medicine is necessary to heal, so going to the doctor is not useless. Therefore, the relevant "idle argument" is fallacious.

While Origen praises Plato's dialectic, even claiming its presence in the Bible, and uses diairetic dialectic especially, he contests an example of Stoic logic. But elsewhere he does use Stoic logic, depending on his own argumentative needs. Origen was acquainted with Chrysippus, cited his ideas on several occasions, and called him "honourable philosopher" (*Cels.* 4.48), but he disagreed with his physical kind of allegoresis (*ibid.*), as opposed to Origen's own spiritual, noetic allegoresis, and in general with Stoic materialism and immanentism, as well as with the Stoic theory of apokatastasis. As was recorded in Origen's day, Chrysippus "became so renowned in dialectics that most people thought that, if there were dialectics among the deities, this would be precisely that of Chrysippus".[109] The Stoics were reported to have considered dialectics or logic as a branch of philosophy

---

107 Ramelli (2020b).
108 Cicero, *Fat.* 28–30. A comparison between Cicero's report of the Stoic ἀργὸς λόγος and Origen's use of it in this passage may be found in Bobzien (1989), pp. 207–209.
109 Diogenes Laertius 7.180.

along with physics and ethics and as able to protect securely the other two branches, or the intellect.[110]

In *Cels.* 7.12–15, Origen applies logic to the defense of Christian prophecy[111] and explicitly mentions the Stoics as antecedents for his argument. Celsus maintained that the prophets could not have foreseen Christ's Passion (against, e.g., Isaiah's prophecy of the suffering servant of the Lord[112]). Origen replies that Celsus' reasoning is mistaken, since it would make hypothetical premises result in contradictory conclusions. 1) If the prophets state that God will be a slave or fall ill or will die, this will happen, because the prophets necessarily tell the truth. 2) But if the prophets foresee that, and such things are intrinsically impossible, what the prophets foresee about God will not come to pass. Origen here traces Celsus' argument back to the argument of the "two conditionals" or "two propositions" or "two figures" (διὰ δύο τροπικῶν), which is again Stoic.[113] He argues that "when two hypothetical premises result in contradictory conclusions, by the syllogism known as that διὰ δύο τροπικῶν, the antecedent of the two premises is denied", namely that God will be a slave, or fall ill, or die. Origen's conclusion is therefore that the prophets did *not* foretell that God will be a slave, or fall ill, or die. Immediately afterwards, Origen clarifies further the Stoic argument he is using in his refutation of Celsus' anti-Christian arguments: "The argument runs as follows: If A is true, B is also true; if A is true, B is not true; therefore, A is not true".

Origen, after using their argument διὰ δύο τροπικῶν, explicitly cites the Stoics' examples: "The Stoics offer the following, concrete instance of this, when saying: 1) if you know that you are dead, you are dead; 2) if you know that you are dead, you are not dead; 3) it follows that you do not know that you are dead". After paraphrasing this reasoning, he concludes: "The same kind of argument is implicit in Celsus' assumption" about the prophets' foreseeing of the Passion. As earlier, Origen is here refuting a Stoic argument used by Celsus. Thus, he uses Stoic logic to support his own arguments as well as to reject those of his opponents. Origen explains the first two verses of the Johannine Prologue by means of predicative logic. In *C.Io.* 2.11.65, he uses both Stoic and Aristotelian logic terminology, calling propositions respectively ἀξιώματα and προτάσεις. Commenting on John 8:42, Origen uses Stoic propositional logic, and particularly Chrysippean anapodictical arguments, of the following format:[114] if A, then B; but non-B is the case; therefore,

---

110 Sextus Empiricus *Adv.Math.* 7.22–23; Philo *Agricult.* 15–16.
111 Ramelli (2017b).
112 Ramelli (2016), pp. 81–82.
113 Attested in SVF 2.248 (from Galen, *Plac. Hipp Plat.* 2.3) and by Sextus Empiricus, *Pyrrh.Hypot.* 2.3 (64 Mutschmann).
114 *C.Io.* 20.17.135–140, SC 290.224–226.

non-A:[115] he applies it to Jesus' words: if God were your father (A), then you would love me (B); but you do not love me (non-B): therefore, God is not your father. Origen uses Stoic logic to support his own theology, but also refutes it repeatedly, to deconstruct the arguments of his opponents. By contrast, he never attacks Plato's dialectics, and even claims to find it in Scriptures.

7 Dialectics, "Zetesis", and the Interrelation between Dialectics and Theology

As anticipated, Origen's "zetetic" method can be considered an expression of dialectics. Origen systematically pursued philosophical investigation, ζήτησις, dialectically in his exegesis and philosophical theology, like Plato and Plotinus. Like Plato, who pursued "paideia rather than didaskalia",[116] Origen was a "zetetic" and maieutic teacher, wanting his disciples and readers to practice zetesis in turn.[117] Ζητέω, ζήτησις, ζήτημα, and ζητητικός are technical terms for philosophical investigation; Plato first used them in this technical sense. Ζήτησις specifically designates philosophical research.[118] Ζητητικός is the philosopher who investigates.[119] Ζήτημα is philosophical research, the object of investigation.[120] Occurrences of ζητέω are frequent in Greek philosophers, especially Plato himself.[121] In Platonism, "zetetic" was a category of Plato's dialogues: some were labelled "instructional" (ὑφηγητικοί), others "investigative" or "zetetic" (ζητητικοί), for "exercise, argument, and refutation".[122]

Origen's "zetetic" philosophizing was similar to that of his fellow-disciple Plotinus, who encouraged his followers to investigate (ζητεῖν) issues by themselves and ask questions,[123] and defended the dialectic, question-and-answer method,[124] encouraging ἀπορίαι in his teaching and including some in his *Enneads*. Longinus and Porphyry even applied the word ζητητικοί to philosophers tout court: Longi-

---

115 SVF 2.242, from Sextus Empiricus, *Adv.Math.* 8.225.
116 Press (2015), p. 195.
117 A full treatment will be devoted to Origen as "zetetic".
118 In *Apol.* 29; *Phaed.* 244; *Crat.* 406; *Resp.* 336;368; *Tim.* 47.
119 *Men.* 81; *Resp.* 528.
120 *Crat.* 421; *Soph.* 221; *Leg.* 631;891.
121 *Phil.* 27; *Men.* 79; *Theaet.* 201.
122 Albinus, *Isag.* 3, mid-2nd century CE.
123 Porphyry, *V.Plot.* 3.35–37.
124 Porphyry, *V.Plot.* 13.10–17.

nus addressed a letter[125] to Porphyry requesting copies of Plotinus' treatises, which he thought "philosophers" (ζητητικοί) should consider among the most remarkable. Origen inherited Plato's "zetetic", heuristic method, as Plotinus did. Both Origen and Plotinus, despite being better acquainted with the dogmatic Plato than the Plato of the skeptic Academy,[126] both worked from a "zetetic" perspective—because this was Plato's own method. For instance, both Origen and Plotinus introduce the treatment of the soul by the same "zetetic" programmatic list: Origen in *C.Cant.* 2.5.21–28, in a zetetic fashion, lists the main issues of philosophical psychology, and Plotinus in *Enn.* 4.3.9.1 introduces the problem "how a soul ends up in a body": before discussing it, he states that this issue must be investigated, ζητητέον. This word is one of Origen's favorite expressions concerning the necessity of investigating philosophical problems, which in his case were applied to theology and exegesis (the latter leads to philosophical theology).[127]

The above-mentioned Gregory Thaumaturgus, who may have inspired Porphyry's *Life of Plotinus*,[128] describes Origen's "zetetic" teaching as Socratic teaching in a Platonic vein. Origen purified his disciples' souls "with his enquiries and restraints, grappling with us *in argument*, and at times overthrowing us *in a really Socratic way*" (7).[129] He wanted his disciples to "philosophize" with all the schools of philosophy, "whether Greek or Barbarian", apart from the atheists: he led his disciples "by hand" in an "exploration"—exactly a "zetesis"—of all the Greek philosophical doctrines, evaluating them and warning his students if anything false, twisted, or dull should appear there, but "retaining in each philosopher what is useful and true, and reject what is false, especially concerning human piety" (13–14). The "zetetic" attitude here is to take into consideration all philosophical ideas and evaluate them. This is also what Origen did with metensomatosis, although in the end he dismissed it, or rather transformed it into ensomatosis (ἐνσωμάτωσις), which fits better with the Biblical paradigm of the end of the world.[130]

For his zetetic method, just as for the structure, plan, and title of *First Principles*,[131] Origen looked to Greek philosophy. Indeed, apart from some occurrences of

---

125 *Ap.* Porphyry *V.Plot.* 19.
126 On these two receptions, see, e.g., Trabattoni (2016).
127 Philosophical and scriptural zetesis are one and the same thing. Thus, it comes as no surprise that the typical philosophical formula ζητητέον, "we must investigate", occurs not only in directly philosophical contexts, but also in exegetical ones, e.g., *H.Luc.* fr. 83.14 in Greek; *C.Cant.* prol. 4.15 in Latin: "we can investigate [*requirere*] why Solomon...".
128 Argument in Edwards (2023).
129 See the analyses of Gregory's testimony in Ramelli (2009a) and Meeren (2019).
130 Ramelli (2022b).
131 Ramelli (2009a).

ζητέω in reference to philosophical investigation in Clement (mostly in quotations from Plato or references to Greek philosophers),[132] there is no such use in Christian authors prior to Origen, whereas this use is extremely widespread and technical in Greek philosophers, especially those—belonging to the Platonic and Pythagorean tradition—with whom Origen was particularly acquainted: Plato, Philo, Numenius, Longinus, Nicomachus, in addition to the Stoic allegorists Chaeremon and Cornutus.

Dialectics, in the form of philosophical zetesis (as well as in other ways), informed Origen's thought. He expressly connected zetesis to life devoted to contemplation and philosophical enquiry, as opposed to life devoted to pleasure or political life (*Princ.* 2.11.1). Origen here insists on the lexicon of zetetic toil: "work hard in wisdom and knowledge" (*sapientiae ac scientiae operam navet*), "study" and "toil" (*studia, industriam*), "through the investigation of truth, learn the causes and rationale/logos of things" (*inquisita veritate rerum causas rationemque cognoscere*). The last definition, in Rufinus' version, recalls Lucretius' and especially Virgil's definitions of philosophy,[133] and Persius' invitation to philosophy: *discite...et causas cognoscite rerum.*[134] A "very philosophical life" aims at the search for truth, and Origen lived it out,[135] and did so consciously at that, as a Christian philosopher. He repeated the same definition of philosophy shortly after: "zeal for the research for truth" and "learn the causes of things" (*inquirendae veritatis studium, rerum causas noscendi*).[136] Here he defends his Christian philosophy—as he also does in a letter[137]—explaining that God confers upon humans the desire for truth: philosophical investigation, the search for truth, is blessed by God. It does not respond to an empty craving, but will find its full achievement in the afterlife: philosophy is a preparation for the eventual beatific knowledge.[138]

---

[132] Forms of ζητέω occur 247 times in Clement, though only a part of these in the technical zetetic sense.
[133] *Naturam student cognoscere rerum*, RN 3.1072; *quibus id fieret cognoscere causis*, RN 5.1185; *rerum cognoscere causas*, Georg. 2.490–2.
[134] *Sat.* 3.66.
[135] Pamphilus *Apol.* 9: *vitam...valde philosopham*; Eusebius HE 6.3.9: φιλοσοφώτατος βίος.
[136] *Princ.* 2.11.4.
[137] *Ap.* Eusebius HE 6.19.12–14.
[138] "Our mind has a proper, natural [probably οἰκεῖον] desire to know the truth about God and learn the causes of things [*sciendae veritatis Dei et rerum causas noscendi*]. We received this desire from God...Love for truth [*amor veritatis*] was infused into our mind by God the Creator... Those who in this life will toil with the utmost strain in pious, religious studies [*summo labore piis studiis ac religiosis operam dederint*], occupying their mind and intellect in this...make them eager to search for truth [*ad inquirendae veritatis studium amoremque convertunt*] and make them readier to re-

God wants humans to philosophize through rational investigation and organized the universe so as to promote this research. Origen grounds the necessity of dialectic as rational enquiry in the Platonic and Biblical twofold scheme of reality: intelligible-sensible, invisible-visible.[139] Humans on earth must reach the knowledge of spiritual realities through material realities—like shadows of the former, according to a Platonic worldview: Paul "teaches that God's invisible things must be learnt from the visible ... by rational analogy [*ratione et similitudine*], showing that this visible world teaches us the invisible ... each visible thing has some rational analogy [*similitudinis et rationi*] to the invisible. Since for enfleshed humans it is impossible to learn anything hidden and invisible, unless by forming an analogous image on the basis of the visible things, for this reason, I think, God, who 'created everything in Wisdom',[140] infused in each visible species on earth the capacity to teach the invisible and heavenly, that through them the *human mind might ascend to the understanding of the spiritual, and search for the causes of things* [*rerum causas*, philosophy's object!] *in the heavenly realm*". This is why Origen blamed his unphilosophical opponents, enemies of allegoresis, for thinking that "truth cannot be found but on earth".[141]

God wants humans to work hard, uninterruptedly, in research. Therefore, Origen exhorts his public to do the same, always investigating and craving to learn: "You also, if you always explore [*scruteris*] the prophetic visions, always investigate [*inquiras*], always desire to learn [*discere cupias*], and meditate and reflect upon them, then you will also receive the Lord's blessing".[142] Origen strongly encourages his audience, including uneducated people, to investigate Scriptures by themselves and produce personal interpretations.[143] Origen often asks the readers of his learned works to find a solution or a better explanation, in a "zetetic" spirit, for example: "Which of the two explanations seems to fit better the verse at stake, you reader will also examine [*probabis*]";[144] "The reader will judge which interpretation can adapt in the worthier way to Scripture's mystical words".[145] Further examples, for instance from *First Principles* or the Commentary on Matthew, abound.

---

ceive instruction in the next world, having already in this life some form of truth and knowledge [*veritatis et scientiae*], to which in the next the beauty of the perfect image will be added".
139 *C.Cant.* 3.13.9–17.
140 Ps. 103:24.
141 *H.Gen.* 13.3; Ramelli 2019.
142 *H.Gen.* 11.3.
143 See, e.g., *H.Gen.*12.5.
144 *C.Cant.* 2.6.13.
145 *C.Cant.* 3.14.34.

Every single letter of Scripture should undergo "scrutiny" and "examination", in an application of zetesis to exegesis.[146]

In a dialectic expression of his "zetetic" method, Origen prospected objections and offered replies in his philosophical exegesis. For instance, in *C.Io.* 2.177, he warns that there is an objection or response (ἀνθυποφορά) to the interpretation he has just expounded, an objection that cannot be effortlessly despised qua easy to reject (οὐκ εὐκαταφρόνητος). Sometimes Origen acknowledges that his answer is only probable (εἰκὸς ἀποκρίνεσθαι),[147] sometimes he thinks he has a particularly striking argument or supposition to propose for discussion (ἔτι δὲ ἐκπληκτικώτερον ὑπόθεσιν).[148] He also propounds confirmations to gain assent (συγκατάθεσιν, a Stoic technical term) regarding the deeper sense of Scripture that he has ferreted out (βαθύτερον ὑπονοούμενον).[149]

Origen made the most of Platonic dialectics and, as I have pointed out, praised it, even stating that it was present in Scripture—like many other teachings, mostly by Plato and ancient philosophy, including the Sages' maxim "Know Yourself".[150] In his system, dialectics explicates itself in theology and serves exegesis and theology, as is clear in the *Commentary on John*.[151] In Plato's *Republic*, dialectic is the supreme science, which unifies the first principles of all individual disciplines under a single principle, the Good. Origen was aware of this overall scheme.

In the prologue to his Commentary on the Song of Songs, Origen theorizes about the division of sciences and seems to suggest that dialectics is an instrument of all the sciences—and especially of theology, which he calls "epoptics", the science of God, which, in the person of the Son, includes the Logos.[152] The Logos is the pivotal nexus between *dialectics* and *theology*: both are the science of the Logos: λόγος qua rational argument and Λόγος qua Christ. Ontologically, the former depends on the latter.

Expounding the division of philosophy into ethics, physics, "epoptics", and (for some schools) logic, Origen posits epoptics as the crowning of philosophy, and reflects on the double status of dialectic-logic, (1) as a specific part of philosophy or (2) as working with all parts. He first lists the tripartition of philosophy: "The general disciplines [*generales disciplinae*] from which one can arrive at the knowledge

---

146 *Philoc.* 10.2.
147 *C.Io.* 2.179.
148 *C.Io.* 2.180.
149 *C.Io.* 2.179.
150 Investigation in Ramelli (forthcoming c). Origen here inspired Gregory of Nyssa regarding the convergence between Greek philosophy and Scripture in relation to this maxim.
151 Heine (1993).
152 See Ramelli (forthcoming d).

of things [*rerum scientiam pervenitur*] are three: what the Greeks called ethics, physics, and epoptics [*ethicam, physicam, epopticen*]; we can call them moral, natural, and inspective [*moralem, naturalem, inspectivam*]": the latter sentence is a clarification by Rufinus.[153] Then he mentions two different positions regarding logic-dialectics: "(1) Some among the Greeks also posited logic [*logicem*], which we can translate 'rational' [*rationalem*], as the fourth discipline [*quarto in numero posuere*]. (2) Others did not posit it as a separate discipline [*extrinsecus*], but as intertwined with the three disciplines mentioned above [*per has tres…disciplinam innexam*], and as interwoven with the whole corpus [*consertamque per omne corpus*]" of philosophy.[154]

Origen then adds some reflections on the nature of logic-dialectics, which pertains to position 2: "Logic [*logice*], indeed, or, as we say, rational science [*rationalis*], is the discipline which appears to contain the rationale [*rationes*] of words and arguments [*verborum dictorumque*], and their proper or improper use [*proprietates et improprietates*], their genera and species, and teach the figures of the single arguments [*figuras singulorum quorumque edocere dictorum*]". Origen's own conclusion favors solution 2: "such a discipline should not be separated from the others but should be integrated and interwoven with the others" (*disciplinam non tam separari quam inseri caeteris convenit et intexi*). It is indeed the science of the Logos that governs all words and reasonings and paradigmatic Ideas, the λόγοι, which are "parts of a whole or forms of the species of the Logos, who was God the Logos in the beginning" and are the ideal paradigms of all creatures in a Platonist exemplaristic fashion (with a typical ontological-epistemological co-extension, which will be inherited by the Christian Platonist Eriugena).[155] Their highest expression is theology-epoptics. This makes clear what I stated earlier: that dialectics and theology are both the science of the *logos:* of human rationality and of its fountainhead, Christ-Logos (also the seat of all paradigmatic Ideas-λόγοι of all creatures).

Logic-dialectics is the science of the Logos and serves philosophy-theology (physics, ethics, and epoptics—the latter crowning all of philosophy), which are structurally based on the Logos, which is Christ. Epoptics is theology, glossed as the science "concerning divine and heavenly things" (*de divinis et caelestibus*), which Origen considers part and parcel of philosophy, insisting that theology cannot be studied without philosophical bases (*C.Cant.* prol. 3.1–4). Here, Origen inserted this treatment of the partitions of philosophy, zetetically (*requirere* likely trans-

---

153 *C.Cant.* prol. 3.1.
154 *C.Cant.* prol. 3.1–2.
155 *Cels.* 5.22; Ramelli (2021e). On Eriugena's co-extension, see below.

lates ζητεῖν), in a major exegetical treatise. The three main branches of Greek philosophy are superimposed onto the Biblical books traditionally ascribed to Solomon: Proverbs (ethics), Ecclesiastes (physics), and the Song of Songs (epoptics-theology).

Origen also calls epoptics θεολογία—the same domain in which Plato excelled, according to Celsus: Plato was the "master of things pertaining to theology".[156] "Ethics, physics, and theology" are the components of philosophy also in *Philoc.* 14.2. Basil identified epoptics with metaphysics,[157] as Aristotle was believed to have done. Porphyry divided Plotinus' *Enneads* into ethics (1), physics (2–3), and epoptics (4–6)—without logic: was this because dialectics was intertwined with all philosophy, and especially with theology, as in Origen? Proclus, who regarded dialectics as "the topmost science" (θριγκὸς τῶν ἐπιστημῶν, with an echo of Plato, *Resp.* 534E2) and the "real science" (ὄντως ἐπιστήμη),[158] deemed Plotinus, Porphyry, and Iamblichus "the exegetes of Platonic epoptics".[159] This also embraced their notion of prayer.[160] Indeed, according to Plotinus, too, philosophy included the investigation of the divine and the divine realm: metaphysics at its highest level. Aristotle himself, as mentioned, treated theology as a synonym of metaphysics.[161] Thus, Plotinus' discourse on the One is both protological (ἕν = first principle) and theological (ἕν = supreme deity), but theology can only be attempted, suggestive, and hinted at.[162]

Origen applied to Christianity the partition of philosophy, not only in the above-examined *C.Cant.* prol. 3, but also in *C.Io.* 1.91–94: philosophy's classification into "practical" and "theoretical" is also the classification of Christianity. Gregory Thaumaturgus attests that "theology" was the culmination of Origen's *cursus studiorum* in his Caesarea university.[163] In his now fragmentary Commentary on Genesis, Origen claimed that ethical, physical and theological issues cannot be addressed "without precise knowledge of how to explain their meaning and without elucidating them according to the logical part".[164] This is exactly the task of dialectics and shows the philosophical basis of theology in Origen's view, as well as the irreducibility of dialectics-logic merely to language logic, in Origen's

---

156 Τῶν θεολογίας πραγμάτων, *Cels.* 7.42.
157 *H.Ps.* 32.341a.
158 *Comm.Remp.* 1.383.13–14. On Proclus' attitude to dialectics, see Opsomer (2022).
159 *Th.Pl.* 1.1.
160 On Neoplatonic theories of prayer and links to contemplation, see Timotin (2017) and my review, *BMCR* 17 (April 2020), https://bmcr.brynmawr.edu/2020/2020.04.32/ (last accessed on November 7, 2022), as well as Dillon (2018), from Plotinus to Proclus.
161 Aristotle, *Met.* 1026a18.
162 See Ramelli (2009b).
163 *Pan.* 13.150.
164 Passage preserved in *Philoc.* 14.2. See also Somos (2015), pp. 22–23.

thought. Indeed, for Origen it is dialectics, instead of language logic, that is theology. Origen's Homilies on Genesis, which simplify the Commentary on Genesis, at 14.3 further emphasize the link between dialectics and theology: "Logic is the part of philosophy which confesses God, the Father of all". Origen in this passage cited the traditional (Stoic) division of philosophy into logic, physics, and ethics, but interestingly ascribed to logic the realm of metaphysics and theology as well. The incongruence results from the fact that the tripartite division of philosophy was Stoic, and in Stoic materialistic immanentism both metaphysics and theology were reduced to physics. There was no transcendent plane. But Origen, who was transcendentalist, could by no means accept such a reduction.

According to Plato, dialectics was coextensive with philosophy, and it was the fullest form of philosophy according to some Stoics, as I have pointed out. Origen used dialectic in his philosophical theology, in the service of theology, and considered the latter, which he calls epoptics, "the art of arts and the science of sciences" (ἡ τέχνη τῶν τεχνῶν καὶ ἐπιστήμη τῶν ἐπιστημῶν).[165] It must be noted that imperial Platonic representations of Plato have a similar view of logic in the service of theology. Alcinous, in particular, interprets dialectical syllogisms—categorical, hypothetical, and mixed syllogisms[166]—as the way through which Plato speaks of the One-Good, the highest level of his protology, which was identified with God.[167] The perspective is the same in Plato as interpreted by imperial, pre-Plotinian Platonists probably known to Origen[168] and in Origen himself.

No less important in this connection is Clement—with whom Origen was well acquainted. Besides presenting dialectics as the means for the "gnostic" or perfect Christian to come to know the first realities, as seen above, he also regarded dialectics-logic as the path to theology. He uses the Christological metaphor of the vine (Christ) to argue that, just as pruning allows people to obtain the best fruit of the vine, so does logic-dialectic allow people to grasp the revelation of Christ and defend it from sophisms.[169]

---

165 C.Io. 13.46.303–304, SC 222.196–200.
166 The distinction between categorical and hypothetical syllogisms is Aristotelian and goes back to Theophrastus, as is attested by the older contemporary of Origen, Alexander of Aphrodisias, In Analytica Priora 45b19 (CAG 2/1, 326.20–328.5).
167 Didaskalikos 6.158–159 (Whittaker, pp. 12–13). Alcinous lists specific Platonic texts among categorical, hypothetical, and mixed syllogisms, especially from Parmenides, but also from Phaedo and I Alcibiades.
168 Although Alcinous is not cited in the list of Origen's favorite readings by Porphyry or Jerome. See Ramelli (2011b).
169 Strom. 1.9.39–44.

Plato himself arguably inspired Origen's connection between dialectics and theology, since, as mentioned at the beginning, dialectic is the art that aims at the knowledge of the Form of the Good and the first principles. Dialectics leads to theology. This is also what Origen seems to have thought. In *Cels.* 6.9 Origen, as often, reads Christian theories in Plato (just as he reads Platonic theories in Scripture): he identifies the factors of the "True Logos" in *Seventh Letter* 342ab (which perhaps inspired Celsus' title[170]) with John, Jesus, and Christ-Logos qua Christ-knowledge. Plato was speaking of *them*. Plato's dialectic is that of the Logos-Christ, being ultimately coterminous with theology.

Origen's allegoresis of the Logos clarifies that Christ-Logos is the source of dialectics and represents it. The Logos is the source and instrument of dialectics and, with Wisdom, Hebrews, and Revelation,[171] is represented by Origen as a "sharp sword", in reference to Platonic diairetic logic. In Rev 1:16, from the Child of the Human Being's mouth "came out a sharp two-edged sword", ῥομφαία δίστομος ὀξεία. Hebr. 4:12 described God's Logos (λόγος τοῦ θεοῦ) as "sharper [τομώτερος] than any two-edged sword", "piercing in division" (δίστομος καὶ διϊκούμενος) and "capable of discerning" (κριτικός).[172] In *C.Io.* 2.6.51 and 2.7.54–57, Origen is referring to Revelation. He has to explain away the violence and destruction that are attributed to Christ therein. Thus, Origen depicts Christ as the divine Logos (ὁ λόγος τοῦ θεοῦ), whose task is "to discern" in justice: ἐν δικαιοσύνῃ κρίνει (51). This is a dialectic diairetic activity. Christ-Logos, being the Truth, judges in justice and fights against what is irrational (τὰ ἄλογα), and thereby anti-dialectic. He chases these components from the souls of those who are conquered and made prisoners by him for their own salvation (54). The Logos will fight his eschatological war (πολεμεῖν) against falsity, irrationality, and ignorance (54–56). Origen finally expounds, consistently with this, the reason why Christ's eyes are said in Rev 19:12 to be like a flame of fire: because Christ-Logos' fiery eyes will consume all thoughts that are too material and crass (ἀφανίζουσι τὰ ὑλικώτερα καὶ παχύτερα τῶν νοημάτων, 57) in a dialectic refinement. Elsewhere as well, Origen attributes to the Logos the actions of diairetic logic: διαιρέω, τέμνω, and διχοτομέω.[173] Therefore, in Origen's view, Christ-Logos applies and even personifies Plato's dialectics. Dialectics is

---

170 Vs. Christ-Logos: see Ramelli (2017c).
171 Origen could include Revelation in the sacred Scripture by offering an entirely allegorical interpretation of it (see Ramelli 2011c and Hedley 2016, n. 133). Origen's exegesis of Revelation survives both in fragments—the scholia or a part thereof—and scattered throughout other works of his.
172 Another of Origen's sources here is, I suspect, Wisdom, a Hellenistic Jewish book that he applied directly to Christ-Logos-Wisdom. In 18:15–16, God's Logos is represented as a warrior with a "sharp sword" (ξίφος ὀξύ).
173 *Protr.Mart.* 37.20; *Fr.Eph.* 35.24.

in the service of theology in Plotinus' *Enneads* as well,[174] although in Origen the coextensivity appears even stronger.

Likewise, the Origenian Eriugena, the last Patristic Platonist and a great and outspoken admirer of Origen, considered dialectics as coextensive with reality—which proceeds from God to God[175]—and arguably constructed his dialogue *Periphyseon* as a dialectical exercise grounded in the Hexaemeron's creation account and finalized at the return to God, an ἀποκατάστασις-ἐπιστροφή (concepts that he links, as Origen, Eriugena and Ps.Dionysius did). Eriugena applies his notion of knowledge to God, whose self-knowledge is transmitted to humans, in *Vox spiritualis aquilae*, his homily on the Johannine Prologue: "it is not you (human) who know Me, but I Myself know Me in you (human) through my Spirit, because you are not substantial Light, but the participation of the Light that subsists in itself".[176] "Substantial Light" is analogous to "substantial knowledge", which is a technical designation of God the Trinity in Evagrius. Also, the notion that the Son and the Spirit transmit the divine knowledge to humans is Evagrian. Eriugena calls God "the Intellect of all", in the sense of the noetic capacity, and "Gnostic Virtue" or the ability to know, which is the divinity itself and also the means through which all knowers know (*intellectus omnium est omnia et ipsa sola intelligent omnia ... gnostica virtus est ipsa quae cognovit omnia: extra se non cognovit omnia, quod extra ea nihil est*).[177]

For Eriugena, in God, knowledge and ontology are one and the same thing, because "to know and to make are one in the Godhead, since by knowing it makes, and by making it knows".[178] This structural correspondence between the ontological and the logical plane is declared at more length by Eriugena in *Periphyseon*.[179] Within this demonstration, he argues that the exercise of dialectics (*dialectica*) leads to truth, because this discipline itself (called here by its Greek, Platonic name) was not invented by humans, but was created by God, who is presented as the author of all the liberal arts in *Periph.* 4.749 A.[180] The human mind, thanks to the seeds of the liberal arts that are embedded in it and are the foundation of

---

174 See Haig (2022).
175 Ramelli (2016b and 2022c).
176 *Jean Scot. Homélie sur le Prologue de Jean*, SC 151, 266.23–26.
177 *Periph.* 3 632D–633 A.
178 *Periph.* 2.559B.
179 See, e.g., 2.559B and, on the basis of Ps.Dionysius, 4.749AC: "Intellectus enim rerum veraciter ipsae res sunt, dicente sancto Dionysio: 'Cognitio eorum quae sunt, ea quae sunt est'... cognitio intellectualis animae praecedit omnia quae cognoscit et omnia quae praecognoscit, ut in divino intellectu omnia causaliter, in humana vero cognitione effectualiter subsistant".
180 *Periph.* 4.749 A: "Διαλεκτική... non ab humanis machinationibus sit facta, sed in natura rerum, *ab auctore omnium artium*, quae vere artes sunt, condita".

theology, also becomes creative: human nature, which is "that in which everything could be found [*inerat*]", becomes "that in which everything is/was created [*condita est*]".[181] As the intelligent disciple (*Alumnus*) states, within a dialectic dialogue with his master (*Nutritor*), the objects of knowledge are "in a way *created in me*". For "when I imprint their phantasms in my memory, and when I treat these things within myself, I divide, I compare, and, as it were, I collect them into a certain unity, I perceive a certain knowledge of the things which are external to me *being created within me*".[182] This is, in a way, a strong form of intentionality.

In his preface to his translation of the Dionysian corpus, Eriugena focuses on theology as the essence of Dionysius' enterprise: "This book is divided into the two main parts of the discipline of logic: affirmative (cataphatic) theology and negative (apophatic) theology, namely, being and non-being. It employs the rules of dialectic analysis and very clearly advises us that we must apply the removal [*privatio*] of whatever can be said or thought to attain the Truth, which is the cause of all creatures, by means of the eminence of its essence", namely, the essence of Truth, which is God.[183]

Eriugena, like Origen, treats theology as coextensive with dialectics or the logical discipline: the rational structure of reality depends on God and divine λόγοι: from God and the λόγοι all derives, and all returns to the λόγοι and God.[184] This is also why Eriugena states that the Truth indicates nothing else but God, in a coincidence of the epistemological and theological realms: "the Truth will indicate you nothing but the One who created all. Besides this, you will have nothing to contemplate, since this [God] is everything".[185]

The genre of *Periphyseon* as a dialogue and thereby as an exercise in dialectics is crucial. Eriugena observes that dialectics in dialogues, like knowledge, creates one mind in the other and vice versa: "For when we enter upon a discussion together, the same thing happens: *each of us is created in the other:*[186] for when I understand what you understand, I am made your understanding, and in a certain way that cannot be described, *I am created in you*. In the same way, when you clearly understand what I clearly understand, *you are made my understanding* and two understandings are made one, formed from that which we both clearly

---

[181] *Periph.* 4.807 A.
[182] *Periph.* 4.765C, version V.
[183] PL 122.1035 A–1036 A.
[184] See Ramelli (2022c).
[185] *In Ioh.Prol.* 289C, Jeauneau (1969), p. 254.
[186] Note the correspondence with the previous quotation (referred to in n. 178), about things external "created within me" (I italicized both expressions in the quotations).

and without doubt understand".[187] Eriugena in his preface to his translation of Maximus the Confessor's *Ambigua* joins dialectic to theology, using "analytics", Plato's dialectical terminology, for the procession (πρόοδος) of all things from God to the cosmos, and θέωσις, the opposite movement, for deification, the return, explicitly called "reversion" [ἐπιστροφή], of all beings to God: "The aforementioned divine procession towards all things is called 'analytics' [ἀναλυτική], meaning 'resolution', and the reversion θέωσις, meaning 'deification'".[188]

The goal of the dialogue *Periphyseon* is reaching God. The dialogue is a performative exercise of dialectics and philosophical theology at the same time: the master and the disciple do not just expound the *exitus* and *reditus* of creation by way of the division and analysis of the genus *natura*, but attempt to perform it, in an ontologico-logical co-extension. For Eriugena, the dialogue is dialectics, coextensive with theology and therefore reality.

Both Origen and his great admirer Eriugena, indeed, conceived of dialectics as coextensive with ontology and in the service of theology, finalized as it is at a theological goal: the knowledge of reality as created by God and, ultimately, the return to God. Not accidentally, both Origen and Eriugena are among the most significant supporters of the theory of apokatastasis.[189] One should not just believe, according to Origen, but provide *reasons* for belief (λογισμὸν ὑπέχειν τῶν πιστευομένων)—reasons which ultimately embrace all reality and God; thus, he exhorted the simple to merely believe, but offered dialectics to the intelligent: "rational arguments by questions and answers", remarkably quoting Plato's *Seventh Letter* 344b in *Cels.* 6.10, which explicitly cites Plato and his *Timaeus* as well. Origen intended to construct a Christian philosophy—particularly Christian Platonism—for the philosophically demanding elites, applying the philosophical, dialectic method indicated by Plato in addition to spreading the Christian faith among the "simple".

---

[187] *Periph.* 4.780BC.
[188] *Proem.* in Maximus the Confessor (1988): *Ambigua ad Iohannem iuxta Iohannis Scotti Eriugenae latinam interpretationem* Édouard Jeauneau (Ed.). CCG 18. Turnhout: Brepols, p. 4.
[189] Ramelli (2013b), pp. 137–214 and 773–815, and Ramelli (2022c). This project has benefitted from a Research Professorship in Patristics and Church History (KUL) I have been awarded as part of the "Initiative of Excellence" program, #028/RID/2018/19.

Bibliography

Ambury, James M. and German, Andy (Eds.) (2019): *Knowledge and Ignorance of Self in Platonic Philosophy.* Cambridge: Cambridge University Press.

Baltzly, Dirk (1999): "Aristotle and Platonic Dialectic in Metaphysics Gamma". In: *Apeiron*, 32. No. 4, pp. 171–202.

Banner, Nicholas (2018): *Philosophic Silence and the "One" in Plotinus.* Cambridge: Cambridge University Press.

Bénatouïl, Thomas and Ierodiakonou, Katerina (Eds.) (2019): *Dialectic after Plato and Aristotle.* Cambridge: Cambridge University Press.

Bobzien, Susanne (1989): *Determinism and Freedom in Stoic Philosophy.* Oxford: Oxford University Press.

Boeri, Marcelo and Salles, Ricardo (2014): *Los filósofos estóicos.* Sankt Augustin: Academia Verlag.

Broadie, Sarah (2020): *Mathematics in Plato's Republic.* Milwaukee: Marquette University Press.

Cameron, Averil (2016): *Arguing it Out: Discussion in Twelfth-Century Byzantium.* Budapest: Central European University, 2016.

Cameron, Averil, and Niels Gaul, eds. (2018): *Dialogues and Debates from Late Antiquity to Late Byzantium.* Milton Park: Routledge, 2018.

Castagnoli, Luca (2010): "How Dialectical Was Stoic Dialectic?" In: Nightingale, Andrea, and Sedley, David (Eds.): *Ancient Models of Mind: Studies in Human and Divine Rationality*, Cambridge: Cambridge University Press, pp. 153–179.

Clark, Gillian (2008): "Can we talk? Augustine and the possibility of dialogue". In: Goldhill, Simon (Ed.): *The End of Dialogue in Antiquity.* Cambridge: Cambridge University Press, pp. 117–134.

Dillon, John (2018): "Prayer and Contemplation in the Neoplatonic and Sufi Traditions". In: Pachoumi, Eleni and Edwards, Mark (Eds.): *Praying and Contemplating in Late Antiquity: Religious and Philosophical Interactions.* Tübingen: Mohr Siebeck, pp. 7–22.

Dutsch, Dorota (2020): *Pythagorean Women Philosophers: Between Belief and Suspicion.* Oxford: Oxford University Press.

Edwards, Mark (2023): "What Porphyry Learned from Gregory Thaumaturgus". In: DePalma Digeser, Elizabeth, Heidi Marx, and Ilaria L.E. Ramelli, *Problems in Ancient Biography: The Construction of Professional Identities in Late Antiquity.* Cambridge: Cambridge University Press.

Fattal, Michel (2018): "Raison critique et crise chez Parménide d'Élée". In. Pulpito, Massimo and Spangenberg, Pilar (Eds.): *ὁδοὶ νοῆσαι—Ways to Think: FS Néstor-Luis Cordero.* Bologna: Diogene, pp. 113–120.

Frankfurter, David (2017): *Christianizing Egypt: Syncretism and Local Worlds in Late Antiquity.* Princeton: Princeton University Press.

Geljon, Albert (2005): "Divine Infinity in Gregory of Nyssa and Philo". In: *Vigiliae Christianae* 59, pp. 152–177.

Gonzalez, Fransisco J. (1998): *Dialectic and Dialogue: Plato's Practice of Philosophical Enquiry.* Evanston: Northwestern University Press.

Gordon, Richard (2016): "Mysteries". In: *Oxford Classical Dictionary.* https://www.doi.org/10.1093/acrefore/9780199381135.013.4318, last accessed on November 7, 2022.

Gourinat, Jean-Baptiste and Lemaire, Juliette (Eds.) (2016): *Logique et dialectique dans l'Antiquité.* Paris: Vrin.

Guiu, Adrian (2021): "Philosophical Dialogue and Contemplation of the Cosmos in Augustine, Boethius, Eriugena". In: Ramelli, Ilaria L.E. (Ed.): *Eriugena's Christian Neoplatonism and Its Sources*. Leuven: Peeters.
Hadot, Pierre (1989): *Simplicius: Commentaries sur les Catégories*. Leiden: Brill.
Hägg, Henny F. (2006): *Clement of Alexandria and the Beginnings of Christian Apophaticism*, Oxford: Oxford University Press.
Haig, Albert R. (2022): "Dialectic as Ostension Towards the Transcendent: Language and Mystical Intersubjectivity in Plotinus' *Enneads*". in *The International Journal of the Platonic Tradition* 16, pp. 1–22. doi:10.1163/18725473-bja10016
Halfwassen, Jens (2021): *Plotinus, Neoplatonism and the Transcendence of the One*. Steubenville: Franciscan University Press.
Havrda, Matyáš (2016): *The So-Called Eighth Stromateus by Clement of Alexandria: Early Christian Reception of Greek Scientific Methodology*. Leiden: Brill.
Hedley, Douglas (2016): *The Iconic Imagination*. London: Bloomsbury.
Heine, Ronald (1993): "Stoic Logic as Handmaid to Exegesis and Theology in Origen's Commentary on the Gospel of John". In: *Journal of Theological Studies* 44, pp. 90–117.
Hoffmann, Manfred (1966): *Der Dialog bei den christlichen Schriftstellern der ersten vier Jahrhunderte*. Heidelberg: Akademie.
Ierodiakonou, Katerina (2019): "Dialectic as a Subpart of Stoic Philosophy". In: Bénatouïl, Thomas and Ierodiakonou, Katerina (Eds.) (2019): *Dialectic after Plato and Aristotle*. Cambridge: Cambridge University Press, pp. 114–133.
Irani, Tushar (2017): *Plato on the Value of Philosophy: The Art of Argument in the Gorgias and Phaedrus*. Cambridge: Cambridge University Press.
Jeauneau, Edouard (Ed.) (1969): *Jean Scot Érigène: Homélie sur le Prologue de Jean*. Sources Chrétiennes 151. Paris: Cerf.
Lavalle Norman, Dawn (2022): *Early Christian Women. Elements on Women in the History of Philosophy*. Cambridge: Cambridge University Press.
Mansfeld, Jaap (2021): "An Early Greek Epic: Narrative Structures in Parmenides' Poem and the Relation between Its Main Parts". In: *Mnemosyne* 74, pp. 200–237.
Mühlenberg, Ekkehard (1966): *Die Unendlichkeit Gottes bei Gregor von Nyssa*. Göttingen: Vandenhoek & Ruprecht.
Mühlenberg, Ekkehard (2015): "Zur Herkunft des Gedankens der Unendlichkeit Gottes". In: Wentz, Gunther (Ed.): *Eine neue Menschheit darstellen*. Götingen: Vandenhoek & Ruprecht, pp. 141–175.
Müller-Jourdan, Pascal (2018): "Anonymus, Prolegomena in Platonis philosophiam". In: Riedweg, Christoph, Horn, Christoph and Wyrwa, Dietmar (Eds.): *Philosophie der Kaiserzeit und der Spätantike. Grundriss der Geschichte der Philosophie*. Basel: Schwabe, pp. 2118–2224.
Opsomer, Jan (2022): "Proclus' *Elements of Theology* and Platonic Dialectics". In: Calma, Dragos (Ed.): *Reading Proclus and the Book of Causes*, 3, *On Proclus and the Noetic Triad*. Leiden: Brill, pp. 18–36.
Press, Gerald (2015): "Changing Course in Platonic Studies". In: Nails, Debra Nails and Tarrant, Harold in Collaboration with Kajava, Mika and Salmenkivi, Eero (Eds.): *Second Sailing: Alternative Perspectives on Plato*. Helsinki: Societas Scientiarum Fennica, pp. 187–196.
Ramelli, Ilaria (2007a): *Gregorio di Nissa sull'Anima e la Resurrezione*. Milan: Bompiani-Catholic University of the Sacred Heart.

Ramelli, Ilaria L.E. (2007b): "Christian Soteriology and Christian Platonism. Origen, Gregory of Nyssa, and the Biblical and Philosophical Basis of the Doctrine of Apokatastasis". In: *Vigiliae Christianae* 61, pp. 313–356.

Ramelli, Ilaria L.E. (2009a): "Origen, Patristic Philosophy, and Christian Platonism: Re-Thinking the Christianisation of Hellenism". In: *Vigiliae Christianae* 63, pp. 217–263.

Ramelli, Ilaria L.E. (2009b): "The Divine as Inaccessible Object of Knowledge in Ancient Platonism: A Common Philosophical Pattern across Religious Traditions". In: *Journal of the History of Ideas* 75. No. 2, pp. 167–188.

Ramelli, Ilaria L.E. (2009c): *Bardesane di Edessa Contro il Fato—Liber legum regionum*. Bologna: ESD.

Ramelli, Ilaria L.E. (2009-2019): *Bardaisan of Edessa: A Reassessment of the Evidence and a New Interpretation, Also in the Light of Origen and the Origenian Fragments from De India*. Piscataway: Gorgias and Berlin: De Gruyter.

Ramelli, Ilaria L.E. (2010): "'Maximus' on Evil, Matter, and God: Arguments for the Identification of the Source of Eusebius *PE* VII 22". In: *Adamantius* 16, pp. 230–255.

Ramelli, Ilaria L.E. (2011a): "Origen's Anti-Subordinationism and Its Heritage in the Nicene and Cappadocian Line". In: *Vigiliae Christianae* 65, pp. 21–49.

Ramelli, Ilaria L.E. (2011b): "Atticus and Origen on the Soul of God the Creator: From the 'Pagan' to the Christian Side of 'Middle Platonism'". In: *Jahrbuch für Religionsphilosophie* 10, pp. 13–35.

Ramelli, Ilaria L.E. (2011c): "Origen's Interpretation of Violence in the Apocalypse: Destruction of Evil and Purification of Sinners". In: Verheyden, Joseph, Merkt, Andreas, and Nicklas, Tobias (Eds.): *Ancient Christian Interpretations of "Violent Texts" in the Apocalypse*, Göttingen: Vandenhoeck & Ruprecht, pp. 46–62.

Ramelli, Ilaria L.E. (2012-2013): "The *Dialogue of Adamantius*: A Document of Origen's Thought?" In: *Studia Patristica* 52, pp. 71–98 and *Studia Patristica* 56. No. 4, pp. 227–273.

Ramelli, Ilaria L.E. (2013a): "Stoic Cosmo-Theology Disguised as Zoroastrianism in Dio's *Borystheniticus?* The Philosophical Role of Allegoresis as a Mediator between *Physikē* and *Theologia*". In: *Jahrbuch für Religionsphilosophie* 12, pp. 9–26.

Ramelli, Ilaria L.E. (2014a): "Alexander of Aphrodisias: A Source of Origen's Philosophy?". In: *Philosophie Antique* 14, pp. 237–290.

Ramelli, Ilaria L.E. (2014b): "Decadence Denounced in the Controversy over Origen: Giving Up Direct Reading of Sources and Counteractions". In: Fuhrer, Therese and Formisano, Marco (Eds.): *Décadence: "Decline and Fall" or "Other Antiquity"?* Heidelberg: Winter, pp. 263–283.

Ramelli, Ilaria L.E. (2016a): *Social Justice and the Legitimacy of Slavery: The Role of Philosophical Asceticism from Ancient Judaism to Late Antiquity*. Oxford: Oxford University Press.

Ramelli, Ilaria L.E. (2016b): "The Reception of Origen's Ideas in Western Theological and Philosophical Traditions". Main lecture at *Origeniana Undecima*, Aarhus University, August 2013. Anders-Christian Jacobsen (Ed.). Leuven: Peeters, pp. 443–467.

Ramelli, Ilaria L.E. (2017a): "Origen and the Platonic Tradition". In: Smith, J. Warren (Ed.): *Plato and Christ: Platonism in Early Christian Theology. Religions* 8(2). No. 21, pp. 1–20. https://www.doi.org/10.3390/rel8020021, last accessed on November 7, 2022.

Ramelli, Ilaria L.E. (2017b): "Prophecy in Origen: Between Scripture and Philosophy". In: *Journal of Early Christian History* 7. No. 2, pp. 17–39.

Ramelli, Ilaria L.E. (2017c): "Contra Celsum". In: *Journal of Theological Studies* 68, pp. 348–350.

Ramelli, Ilaria L.E. (2018a): "Apokatastasis and Epektasis in *Hom. In Cant.*: The Relation between Two Core Doctrines in Gregory and Roots in Origen". In: Maspero, Giulio, Brugarolas, Miguel, and Vigorelli, Ilaria (Eds.): *Gregory of Nyssa: In Canticum Canticorum. Commentary and Supporting*

Studies. Pro^(ce)edings of the 13^(th) International Colloquium on Gregory of Nyssa (Rome, 17–20 September 2014). Supplements to Vigiliae Christianae 150. Leiden: Brill, pp. 312–339.

Ramelli, Ilaria L.E. (2018b): "Origen". In: Cartwright, Sophie and Marmodoro, Anna (Eds.): *A History of Mind and Body in Late Antiquity.* Cambridge: Cambridge University Press, pp. 245–266.

Ramelli, Ilaria L.E. (2018c): "Gregory of Nyssa on the Soul (and the Restoration): From Plato to Origen". In: Marmodoro, Anna and McLynn, Neil (Eds.): *Exploring Gregory of Nyssa: Historical and Philosophical Perspectives.* Oxford: Oxford University Press, pp. 110–141.

Ramelli, Ilaria L.E. (2019): "Autobiographical Self-Fashioning in Origen". In: Niehoff, Maren and Levinson, Joshua (Eds.): *Self, Self-Fashioning and Individuality in Late Antiquity: New Perspectives.* Tübingen: Mohr Siebeck, pp. 273–292.

Ramelli, Ilaria L.E. (2020a): "The *Dialogue of Adamantius:* Preparing the Critical Edition and a Reappraisal". In: *Rheinisches Museum* 163, pp. 40–68.

Ramelli, Ilaria L.E. (2020b): "Epicureanism and Early Christianity". In: Mitsis, Phillip (Ed.): *Oxford Handbook to Epicurus and Epicureanism.* Oxford: Oxford University Press, pp. 582–612.

Ramelli, Ilaria L.E. (2021a): "Unity around a Teacher: Clement and Origen of Alexandria". In: Ramelli, Ilaria L.E., McGuckin, John A., and Ashwin, Piotr (Eds.): *T&T Clark Handbook to the Early Church.* London: T&T Clark Bloomsbury Academic, pp. 191–223.

Ramelli, Ilaria L.E. (2021b): "Mysticism and Mystic Apophaticism in Middle and Neoplatonism". In: Wilke, Annette, Stephanus, Robert, and Suckro, Robert (Eds.): *Constructions of Mysticism as a Universal. Roots and Interactions Across the Borders.* Wiesbaden: Harrassowitz, pp. 29–54.

Ramelli, Ilaria L.E. (2021c): "Origen on the Unity of Soul and Body in the Earthly Life and Afterwards and His Impact on Gregory of Nyssa". In: Ulrich, Jörg, Usacheva, Anna, and Bhayro, Siam (Eds.): *The Unity of Soul and Body in Patristic and Byzantine Thought.* Leiden: Brill, pp. 38–77.

Ramelli, Ilaria L.E. (2021d): "Origen's Philosophical Exegesis of the Bible against the Backdrop of Ancient Philosophy and Hellenistic and Rabbinic Judaism". Main lecture at the conference *The Bible: Its Translations and Interpretations in the Patristic Time*, KUL, October 16–17, 2019. In: *Studia Patristica* 103, pp. 13–58.

Ramelli, Ilaria L.E. (2021e): "The Logos/Nous One-Many between 'Pagan' and Christian Platonism: Bardaisan, Clement, Origen, Plotinus, and Gregory of Nyssa". In: *Studia Patristica* 102, pp. 11–44.

Ramelli, Ilaria L.E. (2022a): "The Question of Origen's Conversion and His Philosophico-Theological Lexicon of *Epistrophē*". Main lecture at the international conference *Religious and Philosophical Conversion*, Bonn University, September 25–27, 2018. In: Wallace, James B. and Despotis, Athanasios (Eds.): *Greek and Byzantine Philosophical Exegesis.* Leiden: Brill, pp. 45–80.

Ramelli, Ilaria L.E. (2022a): "The Strategy and Functions of Philosophical Exegesis in Origen of Alexandria". In: Wallace, James B. and Despotis, Athanasios (Eds.): *Greek and Byzantine Philosophical Exegesis.* Leiden: Brill, pp. 81–108.

Ramelli, Ilaria L.E. (2022b): "The Soul-Body Relation in Origen of Alexandria: Ensomatosis vs. Metensomatosis". Invited lecture at the international congress *Early Christian Mystagogy and the Body*, Utrecht University, August 30–September 1, 2017. Paul van Geest and Nienke Vos (Eds.). Leuven: Peeters, pp. 97–119

Ramelli, Ilaria L.E. (2022c): "From God to God: Eriugena's Protology and Eschatology against the Backdrop of His Platonic Patristic Sources". In: Ramelli, Ilaria L.E. (Ed.): *Eriugena's Christian Neoplatonism and its Sources in Patristic Philosophy and Ancient Philosophy.* Leuven: Peeters, pp. 99–123.

Ramelli, Ilaria L.E. (forthcoming a): "Parmenides' Philosophy through Plato's *Parmenides* in Origen of Alexandria". In: Cornelli, Raffaele, Kurfess, Christopher and Motta, Anna (Eds.): *Eleatic Ontology: Origin and Reception.* Vol. I. Tome 6: *Eleatic Ontology in the Hellenistic Period to Late Antiquity.*

Ramelli, Ilaria L.E. (forthcoming b): "Generic Innovation and Christian Identity". In: Kelly, Gavin and Pelttari, Aaron (Eds.): *The Cambridge History of Later Latin Literature.* Cambridge: Cambridge University Press.

Ramelli, Ilaria L.E. (forthcoming c): "'Know Yourself' in Origen and Gregory of Nyssa: A Maxim of Greek Philosophy Found in Scripture?". In: Filtvedt, Ole Jakob and Schröter, Jens (Eds.): *"Know Yourself" from Paul to Augustine: Exploring the Delphic Maxim in Christian and Non-Christian Sources from the First Centuries.* Berlin: de Gruyter.

Ramelli, Ilaria L.E. (forthcoming d): "Epopteia, epoptics in Platonism, 'pagan' and Christian". In: Tarrant, Harold (Ed.): *The Language of Inspiration and Divine Diction in the Platonic Tradition.* Sedbury: Prometheus Trust Press.

Rigolio, Alberto (2019): *Christians in Conversation: A Guide to Late Antique Dialogues in Greek and Syriac.* Oxford: Oxford University Press.

Rist, John (1981): "The Importance of Stoic Logic in the Contra Celsum". In: Blumenthal, Henry J. and Markus, Robert (Eds.): *Neoplatonism and Early Christian Thought. FS A.H. Armstrong.* London: Variorum, pp. 64–78.

Scherer, Jean (1960): *Entretien d'Origène avec Héraclide, Sources Chrétiennes* 67. Paris: Cerf.

Sedley, David (2015): "Dialectic". In: *Oxford Classical Dictionary,* https://www.doi.org/10.1093/acrefore/9780199381135.013.2140, last accessed on November 7, 2022.

Sedley, David (2019): "Epicurus on Dialectic". In: Bénatouïl, Thomas and Ierodiakonou, Katerina (Eds.): *Dialectic after Plato and Aristotle.* Cambridge: Cambridge University Press, 82–113.

Sferlea, Ovidiu (2010): *Aoristos: le thème de l'infini chez Grégoire de Nysse.* PhD Dissertation. Paris: École pratique des hautes études.

Sferlea, Ovidiu (2013): "L'infinité divine chez Grégoire de Nysse". In: *Vigiliae Christianae* 67, pp. 137–168.

Somos, Robert (2013a): "Is the Handmaid Stoic or Middle Platonic?". In: *Studia Patristica* 56. No. 4, pp. 29–40.

Somos, Robert (2013b): "Homonymy as a Logical Term in Origen". In: *Acta Academiae Hungaricae* 53, pp. 409–421.

Somos, Robert (2015): *Logic and Argumentation in Origen.* Münster: Aschendorff.

Spranzi, Marta (2011): *The Art of Dialectic Between Dialogue and Rhetoric: The Aristotelian Tradition.* Amsterdam and Philadelphia: Johns Benjamins.

Stavru, Alessandro, and Moore, Christopher (Eds.) (2018): *Socrates and the Socratic Dialogue.* Leiden: Brill.

Stock, Brian (2010): *Augustine's Inner Dialogue. The Philosophical Soliloquy in Late Antiquity.* Cambridge: Cambridge University Press.

Stock, Brian (2012): "The Soliloquy. Transformations of an Ancient Philosophical Technique". In: Bochet, Isabelle (Ed.): *Augustin philosophe et prédicateur. Hommage à G. Madec.* Paris: Institut d'Études Augustiniennes, pp. 315–347.

Szlezák, Thomas (2019): *Aufsätze zur griechischen Literatur und Philosophie.* Baden-Baden: Akademie Verlag.

Szram, Mariusz (2020): "Origen's Castration. Solely a Spiritual Phenomenon?" *Gregorianum* 101, pp. 23–36.

Timotin, Andrei (2017): *La prière dans la tradition platonicienne, de Platon à Proclus.* Turnhout: Brepols.

Trabattoni, Franco (2016): *Essays on Plato's Epistemology.* Leuven: Leuven University Press.

Trapp, Michael (2017): *Aelius Aristides: Orations.* Vol. I. Cambridge: Harvard University Press.

Trapp, Michael (2020): "With All Due Respect to Plato". In: *TAPA* 150, pp. 85–113.
Van Der Meeren, Sophie (2019): "Μάλα σωκρατικῶς: formes, fonctions et représentations du 'discours' philosophique chez Origène, d'après le Discours de remerciement". In: Aubert-Baillot, Sophie, Guérin, Charles, and Morlet, Sébastien (Eds.): *La philosophie des non-philosophes dans l'Empire romain du Ier au IIIe siècle.* Paris: Boccard, pp. 287–305.
Wider, Kathleen (1986): "Women Philosophers in the Ancient Greek World: Donning the Mandle". In: *Hypatia* 1. No. 1, pp. 21–62.

Michael Griffin
Exegesis as Philosophy: Notes on Aristotelian Methods in Neoplatonic Commentary

> The investigation (*theōria*) of the truth is in one way hard, in another easy. An indication of this is found in the fact that no one is able to attain the truth adequately (*axiōs... thigein*), while, on the other hand, no one fails entirely, but every one says something true about the nature of things, and while individually they contribute little or nothing to the truth, by the union of all a considerable amount is amassed
> (From Aristotle, *Metaph.* α.1, 993a27–b11, trans. Ross).

1 Introduction

In Mediterranean Late Antiquity, philosophy in Greek was often practiced through exegesis. The writers who we call "Neoplatonists"[1] found their voices in a distinctive interpretation of authors like Plato, Aristotle, and Homer, alongside corpora regarded as Egyptian, Orphic, and Chaldaean. They paid special attention to tracing the consistency or "harmony" of an underlying network of ideas, apparent only to the interpreter who plunges beneath their sources' words (*lexis*) or even mythological symbolism in order to arrive at a shared layer of meaning or intent (*nous*).[2] This exegetical methodology was not a constraint on creative philosophy, or at least it was not *only* a constraint; it led to novel philosophical and scientific conclusions,[3] and the exercise of copying and interpreting a text constituted an important feature of the philosophical "ways of life" adopted in the later Roman Empire.[4]

In this paper, I would like to explore one tendril in the complicated roots of this practice. Loosely, my focus is the methodological value placed on the careful interpretation of diverse, thoughtfully motivated views drawn from philosophy's

---

1 For the problematic limitations of the label *Neuplatonismus*, see Catana (2013) and Gerson (2013).
2 See, for example, Fazzo (2004) and Baltussen (2007 and 2008). Boys-Stones (2001) offers a creative account of the origins of this tendency in post-Hellenistic philosophy. Karamanolis (2006) remains a classic account of the "harmony thesis"; for useful skepticism about its extent and universality in Late Antiquity, see Golitsis (2018). For the Neoplatonist treatment of Plato more broadly, see recently Motta (2021). For the framing of the interpreter's project as accessing the spirit rather than letter, see Simplicius *in Cat.* 7,23–29; for the philosopher's role in grasping the reality underlying myths, see, for example, Olympiodorus *in Gorg.* § 47.
3 See for example Sorabji (2016a), pp. vii and xii as well as Tarrant (1997, 2000a, and 2000b).
4 See, e.g., Hoffmann (2012).

history, seeking to understand why every view *appealed* to its originators, while acknowledging that each position, and each interpreter, is also bounded by their own situation and horizons. The part of the story I will take up here starts with Aristotle (though the road winds back much earlier and in other directions);[5] it meets the Stoics and Plotinus on the way before settling on the Neoplatonist commentators of Late Antiquity. I recognize that is a very broad topic to survey, and excellent work has already been done on these lines:[6] I hope that there is some value in a longitudinal exploration of metaphilosophical and pedagogical concepts that supported "exegesis as philosophy" in Late Antiquity.[7] In conclusion, I will try to point toward some implications of this approach for its later reception.[8]

First (2.1), I attempt to bring together some of the methodological passages from Aristotle that are particularly relevant, focusing on why Aristotle ascribes value to the exercise of comparing multiple viewpoints with philosophical "currency", including earlier philosophy and myth. Second, (2.2) I turn to comparable attitudes in post-Hellenistic witnesses and Plotinus, noting their growing respect for *ancient* or long-enduring ideas as attractive for philosophical investigation and dialectical comparison. Next, (3.1) I focus on the later, systematic Neoplatonist treatment of philosophical authorities like Plato, emphasizing that the authority of a view invites study, but is still no substitute for argument and demonstration, or for the authority of one's own preconceptions in one's conscience, which can be uncovered in the process. In closing (3.2), I explore how this use of ancient sources is situated in Neoplatonic psychology and pedagogy, especially a teacher or text's creative use of language to elicit a student's own "preconceptions".[9]

---

[5] The early study of Homeric "problems" and related grammatical projects, for instance, is already illustrated in the characters' discussion of Simonides in Plato's *Protagoras* (338e–347b), and Socrates' characterization of a rhapsode's practice in the *Ion*. See also François Renaud's study in this volume and Hunter (2011) as well as, on the relationship of Neoplatonic commentaries to these strands, Menn (forthcoming), § 3.2. I am grateful to Renaud for discussion on this point.

[6] See the previous notes for some recent literature; see also Magrin (2016). Barney (2009) shows how, in particular, Aristotle's *Metaphysics* 1 and *Physics* 1.5 guide Simplicius' practice of the history of philosophy. While Simplicius is perhaps the most obvious later commentator to study through the lens of Aristotelian doxographical methodology, Greig (2021) offers an excellent and careful case-study in Damascius' use of Aristotelian aporetic method. A longitudinal arc like the one I explore here, particularly in the post-Hellenistic period, is traced by Boys-Stones (2001), which I have found very helpful.

[7] This can be seen as sketching early roots of a broader "exegetical posture" in philosophy; see, for example, Hadot (1995), which is discussed further below, and Gadamer (1989).

[8] See Futter (2016).

[9] See, for instance, Griffin (2016 and 2020).

2 Background

2.1 Aristotle

The contemporary debate concerning Aristotle's own methods, particularly the so-called "endoxic" method, is fast-moving. There is advancing skepticism about whether Aristotle outlines or practices a cogent version of any such methodology at any point.[10] But there is perhaps clearer progress on Aristotle's domain-specific procedure in ethical inquiry, exploring whether Aristotle practices ethics as a science (*epistēmē*) whose propositions can be demonstrated; and there is also a case for bearing this methodology in mind, at least, when we study Aristotle's procedures in natural science.[11]

Aristotle's later Neoplatonist exegetes display a consistent view about his method in ethics and in other subjects. They regard him, more or less consciously, as an ally of and exemplar for their own philosophical practice, including their interest in studying the history of philosophy, though they recognize that he, too, could fall short of that practice, for example in polemic against the Academy.[12]

In this section, I will attempt to outline some of the textual sources of the Neoplatonists' methodological Aristotle. I would like to emphasize three points that emerge from this survey, again tracing a reading that I will broadly attribute to the Neoplatonists later (§ 2, below).

(1) That the philosopher investigating a topic ought to set down the *phainomena* or appearances concerning that topic (whether they are empirical observations or views with some "currency");[13]
(2) That every credible witness contributes jointly to finding the truth;
(3) That historical witnesses, including very ancient thinkers and writers, count as credible witnesses to the truth—and sometimes as especially credible witnesses.

---

10 See, for example, Frede (2012), Devereux (2015), DaVia (2017), Karbowski (2018 and 2019), and Falcon (2019).
11 The thesis applied to ethics is effectively defended by Karbowski (2019). Alongside the skepticism noted above, I appreciate Melina Mouzala stressing to me that endoxic approaches are hardly alien to Aristotle's physical inquiry, even if we have to be cautious about seeking a single program of approach; Mouzala points out, for instance, the overlaps between the method described in *EE* 1.6, 1216b26–32 and *Physics* 1.1, 184a15–21.
12 Syrianus' commentaries on *Metaphysics* M and N, and B and Γ, illustrate this combination of the Platonists' respect for Aristotle's general methodology and criticism when he is perceived to fall short of it: see Dillon and O'Meara (2006) and O'Meara and Dillon (2008), with introductions.
13 "Currency", Brunschwig (1967), pp. xxxiv–xxxviii.

Much of this discussion can be amplified from explicit and implicit sketches that occur in Aristotle's *Topics*, and the general study of Aristotle's dialectic;[14] I will focus here primarily on Aristotle's methodological remarks in the ethical works.[15]

(a) Setting down the *phainomena*, and identifying *aporiai* (*NE* 7.1)

In the seventh book of the *Nicomachean Ethics*, while preparing the way for his analysis of *akrasia*, Aristotle offers this brief and much-debated methodological trailer:

> **T1.** Aristotle, *Nic. Eth.* 7.1, 1145b2–7. We must, as in the other cases (*epi tōn allōn*), set down the phenomena (*tithenai ta phainomena*) and, after first discussing the difficulties (*diaporēsantas*), prove, if possible, the truth of all the reputable opinions (*endoxa*) about these conditions—or, failing to do this, of the greater number and the most authoritative (*kuriōtata*). For if we both resolve the difficulties and leave the reputable opinions undisturbed, we shall have proved the case sufficiently (trans. Frede 2012, slightly modified).

The force of nearly every phrase in this excerpt has been debated.[16] It appears, at face value, to announce a kind of philosophical practice that begins from "setting down the *phainomena*"[17]—recording "appearances", understood not as "empirical observations", but as "reputable opinions" (*endoxa*)[18] on a topic. For example:

> **T2.** In effect, then, [Aristotle's] idea is that the first thing we must do, when we investigate a subject, is to pay careful attention to what seems to be the case to either everyone, or to most people, or to a special and smaller group—those who have studied the subject (Kraut 2006, p. 78).

---

[14] See Brunschwig (1967, 1986, and 1990), Moraux (1968), Irwin (1981), Devereux (1990), and the overview in Devereux (2015). For an historical overview of the development of dialectic in the 4th century, see Fink (2012). Aristotle's introductions to the *Topics* and *Prior* and *Posterior Analytics* provide a basic framework for several conceptions of dialectic, including the famous description of it as "a line of inquiry whereby we shall be able to reason from reputable opinions about any subject presented to us, and also shall ourselves, when putting forward an argument, avoid saying anything contrary to it" (*Topics* 1.1, 100a20–24); see also *Pri. An.* 1.30–31.

[15] For recent discussions and bibliographies, see Falcon (2019) and Karbowski (2018 and 2019).

[16] See below as well as the helpful "eclectic doxography" in Frede (2012), pp. 200–208.

[17] In an apparent application of the procedure, Aristotle goes on to present various current views (*legomena*, 1145b20) about *akrasia*, then articulates Socrates' denial of *akrasia*, and observes that it is "at variance with the *phainomena*", before proceeding to offer his own view.

[18] Barnes' translation (1980).

Indeed, *NE* 7.1 has sometimes been codified into a set procedure for handling any philosophical topic, or else any ethical topic.[19] For now, we can follow some typical practice by labelling this a "standard account" of an "endoxic method":
(1) Set down the *phainomena* about the subject, by enumerating the *endoxa* (commonly held views) concerning it;
(2) Discuss the difficulties or inconsistencies (*aporiai*) presented by the *endoxa*;
(3) Resolve the difficulties so as to prove the truth of the most or most authoritative *endoxa* (which belong to the set of *endoxa* enumerated at the outset).
(4) The result should be a consistent set of the most or most authoritative *endoxa* concerning the subject.

Many careful readers of Aristotle, including Owen (1961), Barnes (1980), Nussbaum (1982), and Irwin (1988), have developed accounts of some "endoxic method" similar to this, though it is often reconstructed with more flexibility.[20] Others, including Dorothea Frede (2012) and Carlo DaVia (2017), have recently stressed Aristotle's failure to *follow* such a program anywhere, with the possible exception of *NE* 7 itself, and have argued persuasively against the existence of any such wide-ranging and robust endoxic method. We might for now just flag this point: "the endoxic method seeks truth about ethical matters by purging endoxa of their epistemic shortcomings via aporetic investigation",[21] which begins from some sort of survey of the *phainomena*, often including views exhibiting some "currency".

(b) Relating diverse contributions to the truth (*EE* 1.6, 7.1–2; *Metaph. α* 1)

A slightly different, but parallel, articulation of a similar view occurs early in the *Eudemian Ethics*:

---

19 See DaVia (2017), p. 384, with bibliography, for an example.
20 For example, Owen (1961, 1986) suggested that Physics 4 contains a similar study of the phainomena about place—where phainomena in this particular case stand for "commonly held views" (contrasting the empirical observations of astronomy: Prior Analytics 1.30, 46a17–22). Barnes (1980) broadly agreed that phainomena, legomena, and endoxa could point to "things that seem to be the case" in a wide sense, citing EE 7 (above); Barnes also concluded that "the language we use betrays our latent opinions" (Barnes 1980, p. 501), and the method can make these opinions explicit. Irwin (1988) developed a rich analysis of the role that endoxa might play in establishing first principles, while Nussbaum (1982) advanced a creative interpretation of the "method" as a kind of early internal realism. See Frede (2012), pp. 200–208—"a short and eclectic doxography".
21 Karbowski (2016), p. 196.

> **T3a.** Aristotle, *Eud. Eth.* 1.6, 1216b26–32. About all these matters [involving *eudaimonia*] we must try to get conviction by arguments, using the *phainomena* as evidence and illustration (*marturiois kai paradeigmasi*). It would be best that all people should clearly concur with what we are going to say, but if that is unattainable, then that all should in some way at least concur. And this if converted (*metabibazomenoi*) they will do, for every person has some contribution to make to the truth (*echei gar hekastos oikeion ti pros tēn alētheian*), and with this as a starting-point we must give some sort of proof about these matters. For by advancing from true but obscure (*ou saphōs*) judgements he will arrive at clear ones (*saphōs*), always exchanging the usual confused statement for more real knowledge (trans. Solomon in Barnes 1984).

This passage has also earned a scholarly industry in its own right. Donald J. Allan compared the "Eudemian method" to the "mathematical pattern of deduction" of Euclid's *Elements*.[22] Recently, Andrea Falcon and Joseph Karbowski have treated this "Eudemian method" in much more depth, in the context of exciting new monographs on Aristotle's method in ethics more broadly (2019).[23] The key points that I would like to note for our purpose here are the following: in the course of summoning the *phainomena* (so to speak) to the stand as witnesses (a metaphor whose force might sometimes be underplayed), Aristotle stresses that "every person has some contribution to make to the truth".

This method is not usually presented as a demonstration (*apodeixis*), but as a reconstruction of a picture from basic evidence, where the more witness evidence can be offered, the rounder the picture of the truth that appears. In a way, it contributes to the "excavation" or "articulation" of what the view *is* before it is examined in more detail.

This picture can also be juxtaposed with another methodological passage from *EE* 7.1–2.

> **T3b.** Aristotle, *Eud. Eth.* 7.1–2, 1235a4–b18. Many questions are raised (*aporeitai*) about friendship (*philia*)... [for example, among those who apply the term to the external world, some maintain that similar things are friends, others that opposite things are friends]. There are, then, these two views about friendship; and they are too general and far removed. There are other views that come nearer to and are more suitable to the *phainomena*... [for example, some maintain that only good people can be friends to one another.]
>
> We must, then, find a method (*tropos*) that will best explain the views held on these topics (*dokounta... malista apodōsei*), and also put an end to difficulties and contradictions (*aporiai... enantiōseis*). And this will happen if the contrary views are seen to be held with some show of reason (*eulogōs*); such a view will be most in harmony (*homologoumenos*) with the *phainomena*; and both the contradictory statements will in the end stand, if what is said is true in one sense but untrue in another (trans. Solomon in Barnes 1984).

---

22 See Allan (1980).
23 Karbowski (2018); see also Falcon (2019).

Setting aside for now the question of whether this is a *different* method than that described in *EE* 1, we can find some joint conclusions supported by the intersection of both texts. Briefly, this passage implies that *agreement* (*homologia*, or harmony) among the "witnesses" and *phainomena* about a given proposition can be seen as a guide of guide toward the promise that proposition holds of being true. A similar notion might be read into the opening of the second book of the *Metaphysics*:

> **T4.** Aristotle, *Metaphysics* α 1, 993a27–b11. The investigation (*theōria*) of the truth is in one way hard, in another easy. An indication of this is found in the fact that no one is able to attain the truth adequately (*axiōs... thigein*), while, on the other hand, no one fails entirely, but every one says something true about the nature of things, and while individually they contribute little or nothing to the truth, by the union of all a considerable amount is amassed. Therefore, since the truth seems to be like the proverbial door, which no one can fail to hit, in this way it is easy, but the fact that we can have a whole truth and not the particular part we aim at shows the difficulty of it.
>
> Perhaps, as difficulties are of two kinds, the cause of the present difficulty is not in the facts but in us. For as the eyes of bats are to the blaze of day, so is the reason (*nous*) in our soul to the things which are by nature most evident of all.
>
> It is just that we should be grateful, not only to those whose opinions we may share, but also to those who have expressed more superficial (*epipolaioteron*) views; for these also contributed something, by developing before us the powers of thought (trans. Ross in Barnes 1984).

Here is another version of the idea that every witness offers a slice of the truth. We might amplify the spirit of the notion through the well-known Indian metaphor, found in early Buddhist and Jain literature, of the blind men and the elephant.[24] Indeed, there are particularly interesting parallels in the current literature on exegetical "inclusion" or harmony in Indian philosophical and contemplative commentary traditions.[25]

But returning to Aristotle's case, in what sense does each witness offer some aspect of the truth? As David McNeill has argued (2018), particularly in the context of *EN* 7.1, studying the *endoxa* and so "going through perplexity well" (McNeill 2018, p. 259) is not a matter of comparing "isolated practical or theoretical positions or propositions that one could endorse", but of encountering "phenomenally concrete modes of human orientation to a shared practical world", making the *endoxa* "cognitively and experientially salient" (McNeill 2018, p. 265). This point seems particularly clear when Aristotle stresses the value of encountering historical views

---

[24] An early example of the parable is preserved in Udana 6:4 in the Buddhist Pāli canon (Tittha Sutta).
[25] Paul Hacker's notion of "inclusivism" in Indian philosophy (see Hacker 1983) is often commented on in this connection.

taken by those *with experience*—and hence the value of studying the history of philosophy.

(c) Finding witnesses in the past (*NE* 6.11, *Metaph.* Λ 8, fr. 13)

In searching for witnesses, we should not limit ourselves to the newest opinions. Aristotle brings home this advice in *Nicomachean Ethics* 6:

> **T5.** Aristotle, *Nic. Eth.* 6.11, 1143b11–14. One should pay attention (*prosechein*) to the undemonstrated sayings and opinions (*anapodeiktois phasesi kai doxais*) of those who have experienced and are older (*tōn empeirōn kai presbuterōn*), or to those who have practical wisdom (*phronimōn*), no less than to demonstrations. For, because they have an eye that derives from their experience, they see rightly (Trans. Frede 2012).

In context, this helps to illustrate the value of practical experience in certain domains of philosophy. An inexperienced and childlike person (figuratively or literally) might be a mathematical prodigy—but it requires experience to be a good judge of political science (*Nic. Eth.* 1.3). In a parallel fashion, perhaps, the witness of very ancient (*palaioi*) historical views that have stood the test of time might be particularly valuable, due to experience.

This is a point that I would suggest (following Boys-Stones 2001, among others) is particularly important for understanding the progress of ancient exegesis or commentary. As the following passages illustrate, ancient or historical wisdom sometimes takes an unusual form—for example, the form of myths, or highly condensed sayings (*paroimia*) that require unpacking. The next text, preserved by Synesius, reports Aristotle's view that some particularly valuable ancient wisdom is preserved in abbreviated sayings:

> **T6.** Aristotle, fr. 13 Rose. Whether a "saying" (*paroimia*) counts as something wise: Why not? Aristotle says about them that they are remnants of an ancient philosophy (*palaias eisi philosophias... egkataleimmata*) saved by their brevity and acuity when it was lost in the great destructions of mankind *(Ap.* Synesium *calvit. encom.* 22, trans. Boys-Stones 2001).

And as *Metaphysics* Λ suggests, sayings of this kind invite interpretation, or exegesis, due to their form.

> **T7.** Aristotle, *Metaphysics* Λ 8, 1074a38–b14. Fragments of the thought of the ancient, the very ancient thinkers have been handed down to us in the form of a myth, to the effect that these [se. the planets] are gods, and that the divine embraces the whole of nature. The rest has been added later in a mythological form to influence the beliefs of the vulgar, for the benefit of the laws, and for pragmatic reasons. And they say that these [gods] are human in shape or are

like some other animals, and add other things consequent on and similar to what has been said.
> If we separate out and take only the first claim, that they thought the first substances to be gods, we must regard it as a divine saying, and reflect that, since every art and philosophy has probably been repeatedly developed to the utmost and has perished again, these beliefs of theirs have been preserved as relics. To this extent only, then, can we see the opinion of our ancestors and of the first men (trans. Boys-Stones 2001).

These exegetical or metaphilosophical views in Aristotle are shared, to some extent, with Plato,[26] and anticipate one current of classical and Presocratic thought: to articulate and discuss the views of predecessors within a framework that explains their motives, and the performance of this explanation, illustrates its own greater *inclusiveness*. The rhetorical and exegetical function of the four-causal schema early in Aristotle's *Physics* and *Metaphysics*, respectively, illustrates this practice neatly (e.g., *Metaph.* 1.3, 983a33–b4):[27] the conjoint validity of all four modes of explanation explains why, for instance, the Milesians (in focusing on material causation) and the Pythagoreans (in focusing on formal causation) are both partly right, even as it explains why their views are partial, and illustrates the greater, and more inclusive, explanatory power of the four-causal schema itself. The Neoplatonists use the same sort of strategy with Aristotle himself, applying a typically *six*-causal schema that illustrates the Platonic causes he omitted.[28]

2.2 The 1st Centuries, the Stoics, and Plotinus

(a) The 1st centuries BCE and CE

The project of unpacking the "witnesses" of the *phainomena* with a special emphasis on the history of philosophy was flourishing during the 1st century BCE—not only by way of myths and *paroimia*, but now engaging in extended, sometimes line-by-line exegesis of poetic and philosophical texts. This practice breathed the air of scholarly and philological centers like Alexandria and Pergamon, and especially the Mouseion, itself an institution with Aristotelian roots. Its roots are deeply

---

[26] Compare the proem of the *Timaeus*: the Egyptian priest makes a similar point to Solon (21e–25d).
[27] But compare Frede (2012), pp. 192–193, who suggests that Aristotle relies on this schema "not... to establish the opinions of his predecessors, but rather to point out their weaknesses". We could also argue, as above, that he can deploy the schema to illustrate the strengths *as well* as weaknesses of each theory.
[28] Simplicius, *in Phys.* 3,13; 6,31.

intertwined with the practice of criticism of Homer pioneered by figures like Aristarchus.[29]

The early record of word-for-word philosophical commentary and exegesis in the Mediterranean has been well treated.[30] Here is one way of articulating its development briefly. As David Sedley wrote in 2012:

> [C]ut adrift from the historic institutions which had linked them to their revered founders, the major philosophies shifted their efforts increasingly onto the study of their foundational texts[31] (Sedley 2012, pp. 1–2).

Mediterranean philosophers gradually fostered a new value for working in text; at the same time, they began to cultivate a new respect for the authority of ancient or "classical" voices. These two developments were not necessarily symbiotic.[32] Within a few centuries, however, philosophy had come to consist primarily in exegesis of canonical texts in a set curriculum.

---

[29] For the development of that practice, see recently Schironi (2018).
[30] Recently, see Myrto Hatzimichali's contribution published in *Aristotle Re-Interpreted* (ed. Sorabji 2016b); surveys can be found in Richard Sorabji's updated introductions to *Aristotle Transformed* (ed. Sorabji 2016a) and in the recent *Brill Companion to the Reception of Aristotle* (ed. Andrea Falcon 2016) and *Brill Companion to the Reception of Plato* (eds. Tarrant and Layne 2018). For this development in general, see recently Malcolm Schofield (Ed.): *Aristotle, Plato and Pythagoreanism in the First Century BC: New Directions for Philosophy*, Cambridge: Cambridge University Press, 2013.
[31] See Sedley (2012), pp. 1–2.
[32] Consider for instance Epictetus, whose school cultivated the close reading of philosophical texts. Epictetus himself found value in philosophical reading, provided that the student was capable of digesting the material and apply it in practice (Diss. 1.26.13–18). Epictetus also recommended writing as a therapeutic exercise: for this theme in general, see P. Hadot (1995). In this role, writing serves as a replacement for the cross-examination of one's own soul when no conversation partners are at hand, as his presumably anachronistic ascription of a kind of "writing for oneself" to Socrates illustrates (Diss. 2.1.31–32). But Epictetus prided himself on not being a classicist, not a (mere) philologue or antiquarian. He vigorously critiqued philosophers who advertised themselves primarily as exegetes or commentators on the greats (Diss. 3.21.6–7), and in the passage cited above, he gently corrects an advanced student who sets a junior student a passage to absorb just "as a reader" (ὡς ἀναγνώστῃ, 1.26.13), before evaluating the younger man's ability to digest its meaning. Criticisms of philosophy as philology are echoed by Seneca (Letter 108.23; cf. Frede (1999), pp. 771–797 and 785, and famously Plotinus (Porphyry, Life 14).

(b) The post-Hellenistic Stoics

The development—toward a reverent, exegetical attitude to the "ancients"—was, as Boys-Stones argued,[33] already exemplified by the Stoic Cornutus in the later 1st century CE, building on Posidonius' notion (cf. Seneca, *Letter* 90.5) that self-conscious sages flourished in the golden age. So Cornutus:

> **T8.** Cornutus, *Introduction* 35, 75.18–76.5 Lang. [T]he ancients were no ordinary people, but capable of understanding the nature of the cosmos, and inclined to use symbols and riddles (*dia symbolōn kai ainigmatōn*) in their philosophical discussions of it (trans. Boys-Stones 2001, p. 53).

Diogenes Laertius' *Lives* 1.1–12 also illustrates how the sages of various cultures, including Egyptian (1.10), Celtic, and Indian (1.6), operate by way of philosophical riddles (αἰνιγματωδῶς ἀποφθεγγομένους φιλοσοφῆσαι, 1.6). A concept began to take root that ancient wisdom not only *could* be reconstructed by a good interpreter, but that it *had* been so reconstructed, by Plato, whose philosophy—rightly interpreted by a competent interpreter (*exēgētēs*)—could furnish a direct and immediate guide to the wisdom of the golden age[34] and so paint a trail to the human good life and a deeper understanding of I world around us.

(c) Plotinus: Early Neoplatonism

This development lent momentum to commentary and exegesis in the Platonic tradition.[35] We might illustrate this the following passages from Plotinus. The first introduces Plotinus' doxography of views maintained by predecessors about the nature of time (in which Plotinus keeps fairly close to the general structure of Aristotle's doxography at *Physics* 4.10, 218a30–b20). Here Plotinus explains why it is necessary to review such prior views on the subject:

---

[33] *Post-Hellenistic Philosophy* (2001).
[34] Boys-Stones (2001), 115.
[35] Already in the early empire, Aristotelianism was also beginning to operate primarily as a commentary tradition (see Fazzo 2004, pp. 1–19, particularly p. 3; cf. Sharples 2010, p. 331), and by the later 3rd century CE, it is possible to speak of a "philosophy of the commentators" (see Sorabji 2004 and Tuominen 2009). This development toward philosophy as commentary on wise teachers of the past sets the stage for Simplicius. See, for example, Sedley 2012, Introduction, and Boys-Stones (2001), especially pp. 99–150; and for the Neoplatonic maturity of the form, see Mansfeld (1994), Hoffmann (1998), pp. 228–40, Hoffmann (2008), 601–602, and Baltussen (2008). (1994) and 601–602

T9. Plotinus, *Enneads* 3.7.7. Now if the blessed persons (*makares*) of antiquity had said nothing about time, it would be necessary for us to take eternity as our starting point and connect it with our subsequent account of time, endeavoring to fit (*epharmozein*) the [15] opinion of it that we state (*ta dokounta*) with the conception (*ennoia*) of it that we possess. But as it is, it is necessary first to take the most notable statements that have been made about time (*ta malista axiōs logo eirēmena*), and to consider whether our own account will be in agreement with any of them (trans. McGuire and Strange 1994).

T10. Plotinus, *Enneads* 3.7.1. When we say that eternity and time are different things, and that eternity pertains to the eternal nature, while time pertains to what comes to be and to this universe, we immediately think, as we do in the case of more cursory conceptual apprehensions (*athroōterais epibolais*), that we possess [5] a clear (*enarges*) impression of them in our souls, since we are always talking about them and referring to them everywhere.

But when we try to go on to examine them and, as it were, get close to them, we once again find ourselves at a loss what to think (*aporountes*): different ones of us fix upon different declarations of the ancient philosophers about them, and perhaps [10] even disagree about how to interpret these statements. So we stop here, and deem it sufficient if when asked we can state their views about them. Content with this, we give up inquiring any further about these matters.

Now we must indeed think that some of the ancient and blessed philosophers have found the truth. But who among them [15] most attained to it, and how we might gain an understanding (*sunesis*) of these things for ourselves, needs to be investigated (trans. McGuire and Strange 1988).

The methodological implications of these texts have stirred considerable discussion.[36] For example, one team of scholars (commenting on *Enn.* 3.7.1 and 3.7.7), suggest that:[37]

T11 Plotinus accepts as conditions on the adequacy of a philosophical theory (1) that it be consistent with the texts of the most authoritative of the *palaioi*, the ancient philosophers, on the matter in question, (2) that it be in agreement with our common conceptions (*koinai ennoiai*) about the subject... But we must ensure that these conceptions are clear (*enarges*) by careful

---

36 With respect to the second of these two criteria, debate has focused on Plotinus' assumptions about the nature of our "prephilosophical" common notions, and the extent of his debt to earlier Middle Platonic reception of Stoicism. In particular: does he regard the koinai ennoiai as confused and in need of correction (cf. Phillips 1987, pp. 40–41, as well as McGuire and Strange 1988, p. 266, n. 13)? Might he accept two classes of common notions, (a) those that arise from the physical world, rather like Epicurean prolēpseis, and coincide with the ordinary meaning of words (illustrated by the remark that "we possess a clear impression of [eternity and time] in our souls, since are always talking about them and referring to them everywhere"), but capture only accidents; and (b) those that arise from direct contact with the intelligible world, and do in fact offer a grasp of the essential nature of things? For this bipartition, see Van Den Berg (2005), p. 120. For the role of concept formation in Middle Platonism more broadly, see also Helmig (2012).

37 McGuire and Strange (1988).

philosophical examination of them, since our initial unreflective ("cursory") formulations of them can be confused and misleading... (McGuire and Strange 1988, p. 266, n. 13).

For our purposes here, I just want to note that the familiar pattern of comparing witness testimony—setting down the *phainomena*, identifying *aporiai*, searching for harmony, and highlighting the particular testimony of the ancient or historical—has taken on a slightly novel coloring, flavored by the philosophy of language.

Following Stoic and Epicurean predecessors, Plotinus maintains that we all have a body of more or less vague preconceptions (*prolēpseis* in Greek) that enable us, among other things, to converse in ordinary speech. But they are inexact, "cursory". When we challenge them, we run into puzzles or inconsistencies—in Greek, *aporiai*. They need to be hardened, excavated—a process that contemporary philosophers called *diarthrōsis*, "articulation". What is excavated is in the end our shared universal conceptual grammar—in Greek, our common concepts or *koinai ennoiai* (already a problematic before Plotinus, but I will leave that to one side here). What is the process? We entertain as wide a corpus of dialectically and conceptually permissible accounts of the real meaning of our words as possible—as many as "make sense" and have stood the test of time. Then we look at which ones really "feel" right: and the "feel" comes from the *koinai ennoiai*.

3 The Neoplatonist Exegetes

Now we come to the *exēgētai* of our period in Late Antiquity. By this time, reading a philosophical classic with a teacher and a commentary could function as a kind of exercise in personal development for the student reader.[38] Even the deceptively

---

[38] To compose a commentary was a primary vehicle of "research" or academic output, a centerpiece of the professional philosopher's working day (cf. Marinus, *Life of Proclus* 22), and a spiritual exercise in its own right (see I. Hadot 1978, pp. 147–65 and 1996, pp. 51–60, as well as P. Hadot 1995). To read Aristotle's *Physics* and *De Caelo* with an able teacher invites the student to "unite" (*henōsis*) with the mind of the world (Hoffmann 2000). To read Epictetus' *Handbook* with understanding might exercise and rehabilitate the soul of a student of innate but untutored talent (*phusikē aretē*) to achieve what Neoplatonists after Iamblichus dubbed "habituative" excellence (*ēthikē aretē*) (Simplicius, *in Ench.* pr. 87–90). That achievement, in turn, would prepare the reader to tackle a philosophical *scala virtutum* (Dillon 1996 and Edwards 2000), climbing from "constitutive" or "social" excellence (*politikē aretē*), available through the close reading of a Platonic text like the *Alcibiades I* or *Gorgias* with an instructor, to "cleansing" or "purifying" excellence (*kathartikē*) via a dialogue like *Phaedo*, to "contemplative" (*theoretikē*) excellence via a dialogue like *Phaedrus*, and on to a loftier, "emblematic" (*paradeigmatikē*) excellence achieved through more advanced readings and theurgical practices. Reading an authentic Platonic dialogue, itself a kind of "model" of

simple practice of copying a classical text out could offer educational and spiritual insight. Thus, Simplicius counts his annotated, condensed revising of the major commentary on Aristotle's *Categories* by his predecessor Iamblichus as a contemplative exercise (*in Cat.* 3,2–9, 438,33–36), an effort to ascertain the truth and to discover Iamblichus' spirit or meaning (νοῦς) as well as a productive and creative project designed to help students and οἱ πολλοί on their way to more advanced texts and lessons (3,13–17):

> **T12.** Simplicius *in Cat.* 3,2–9. My goal (*skopos*) in making this copy (*apographē*) was, in the first place, to obtain, through the act of writing, as accurate a comprehension (*katanoēsis*) as possible of what had been said. At the same time, I wished to reduce this man's lofty spirit (*hupsēlos nous*), inaccessible to the common people, until it was more clear and commensurate [with the common understanding] (trans. Chase 2003).

3.1 In Relation to Authority

This may look like a development towards deference to past authority, especially in a particular doctrinal school or tradition. And indeed, already in the 3rd century, the doctor-philosopher Galen complains of frustration with contemporaries who have become "slaves of their schools" (οἱ ταῖς αἱρέσεσι δουλεύοντες, *On Natural Capacities* 35,6).[39] But the situation is more complex than merely becoming slaves to the past or "the greats". Simplicius' contemporaries, the ancient commentators, do not consider Platonic authorship as a guarantee of truth in itself (Ammonius ap. Olympiodorum *in Gorg.* 41.9):

> **T13.** Olympiodorus *in Gorg.* 41.9. Concerning Aristotle we must point out that in the first place he in no way disagrees (*diaphōnei*) with Plato, except in appearance (*kata to phainomenon*). In the second place, even if he does disagree, that is because he has benefited from Plato. For [Plato] says in the *Alcibiades* "Unless you hear yourself speaking, do not put your trust in the words of another" [114e]. And again in the *Phaedo* he says "Care little for Socrates, but greatly for the truth" [91c].

---

the cosmos (Proclus *in Alc.* 10,3–18), offered the opportunity to foster such excellences through the emulation of central characters (*mimēsis:* see the anonymous *Introduction to Platonic Philosophy*, § 4, 15, Westerink, pp. 1–19): when we grasp the reasons for Socrates' actions, for instance, we will be drawn to pursue his knowledge (Proclus *in Alc.* 21, 5–8), and this applies more broadly to the emulation of all good people (*in Tim.* I, 16, 6–12).

39 Or more recently Dodds' critique of the "museum of metaphysical abstractions" that comprised Proclus' allegorical exegesis of the gods (Dodds 1963, p. 260). As Barnes (2003, p. 29) summarizes the charge, late ancient Platonists are sometimes criticized as surrendering "reason to trust and proof to authority", paving the way for a less rational and more religious sensibility in philosophy (compare Galen, *On Types of Pulse* 8.579, Dodds 1951, pp. 327–340, and Cooper 2012, p. 387).

So Plato himself urges us not to believe him indiscriminately (*haplōs kai hōs etuchen*), but to inquire (*zētein*) [for ourselves]. That is surely why the philosopher Ammonius says "I may have acted wrongly, but when someone once said something and declared "Plato said so" (*Platōn ephē*), I answered "He did not mean it like that, and in any event—may Plato forgive me—even if did mean it like that I am not persuaded, unless he added a demonstrative argument (*meta apodeixeōs*)"" (trans. Jackson, Lycos and Tarrant 1998).

I would like to draw out core points here in particular:
(1) The importance of "harmony" (*homologia, symphonia*) or agreement between Aristotle and Plato (see Karamanolis 2006 for discussion of the origins of this view)—and compare Aristotle himself;
(2) The role of demonstration (*apodeixis*) in the search for truth: compare already Aristotle's stress that the "witnesses" consulted may not be *demonstrative*, but are nonetheless useful (*NE* 6.11, 1143b11–14, **T7** above).

Simplicius himself critiques those who deploy quotations from the ancients as *proofs* of their arguments (*in Phys.* 1318,10–15, on *Phys.* 265b17–266a5), and praises the "impartiality" of a good commentator:[40]

> **T14.** Simplicius *in Phys.* 1318,10–15, ad *Phys. 265b17–266a5*. Aristotle's practice after his demonstrations is to introduce the testimonies of his predecessors as agreeing with his demonstrations, in order on the one hand to teach and compel his readers through his demonstrations, and on the other to make the belief more certain in his hearers through the testimony of the others; he does not employ the testimony of his predecessors [with Diels' emendation] as a demonstration, as is the practice of more recent writers. This is what he does now, confirming that motion in place is the primary kind of motion also from the fact that everyone who had declared his opinion on these matters was brought to this view as if guided by nature itself, as he himself says elsewhere[41] (trans. McKirahan 2001).

This brings us to the ecumenical posture that Simplicius recommends in the study of Aristotle:

---

40 See Barnes (2003), p. 29: "Yet if there were some intellectual slaves—and perhaps some happy slaves—slavery was not a common condition among philosophers". For Simplicius' remarks on "mpartiality" in the *Categories* commentary, see also Barney (2009), p. 108. Perhaps Simplicius himself placed more weight on some voices from the past, but it seems clear that he himself, like Ammonius, would have denied a claim that any proposition was true *because Plato* (rightly interpreted) *had said it*, even if we are sometimes inclined to attribute such a view to Simplicius. See Baltussen (2002), p. 181. Baltussen (2002, p. 187) speculates that Simplicius even uses direct quotation as a kind of living "voice from the past".
41 Perhaps *Phys.* 1.5, 188b29–30, *Metaph.* 1.3, 984a18–19, b9, and *PA* 1.1, 642a19.

T15. Simplicius *in Cat.* 7,23–29. [T]he worthy exegete of Aristotle's writings must not fall wholly short of the latter's greatness of intellect. He must also have an experience of everything the philosopher has written, and must be a connoisseur (*epistēmōn*) of Aristotle's stylistic habits. His judgement must be impartial (*adekastos*), so that he may neither, out of misplaced zeal, seek to prove something well said to be unsatisfactory, nor, if some point should require attention, should he obstinately persist in trying to demonstrate that [Aristotle] is always and everywhere infallible, as if he had enrolled himself in the philosopher's school. [The good exegete] must, I believe, not convict the philosophers of discordance by looking only at the letter (*lexis*) of what [Aristotle] says against Plato; but he must look towards the spirit (*nous*), and track down (*ankhineuein*) the harmony which reigns between them on the majority of points (trans. Chase 2003).

Simplicius' value for a certain impartiality or open-mindedness does not stop here. At another point in his voluminous commentary on the *Physics*, in discussion of notions of "place" or *topos* (and incidentally commenting on the same passages that motivated Gwilym Ellis Lane Owen's treatment of *Physics* 4 in terms of endoxic method), he writes:

T16. Simplicius, *in Phys.* 640,12–18. Why, then, should we say that so many great people (*tēlikoutoi*) were mistaken in their opinions about place, putting forward our problems (*aporiai*) as an unfortunate feast for those who are accustomed to take pride in the apparent contradictions (*hai dokousai... enantiologiai*) of the people of old (*hoi palaioi*)? Should we not rather follow up (*parakolouthountes*) each of those who wrote about place and show that none of them missed the truth about place? But, since place has many aspects (*polyeidous*), we should show that each person has seen and revealed a different aspect of it (*kat' allo ti tōn eidōn tou topou theasasthai te kai ekphēnai*). But now, in order that we may both see the common conception (*ennoia*) of the whole of place and observe from that standpoint the above-mentioned differences between the philosophers, I shall continue with a fresh start as follows (trans. Urmson-Siorvanes, 1992).

Simplicius is here articulating a view similar, for example, to Aristotle's methodological remarks in *Metaphysics* α cited earlier. Like Aristotle, Simplicius and his contemporaries see themselves atop a kind of peak of philosophical achievements, from which the bounded achievements of the past could be surveyed.[42] Part of the task is to see how earlier thoughtful writers *partially* hit the truth. In one particularly interesting stretch of commentary on the *Categories*,[43] Simplicius explores

---

[42] Fr. 53 Rose = Cicero, *Tusculan Disputations* 3.69; cf. Boys-Stones (2001), pp. 26–27.
[43] The passage excerpted from his "corollary on space" above (*in Phys.* 640,12–18) offers an example of this procedure. Another extended example of this kind of effort can be found in a long doxographical passage early in the *Categories* commentary, where past interpretations of the *skopos*—the goal or target—of the *Categories* are subdivided into verbal, conceptual, and ontological readings, which are then unified under the more "complete" banner of a semantic understanding that

how ancient readers of that important text had all been like the blind men who touch a part or organ of the elephant (without this analogy): some caught the treatise's import for the study of language, others for the study of concepts, others again for the study of beings—but Simplicius and his teachers can see how *all* the ancient views fit together. Studying many of these different positions will help us to articulate a more holistic picture. (And incidentally, studying the *Categories* will also be also the first step in making our words accurately fit our concepts and reality).

3.2 Ennoiai

In this short detour, I would like to argue that the position we have been exploring has more or less explicit roots in Neoplatonic philosophy of language and philosophy of education.

Here is a passage where Simplicius articulates the need for a *teacher* in philosophy. In this text he addresses how verbal expressions always address and reflect reality to a limited degree, but the "common notions", our universal conceptual grammar, are not fully "stirred" until a more expert concept-worker teaches us. (We might again compare McNeill's point about Aristotle's method in *EN* 1.7, that studying experiential *endoxa* "reanimat[es] ways of questioning that experience" for the would-be learner; 2018: 265). Simplicius borrows compelling language of "stirring up" (*anakinein*) a common notion, beginning from examples and sketches (*hupographai*) in the use of a word (Simplicius *in Cat.* 159,10–15):[44]

> **T17.** Simplicius *in Cat.* 12,10–13,4. [N]either are significant expressions wholly separate from the nature of beings, nor are beings detached from the names which are naturally suited

---

integrates verbal, conceptual, and ontological components. Among earlier philosophers, Simplicius says, "each one had an incomplete grasp of the goal" (2,26). The "full" view is this: "It is thus clear from what has been said that [those who have a more complete conception: Alexander, Porphyry, Iamblichus] do not define the goal as being about mere words, nor about beings themselves in so far as they are beings, nor about concepts alone... it is about simple words and expressions qua significant of primary and simple beings... It is thus clear from the preceding considerations that the goal (*skopos*)... is about simple, primary, and generic words (*phōnai*), in so far as they are significant of beings... and about concepts..." (11,30–13,19) The question at hand is what *kind of thing* a category or genus like "the relative" or "quality" might turn out to be—a word, or a concept, or a real being. In Plotinus' terms, Simplicius has explored the views of thoughtful, past philosophers about the nature of these genera, and explored *aporiai* about them, in order to discover which, if any, matches with our *ennoia*, along the way trying to excavate and clarify what that concept is.

44 See also Hoffmann (1987).

to signify them. Nor, finally, are intellectual concepts extraneous to the nature of the other two; for these three things were previously one, and became differentiated later. For Intellect (*nous*), being identical with realities and with intellection (*noēsis*), possesses as one both beings and the intellectual concepts of them, by virtue of its undifferentiated unity (*adiakritos henōsis*), and there [sc. in the intelligible world] there is no need for language. [...]

[The soul] needs someone who has already beheld the truth, who by means of verbal expression (*phōnē*) uttered forth from the common concept (*ennoia*) also moves the concept within [the soul of the student] which had until then grown cold. This, then, is how the need for *phōnē* came about... *noēseis* join the learner's concepts to those of the teacher... When *noēseis* are set in motion in the appropriate way, they adjust themselves to realities, and thus there comes about the knowledge of beings, and the soul's spontaneous *erōs* is fulfilled (trans. Chase 2003, modified, ACA).

For Plotinus and Simplicius, we all come with the basic conceptual structure of reality baked in, but it needs some stimuli to "stir it up"—and the right stimuli, offered by a good teacher, will teach us the real "grammar". The following are also examples:[45]

T18. Simplicius *in Cat.* 159,10–15. It was not feasible to give definitions of the primary genera... But it was possible, by means of a general description (*hupographēs*), to actuate (*anakinein*) our conception (*ennoian*) that fits (*sunarmozousan*) with the relative. He does this by following Plato according to the first definition, as Boethus [of Sidon] tells us... (trans. Fleet 2002, slightly modified, ACA).

T19. Simplic. *in Cat.* 163,28–29. [Boethus of Sidon] himself goes on to claim in his defense that it is necessary to present sketch accounts (*hupographas*) of the primary genera by means of the things which are posterior to them as well as themselves (trans. Fleet 2002, ACA).

T20. Ammonius *in Cat.* 93,9–12. Aristotle's aim here is to teach us about the words mentioned in the doctrine of categories that require some articulation (*diarthrōsis*) and explanation... (trans. Matthews and Cohen 1991, ACA).

T21. Herminus ap. Porph. *in Cat.* 59,17-33. [T]he subject of the [Categories] is not the primary and highest genera in nature, for instruction in these is not suitable for young persons, nor the issue of what the primary and fundamental differentiae of things said are, since in that case the discussion would seem to be about the parts of speech. Rather it is about the sort of predication that will properly belong to what is said in each of the genera of being. Hence it also became necessary to touch in some way upon the genera to which the predications in question correspond, for it is impossible to recognize the kind of signification that is proper to each genus without some preconception (*prolēpsis*) of it. This also accounts for the title

---

45 Thus, the *Categories* itself is an example of how we might be helped in this project—before we are able to understand a full "definition" of a genus like the relative, Aristotle's text can offer us a sketch account or *hupographē* that will "stir up" the appropriate conception. See Long and Sedley (1987), 26H, for Stoic examples of the Hellenistic use of the Aristotelian concept of the "outline account".

Predication (*katēgoria*), which means "the proper mode of signification connected with each genus" (trans. Strange 1992, slightly modified, ACA).

In at least some respects, the Neoplatonists believe that mastering language just *is* mastering philosophy, or at least the first stage of philosophy: it is the task of amending our referring terms to carve reality at the joints. Thus, at the outset of their exegeses of Aristotle's *Categories*, Simplicius and his predecessors describe how our ordinary referring terms almost but not perfectly map to real beings—we are given the beginnings by nature and nurture, but need to train further to get it exactly right.[46] And ancient philosophers carefully coined new terms to capture reality more precisely: thus, we may study all the ancients to learn this tongue.

From this standpoint, the activity of recording and reflecting upon philosophical sources that have lasted a long time is almost like immersion in a broad corpus of speech. As we engage in this exercise, though, some of those speech-acts may start to seem especially "good" or grammatical; and we will learn ever richer ways of deploying conceptual language by spending time in dialogue with exemplars. So, on this view, we might read Simplicius' sprawling but introductory commentary on Aristotle's *Categories* or *Physics*, brimming over with examples from the history of the philosophy of language and nature respectively, as texts analogous to an elementary rhetorical treatise like Longinus' on the sublime: full of descriptive samples of great and not so great figures of speech, plus some rough prescriptive rules for which are best. But here it is not diction but conceptual analysis to which we are invited to attend.

Impressionistically, on this interpretation, Neoplatonists take ordinary language as a vaguer form of an ideal language that precisely carves reality and discloses the truth: Plato may be an outstanding user of this language, but many ancient philosophers use something like it, coining better words and clarifying the initially fuzzy lines of concepts behind our ordinary words, since speech is also, in our pre-philosophical condition, comprised of imperfect representations of the ἔννοιαι that have "grown cold" in the person's mind.[47] (The Platonists have an aesthetically thoughtful way of putting this: they argue that language (*logos*) is generated by the soul in the course of its descent from the true, intelligible

---

[46] See Griffin (2012).
[47] Here are the baser or lower concepts of Plotinus, those that are partial and not guaranteed to be accurate, that need dialectical work to make "clear" and usable for philosophical activity. In the context of this passage in the Categories commentary, it is apparent that language, at this level, is the kind of sloppy "ordinary language" that is used by the student who is coming to philosophy for the first time at the level where all words refer to sensible objects of ordinary experience (cf. in Cat. 74,4, 82,15–20).

realm outside Plato's "cave", and philosophy, which leads through and past language, is a road home).[48] So, the study of language and philosophy are steps on one ladder to the truth.

4 Conclusion

To return to the central thread: for the Platonists, exegesis or commentary (*exēgēsis, hermēneusis*) on a range of authoritative witnesses, cultivating a recognition of their harmony or agreement as witnesses (*homologia*), can constitute creative philosophical work. I would briefly add that not only textual witnesses, but also *agalmata*, sacred works of art, and oracles invite such a treatment from the Neoplatonists.[49] Our short survey has traced one vision of this kind of view, back from the commentators through Plotinus and the Stoics to, in some sense, several methodologies employed by Aristotle himself.

This approach to philosophy sometimes contrasted with the methods of modern analytic thought. For example, Pierre Hadot writes:

> T22. Philosophers of the modern era, from the seventeenth to the beginning of the nineteenth centuries, refused the argument from authority and abandoned the exegetical mode of thinking. They began to consider that the truth was not a ready-made given, but was rather the result of a process of elaboration, carried out by a reason grounded in itself (P. Hadot 1995, p. 76).

And indeed, in contrast to this, we might notice some benefits of engaging in philosophical commentary with respect for our sources. As Sorabji (2016a) has stressed, the ancient commentators were bound to generate highly creative ideas in order to maintain their thesis of *harmony*. Again, Dylan Futter has recently developed the idea that philosophical commentary traditions encourage, rather than discouraging, philosophical conversation to continue:

---

[48] Meaning in speech is "the most authentic core of human language" (Hoffmann 1987, p. 90). Hoffmann explains that "La structure de la signification, telle que nous la percevons dans l'analyse de notre langage, c'est-à-dire comme un système liant les mots aux réálités par la médiation des notions, est le produit de la séparation effectuée par l'âme, et tel est l'horizon métaphysique dans lequel est conçue la doctrine des catégories. Pour Simplicius, comme pour les autres commentateurs néoplatoniciens, celles-ci sont les mots les plus génériques de tous, institués dès l'origine pour signifier les genres suprêemes. Ces mots simples et premiers sont le noyau le plus authentique du langage humain, et leur étude est à juste titre le <viatique> qui conduit l'âme <vers de plus hautes connaissances> dans son cheminement en direction du premier principe. Le σκοπός du traité d'Aristote s'ordonne à cette fin ultime" (Hoffman 1987, p. 90).
[49] See, for example, Addey (2014).

T23. I shall refer to this hermeneutical standpoint as "philosophical anti-authoritarianism" (PA)... No reader is required to defer to any philosophical author on any philosophical claim. [...]

In some commentary traditions [by contrast]... the reader... was not entitled to judge that the author had made a mistake... Since [such a reader] is required to defer to the author, he must assume that his own judgement is incorrect, or that he has failed to grasp the author's meaning. Hence he cannot move past the claim or argument. He cannot accept or believe it because it seems to him to be false. But he cannot reject it... He is in *aporia*. If [this reader] is to dissolve the perplexity, he must continue to philosophise. The fact that he is not permitted to reject the apparently false claim of the philosophical authority forces him to rethink the matter... He must ask... questions... Hence [his hermeneutical approach] generates philosophical activity of the sort that can improve understanding (Futter 2016, pp. 1333–1334 and 1345).

As we have seen among the Platonists, this obstacle to rejecting a claim on first encounter does not need to imply a sacrifice of argumentative rigor. Nor, I think, should it imply that we as readers are obliged to engage seriously with views that we find reprehensible: after all, what is being "excavated" is the notions or values, the *ennoiai* or *prolēpseis*, that guide our deepest moral and aesthetic intuitions. The commentators likewise maintain the essential value of demonstration (*apodeixis*) and the articulation of our common notions (*ennoiai*) apart from ancient authority. Ancient philosophical resources act as inspirations, kindling the spark at the beginning of the philosophical journey, and providing a kind of companionship and company at its end, when different words or concepts are set aside and a shared meaning persists—as some, at least, of the Platonists believe it must.[50]

Bibliography

Addey, Crystal (2013): *Divination and Theurgy in Neoplatonism: Oracles of the Gods*. London and New York: Routledge.

Allan, Donald J. (1961): "Quasi-mathematical method in the Eudemian Ethics". In: Mansion, Suzanne (Ed.): In: Mansion, Suzanne: *Aristote et les problèmes de méthode. Communications présentées au Symposium Aristotelicum tenu à Louvain du 24 août au 1er septembre 1960*. Louvain: Publications Universitaires, pp. 303–318.

Baltussen, Han (2002): "Philology or Philosophy? Simplicius on the Use of Quotations". In: Worthington, Ian and Foley, John (Eds.): *Epea and Grammata. Oral and Written Communication in Ancient Greece*. Leiden: Brill, pp. 173–189.

---

50 So Olympiodorus *in Gorg.* §§ 4 and 47 and the sources cited above.

Baltussen, Han (2007a): "Dialectic in Dialogue: The Message of Plato's Protagoras and Aristotle's Topics". In McKay, Anne (Ed.): *Orality, Literacy, Memory in the Ancient Greco-Roman World*. Leiden: Brill. pp. 201–224.

Baltussen, Han (2007b): "From Polemic to Exegesis: The Ancient Philosophical Commentary". In: *Poetics Today* 28, pp. 247–281.

Baltussen, Han (2008): *Philosophy and Exegesis in Simplicius: The Methodology of a Commentator.* London: Duckworth.

Barnes, Jonathan (1980): "Aristotle and the Methods of Ethics". In: *Revue Internationale de Philosophie* 34, pp. 490–511.

Barnes, Jonathan (2003): *Porphyry's Introduction. Translation of the "Isagoge" with a Commentary.* Oxford: Oxford University Press.

Barney, Rachel (2009): "Simplicius: Commentary, Harmony and Authority". In: *Antiquorum Philosophia* 3, pp. 101–120.

Blondell, Ruby (2002): *The Play of Character in Plato's Dialogues.* Cambridge: Cambridge University Press.

Bonazzi, Mauro and Celluprica, Vincenza (2005): *L'eredità Platonica. Studi Sul Platonismo Da Arcesilao a Proclo.* Naples: Bibliopolis.

Boys-Stones, George (2001): *Post-Hellenistic Philosophy: A Study of Its Development from the Stoics to Origen.* Oxford: Oxford University Press.

Boys-Stones, George (2005): "Alcinous, Didaskalikos 4: In Defence of Dogmatism". In: Bonazzi, Mauro, and Celluprica, Vincenza (Ed.): *L'eredità platonica. Studi sul platonismo da Arcesilao a Proclo.* Naples: Bibliopolis, pp. 203–234.

Brittain, Charles (2001): *Philo of Larissa: The Last of the Academic Skeptics.* Oxford: Oxford University Press.

Brittain, Charles (2005): "Common Sense: Concepts, Definition and Meaning in and Out of the Stoa". In: Frede, Dorothea and Inwood, Brad (Eds.): *Language and Learning: Philosophy of Language in the Hellenistic Age.* Cambridge: Cambridge University Press, pp. 164–209.

Brunschwig, Jacques (1967): *Aristote: Topics I.* Paris: Les Belles Lettres.

Brunschwig, Jacques (Trans.) (1967): *Aristote: Topiques Livres I–IV.* Paris: Les Belles Lettres.

Brunschwig, Jacques (1986): "Aristotle on arguments without winners or losers". In: *Wissenschaftskolleg-Jahrbuch* 1984–1985. Berlin: Siedler, pp. 31–40.

Brunschwig, Jacques (1990): "Remarques sur la communication de Robert Bolton". In: Devereux, Daniel and Pellegrin, Pierre (Eds.): *Biologie, logique et métaphysique chez Aristote.* Paris: CNRS, pp. 237–262.

Catana, Leo (2013): "The Origin of the Division Between Middle Platonism and Neoplatonism". In: *Apeiron* 46. No. 2, pp. 166–200.

Chiaradonna, Riccardo (2005): "Plotino e La Corrente Antiaristotelica Del Platonismo Imperiale: Analogie e Differenze". In: Bonazzi, Mauro, and Celluprica, Vincenza (Eds.): *L'eredità platonica. Studi sul platonismo da Arcesilao a Proclo. Elenchos* 45. Naples: Bibliopolis, pp. 235–274.

Chiaradonna, Riccardo (2011): "Plotinus' Account of the Cognitive Powers of the Soul: Sense Perception and Discursive Thought". In: *Topoi* 31, pp. 191–207.

Chiaradonna, Riccardo and Trabattoni, Franco (Eds.) (2009): *Physics and Philosophy of Nature in Greek Neoplatonism: Proceedings of the European Science Foundation Exploratory Workshop* (Il Ciocco, Castelvecchio Pascoli, June 22–24, 2006). Leiden and Boston: Brill.

Cooper, John M. (2012): *Pursuits of Wisdom: Six Ways of Life in Ancient Philosophy from Socrates to Plotinus.* Princeton: Princeton University Press.

DaVia, Carlo (2017): "Aristotle and the Endoxic Method". In: *Journal of the History of Philosophy* 55. No. 3, pp. 383–403.
Devereux, Daniel (1990): "Comments on Robert Bolton". In: Devereux, Daniel and Pellegrin, Pierre (Eds.): *Biologie, logique et métaphysique chez Aristote*. Paris: CNRS, pp. 263–286.
Devereux, Daniel (2015): "Scientific and ethical methods in Aristotle's Eudemian and Nicomachean Ethics". In: Henry, Devin and Nielsen, Karen Margrethe (Eds.) (2015): *Bridging the Gap between Aristotle's Science and Ethics*. Cambridge: Cambridge University Press, pp. 130–147.
Dillon, John and O'Meara, Dominic (2006): *Syrianus: On Aristotle's Metaphysics 13–14*. London: Bloomsbury.
Dodds, Eric Robertson (1951): *The Greeks and the Irrational*. Berkeley: University of California Press.
Dodds, Eric Robertson (1968): *The Elements of Theology: A Revised Text with Translation, Introduction, and Commentary*. Oxford: Oxford University Press.
Dyson, Henry (2009): *Prolepsis and Ennoia in the Early Stoa*. Berlin: de Gruyter.
Falcon, Andrea (2019): "Aristotle's Method of Inquiry in Eudemian Ethics 1 and 2". In: Bonazzi, Mauro, Ulacco, Angela and Forcignanò, Filippo (Eds.): *Thinking, Knowing, Acting: Epistemology and Ethics in Plato and Ancient Platonism*. Leiden: Brill, pp. 186–206.
Fazzo, Silvia (2004): "Aristotelianism as a Commentary Tradition". In: *Bulletin of the Institute of Classical Studies* 47, pp. 1–19.
Frede, Dorothea (2012): "The Endoxon Mystique: What Endoxa Are and What They Are Not". In: *Oxford Studies in Ancient Philosophy* 43, pp. 185–215.
Frede, Dorothea and Inwood, Brad (Eds.) (2005): *Language and Learning: Philosophy of Language in the Hellenistic Age*. Cambridge: Cambridge University Press.
Frede, Michael (1999): "Epilogue". In: Algra, Keimpe, Barnes, Jonathan, Mansfeld, Jaap and Schofield, Malcolm (Eds.): *The Cambridge History of Hellenistic Philosophy*. Cambridge: Cambridge University Press, pp. 771–797.
Futter, Dylan B. (2016): "Philosophical Anti-Authoritarianism". In: *Philosophia* 44. No.4, pp. 1333–1349.
Gadamer, Hans-Georg (1989): *Truth and Method*. 2nd revised ed. Joel Weinsheimer and Donald G. Marshall (Trans.). New York: Crossroad. [1st English ed.: Gadamer, Hans-Georg (1975): *Truth and Method*. William Glen-Doepel (Trans.). John Cumming and Garrett Barden (Eds.). London and New York: Continuum.]
Gerson, Lloyd (2002): "The Study of Neoplatonism Today". In: *Plato—The Internet Journal of the International Plato Society* 2. DOI: 10. 14195/2183–4105_2_2.
Gerson, Lloyd (2013): *From Plato to Neoplatonism*. Ithaca: Cornell University Press.
Gerson, Lloyd (2014): "Plotinus". In: Zalta, Edward N. (Ed.): *The Stanford Encyclopedia of Philosophy* (Summer 2014 Edition) http://plato.stanford.edu/archives/sum2014/entries/plotinus/, last accessed on November 7, 2022.
Gill, Mary Louise, and Pierre Pellegrin (2008): *A Companion to Ancient Philosophy*. Hoboken: John Wiley & Sons.
Golitsis, Pantelis (2018): "Syrianus, Simplicius, and the Harmony of Ancient Philosophers". In: Strobel, Benedikt (Ed.): *Die Kunst der Philosophischen Exegese bei den Spätantiken Platon- und Aristoteles-Kommentatoren*. Berlin: de Gruyter, pp. 69–99.
Greig, Jonathan (2021): "The Aporetic Method of Aristotle's Metaphysics B in Damascius' De Principiis: A Case Study of the First aporia". In: *History of Philosophy & Logical Analysis* 24, pp. 161–209.
Griffin, Michael J. (2012): "What does Aristotle categorize? Semantics and the Early Peripatetic reading of the *Categories*". *BICS* 55, pp. 69–108.

Griffin, Michael J. (2016): "Why Philosophy Begins with the *Categories:* Perspectives from the 1st-century Greek Commentators". In: *Documenti e studi sulla tradizione filosofica medievale* 27, pp. 19–42.
Griffin, Michael J. (2020): "Articulating Preconceptions: A Reconsideration of *Aristotle's Categories in the Early Roman Empire.*" In: *Documenti e studi sulla tradizione filosofica medieval*, forthcoming.
Hacker, Paul (1983): "Inklusivismus". In: Oberhammer, Gerhardt (Ed.): *Inklusivismus*. Vienna: Akademie, pp. 11–28.
Hadot, Ilsetraut (1978): *Le Problème Du Néoplatonisme Alexandrin*. Paris: Études augustiniennes.
Hadot, Ilsetraut (1987): *Simplicius: sa vie, son oeuvre, sa survie*. Berlin and New York: de Gruyter.
Hadot, Ilsetraut (1996): *Commentaire Sur Le Manuel d'Épictète: Introduction Et Edition Critique Du Texte Grec*. Leiden and Boston: Brill.
Hadot, Pierre (1995): *Philosophy as a Way of Life: Spiritual Exercises from Socrates to Foucault*. Hoboken: John Wiley & Sons.
Helmig, Christoph (2012): *Forms and Concepts: Concept Formation in the Platonic Tradition*. Google eBook. Berlin and Boston: de Gruyter.
Henry, Devin and Nielsen, Karen Margrethe (Eds.) (2015): *Bridging the Gap between Aristotle's Science and Ethics*. Cambridge: Cambridge University Press.
Hoffmann, Philippe (1987): "Catégories et langage selon Simplicius—La question du 'skopos' du traité aristotélicien des Catégories". In: Hadot, Ilsetraut (Ed.): *Simplicius: sa vie, son oeuvre, sa survie*. Berlin and New York: de Gruyter, pp. 61–90.
Hoffmann, Philippe (1998): *La Fonction Des Prologues Exégètiques Dans La Pensée Pédagogique Néoplatonicienne*. Paris: Cerf.
Hoffmann, Philippe (2008): "What Was Commentary in Late Antiquity? The Example of the Neoplatonic Commentators". In: Gill, Mary Louise and Pellegrin, Pierre (Eds.): *A Companion to Ancient Philosophy*. Hoboken: John Wiley & Sons, pp. 597–622.
Hunter, Richard (2011): "Plato's *Ion* and the Origins of Scholarship". In: Matthaios, Stephanos, Montanari, Franco and Rengakos, Antonios (Eds.): *Ancient Scholarship and Grammar: Archetypes, Concepts and Contexts*. Berlin, and York: de Gruyter, pp. 27–40.
Irwin, Terence (1981): "Aristotle's methods in ethics". In: O'Meara, Dominic (Ed.): *Studies in Aristotle*. Washington, D.C.: The Catholic University of America Press, pp. 193–223.
Karamanolis, George E. (2006): *Plato And Aristotle in Agreement?: Platonists on Aristotle from Antiochus to Porphyry*. Oxford: Clarendon Press.
Karbowski, Joseph (2015): "Phainomena as Witnesses and Examples: The Methodology of *Eudemian Ethics* 1.6". In: *Oxford Studies in Ancient Philosophy* 49, pp. 193–226.
Karbowski, Joseph (2019): *Aristotle's Method in Ethics*. Cambridge and New York: Cambridge University Press.
Kraut, Richard (Ed.) (2006): *The Blackwell Guide to Aristotle's Nicomachean Ethics*. Hoboken: Wiley-Blackwell.
Long, Anthony Arthur and Sedley, David N. (1987): *The Hellenistic Philosophers*. Vol. I–II. Cambridge: Cambridge University Press.
Mackay, Anne E. (2008): *Orality, Literacy, Memory in the Ancient Greek and Roman World*. Leiden: Brill
Magrin, Sara (2016): "Plotinus' Reception of Aristotle". In: Falcon, Andrea (Ed.): *Brill's Companion to the Reception of Aristotle*. Leiden: Brill, pp. 258–276.
Mansfeld, Jaap (1994): *Prolegomena: Questions to Be Settled Before the Study of an Author or a Text*. Leiden: Brill.

McGuire, James E., and Strange, Steven K. (1988): "An Annotated Translation of Plotinus Ennead III 7: On Eternity and Time". In: *Ancient Philosophy* 7, pp. 251–271.

McNeill, David N. (2018): "Akratic Ignorance and Endoxic Inquiry". In: *Review of Metaphysics* 72. No. 2, pp. 259–299.

Menn, Stephen (forthcoming): *Simplicius on Aristotle's Physics: An Introduction*. London: Bloomsbury.

Moraux, Paul (1968): "La joute dialectique d'après le huitième livre des *Topiques*". In: Owen, Gwilym Ellis Lane (Ed.): *Aristotle on Dialectic. The Topics. Proceedings of the Third Symposium Aristotelicum*. Oxford: Clarendon Press, pp. 277–311.

Motta, Anna (2021): "The Many Voices of a Teacher Without Teachers". In: *Methexis* 33, pp. 170–196.

O'Meara, Dominic and Dillon, John (2008): *Syrianus: On Aristotle's Metaphysics 3-4*. London: Bloomsbury.

Owen, Gwilym Ellis Lane (1961): "*Tithenai ta phainomena*". In: Mansion, Suzanne (Ed.) *Aristote et les problèmes de méthode: communications présentées au Symposium Aristotelicum tenu à Louvain du 24 août au 1er septembre 1960*. Louvain: Publications Universitaires, pp. 83–103. [Reprinted in: Owen, Gwilym Ellis Lane (1986): *Logic, Science, and Dialectic: Collected Papers in Greek Philosophy*. Ithaca: Cornell University Press, pp. 239–251.

Phillips, John F. (1987): "Stoic 'Common Notions' in Plotinus". In: *Dionysius* 11, pp. 33–52.

Quinney, Laura (2008): "Romanticism, Gnosticism, and Neoplatonism". In: Mahoney, Charles (Ed.): *A Companion to Romantic Poetry*. Toronto: Wiley-Blackwell, pp. 412–424.

Reinhardt, Tobias (2007): *Cicero's Topica*. Oxford: Oxford University Press.

Schofield, Malcolm (2013): *Aristotle, Plato and Pythagoreanism in the First Century BC: New Directions for Philosophy*. Cambridge: Cambridge University Press.

Schrenk, Lawrence P. (1994): *Aristotle in Late Antiquity*. Washington, D.C.: The Catholic University of America Press.

Scott, Gary Alan (2000): *Plato's Socrates as Educator*. Albany: SUNY Press.

Scott, Gary Alan (2002): *Does Socrates Have a Method: Rethinking the Elenchus in Plato's Dialogues and Beyond*. University Park: Pennsylvania State Press.

Sedley, David (1997): "Plato's Auctoritas and the Rebirth of the Commentary Tradition". In: Barnes, Jonathan and Griffin, Michael (Eds.): *Philosophia Togata II*. Oxford: Oxford University Press, pp. 110–129.

Sedley, David (2012): *The Philosophy of Antiochus*. Cambridge: Cambridge University Press.

Sharples, Robert William (2010): *Peripatetic Philosophy, 200 BC to AD 200: An Introduction and Collection of Sources in Translation*. Cambridge: Cambridge University Press.

Smith, Robin (1997): *Aristotle: Topics, Books I and VIII*. New York: Oxford University Press.

Sorabji, Richard (2004): *The Philosophy of the Commentators 200-600 AD: A Sourcebook*. London and Ithaca: Duckworth and Cornell University Press.

Stamatellos, Giannēs (2007): *Plotinus and the Presocratics: A Philosophical Study of Presocratic Influences in Plotinus' Enneads*. Albany: SUNY Press.

Strange, Steven K. (1994): "Plotinus on the Nature of Eternity and Time". In: Schrenk, Lawrence P. (Ed.): *Aristotle in Late Antiquity*. Washington, D.C.: The Catholic University of America Press, pp. 22–53.

Strobel, Benedikt (2009): "Von Einem Subjekt Ausgesagt Werden und an Einem Subjekt Vorliegen: Zur Semantik Genereller Terme in der Aristotelischen Kategorienschrift". In: *Phronesis* 54, pp. 40–75.

Tarrant, Harold S. (1997): "Cultural and religious continuity: 2. Olympiodorus and the surrender of paganism". In: Garland, Lynda (Ed.): *Conformity and non-conformity in Byzantium*. Amsterdam: Hakkert, pp. 181–192.

Tarrant, Harold S. (2000a): *Plato's first interpreters*. Ithaca: Cornell University Press.

Tarrant, Harold S. (2000b): *Reason, faith, and authority: Some Platonist debates about the authority of the teacher*. In: *Sophia* 39, pp. 46–63.

Tuominen, Miira (2009): *The Ancient Commentators on Plato and Aristotle*. Berkeley: University of California Press.

Van Den Berg, Robbert M. (2005): "As We Are Always Speaking of Them and Using Their Names on Every Occasion. Plotinus, Enn. III.7 [45]: Language, Experience and the Philosophy of Time in Neoplatonism". In Chiaradonna, Riccardo and Trabbatoni, Franco (Eds.): *Physics and Philosophy of Nature in Greek Neoplatonism: Proceedings of the European Science Foundation Exploratory Workshop*. Leiden: Brill, pp. 101–120.

Watts, Edward (2008): *City and School in Late Antique Athens and Alexandria*. Berkeley: University of California Press.

Worthington, Ian, and Foley, John Miles (2002): *Epea and Grammata: Oral and Written Communication in Ancient Greece*. Leiden: Brill.

Zanker, Paul (1996): *The Mask of Socrates: The Image of the Intellectual in Antiquity*. Sather Classical Lectures. Berkeley: University of California Press.

Ziolkowski, Jan M. (2009): "Cultures of Authority in the Long Twelfth Century". In: *Journal of English and Germanic Philology* 108. No. 4, pp. 421–448.

Sarah Klitenic Wear
Syrianus and the Dialectical Cosmos

1 Introduction

In his *Commentary on the Parmenides* 1001.13–21, Proclus outlines the principle method of dialectic as invented by Zeno, the disciple of Parmenides. He says:

> "Let there be two hypotheses, then. Each of them generates for us three further hypotheses. Previously we took being and not-being to be predicated of something, and so divided our hypothesis in two; now, if we take the predicate three ways, we shall triple each of these two hypotheses. The predicate, then, is to be taken as either being true or not being true, or being both true and not true" (Proclus, *In Parm.* 1001.13–21).[1]

He elaborates a few lines later that if we consider these three propositions in the context of the existence and of the non-existence of the subject, it is plain that we will have tripled the original two hypotheses. The subject is then double: being and not being. Later, he explains that each of these hypotheses must be quadrupled by reason of the varied relationship of the predicate to the subject:

> "Either it is predicated of its subject, or of anything else; and each of these two has two possibilities: each of its subject in relation to itself, or of its subject in relation to others; or of others in relation to themselves, or to it" (Proclus, *In Parm.* 1001.37–42).[2]

What Proclus means by these two passages is that "if a thing is" and "if it is not" is the relationship between the thing itself with respect to others and others with respect to themselves and others with respect to the thing itself. Proclus derives a sequence of twenty-four propositions outlining what a thing is and what it is not. He takes this dialectical exercise with the express purpose of applying this method to his discussion of the One in his *Commentary on the Parmenides*, with a focus on the hypotheses concerning what can be said and cannot be said

---

[1] All translations of Proclus' *Commentary on the Parmenides* used in this article are from Morrow and Dillon (1987); the text used is Steel (2007–2009). Καὶ δύο ἔστωσαν αὗται ὑποθέσεις· ἑκατέρα δὲ τούτων τρεῖς ἡμῖν ὑποθέσεις γεννᾷ· πρότερον μὲν γὰρ τό τινι λαβόντες ὂν καὶ μὴ ὂν ἐδιχοτομοῦμεν τὴν ὑπόθεσιν· νῦν δὲ ἂν τὸ συμβαῖνον τριχῶς λάβωμεν, ἑκατέραν τῶν ὑποθέσεων τριπλασιάσομεν. Τὸ τοίνυν συμβαῖνον, ἢ ὡς ἑπόμενον ληπτέον, ἢ ὡς μὴ ἑπόμενον, ἢ ὡς ἅμα καὶ ἑπόμενον καὶ ἑπόμενον καὶ οὐχ ἑπόμενον.

[2] ἢ γὰρ αὐτῷ τι συμβαίνει, ἢ τοῖς ἄλλοις· καὶ τούτων ἑκατέρῳ διχῶς· ἢ αὐτῷ πρὸς ἑαυτό, ἢ αὐτῷ πρὸς τὰ ἄλλα, καὶ τοῖς ἄλλοις ἢ πρὸς ἑαυτά, ἢ πρὸς αὐτό·

about the One and what can be said about being and intelligible beings as they relate to the One. These questions give rise to a metaphysical discussion of the One and the universe. Proclus understands the universe using the methodology of relation as found in the propositions of the first and second hypotheses of the *Parmenides*. Thus, Proclus creates what he calls "a dialectical cosmos";[3] the first two hypotheses of the *Parmenides* classify intelligible beings according to their relation to other entities in the universe. The first two hypotheses of the *Parmenides* divide and separate the universe, but then collect the cosmos again into a coherent whole; for Proclus, dialectic is an ontological, rather than logical methodology.

Proclus credits his teacher Syrianus with this approach. In his theological interpretation of the *Parmenides*,[4] Syrianus shows how the nature of the One appears in the structure of the dialogue itself. The first hypothesis of the *Parmenides* outlines the primal God, while the intelligible universe is the subject of the second hypothesis, insofar as the intelligible universe is a product of the One.[5] The second hypothesis, thus, consists of a series of distinct propositions that correspond to a level of being within the intelligible world.[6] More specifically, for the propositions "if a thing is"/"if a thing is not", Syrianus creates a series of positive and negative hypotheses, one of which relates to what can be said about the One "in itself" or "in another" (*In Parm.* 1114.25). Here, Syrianus says that "in itself" speaks to the primal One, while the One "in another' speaks to the realm below the One, the intelligible realm. The propositions "in itself"/"in another" speak to how the One relates to itself and to another entity—again, Proclus' cosmic dialectic is relational; the universe is organized according to defining a principle in its own right and what it is in relation to another principle. As with the propositions "if a thing is"/"if a thing is not", this question of how something relates to itself and relates to Ir becomes the primary paradigm by which Syrianus views the One and its relationship to the intelligible realm.

Syrianus applies this understanding of dialectic[7] to express how forms and the monads interact with each other and the rest of the universe. In his *Commentary*

---

3 Proclus, *Theol. Plato.* I, 11, p. 53, 19–22. Cf. Proclus, *In Parm.* 1099, 27–31; Proclus, *Theol. Plato.* I, 11, p. 49, 3–11.
4 Proclus considers the first hypothesis a theological hymn by means of negations to the One (*In Parm.* VII 1191, 34 f); See also *Theol. Plato.* 1968, LXXII, n.2 and Van Den Berg (2001), p. 24. The second hypothesis of the *Parmenides* is also considered a hymn celebrating the generation of all the gods; see *Theol. Plato.* I 7, pp. 31 and 25–27. See also Van Den Berg (2001), p. 24.
5 Wear (2011a), p. 67.
6 Dillon (2009), p. 235.
7 Syrianus' understanding of the dialectical discussion of the first and second hypotheses of the *Parmenides* ultimately finds its origins in the divided line of Plato's *Republic* which defines kinds of knowledge, the object of the knowledge, and the method of the knowledge. For a descrip-

on the Metaphysics (p. 119, 28; Syrianus, *In Metaph.* 1179b33–5), Syrianus follows the same logical structure he applied to the One, this time to the forms. He says that forms are each distinct in themselves and yet they relate to each other when viewed from another perspective. That is they remain in themselves and yet each are contained within each other in the intelligible realm. This logical framework is repeated in fragments from Syrianus' *Commentary on the Philebus* (Damascius, *In Phil.* Sect. 244) where the three monads of the *Philebus* mimic the basic behavior shown in the propositions "in itself" and "in another" in that the monads are differentiated, but still coordinated at the level of Intellect. Thus, for Syrianus, dialectic in the *Parmenides* is essentially a way to explore relationships among metaphysical principles; moreover, for Syrianus, the relationships among principles are not logical, but metaphysical.[8] At the heart of this procedure is Syrianus' desire to show how all things are in all things and all things are in and from the One. This concern proliferates Syrianus' writings to express some of the basic concerns usually attributed to Proclus—primarily, the question of how a manifold universe derives from, participates in, and yet does not detract from a unified prince.

---

tion of Syrianus' description of dialectic and its relationship to knowledge, see Bonelli (2009), p. 429, and Ierodiakonou (2009), pp. 419–421. Lernould (1987, pp. 509–536) outlines Proclus' views of dialectic, which are very similar to those of Syrianus. Both Syrianus and Proclus use the divided line from *Republic* VII 534 e 2–3 as a division of the sciences concerning epistemology; each division defines itself according to its level of dialectical power and by its object (Lernould 1987, p. 516). Proclus interprets the divided line of the *Republic* with an explanation that there are three kinds of knowledge: opinion (a knowledge turned towards practical and material concerns); mathematical sciences (including metaphysics, dialectic, mathematics, and physics). These sciences pertain to the intermediate forms. The third kind of knowledge, intellection, has to do with the intelligible forms; it is only possible through intellection. Dialectic, for Syrianus, is a method of attaining scientific knowledge about intermediate forms through analysis, division, definition, and demonstration (Ierodiakonou 2009, p. 414). With respect to the first and second hypotheses of the *Parmenides*, Syrianus uses dialectic to understand the metaphysical structure of the universe. On the divided line of the *Republic* and dialectic, see Proclus, *In Eucl.* 31.4–17; *In Tim.* I, 350.8–20; For Syrianus, see *In Metaph.* 81, 38–83, 11; 2, 18–3, 1; 3, 17–24; on the three different kinds of forms presented in the Line simile of Plato's *Republic*, see Ierodiakonou (2009), p. 406, O'Meara (1989), pp. 128–141, and Syrianus, *In Metaph.* 55, 38–56, 2; and for a discussion of Plato's *Republic* and demonstrative science, see Bonelli (2009), p. 433.
8 Lernould (1987), p. 516.

2 Dialectical Structure in Proclus' *Commentary on the Parmenides*

Lemmata 135d7 through 136e4 treat Parmenides' description of his dialectical method, illustrated by Proclus in *In Parm.* 997.16–1007.34. Proclus understands dialectic as an exploration of contradictions in this world that point to metaphysical truths and syntheses underlying the apparent disjunctions. Proclus reads lemma 135d5: "What form, then, Parmenides", he said, "shall this exercise take?" "The form", he said, "which you heard Zeno employing". To this, Proclus offers the interpretation that the lemma describes a dialectical procedure that does not involve argumentation (994). Proclus goes on to say that "the whole of the logical procedure and the unfolding of theorems" is the field of the exercise. Again, Proclus explains the dialectical method in his exegesis of *Parm.* 135e: "If you want to be exercised, you must not merely make the supposition that such and such a thing is, and then consider the consequences; you must also make the supposition that the same thing is not". Proclus says that Plato uses the exercise of Zeno who used sets of opposite statements to advance the theorems of Parmenides on the One (997). He credits Parmenides for asking not only whether something is in his dialectical investigations, but whether something is not, and then to consider what follows and what does not follow.[9] Proclus explains that this method allows one to understand "of what the thing postulated is a cause, or what attributes belong to it in and of itself" (999).

Proclus, then, offers a new opinion insofar as he recognizes the dialectical method of the *Parmenides*, but sees the principle aim of that method an explanation of metaphysical truths.[10] This innovative technique arises in the teachings of Syrianus; in a doxography on the *skopos* of the *Parmenides*, Proclus lays out the history of interpretation of the *Parmenides*, ending with the opinion of his teacher. Proclus begins by outlining those of his predecessors who thought that the *skopos* of the *Parmenides* was an exercise in logical method with a polemical aim. The first among these (630.37–633.12) took the *Parmenides* as an antigraphe against Zeno on the intelligibles. The second group argued that it was a practice in the exercise of logical disputation.[11] Proclus argues against this group both here, in the *Commen-*

---

[9] The hypothesis in dialectic is turned, ultimately, towards the principle, which alone is really a principle: Proclus, *In Remp.* I, 292. 6. Lernould notes that thus, the hypothesis is an "actual hypothesis" (τῷ ὄντι ὑπόθεσις), a condition of possibility for returning to the true principle (1987, p. 526).
[10] Beierwaltes (1979), p. 258.
[11] Alcinous, *Didaskalikos* 6, where he finds in the *Parmenides* the ten categories of Aristotle. On the *Parmenides* as a logical exercise, see Albinus, *Isagoge* 3, Diogenes Laertius III, 58, Philoponus,

tary on the Parmenides, as well as in Platonic Theology I.9, where he says that *Parmenides* is a dialogue on Being, not a gymnasia of dialectic.¹² The third group Proclus lists says that *Parmenides* is a metaphysical dialogue: he divides this group into three subcategories¹³—the third being the opinion of Syrianus. Proclus discusses groups who maintain that the subject of the *Parmenides* is metaphysical. The first group (635.31–638.2) say that Parmenides uses logical dialectic to express that Being is the subject of the *Parmenides*. For according to Proclus, no Platonic dialogue has a method as its topic (637.9). Rather, logic is used by Plato to posit the existence and non-existence of Being. The second group (638.14–640.16) also rejects that dialectic is the subject of *Parmenides*; it always thinks that the topic is metaphysical—this group considers that the dialogue is about the One Being and everything that receives its reality from the One. This group held that the topic cannot be the One Being exclusively because everything attributed and denied of the One cannot be applied to the One Being alone.¹⁴ Plotinus, Porphyry, and Iamblichus hold this position.

3 Syrianus' Metaphysical Interpretation of the *Parmenides* and "If a Thing Is"/"If a Thing Is Not"

When Proclus arrives at the opinion of Syrianus, it is clear that Syrianus uses the dialectical structure of the *Parmenides* for a metaphysical purpose: to preserve the transcendence of the One, while still connecting the One to the rest of creation for the unification of the cosmos.¹⁵ The first hypothesis of the *Parmenides* purports

---

In Anal. Pr. 9. 18–19, Alex. Aphr. *In Topica*, 28.23–29.5; Proclus, *Theol. Plato.* I.9 deals with the topic of the *Parmenides*.
12 *Theol. Plato.* I.9, p. 38, 4–7: "We then have a reason to say that the Parmenides does not have logic as its aim, rather it seems to constitute a science of all the first principles as its aim".
13 The first group (635.31–638.2) says the subject of the Parmenides is Being. This group (possibly, Origen the Platonist) said that Plato did not introduce logic as the subject of the Parmenides; rather, its concerns include existence and non-existence of Being. The second group (638.14–640.16) argues that the subject of the Parmenides is One Being, and everything deriving reality from the One. This group (possibly, Plotinus, Porphyry, and Iamblichus) says that all things attributed and denied of the One cannot be applied to the One Being alone. The discussion must concern all things, from the primary cause to the lowest. Cf. Wear (2011b), pp. 217–218.
14 Wear (2011a), p. 65.
15 The dialectical hypothesis is confirmed it its own being by surpassing the dialectical operation; it is connected to the first principle and reveals its self, in another manner as a true hypothesis

that the One "is not"—that is, nothing positive can be said about the One; its ineffability maintains its transcendence and unity. The second hypothesis of the *Parmenides* defends a positive view of the universe, describing the intelligible universe as what "is". This metaphysical understanding of the One, then, emerges throughout the relationship between what a thing is and what it is not as expressed through the dialectical statements of the *Parmenides*. By focusing on the relational statements of the *Parmenides* to describe the One's connection with creation, Syrianus outlines principles of the One and the intelligible universe, expressing what each is and is not with respect to the other. In Proclus' *Commentary on the Parmenides*, Proclus reports that Syrianus—unlike his predecessors, who thought the skopos was exclusively logical—held the view that the subject of the *Parmenides* is metaphysical:[16]

> These are the differences of opinion among the ancients with respect to the purpose of the *Parmenides*. Now we must say that our Master has added to their interpretations. He agrees with those of our predecessors who thought that the aim of the dialogue is metaphysical and dismisses the idea that it is a polemic as implausible" (Proclus, *In Parm.* 640.17–24).[17]

Thus, Syrianus says the following is the "true meaning" of the dialogue:

> Considering such to be the dialogue's purpose, our Master denied that it was about Being, or about real beings alone; he admitted that it was about all things, but insisted on adding, "insofar as all things are the offspring of one cause and are dependent upon a universal cause" (Proclus, *In Parm.* 641.1–4).[18]

Syrianus' primary contribution to a discussion of the skopos of the *Parmenides* was to set the One, rather than Being, as the topic of the first hypothesis of the *Parmenides*.[19] Moreover, he justifies this interpretation by saying that it defends the basic

---

founded in its own being. For Proclus, the hypothesis is a statement about being, not a mathematical term. See Beierwaltes (1979), p. 258, and Lernould (1987), p. 526.

16 Wear (2011b), p. 216.

17 Οἱ μὲν οὖν παλαιοὶ περὶ τῆς τοῦ Παρμενίδου προθέσεως τοῦτον διέστησαν τὸν τρόπον· ὅσα δὲ συνεισήγαγε ταῖς τούτων ἐπιστάσεσιν ὁ ἡμέτερος καθηγεμών, ἤδη λεκτέον. Εἶναι μὲν δὴ καὶ αὐτὸς τὸν σκοπὸν πραγματειώδη τοῦ διαλόγου τοῖς οὕτως ἑλομένοις τῶν πρεσβυτέρων ὡμολόγει, τήν τε ἀντιγραφὴν ὡς ἀπίθανον χαίρειν ἀφείς·

18 περὶ τοῦ ὄντος οὔτε περὶ τῶν ὄντων αὐτῶν εἶναι μόνων διετείνετο· συγχωρῶν δὲ εἶναι περὶ τῶν πάντων, ἠξίου προστιθέναι καθ' ὅσον ἑνὸς πάντα ἐστὶν ἔκγονα καὶ εἰς ἓν ἀνήρτηται πάντων αἴτιον.

19 Wear (2011a), p. 60 and Wear (2011b), p. 216.

premise that all things derive from the One and depend (or participate) upon it.[20] In addition, Syrianus gives a practical reason for why the skopos of the *Parmenides* must not be dialectical method. He says that if Zeno's purpose was to teach a dialectical method, he would not have chosen such a difficult metaphysical topic (640.27–37); for dialectical method is taught to the young, and the audience of the *Parmenides* is not the young but rather those learned in Platonic metaphysics. Rather, the skopos of the *Parmenides* is Being and all beings as the product of the One.

In his *Commentary on the Parmenides* (1061.20–1064.12), Syrianus gives what Proclus calls the "theological interpretation" of the *Parmenides*, something Proclus also credits Syrianus with inventing (641.1; 1061.20.) This is expressed in the internal ordering of the hypotheses such that the higher produce the lower and contain the lower; for instance, Syrianus says that the first hypothesis describes the One in its capacity for generating gods. According to Syrianus:

> All things are presented in logical order, as being symbols of divine orders of being and also that the fact that all those things which are presented positively in the second hypothesis are presented negatively in the first indicates that the primal cause transcends all the divine orders, while they undergo various degrees of procession according to their distinct characteristics (Proclus, *In Parm.* 1062.10–17).[21]

In the lines that follow, he says that the One in the second hypothesis is neither the primal One (for it is complex, being all things) nor it is that which is inseparable from Being. The second hypothesis reveals a multiplicity of divine henads connected to the One-Being. Thus, according to Syrianus' metaphysical interpretation of the dialectical structure of the *Parmenides*, the *Parmenides* outlines the universe according to the relationship between the One and the universe, as described chiefly with the statement "is"/"is not".[22] Moreover, the structure of the universe follows

---

20 The five positive hypotheses show all the meanings of the One, understood as the relationship between the One and Being (1041.22ff.): insofar as it is superior to Being (as One); coordinate with Being (as Intellect); inferior to Being (Soul); moreover, they show how the One is "not Being" in its relationship to others that participate in it; relative to the physical world; absolute nont-being (pure matter).

21 καὶ διὰ τοῦτο παραλαμβάνεσθαι ἐπόμενα πάντα, σύμβολα τῶν θείων ὄντα διακόσμων, καὶ ἐπὶ σύμβολα τῶν θείων ὄντα διακόσμων, καὶ ἐπὶ τούτοις πάντα ὅσα καταφατικῶς ἐν τῇ δευτέρᾳ λέγεται τῶν ὑποθέσεων, ταῦτ' ἀποφάσκεσθαι κατὰ τὴν πρώτην, εἰς ἔνδειξιν τοῦ τὴν μὲν πρώτην αἰτίαν πασῶν ἐξῃρῆσθαι τῶν θείων δια-κοσμήσεων, ἐκείνας δὲ ἄλλως κατ' ἄλλας ἀφωρισμένας ἰδιότητας προεληλυθέναι·

22 Lernould argues that the hypothesis, "if the One is, it is not many" as evidence that dialectic in the *Parmenides*, as interpreted by Proclus, is not interested in demonstration, as is a typical hypothesis in dialectic because the hypothesis itself cannot be demonstrated (Lernould 1987, p. 527).

the logical ordering outlined in the hypotheses; when properly interpreted, the second hypothesis of the *Parmenides* describes how each order of the intelligible universe participates in the order above it so that the entire universe coheres.[23] The organization of the nine hypotheses into five positive and four negative statements unfold the different senses of One and Not-Being. The five positive hypotheses show all the meanings of One, understood as the relationship between the One and Being (1041.22ff.). To these, Syrianus adds four negative hypotheses—if the One does not exist—again, expressed in terms of relationship of the One to itself and another entity.[24] Thus, Proclus gives us Syrianus' proposal for the nine hypotheses of the *Parmenides* is that which is systematically denied of the One in the first hypothesis is affirmed of the One in the second. Each positive attribute corresponds in order to the preceding negation. From the propositions, "If a thing is"/ "if it is not", Syrianus outlines the attributes of the intelligible universe, level by level, according to the relationship between higher and lower principles. This interpretation of the *Parmenides* is the organizing principle of both Proclus' *Commentary on the Parmenides* and his *Platonic Theology*. It is primarily a theological interpretation, moreover, as the attributes of being described by the One-that-is are really attributes of a god.[25]

4 In Itself and In Another: *Parmenides* 138 A[26]

Syrianus uses the propositions "in itself/in another" to describe the participation of intelligible entities and to describe the nature of the One.[27] In this way, he applies

---

[23] Lernould surmises that the fact that the Procline hypotheses are not mere deductions; the Procline dialectic is unhypothetical because unlike the inferior sciences, Procline dialectic is not hypothetical (Lernould 1987, p. 526). In the sciences other than dialectic, Lernould explains, the principles and the conclusions are not actually known. The dialectician bases his demonstrations on principles not demonstratable.

[24] The nine hypotheses are: 1) relationship of the One superior to Being to itself and other things; 2) One coordinates with Being; 3) One inferior to Being to itself and other things; 4) the relationship of others which participate in the One to themselves and to the One; 5) the relations the others which do not participate in the One have to themselves and to the One; 6) the relations of the One (if it does not exist), in the sense of existing in one way and not in another, to itself and to other things; 7) the relations of the One (if it does not exist), in the sense of absolute non-existence, towards itself and others; 8) the relations of the orders to themselves and to the One (when taken as non-existent) in the sense of existing in one way and not in another; 9) the relations of the others to themselves and to the One (when taken as absolutely non-existent.)

[25] Van Campe (2009), p. 250.

[26] Proclus, *In Parm.* 1142.10–1143.39

the dialectical principle established earlier in the commentary to describe the relationship between the One and the intelligible realm. He begins with the second hypothesis: everything denied of the One in the first hypothesis has an analogue to the second hypothesis where it is positively asserted of the intelligible realm.[28] More specifically, the positive elements of the second hypothesis describe the noetic triads; the noetic-noeric triads; and the noeric hebdomad.[29] Again, Syrianus does this by using the second hypothesis of the *Parmenides* to describe the levels of gods within the intelligible realm, something Proclus attributes to him in *Platonic Theology* book III.[30] In 1142.10, Syrianus says that the propositions "in itself"/"in another" must refer to the monad at the intellectual level; this monad relates back to its source in the intelligible level while simultaneously participating in the intelligible/intellectual gods. Thus, at the intellectual level it rests in its cause, while participating in other gods. The intelligible characteristics remain attributed to the One, while the One transcends the intellectual gods assigned to One-Being in Syrianus' understanding of the second hypothesis of the *Parmenides* (145B6). Van Campe points out that the diversity of properties within the henadic realm are unfolded by the structure of Parmenides' argument: "insofar as the premise remains the same each time ("if the One is"), it refers to their unity; insofar as each conclusion, by contrast is different ("it is a whole", "it is in itself", etc.), their distinctive prop-

---

[27] For Porphyry, Iamblichus, and Syrianus, "in another" signifies "that which remains in its cause" (Saffrey and Westerink 1974, p. 129).
[28] Steel (2009), p. 355. *Parmenides* 136B 6–c 5: "concerning whatever you might hypothesise as being and as not being and as not being and as having any other property, you must examine the consequences for it in relation to itself and in relation to each one of the others, whichever you select, and in relation to several of them, and to all of them in the same way; and in turn, you must examine the others, both in relation to themselves and in relation to any other thing you select on every occasion, whether you hypothesise you you hypothesise as being or as not being" (trans. Gill and Ryan 1996).
[29] Van Campe (2009), p. 268; see also Proclus, *Theol. Plato.* 1, 11, p. 50, 2–3. Van Campe shows that Syrianus understands different attributes of Being as attributes of different gods; "the One that is" is at the highest level of intelligible Being, "the One Being", comprehending all attributes in a unified way while simultaneously it is the procession of beings into multiplicity. See Wear (2011a), p. 75.
[30] "This, then, will be my procedure. I will take each of the conclusions separately, and will endeavor to refer it to the corresponding class of gods, following closely in this case also the inspired insights of my Master, that divine man with whom we have entered into the ecstasy of the study of the Parmenides, as he revealed to us these sacred paths, which has truly roused us up from our sleep to the ineffable initiation into its mysteries" (Proclus, *Theol.Plato.* p. 83, 10–18). Saffrey-Westerink.

erties become visible".[31] Thus, Syrianus uses dialectic to describe the metaphysical structure of Being in a theological framework.

In his treatment of the lemma, Syrianus denies being to the One in order to assert that the One is cause of intellectual realm, while simultaneously transcending it. Thus, Syrianus makes the One distinct from the summit of the intellectual realm, the highest monad of the noeric gods. Proclus adopts and repeats this theory in *Platonic Theology* II, 12, p. 68, 6–22, also basing his organization of the universe on the Parmenidean statements "in itself"/"in another". What is at stake for Syrianus' interpretation of the *Parmenidean* statement is the nature of cause. A cause is participated in by particular elements in a variety of ways; the One, however, as ultimate cause is not comprehended by any entity.

5 Wholes and Parts and In Itself/In Another: Language for the One

Syrianus applies the logical relationship articulated above in his exegesis of *Parmenides* 137D, a question of whether the One has parts and whether these parts would mean that the One would have a beginning, middle, or end. This lemma (paired with "in itself"/"in another") when applied to the One speaks to the One's relationship with the created world and to the One's relationship to itself. In Proclus' *Commentary on Parmenides* 1114.25, Syrianus addresses *Laws* IV 715E where the Athenian Stranger says that God has beginning, middle, and end and *Parmenides* 137D where god has no parts; these two texts seem to contradict one another. Referring to the passage in the *Parmenides*, Syrianus says that Plato's discussion describes how the One relates to itself as a monad (and hence denies a beginning, middle, and end), to maintain its unity while simultaneously interacting with a universe full of multiplicity.[32] Conversely, the discussion in the *Laws* addresses an external relationship, that is, how the One relates to the world or "in another". For Syrianus, because beginning, middle, and end exist in the universe, they must exist in the One. Citing Plato, *Ep.* 2 312 E, Syrianus says that the One acts as beginning because it is the source of all things, the center because all things are established in it, and the end, because it is the ultimate goal of all things. The One itself, however, does not consist of parts; rather it is a total entity. Proclus elaborates upon Syrianus' distinction in *Platonic Theology* II, 12, where the second triad of the intelligible-intellectual gods is associated with whole and parts. Here, Proclus goes into de-

---

31 Van Campe (2009), p. 268. Cf. *In Parm.* 1050, 5–26.
32 Wear (2017), p. 219.

tail as to how the dialectic of *Parmenides* outlines the relationship between One and intelligible being; through diairesis, he makes a division into beginning, middle, and end.³³ For Proclus, the issue of wholes and parts describes the relationship between the One primal God (as described in the first hypothesis) and the One-Being (as described in the second hypothesis.) In *PT* II, 12, Proclus says that intelligible triads of the *Parmenides* correspond to the conclusions of the second hypothesis, with the first intelligible triad corresponding to the first conclusion (if the One is, it participates in Being: *Parm.* 142 B 5–7). He further specifies, again using the dialectic of the *Parmenides* to delineate intelligible levels as first conceived by Syrianus. For Proclus, the second triad corresponds to the second conclusion: if the One is, it is a totality and it has parts (*Parm.* 142 C 7–D); and the third triad corresponds to the third conclusion; if the One is, it is an infinite multiplicity of parts (*Parm.*, 142D 9–143 A 3). In his reading of *Parmenides* 138a3–6, Syrianus interprets "in itself" and "in another" to describe the One in the first and second hypotheses as having contact with many points, but containing no parts;³⁴ both of these attributes he denies of the One. Likewise, the beginning of the second hypothesis, "if the One is" is a discussion on Being.

6 Is/Is Not and In Itself/In Another: Language for Forms

Syrianus' description of the forms uses the dialectical language found in his description of the One and its relationship to the intelligible world. Namely, forms are defined with respect to what they are not, according to the dialectic of the *Parmenides*.³⁵ At the intelligible level, each form has its own identity so that we can say "is not" when speaking of rest, because it is not movement, or identity or difference. Each form is different and specific, but still communicates with other forms.³⁶ The forms and the One mark a significant difference from statements relating to "it is not" at the sensible level; at the sensible level, a negative proposition implies a reference to some reality for which we are denying the attributes. Unlike intelligibles, which participate in each other, sensibles are not in another because

---

33 Glasner (1992), p. 198. Cf. Proclus, *In Eucl.* 98.13–14.
34 Wear (2017), p. 221; See also Saffrey and Westerink (1974), p. 129.
35 Hermias says something similar in his *Commentary on the Phaedrus* (248.4–8) where dialectic concerns the forms. On this passage, Gary Gabor notes that the dialectic "has the potential to divide the noetic forms into many and lead them back into one, or what is the same, to distinguish each of the ideas from each other" (Gabor 2020, p. 54).
36 Steel (1999), p. 356.

they are particular beings. Syrianus gives the example Socrates, being a man as a particular thing, is not a horse or a lion; he lacks the characteristics of lion and horse. Hence, a negation at the sensible level does not specify (like the forms, where negations maintain their identity) but negates. The dialectical structure of the *Parmenides* defines one thing with respect to another.

The structure of the *Parmenides* defines the intelligibles according to their relationship with each other. That is, one specific form is a form because it is not the others, and yet they inter-relate. In *In Parm.* 997, a section on the dialectical method of the dialogue, Proclus reads lemma 13e, that things are both like and unlike, a passage he understands in light of Socrates' teaching on the mixture of forms. Here Proclus says that Socrates asks to see how the forms at the intelligible level are mixed and he wonders as to what is the method of communion. In Syrianus' *Commentary on the Metaphysics*, (p. 119, 28; Syrianus, *In Metaph.* 1179b33–5) the method of communion of forms is described. Forms interpenetrate one another purely and without mingling:

> "Actually, as a general principle, the divine, intellectual forms might be said to be united with one another, and to interpenetrate one another purely and without mingling, but they would never be said to participate in one another in the way that secondary and lower natures participate in them" (Syrianus, *In Metaph.* 1179.33–5).[37]

These forms—existing as unities in the realm of Intellect—provide unity and characteristics to entities below them. Proclus, likewise, describes the forms as "unconfused" in *Elements of Theology* (*ET*) 176, where "all intellectual forms are both implicit in each other and severally existent". He explains that the lack of confusion allows the forms to be participated in individually; this individuality is maintained in inferior principles precisely because it exists at the level of the form.[38] On the other hand, he adds, the unity of the forms appears as "the undivided substance and unitary existence of the intelligence which embraces them". For Syrianus, the propositions "in itself"/"in another" outlines at once a dialectical structure, as well as the metaphysical reality of forms which are at once in themselves

---

37 Trans. Dillon and O'Meara (2008). ὅλως δὲ ἡνῶσθαι μὲν ἀλλήλοις καὶ χωρεῖν δι' ἀλλήλων καθαρῶς καὶ ἀσυγχύτως τὰ θεῖα εἴδη καὶ νοερὰ λέγοιτ' ἄν, μετέχειν δὲ καθάπερ αἱ δεύτεραι καὶ πολλοσταὶ φύσεις αὐτῶν μετέχουσιν οὐδαμῶς ἂν ἀλλήλων. (ed. Kroll and Reimer 1902)

38 Proclus, *ET* 176 (10–16), trans. and ed. Dodds (1963): δηλοῖ δὲ τὸ μὲν ἀσύγχυτον τῶν νοερῶν εἰδῶν ἡ τῶν ἑκάστου διακεκριμένως μετεχόντων ἰδιάζουσα μέθεξις. εἰ μὴ γὰρ τὰ μετεχόμενα διεκέκριτο καὶ ἦν χωρὶς ἀλλήλων, οὐδ' ἂν τὰ μετέχοντα αὐτῶν ἑκάστου μετεῖχε διακεκριμένως, ἀλλ' ἦν ἂν πολλῷ μᾶλλον ἐν τοῖς καταδεεστέροις ἀδιάκριτος σύγχυσις, χείροσιν οὖσι κατὰ τὴν τάξιν· πόθεν γὰρ ἂν ἐγίνετο διάκρισις, τῶν ὑφιστάντων αὐτὰ καὶ τελειούντων ἀδιακρίτων ὄντων καὶ συγκεχυμένων;

and in each other. Forms are defined with respect to what they are and what they are not (and what they are not speaks to the relationship they hold with other forms.)

7 "In Itself/In Another" and "It Is/It Is Not": Parmenidean Dialectic in the *Commentary on the Philebus*

Above, we saw that Syrianus understands the One with respect to its relationship to the intelligible world. Likewise, he describes forms using the propositions "in itself"/"in another" and "it is"/it is not"; these are dialectical statements indicating the way forms interact with each other and how they are defined contradistinguished with one another. Even when Syrianus does not use this statement explicitly, one may see the dialectical structure of the *Parmenides* at work in other areas of Syrianus' metaphysical thought. For instance, Syrianus, in testimonia from Damascius' *Commentary on the Philebus*, places the three monads in the realm of One-Being (that is, the metaphysical principle existing after the One, and the pair *peras* and *apeiria*, but before the Intellectual triad). Syrianus equates each monad with one of the three hypostases of Being, Life, and Mind. Syrianus treats the three monads individually and yet they are defined with respect to the other monad: that is, I argue, there is an application of the statements "in itself"/"in another" and "it is"/ "it is not" in Syrianus' explanation of the three monads of the *Philebus*. In testimonia from Damascius' *Commentary on the Philebus*, Syrianus is reported to have taught the following:

> Syrianus separates the three [monads] and regards Truth as first revealed in Being, inasmuch as it is entirely permeated with its own essence and absolutely incapable of non-existence; Beauty as first present in Life, because Beauty is prolific and delights in self-development, for after the completely undifferentiated it is Life that carries in it the seeds of differentiation; Proportion in Intelligence, for here forms are first differentiated and harmoniously coordinated (Damascius, *In Phil.* Sect. 244).[39]

---

[39] Trans. and ed. Westerink (1959). The Greek for the entire fragment of Syrianus runs as follows: "Ὅτι Συριανὸς μερίζει· καὶ τὴν μὲν ἀλήθειαν ἐν τῷ ὄντι πρώτῳ θεωρεῖ ὡς πάντη διακορεῖ τοῦ εἶναι ὅ ἐστι καὶ οὐδαμῇ οὐδαμῶς τὸ μὴ ὂν προσδεχομένῳ· τὸ δὲ κάλλος ἐν τῇ ζωῇ ὡς γόνιμόν τε καὶ προόδοις χαῖρον, μετὰ γὰρ τὸ πάντη ἀδιάκριτον οἷον ὠδῖνα περιφέρει διακρίσεως ἡ ζωή· τὴν δὲ συμμετρίαν ἐν τῷ νῷ ἅτε πρῶτον ἐν αὐτῷ τῶν εἰδῶν διακρινομένων καὶ ἐναρμονίως συντεταγμένων. Διέλοις δὲ ἂν καὶ εἰς τὰς ἀρχὰς μετὰ τὴν μίαν· τῷ μὲν γὰρ πέρατι τὴν ἀλήθειαν, τῇ δὲ ἀπειρίᾳ τὸ κάλλος διὰ τὸ προοδικόν, τῷ δὲ μικτῷ τὴν συμμετρίαν δικαίως ἂν ἀποδοίης. ἀλλὰ καὶ εἰς τὰς τρεῖς

In this quotation, Syrianus shows that Truth is first revealed in Being, inasmuch as it has its own essence—he describes it as "absolutely incapable of non-existence". Syrianus says that Beauty is first present in Life, because Beauty is prolific and "delights in self-development". Finally, proportion (*symmetria*) is found in Intelligence, for here forms are first differentiated and harmoniously coordinated. Thus, each monad is in itself as well as in another. Proclus elaborates upon these relationships, as described by Damascius throughout the *Commentary on the Philebus*. Moreover, Proclus uses Syrianus' metaphysical interpretation of dialectic in the *Parmenides* to understand the relationship between monads and the One.[40] That is, Proclus connects the monads to the One using the basic premises found in his discussion of the first and second hypotheses of the *Parmenides* and the statement "if the One is"/"If the One is not".

8 Conclusion

This paper is an examination of how Syrianus understands certain dialectical statements from the *Parmenides* and how he uses these statements as descriptions of the whole universe, particularly how one entity relates to another.[41] Namely, we see that he reads "if a thing is/if it is not" to describe the relationship between the One and the intelligible universe; moreover, "in itself/in another" also speaks to the relationship between the One and the entities below the One. These two statements become the paradigm whereby Syrianus comes to understand other entities, such as the forms in his *Commentary on Metaphysics*, or the monads, as reported

---

ἀρχικὰς ὑποστάσεις μερίσειας ἂν τὰς μονάδας, θεὸν νοῦν ψυχήν. καὶ μὴν καὶ εἰς τὰς οὐσίας, μεριστὴν ἀμέριστον μέσην· δεῖται γὰρ συμμετρίας ἡ μὲν μεριστὴ ὡς πάντῃ διαστᾶσα, ἀληθείας δὲ ἡ ἀμέριστος ὡς ὄντως οὖσα, κάλλους δὲ ἡ μέση ὡς οἷον ἐν συγκράσει θεωρουμένη τοῦ μεριστοῦ καὶ κάλλους δὲ ἡ μέση ὡς οἷον ἐν συγκράσει θεωρουμένη τοῦ μεριστοῦ καὶ ἀμερίστου".

40 In Damascius' *Commentary on the Philebus*, Proclus is reported to have taught that the three monads preserve the invioability of the One, allowing it to remain in itself as "the One is not". Moreover, in his *Commentary on the Republic* 1, p. 295.19–24, Proclus shows how we use the three monads to come to the good because of their affinity to the Good. The monads have a certain affinity to the Good, without acting as attributes of the One itself. Instead, the three monads characterizes the way in which the One affect us—now, Proclus' emphasis is on the One's relation to creation, as governed by its point of perspective.

41 In his *Commentary on the Phaedrus* 261C, Hermias notes that the highest task of philosophy is to know the similarities and differences between things; the highest kind of similarities and differences are between the forms. See Gabor (2020), p. 65. David Butorac points out that for Proclus, one of the problems of dialectic is that it divides that which it treats Platonic dividing and combining or the Aristotelian definition, division, demonstration, and analysis (Butorac 2017, p. 130).

in Damascius' *Commentary on the Philebus*. Thus, these two statements act as the lens by which Syrianus understands participation by intelligible entities, primarily following the example of the intelligible realm's participation in the One. At the heart of the propositions "if a thing is/if it is not" and "in itself/in another" is a concern for preserving the unity of the One while showing how it gives rise to an elaborate, inter-connecting universe that coheres in its cause.

Bibliography

Primary Sources

Proclus

Diehl, Ernestus (Ed.) (1903–1906, 1965): *Procli Diadochi In Platonis Timaeum commentaria*. Leipzig: Lipsiae [reprint Amsterdam: Hakkert].
Dodds, Eric Robinson (Ed. and Trans.) (1963): *Proclus. The Elements of Theology*. 2nd ed. Oxford: Oxford University Press.
Friedlein, Gottfried (Ed.) (1873, 1967): *Procli Diadochi in primum Euclidis elementorum librum commentarii*. Leipzig: Teubner [reprint Hildesheim: Olms].
Kroll, Wilhelm (Ed.) (1899–2001): *Procli Diadochi in Platonis rem publicam commentarii*. 2 Volumes. Leipzig: Teubner.
Lucarini, Carlo M. and Moreschini, Claudio (Eds.) (2012): *Hermias Alexandrinus, In Platonis Phaedrum Scholia*. Berlin: de Gruyter.
Morrow, Glenn R. and Dillon, John M. (Trans.) (1987): *Proclus' Commentary on Plato's* Parmenides. Princeton: Princeton University Press.
Saffrey, Henri D. and Westerink, Leendert G. (Eds. and Trans.) (1968–1997): *Proclus: Théologie platonicienne*. 6 Vols. Paris: Les Belles Lettres.
Steel, Carlos, Macé, Caroline, d'Hoine, Pieter, Gribomont, Aurélie, and Van Campe, Leen (Eds.) (2007–2009): *Procli in Platonis Parmenidem commentaria*. Vols. I–VII. Oxford: Oxford University Press.
Wallies, M. (Ed.) (1905): *Philoponus In Aristotelis Analytica Priora commentaria*. CAG XIII.2. Berlin: Reimer.
Westerink, Leendert G. (Ed.) (1954): *Proclus Diadochus. Commentary on the First Alcibiades of Plato*. Amsterdam: North Holland Publishing Company.
Westerink, Leendert G. (Ed.) (1959): *Damascius. Lectures on the Philebus, wrongly attributed to Olympiodorus. Text, translation, notes and indices*. Amsterdam: North-Holland Publishing Co.
Westerink, Leendert G. and Combès, Joseph (Eds.) (1986–1991): *Damascius. Traité des Premiers Principes. Vol. III, texte établi, traduit et annoté*. Paris: Les Belles Lettres.

Syrianus

Dillon, John and O'Meara, Dominic (Eds.) (2006): *Syrianus: On Aristotle* Metaphysics *3–4*. London: Bloomsbury.

Dillon, John and O'Meara, Dominic (Eds.) (2008): *Syrianus: On Aristotle* Metaphysics *13–14*. London: Bloomsbury.
Kroll, Wilhelm (Ed.) (1902): *Syrianus, In Aristotelis 'Metaphysica' commentaria*. CAG VI, I. Berlin: Reimer.

Secondary Sources

Beierwaltes, Werner (1979): *Proklos. Grundzüge seiner Metaphysik*. Frankfurt am Main: V. Klostermann.
Bonelli, Maddalena (2009): "Dialectique et Philosophie Première: Syrianus et Alexandre D'Aphrodise". In: Longo, Angela (Ed.): *Syrianus et La Métaphysique de l'Antiquité Tardive. Actes du colloque international. Universitéde Genève, 29 Septembre-1 Octobre 2006*. Naples: Bibliopolis, pp. 423–438.
Butorac, David (2017): "Proclus' Aporetic Epistemology". In: Butorac, David D. and Layne, Danielle A. (Eds.): *Proclus and His Legacy*. Berlin: de Gruyter, pp. 123–136.
Dillon, John M. (2009): "The Architecture of the Intelligible Universe Revealed: Syrianus' Exegesis of the Second Hypothesis of the *Parmenides*". In: Longo, Angela (Ed.): *Syrianus et La Métaphysique de l'Antiquité Tardive. Actes du colloque international. Universitéde Genève, 29 Septembre-1 Octobre 2006*. Naples: Bibliopolis, pp. 253–246.
Gabor, Gary (2020): "Hermias on Dialectic, the Technē of Rhetoric, and the True Methods of Collection and Division". In: Finamore, John F., Manolea, Christina-Panagiota, and Wear, Sarah Klitenic (Eds.): *Studies on Hermias' Commentary on Plato's* Phaedrus. Leiden: Brill, pp. 50–67.
Glasner, Ruth (1992): "Beginning, Middle, and End in Proclus' Commentary on Plato's Parmenides". In: *Hermes* 120. No. 1, pp. 194–204.
Ierodiakonou, Katerina (2009): "Syrianus on Scientific Knowledge and Demonstration". In: Longo, Angela (Ed.): *Syrianus et La Métaphysique de l'Antiquité Tardive. Actes du colloque international. Universitéde Genève, 29 Septembre-1 Octobre 2006*. Naples: Bibliopolis, pp. 401–422.
Lernould, Alain (1987): "La Dialectique Comme Science Première Chez Proclus". In: *Revue des Sciences philosophiques et théologiques* 71. No. 4, pp. 509–536.
O'Meara, Dominic (1989): *Pythagoras Revived: Mathematics and Philosophy in Late Antiquity*. Oxford: Oxford University Press.
Steel, Carlos (1999): "'Negatio Negationis': Proclus on the Final Lemma of the First Hypothesis of the *Parmenides*". In: Cleary, John (Ed.): *Traditions of Platonism. Essays in Honour of John Dillon*. Aldershot: Ashgate, pp. 351–368.
Van Campe, Leen (2009): "Syrianus and Proclus on the Attributes of the One in Plato's *Parmenides*". In: Longo, Angela (Ed.): *Syrianus et La Métaphysique de l'Antiquité Tardive. Actes du colloque international. Universitéde Genève, 29 Septembre-1 Octobre 2006*. Naples: Bibliopolis, pp. 247–280.
Van Den Berg, Robbert (2001): *Proclus' Hymns: Essays, Translation, and Commentary*. Leiden: Brill.
Wear, Sarah Klitenic (2011a): "The One in Syrianus' Teachings on the *Parmenides* 137d and 139a1". In: *The International Journal of the Platonic Tradition* 4, pp. 55–84.
Wear, Sarah Klitenic (2011b). *The Teachings of Syrianus on Plato's Timaeus and Parmenides*. Leiden: Brill.
Wear, Sarah Klitenic (2017): "Pseudo-Dionysius and Proclus on *Parmenides* 137d: On Wholes and Parts". In: Butorac, David D. and Layne, Danielle A. (Eds.): *Proclus and His Legacy*. Berlin: de Gruyter, pp. 219–232.

Dirk Baltzly
Proclus on Plato's Dialectic: Argument by Performance

1 Introduction

Both contemporary interpreters of Plato and the Neoplatonists faced the question of reconciling the dialogues' various discussions of *dialectic*.[1] The *Republic* makes dialectic *the* distinctive philosophical method through which the Guardians know the Forms and especially the Form of the Good (*Rep.* VI 510c–511c; VII 534b–c).

In the *Republic*, Plato famously describes dialectic by its contrast with mathematical reasoning. The dialectician is able to "give an account of the essence of each thing". His cognitive state is distinguished from that of the mathematician in its nature (*Rep.* VI 511a–c). The dialectical philosopher enjoys intellection or *noêsis*, as opposed to the mathematician who seems to have merely discursive thought or *dianoia*.[2] In addition, the movement or direction of thought involved in dialectic is also distinguished from that involved in mathematical thinking. We are told by Socrates that while dialectic treats hypotheses as hypotheses, using them as steps to ascend to an unhypothetical first principle, mathematicians proceed from hypotheses to a conclusion (510b4–9, 511b3–c2). Dialectic, however, "does away with" its hypotheses (533c8). Finally, while mathematical thinking involves the use of images, dialectic does not (510d5–511a1, 511b8–c2).

Much of this is puzzling if considered merely in the context of the single dialogue. It is an important puzzle because dialectic plays a key role in Plato's *Republic*. The authority of the Guardians to lead the ideal *polis* is grounded in the psychic consequences—both in terms of knowledge and character or motivation—that result from their acquaintance with, and love of, the Forms. Dialectic, in turn, is the method through which they gain this acquaintance. But in spite of this key role, Socrates' remarks on the nature of dialectic are far from clear. The reader is left

---

[1] For a sketch of the contemporary landscape around the topic of dialectic over all the dialogues, see Benson (2006) or, more briefly, Baltzly (2012). Other reference works treat dialectic in the *Republic* in separate entries from dialectic in, e.g., the *Philebus*. Cf. Fine (2008).
[2] It remains a subject of hot dispute whether the objects of these distinct cognitive attitudes are similarly distinct. This, of course, is the problem of whether the segments of the doubly-divided line each have their own distinctive objects and whether there is a distinct class of "mathematicals" intermediate between sensibles and Forms. For the state of the question and a negative verdict on the intermediates, see Franklin (2012).

to wonder in what, exactly, it consists. The explanation of the nature of dialectic we are given does not seem proportionate to the argumentative weight resting on the Guardians' distinctive method and its results in the dialogue.

Things become even more puzzling when we look to dialogues other than the *Republic*. There are, of course, remarks about the method of dialectic in the *Phaedrus*, *Sophist* and *Philebus* that do not resemble closely the method of dialectic described in *Republic*. Here it is the method of *collection and division* that is identified with dialectic.³ At no point do these various descriptions of collection and division advert to an unhypothetical starting point or stipulate that the method eschews the kinds of images or figures used by geometers. Thus, the relation between *Republican* dialectic, on the one hand, and the methods (including collection and division) that are called dialectic in other dialogues is a matter of deep dispute in current Plato scholarship.⁴ Readers of Plato who take a chronological approach to the dialogues may find this final puzzle less problematic. Perhaps Plato just changed his mind or refined his views. But chronological approaches to the interpretation of such divergences between Plato's dialogues have now fallen somewhat out of favor. What should readers who suppose that Plato's dialogues present (at least in some sense) a reasonably unified Platonism think of the idea of dialectic as it appears in the *Republic*, on the one hand, and in *Phaedrus*, *Sophist* and *Philebus*, on the other? One way to approach this question is to see if there is any insight that can be gained from post-Hellenistic attempts to assemble—largely but exclusively—from Plato's dialogues a doctrinaire version of Platonism.

2 Post-Hellenistic Platonists on the Unity of Dialectic

Alcinous' *Handbook of Platonism* is our most complete representative of Middle Platonic attempts to systematically articulate a doctrinaire or dogmatic version of Platonism. The *Handbook*'s Chapter 5 on dialectic nicely illustrates a strategy that declines to attempt to unify the various things that the dialogues characterize

---

3 Perhaps most explicitly at *Phdr.* 266b3–c1: Τούτων δὴ ἔγωγε αὐτός τε ἐραστής, ὦ Φαῖδρε, τῶν διαιρέσεων καὶ συναγωγῶν, ἵνα οἷός τε ὦ λέγειν τε καὶ φρονεῖν.... Καὶ μέντοι καὶ τοὺς δυναμένους αὐτὸ δρᾶν εἰ μὲν ὀρθῶς ἢ μὴ προσαγορεύω, θεὸς οἶδε, καλῶ δὲ οὖν μέχρι τοῦδε διαλεκτικούς.
4 For a good overview of the range of procedures variously called "dialectic" in the Platonic dialogues, see Dorter (2006), pp. 10–18. Dorter is, in the end, a kind of "unitarian" about dialectic, but the unity he discovers behind the various senses of dialectic is one of purpose where the activities so described "converge" but do not exactly coincide.

as "dialectic". Alcinous attributes a unified *purpose* to dialectic: to examine the essences of things and, secondarily, their accidents. But he immediately concedes that dialectic can do this in two different ways—either "from above" by means of division and definition or "from below" by means of analysis.[5] In the end, Alcinous enumerates five components from which dialectic is comprised: division, definition, analysis (i.e., collection), induction, and syllogistic. The first four of these are discussed in Chapter 5, with syllogistic being given a separate treatment in Chapter 6. But even within the context of Chapter 5, no clear connections emerge between the first four components. Instead, the exegesis is exclusively occupied with further subdivisions of division and analysis. It is only in Alcinous' third form of analysis that one can perhaps hear some echoes of *Republic* VI 510c–511c or VII 534b–c since he describes this as a method by which one can "advance upward from a hypothesis to an unhypothetical first principle" (5.4).[6] But the explanation of this in in 5.6 is disappointingly opaque and, more importantly for our purposes, in no way connected with the any kind of collection or division.[7] So Alcinous' *Handbook* treats Plato's dialectic as a toolkit composed of distinct tools with distinct instructions for their use. Nor does any characterization of any of the components contained within dialectic greatly illuminate the *Republic*'s description of dialectic.

Plotinus dedicates a short treatise, *Ennead* I.3, specifically to the subject of dialectic. While Alcinous treats dialectic's function as principally epistemological—to know the essences and accidents of things—Plotinus' characterization of it is principally soteriological. Dialectic is the manner (*tropos*) in which some souls—those that are philosophical, as opposed to erotic or musical (*Enn.* I.3.1, 9–10; cf. *Phdr.* 248d)—ascend from their foothold in the intelligible realm to the "end of their journey" (*Enn.* I.3.1, 16–17; cf. *Rep.* VII 532e3) at the limit of the intelligible. These "philosophical souls" are said by Plotinus to be prepared for the grasp of the incorporeal by their studies of mathematics (*Enn.* I.3.3, 5–10; cf. *Rep.* VII 531d–35a). In thus juxtaposing phrases and ideas from the *Phaedrus* and the *Republic*, Plotinus

---

[5] Alcinous' terminology is Aristotelian since he calls this "analysis" rather than Plato's more usual term *synagôgê* but such appropriations of Aristotelian or Stoic terminology for methods or concepts that Alcinous attributes to Plato is common.

[6] Ἀναλύσεως δὲ εἴδη ἐστὶ τρία· ἡ μὲν γάρ ἐστιν ἀπὸ τῶν αἰσθητῶν ἐπὶ τὰ πρῶτα νοητὰ ἄνοδος, ἡ δὲ διὰ τῶν δεικνυμένων καὶ ὑποδεικνυμένων ἄνοδος ἐπὶ τὰς ἀναποδείκτους καὶ ἀμέσους προτάσεις, ἡ δὲ ἐξ ὑποθέσεως ἀνιοῦσα ἐπὶ τὰς ἀνυποθέτους ἀρχάς.

[7] Ἡ δὲ ἐξ ὑποθέσεως ἀνάλυσίς ἐστι τοιαύτη· ὁ ζητῶν τι ὑποτίθεται αὐτὸ ἐκεῖνο, εἶτα τῷ ὑποτεθέντι σκοπεῖ τί ἀκολουθεῖ, καὶ μετὰ τοῦτο εἰ δέοι λόγον ἀποδιδόναι τῆς ὑποθέσεως, ἄλλην ὑποθέμενος ὑπόθεσιν, ζητεῖ εἰ τὸ πρότερον ὑποτεθὲν πάλιν ἐστὶν ἀκόλουθον ἄλλῃ ὑποθέσει, καὶ τοῦτο μέχρις οὗ ἂν ἐπί τινα ἀρχὴν ἀνυπόθετον ἔλθῃ ποιεῖ.

implies a continuity in the meaning of "dialectic" between the two dialogues. But it is noteworthy that Plotinus draws more upon the soul's ascent in Socrates' palinode in the earlier part of dialogue (*Phdr.* 243e–57b) than he does upon the discussion of collection and division in the latter part of the *Phaedrus*. This is, of course, consistent with Plotinus' characterization of dialectic principally in soteriological terms: it is a means of ascent and the soul's salvation. Less is said by Plotinus about its epistemic character as a way of knowing that is somehow superior to mathematical knowledge.

When Plotinus does turn to the question "What is dialectic?" in *Enn.* I.3.4 phrases and ideas from *Republic*, *Sophist* and *Phaedrus* are conspicuously invoked. From *Republic* 534b–c, we find the claims that dialectic is the capacity to say *what each thing is* and how it differs from all other things (lines 2–4).[8] Conjoined to this is the idea that dialectic says *how many* things there are (I.3.4, 4–5), thus alluding to the quantitative aspect of collection and division that is described in *Philebus* 16c–17a, where it is crucial to rightly enumerate the divisions among the things that are. In Plotinus' description, dialectic does not seem to be confined to the realm of Forms, since it speaks about both things that are everlasting and those that are not (I.3.4, 7–8).[9] Yet it deals with *epistêmê*, in contrast to *doxa*, and this is perhaps explained by the fact that sensible things are only dealt with at the start of the process, since the dialectician's soul "ceases to wander in the sensible world" (I.3.4, 9–12) when it is nourished upon the *Phaedrus*' "plain of truth" (*Phdr.* 248b) where it uses the method of collection and division to discern the Forms.[10] Perhaps as a consequence, it can say "what each thing is"—knowing the Forms in the first instance and their sensible copies as a consequence of this prior knowledge. This method is also used to know the "primary genera" (I.3.4, 14–15), which are presumably the greatest kinds from *Sophist* 255bff. and the things that are "woven together from these by the use of intellect" (*Soph.* 259e4–6). This seems to be a descent through the intelligibles that result from the weaving together of the greatest kinds by means of synthesis until, picking

---

8 *Enn.* I.3.4, 1–6: Τίς δὲ ἡ διαλεκτική, ἣν δεῖ καὶ τοῖς προτέροις παραδιδόναι; Ἔστι μὲν δὴ ἡ λόγῳ περὶ ἑκάστου δυναμένη ἕξις εἰπεῖν τί τε ἕκαστον καὶ τί ἄλλων διαφέρει καὶ τίς ἡ κοινότης· ἐν οἷς ἐστι καὶ ποῦ τούτων ἕκαστον καὶ εἰ ἔστιν ὅ ἐστι καὶ τὰ ὄντα ὁπόσα καὶ τὰ μὴ ὄντα αὖ, ἕτερα δὲ ὄντων.

9 *Enn.* I.3.4, 6–9: Αὕτη καὶ περὶ ἀγαθοῦ διαλέγεται καὶ περὶ μὴ ἀγαθοῦ καὶ ὅσα ὑπὸ τὸ ἀγαθὸν καὶ ὅσα ὑπὸ τὸ ἐναντίον καὶ τί τὸ ἀΐδιον δηλονότι καὶ τὸ μὴ τοιοῦτον, ἐπιστήμῃ περὶ πάντων, οὐ δόξῃ.

10 *Enn.* I.3.4, 9–12: Παύσασα δὲ τῆς περὶ τὸ αἰσθητὸν πλάνης ἐνιδρύει τῷ νοητῷ κἀκεῖ τὴν πραγματείαν ἔχει τὸ ψεῦδος ἀφεῖσα ἐν τῷ λεγομένῳ ἀληθείας πεδίῳ τὴν ψυχὴν τρέφουσα, τῇ διαιρέσει τῇ Πλάτωνος χρωμένη μὲν καὶ εἰς διάκρισιν τῶν εἰδῶν, χρωμένη δὲ καὶ εἰς τὸ τί ἐστι, χρωμένη δὲ καὶ ἐπὶ τὰ πρῶτα γένη, καὶ τὰ ἐκ τούτων νοερῶς πλέκουσα.

up the analysis again, it arrives back at the *archê* where it subsides into stillness and is no longer occupied with many things (I.3.4, 15–18).[11] So this aspect of Plotinus' description aligns with the upward and downward movement of thought in the description in the *Republic*.

While there is much that is unclear about the description of dialectic that Plotinus offers us in these terse 18 lines of text, the message of the next 5 lines is clear: dialectic is *distinct* from logic.[12] Though it presupposes logical techniques involving premises and syllogisms, it is superior to such matters. So, while Alcinous' *Handbook* is content to assimilate ideas from Aristotelian syllogistic and Stoic logic to dialectic, Plotinus distinguishes genuinely Platonic dialectic from these lesser pursuits. While Plotinus' description of dialectic imputes to it some sort of role in coming to know, its real purpose is to facilitate psychic ascent. While knowing is clearly part of such an ascent, the ascent seems to be a moral or spiritual condition as much as an epistemic one. Given the predominance of the soteriological in Plotinus' thought about dialectic, it is not too surprising that the puzzles we raised about the way of knowing described by collection and division, on the one hand, and the way of knowing that involves the *Republic*'s transcending of hypotheses and images, on the other, do not occupy him much. These are assumed to be *somehow* complementary in the ascent to the One—though Plotinus sheds little light on the nature of this complementarity. Nor does Plotinus show us in detail how Plato's various pronouncements on dialectic fit together. Rather, he invokes images and phrases from all the dialogues that describe dialectic in a compressed description of the method that *assumes*, rather than *argues for*, the unity of Plato's various remarks about it. As we shall see, this pattern is repeated in the work of Proc.

3 Proclus on Dialectic across the Various Dialogues

Proclus similarly takes a view of Platonic dialectic according to which it is one and the same method throughout the Platonic corpus. Commenting on the questioning of Socrates by Parmenides at the opening of the dialogue, Proclus says:

---

11 *Enn.* I.3.4, 15–18: ἕως ἂν διέλθῃ πᾶν τὸ νοητόν, καὶ ἀνάπαλιν ἀναλύουσα, εἰς ὃ ἂν ἐπ' ἀρχὴν ἔλθῃ, τότε δὲ ἡσυχίαν ἄγουσα, ὡς μέχρι γε τοῦ ἐκεῖ εἶναι ἐν ἡσυχίᾳ, οὐδὲν ἔτι πολυπραγμονοῦσα εἰς ἓν γενομένη βλέπει.
12 *Enn.* I.3.4.18–20: τὴν λεγομένην λογικὴν πραγματείαν περὶ προτάσεων καὶ συλλογισμῶν, ὥσπερ ἂν τὸ εἰδέναι γράφειν, ἄλλῃ τέχνῃ δοῦσα·

When Socrates shows his bewilderment in the face of these problems, Parmenides advises him, if he is really enamoured of the truth about Being, to exercise himself in dialectic before undertaking this larger inquiry—meaning by dialectic that method that Socrates himself teaches us in other works, such as the *Republic*, the *Sophist*, the *Philebus*. When Socrates asks what this method is and shows himself ready to accept these visitors teaching, Parmenides expounds the method whose praises Socrates has already sung on many occasions. In the *Phaedo* (101d), for example, in distinguishing the function of dialectic from eristic, he says that one must at every step assume a hypothesis and continue an inquiry until from many hypotheses we come up to "something adequate", which he calls "the unhypothetical" (*Rep.* VI 510b) (*in Parm.* 622.18–33, trans. Dillon and Morrow).

Proclus seems to regard the unity of Platonic dialectic as obvious. But at 648.1ff. Proclus takes up the arguments of those who assert that Parmenides' method, as it is described in that dialogue, cannot be the same as Plato's dialectic as it is described in the *Republic*. Steel has argued persuasively that the people who suppose that Parmenidean dialectic differs from Platonic dialectic are Middle Platonist readers of the dialogue.[13] Three grounds are offered for this non-identity between Parmenidean and Platonic dialectic. First, in the dialogue that bears his name, Parmenides urges the *young* Socrates to engage in it, but in the *Republic* Socrates recommends against young people being allowed to engage in dialectic too early (*Rep.* 537–39d). Second, Parmenides describes dialectic as an exercise (*askêsis*) and Proclus takes this to mean that it involves arguing both sides of a question. Thus, he claims, it more closely resembles Aristotelian dialectic than Plato's. Proclus then summarizes Platonic dialectic in way that resembles the *Republic*:

> Plato's dialectic is described in the dialogue [sc. the *Republic*?] as leading to the highest and purest stage of knowledge and insight, since its activity is based on intelligible Forms, through which it advances to the very first member of the intelligible world, paying no attention to human opinion but using irrefutable knowledge at every step (649.1–8, trans. Dillon and Morrow).[14]

Finally, while Parmenidean dialectic is characterized as a procedure that resembles "babbling" (*Parm.* 135d) to some, Platonic dialectic is the "capstone of knowledge" (*Rep.* 534e), which is suitable only for genuine philosophers (*Soph.* 253e). This final point suggests that Proclus' Middle Platonic opponents do not sharply distinguish the method of collection and division described in the *Sophist* (and else-

---

13 Cf. Steel (1997).
14 τὴν δὲ Πλάτωνος τὸ ἀκρότατον παρ' αὐτῷ λέγεσθαι καὶ καθαρώτατον νοῦ καὶ φρονήσεως, ἐν τοῖς νοητοῖς εἴδεσι τὴν πραγματείαν ἑαυτῆς ἱδρύσασαν καὶ διὰ τούτων χωροῦσαν ἐπ' αὐτὸ τὸ ἡγούμενον τοῦ νοητοῦ παντός, οὐ πρὸς δόξαν βλέπουσαν ἀνθρώπων, ἀλλ' ἐπιστήμῃ περὶ πάντων ἀνελέγκτῳ χρωμένην·

where) from the dialectic of the *Republic*. Or, at the very least, Proclus has seen fit to put their arguments in ways that elide the difference since both *Republic* and *Sophist* are mentioned in the same breath, as if were obvious that these things were one and the same, or at least closely related.

Let us set aside as not philosophically interesting the tedious special pleading that Proclus does on behalf of the young Socrates (who is, of course, *so* obviously talented that he should enjoy an exemption for youthful participation in Platonic dialectic). The overall thrust of Proclus' reconciliation of the putative differences between Parmenidean and Platonic dialectic is to assimilate collection and division to the method of hypothesis employed in the latter half of the *Parmenides*. In order to do this, Proclus offers a tripartite division of Platonic dialectic: 1) the kind that argues both sides of the question; 2) the kind that exhibits only the truth; and 3) the kind that serves only to refute false beliefs (*in Parm.* 654.11–13).[15] The second kind seems to include both Republican dialectic and the method of collection and division from the *Phaedrus* and the *Statesman*. As Proclus says:

> In another form of its activity, dialectic places the mind at the outset of the region of thought where it is most at home, looking at truth itself, "sitting on a sacred pedestal" (*Phaedr.* 254b7), which Socrates says unfolds before the mind the whole intelligible world, making its way from Form to Form, until it reaches the very first Form of all, sometimes using analysis, sometimes definition, now demonstrating, now dividing, both moving downwards from above and upwards from below until, having examined in every way the whole nature of the intelligible, it climbs aloft to that which is beyond all being. When it has safely anchored the soul there, it has reached its goal and there will no longer be anything greater to be desired. You could say these are the functions of dialectic spoken of in the *Phaedrus* and in the *Sophist*, the former dividing dialectical procedures into two, the latter into four parts (*in Parm.* 653.18–33, trans. Dillon and Morrow).[16]

Steel's new Oxford Classical Text reveals the extent to which this passage approaches a cento of Platonic passages. The quotation from *Phaedrus* 247b8 is of course obvious. Steel rightly notes the connection between Proclus' ἅπαν τὸ νοητὸν ἀνελίττειν, δι' εἰδῶν ἀεὶ πορευομένην and the double use of the same par-

---

15 On the third form of dialectic, see Layne (2009).
16 ἑτέρα δὲ ἀναπαύουσα ἤδη τὸν νοῦν οἰκειοτάτῃ θεωρίᾳ τῶν ὄντων καὶ αὐτὴν ὁρῶσα καθ' αὑτὴν τὴν ἀλήθειαν ἐν ἁγνῷ βάθρῳ βεβῶσαν, ἥν φησιν ὁ Σωκράτης ἅπαν τὸ νοητὸν ἀνελίττειν, δι' εἰδῶν ἀεὶ πορευομένην ἕως ἂν εἰς αὐτὸ καταντήσῃ τὸ πρῶτον, τὰ μὲν ἀναλύουσαν, τὰ δὲ ὁριζομένην, τὰ δὲ ἀποδεικνῦσαν, τὰ δὲ διαιροῦσαν, ἄνωθέν τε καὶ κάτωθεν εἰς τὸ ἀνάντες χωροῦσαν, ἕως ἂν πᾶσαν πάντῃ διερευνωμένη τὴν τῶν νοητῶν φύσιν εἰς τὸ ἐπέκεινα πάντων ἀναδράμῃ τῶν ὄντων, οὗ τὴν ψυχὴν ὁρμίσασα τελέως οὐκ ἔτι ποθήσεται κρεῖττον ἐφετὸν ἐπὶ τέλος ἥκουσα· καὶ ταύτης ἂν εἴποις ἔργα εἶναι τά τε ἐν Φαίδρῳ ῥηθέντα καὶ τὰ ἐν Σοφιστῇ, τὰ μὲν διχῇ διῃρημένα, τὰ δὲ τετραχῇ τῆς διαλεκτικῆς ἔργα.

ticiple at the point where the divine souls survey the realm of Forms in *Phaedrus* 247a8–b3 ἄκραν ἐπὶ τὴν ὑπουράνιον ἀψῖδα πορεύονται πρὸς ἄναντες, ᾗ δὴ τὰ μὲν θεῶν ὀχήματα ἰσορρόπως εὐήνια ὄντα ῥᾳδίως πορεύεται, τὰ δὲ ἄλλα μόγις. But Proclus' phrase δι' εἰδῶν ἀεὶ πορευομένην also echoes the way in which the *Republic* describes dialectic as "moving by means of Forms, through Forms, to its conclusions which are Forms" (511c1–2, trans. Grube ἀλλ' εἴδεσιν αὐτοῖς δι' αὐτῶν εἰς αὐτά, καὶ τελευτᾷ εἰς εἴδη.) Now, in Proclus' passage, this movement extends upward rather than downward. So that instead of τελευτᾷ εἰς εἴδη, dialectic arrives at what is *first* (τὸ πρῶτον). And dialectic's omni-directional capacity is stressed in ἄνωθέν τε καὶ κάτωθεν. So, it descends through Forms to Forms, but also arises through Forms (perhaps by seeing many kinds as unified in their genus) and descends through Forms (dividing a genus into its parts). Arising to the first principle and descending again it "analyzes, defines, demonstrates, and divides" along the way. Analyzing, dividing and defining are plausibly activities associated with the method of collection and division from the later dialogues. But in the next sentence, we find διερευνωμένη τὴν τῶν νοητῶν φύσιν εἰς τὸ ἐπέκεινα πάντων ἀναδράμῃ τῶν ὄντων which clearly recalls the Form of the Good in *Republic* 509b9 (ἐπέκεινα τῆς οὐσίας). The soul is finally "safely anchored" here and Proclus' language in relation to this anchoring *may* function to suggest a confluence between Platonic dialectic and the practice of theurgy.[17] So, Proclus' second form of Platonic dialectic—the kind that μόνον τὸ ἀληθὲς ἐκφαινούσης (*in Parm.* 654.17)—seems to combine what is described in the *Republic* with collection and division and, perhaps, even theurgical practice.[18] So, the truth-revealing kind of dialectic is capacious indeed! Thus, where modern interpreters have found a puzzle about how Plato's seemingly different characterizations fit together, Proclus supposes that these are simply different perspectives on a single movement of thought. But what is his *argument* for this interpretation?

Proclus' passage quoted above is not really an *interpretive argument* that this unified conception of truth-revealing dialectic is present across Plato's dialogues. The rich and rapid-fire string of associations or near quotations seems intended

---

[17] Cf. Van Den Berg (2000), especially the appendix in which Van Den Berg catalogues the extensive use of this image. He relies on Saffrey and Westerink *Platonic Theology*, Vol. IV, p. 147, n. 3, for the connection with the *Chaldean Oracles*. On Van Den Berg's reading, Proclus supposed that only the divine Nous or Intellect—equated with the Demiurge—contemplates the Forms. The human soul's access to the Demiurge is mediated through the Leader Gods of *Platonic Theology* VI. Van Den Berg further argues that humans can use theurgy to bring themselves to the attention of these Leader Gods in order to facilitate access to the Demiurge.

[18] For the view that philosophical and theurgic methods of ascent to the divine should be seen as complementary rather than opposed alternatives, see Baltzly (2006).

to *reveal* a unity of method between *Republic*, on the one hand, and *Phaedrus*, *Statesman* and *Sophist* on the other. An audience who knows Plato's dialogues sufficiently well would literally *hear* the presence of all these dialogues in Proclus' description of this species of dialectic. The interpretive argument for the existence of a species of "truth-revealing dialectic" that includes both the curious procedure of the *Republic* and also the method of collection and division described in the *Phaedrus* is more *performance* than argument. Or perhaps we might call it argument by performance. Much the same can be said of Plotinus' compressed discussion of the nature of dialectic in *Enneads* I.3.4. How should we, who also wish to understand Plato, think about this? Should it engender disappointment?

4 What Kind of Philosophizing Is This?

Some philosophers think that we make Neoplatonic authors relevant just to the extent that we find clues to the proper interpretation of the dialogues in them, thus treating Proclus' *Timaeus Commentary* as one might treat Taylor's or Cornford's.[19] But our ancient Platonists are not merely interpreters of Plato. They study Plato's dialogues together in communities of like-minded Platonists in order to acquire the various gradations of the virtues that they distinguished. This correlation between the gradations of the virtues and the reading order of Plato's dialogues is most explicit in the *Anonymous Prolegomena to Platonic Philosophy*.[20] But within individual commentaries it is implicit in the care that is taken over determining the *skopos* or objective of the dialogue, for the *skopos* is, in turn, related to the various gradations of the virtues. This is perhaps most obvious in the surviving commentaries on the *Phaedo*—a dialogue which correlates with the cathartic or *purificatory* virtues in the reading order.[21] But it is also present with Olympiodorus' *Gorgias Commentary* which serves, in the place of the *Republic*, as a text through which one acquires the *civic* virtues.

Through the possession and exercise of the grades of virtues acquired through the Platonic reading order, the Neoplatonists supposed themselves to be assimilated to the divine. Now, there are two familiar observations about this transformative philosophical project. First, the texts that describe and distinguish the natural, ethical, civic, kathartic, and contemplative gradations of the virtues are hardly

---

[19] Cornford (1957) and Taylor (1928).
[20] *Anon Proleg.* § 26. For discussion, see Westerink (1962), p. xl, as well as Baltzly (2107), pp. 266–270, and Tarrant (2014), pp. 23–25.
[21] Cf. Westerink (1976), p. 28.

transparent in their meaning.²² Second, the connection that as alleged between the stages of the Neoplatonic reading order of the dialogues and the acquisition of these virtues is similarly opaque. While one might plausibly connect the *Republic* with some concept of civic virtue or the *Phaedo* with some notion of *kathartic* virtue, it is far from clear why the *Phaedrus* should be a dialogue from which one learns the contemplative virtues. So, it is not only unclear what the aretaic pay-off of studying Plato's dialogues with a certified Platonist is; it is also unclear how the lectures that form the basis of our commentaries were meant to assist in the acquisition of the aretaic pay-off.

We are left to hypothesize. Here is the hypothesis that I have been defending for some time now: the various gradations of the virtues should be understood in terms of the possession of an improvisational ability I have called "Platonic literacy".²³ It is the capacity to live "in and through the dialogues" which mimics the way in which the educated elite of the late Roman empire lived in and through their notion of *paideia*. The study of grammar and rhetoric that comprised *paideia* was thought to similarly transform the educated man into someone truly remarkable:

> We are recognised to be as much superior to the uneducated, who by the formlessness of their rusticity and the disorder of their untrained speech wound and even maim the purity of language guided by strict rule, and obscure the brilliance of its elegance, which is the fruit of art, as they themselves seem superior to beasts.²⁴

The beneficiaries of this elite education could write and speak in ways that could scarcely be grasped by the ordinary person in the street. An educated person's Greek would be "pure"—like that of the great Attic writers, Plato and Demosthenes. He could produce quotations from poets and readily adapt a line from Homer to any situation in a creative and witty manner. His learned and disciplined use of language was mirrored in self-controlled and measured physical movements. His refined aesthetic sensibilities were continuous with his flawless social decorum. Or at least this was the theory. Now, the *bios* tradition that accompanies members of the Platonic schools in Athens and Alexandria tells us that most of the Platonic philosophers would have had the benefit of the elite education in rhetoric and composition that formed the core of late antique *paideia*. Indeed, we find commentaries on the rhetorical works of Hermogenes that are plausibly from the hand of Syrianus, the teacher of both Hermias and Proclus. So, our Neoplatonic philos-

---

22 Baltzly (2017), pp. 263–265.
23 Baltzly (2018), pp. 33–36.
24 Diomedes, *Ars Grammatica*, 299.19–23, cited and translated in Browning (2000), p. 858.

ophers were well-acquainted with the elite education. My suggestion is that it presented itself as a model that could be adapted to understand both the nature of the grades of virtue and also the means through which such virtues were acquired.

What does the successful rhetorician or the educated person of Late Antiquity know? It is not, I submit, principally a *knowledge-that* concerning any specific subject matter. It is rather a *knowledge-how* that permits the person to do a variety of things: how to combine words in ways that different audiences will find appealing or persuasive; how to allude to canonical texts in ways fitting to the occasion at hand; how to display one's erudition to the similarly erudite in ways that escape the notice of those who are not so erudite; and so on. The content of the educated person's knowledge cannot be given any complete discursive specification, since it is ability to adapt a received body of texts and styles of writing or speaking to an unspecifiable range of social settings.

The contemplative virtues of the Neoplatonists likewise aimed at a simultaneously epistemic and moral condition that would outrun any attempt at complete discursive specification. After all, dialectic aims not at the discursive *dianoia* but rather at *noêsis*—a noetic insight that outruns any attempt to reduce it grasp of a simple body of propositions. This is simply a consequence of the nature of the objects of *noêsis:* Forms. Forms, for the Neoplatonists, form a pluralized unity in the living intelligible creature that serves as the model for the Demiurge in the *Timaeus*. This intelligible living creature not only has an organic unity but even enjoys a non-temporal life. The open-ended ability or knowledge-how that was the product of *paideia* thus forms a suitable model for Platonic *noêsis* and particularly for the higher gradations of the Platonic virtues.

There is a second parallel between the Platonic virtues and *paideia* as well. The traditional education was acquired through the study of certain canonical authors: Homer, Demosthenes, Plato and so on. Moreover, *paideia* was often demonstrated or performed through doing things with these canonical texts. An illustration from Peter Brown makes this point nicely:

> Through a shared paideia, they could set up instant communication with men who were, often, total strangers to them. They signaled, above all, that they were approachable and that they knew the rules of the game. As a "fellow servant of the Muses" no administrator could mistake a compliment or ignore a challenge posed to him as a classical reminiscence. Confronting the legal advisors of a newly arrived governor (who may have grown up in Rome), Libanius posed the crucial question: "How did Odysseus rule when he was king of Ithaca?" "Gently as a father" was the instant reply. The classical phrase set the tone for relations between the governor and town council in the months that followed.[25]

---

25 Brown (1992), p. 40.

My hypothesis is that the virtues the Neoplatonists supposed to be acquired through the study of Plato were similarly manifested in relation to Plato's text. The person who possessed the counterpart to rhetorical knowledge-how in the case of Plato could interpret or weave together Platonic texts in a satisfying manner for any occasion.

There was, however, one important point of contrast, and this difference goes to the reasons for acquiring the philosophical counterpart to *paideia*. Ordinary *paideia* was performed for others as a means of entering into the economy of favors that existed among the "friends of the Muses". Platonic *paideia*, by contrast, was performed for those who similarly identified as philosophers—with all the implications this had for one's social role and self-understanding. The Platonist philosopher was recognized by his ability to creatively improvise on themes in Plato and to bring different aspects of the dialogues into alignment with one another in ways that were sensitive to the context of the setting and occasion of the performance. This virtuosity, however, was taken by participants in this hyper-elite culture to be *the outward sign of an inward condition:* the possession of the virtues attained through the ascent through the Neoplatonic curriculum.

It is within this context that we should understand Proclus' performative argument for the inclusion of collection and division alongside Republican dialectic. A Platonic commentary was not for him what it was for Taylor or for Cornford. It is a showing as much as a saying and may sometimes involve exhibiting rather than arguing. The goal, after all, is not merely the possession of true conclusions about the meaning of Plato's dialogues. It is unity with the divine. Dialectic's pathway to the Good, according to Proclus, lies through subtraction or the stripping away of the plurality of words (*in Remp.* I 285.5–9). If the goal is a non-discursive unification with the Good, then the nature of the dialectic through which such unification is to be achieved may similarly be exhibited in a performative display that produces an intuition or immediate apprehension of a unity among Plato's various descriptions of dialectic. In rhetorical performances, content is inseparable from form and the feelings produced in the audience from their grasp of the orator's conclusion. If the Platonic literacy is a distinctively philosophical extension of rhetorical *paideia* that is directed toward the achievement of an ineffable union with the divine, Proclus' performative demonstration of the unity of Platonic dialectic is just what we should expect.[26]

---

[26] At the point at which it was no longer feasible to make changes to this paper, I had the good fortune to read a draft chapter from Corentin Tresnie on the subject of dialectic in Proclus. Corentin is completing his PhD at KU Leuven and we met during my visiting fellowship there in October of 2022. When his paper on this subject is published – as I am confident it will be – readers will be able to see for themselves just how incomplete and superficial is the picture presented here in my

Bibliography

Baltzly, Dirk (2006): "Pathways to Purification: The Cathartic Virtues in the Neoplatonic Commentary Tradition". In: Tarrant, Harold and Baltzly, Dirk (Eds.): *Reading Plato in Antiquity*. London: Duckworth, pp. 169–184.

Baltzly, Dirk (2012): "Dialectic". In: Press, Gerald Alan (Ed.): *The Continuum Companion to Plato*. London and New York: Continuum, pp. 159–161.

Baltzly, Dirk (2017): "The Human Life". In: Martijn, Marije and D'Hoine, Pieter (Eds.): *All From One: A Guide to Proclus*. Oxford and New York: Oxford University Press, pp. 258–275.

Baltzly, Dirk and Share, Michael (2018): *Hermias: On Plato's Phaedrus 227a–245e. Ancient Commentators on Aristotle*. London: Bloomsbury.

Benson, Hugh H. (2006): "Plato's Method of Dialectic". In: Benson, Hugh H. (Ed.): *A Companion to Plato*. Oxford: Blackwell Publishing, pp. 85–99.

Brown, Peter (1992). *Power and Persuasion in Late Antiquity*. Madison: University of Wisconsin Press.

Browning, Robert (2000): "Education in the Roman Empire". In: Cameron, Averil, Ward-Perkins, Bryan, and Whitby, Michael (Eds.): *The Cambridge Ancient History Volume XIV: Late Antiquity: Empire and Successors, AD 425–600*. Cambridge: Cambridge University Press, 855–883.

Cornford, Francis M. (1957): *Plato's Cosmology*. Library of the Liberal Arts. New York: Liberal Arts Press.

Dorter, Kenneth (2006): *The Transformation of Plato's Republic*. Lanham: Lexington Books.

Fine, Gail (Ed.) (2008): *The Oxford Handbook of Plato*. Oxford: Oxford University Press.

Franklin, Lee (2012): "Inventing Intermediates: Mathematical Discourse and Its Objects in Republic VII". In: *Journal of the History of Philosophy* 50. No. 4, pp. 483–506.

Layne, Danielle A. (2009): "Refutation and Double Ignorance in Proclus". In: *Epoche* 13. No. 2, pp. 347–362.

Morrow, Glenn R. and Dillon, John (1987): *Proclus: Commentary on Plato's Parmenides*. Princeton: Princeton University Press.

Steel, Carlos (1997): "Proclus et l'interprétation 'logique' du *Parménide*". In: Benakis, Linos G. (Ed.): *Néoplatonisme et Philosophie Médiévale*. Turnhout: Brepols, pp. 67–92.

Tarrant, Harold (2014): "Platonist Curricula and Their Influence". In: Remes, Pauliina and Slaveva-Griffin, Svetla (Eds.): *The Routledge Handbook of Neoplatonism*. London and New York: Routledge, pp. 15–29.

Taylor, Alfred Edward (1928): *Commentary on' Plato's Timaeus*. Oxford: Clarendon Press.

Van Den Berg, Robbert (2000): "Towards the Paternal Harbor: Proclean Theurgy and the Contemplation of the Forms". In: Steel, Carlos and Segonds, Alain-Philippe (Eds.): *Proclus et La Théologie Platonicienne*. Leuven and Paris: Leuven University Press and Les Belles Lettres, pp. 425–443.

Westerink, Leendert G. (1962): *Anonymous Prolegomena to Platonic Philosophy*. Amsterdam: North Holland Publishing.

Westernik, Leendert G. (1976): *Olympiodorus: Commentary on Plato's Phaedo*. Amsterdam: North Holland Publishing.

---

own paper. But *ars longa, vita brevis*. So let us all learn to enjoy having our best efforts superseded by those who are younger. (It's what happens anyway, so it's wise to take pleasure in the inevitable.)

Harold Tarrant
Elenchus and Syllogistic in Olympiodorus of Alexandria

1 Introduction

Olympiodorus has left us both Aristotelian and Platonic commentaries, but he was the admired leader of an Alexandrian school that has otherwise left us only the latter in spite of a widespread admiration for Plato. It is therefore natural to be asking whether the term "Neoplatonist" is a satisfactory appellation either for the school or for this one individual. In his case, my impression is that Plato progressively became more central to his endeavors, but uncertainties regarding the internal chronology of his works certainly cloud the picture. Here I shall confine myself to matters of logic, where, if anywhere, one might expect Aristotle to win greater honors, especially for his contribution to syllogistic. However, Platonists were reluctant to concede that Plato lacked knowledge of syllogistic patterns, of sophistic tricks, or of the categories. This can clearly be seen in Alcinous' treatment of Platonic logic in *Didaskalikos* 6, where syllogisms, including hypothetical syllogisms more typical of the Stoa, are detected in demonstrations and refutations from a variety of works, patterns of sophistic argument are exposed in the *Euthydemus*, categories are employed in the *Parmenides*, and etymology is explained in the *Cratylus*. When we come to Proclus, one finds him idealizing Plato's arguments in the Parmenides in particular, always insisting that Plato *never* produces a work intended to teach logic in isolation but rather selects his methods in accordance with the needs of the principal topic.[1]

The Alexandrian school clearly devoted considerable attention to the teaching of Aristotle. The logical works took precedence and a familiarity with Aristotle was assumed prior to the teaching of Plato, but that did not of course make Plato Aristotle's inferior. The *Prolegomena Philosophiae* of both David and Elias discuss various concepts of philosophy, seemingly valuing those associated with Plato at least

---

[1] See Proclus, *in Parmenidem* 637.5 ff.: "Nowhere do we find Plato producing a work which is principally a study of method; but rather we find him employing different methods at different times according to what each subject requires, and always adopting his method for the sake of his object of inquiry. Thus in the *Sophist* he brings in the method of Division not in order to teach his hearers Division (though this is an incidental result), but in order to catch and bind the many-headed sophist. ... A method is a necessary means when we want to exercise it in gaining knowledge of things, but not worthy of attention for its own sake" (trans. Morrow and Dillon).

as highly as those associated with Aristotle.² So, far from being frowned on, Olympiodorus' teaching of Plato was admired by followers in the school. For instance, on three occasions, David's *Prolegomena* refer respectfully to Olympiodorus,³ twice calling him "the philosopher"; Elias does not refer to him, but his own *Prolegomena* (14.9–10) quote verses about Plato's salutary influence that David said were composed by Olympiodorus (32.1–2).⁴

In fact, Plato's works were seen as the crowning element in philosophical education, studied by those who had already had a grounding in Aristotle, and particularly in Aristotelian logic. At the beginning of both the anonymous *Prolegomena to Plato's Philosophy* (1.1–12) and Olympiodorus' *Commentary of the First Alcibiades*,⁵ the beginning of Aristotle's *Metaphysics* is quoted for its remarks about everybody desiring knowledge, and the claim is then made that all normal⁶ men desire the philosophy of Plato *in particular*.⁷ Understanding what Plato has to offer remains the pinnacle of philosophic inquiry, as it had been in Proclus' Athenian school. Consequently Plato, like Aristotle, had to be treated not only as somebody who understood the rules of argument well, but also as a philosopher offering even greater insights through the arguments that he employed. The signs, however, are that Olympiodorus became increasingly aware of what is distinctive about Platonic argument.

---

2 Elias, lecture 5, and David, lectures 7–8. Both sets of *Prolegomena* are now conveniently translated with notes in Gertz (2018).
3 See *Prolegomena* 16.3, 31.34, and 63.32.
4 The titles of Olympiodorus' commentaries on *Alcibiades* and *Gorgias* also refer to him as "the great philosopher". Olympiodorus had himself used the term several times when referring to Ammonius in the late commentary on Aristotle's *Meteorologica*, and Damascius refers similarly to Proclus in *Doubts and Solutions*.
5 For a recent translation into English, with notes, of *in Alc.* see Griffin (2015) and (2016), of which the former is relevant here. For recent translations into French and Italian of anon. *Proleg.*, see Westerink, Trouillard and Segonds (1990) and Motta (2014); Filippi (2017) translates both works into Italian.
6 Anon. *Proleg.* 1.8–12 excepts those individuals whose cognitive powers are maimed and are thus unable to view intelligible truths, rather like bats who cannot look upon the light of the sun.
7 See Olympiodorus *in Alc.* 1.3–9: "Aristotle at the start of his own *Metaphysics* claims that 'All people naturally desire to know', and a sign of this is our delight in the senses, as this is the reason we delight in the senses. I, however, would make this greater claim when introducing Plato's philosophy: that all people desire Plato's philosophy, wanting to draw off benefit from it, eager to be in possession of the flow from it and filling themselves full of Plato's inspirations". On this passage and its sequel, see also Tarrant (2021b).

2 Olympiodorus' Development

In the supposedly immature *Commentary on Plato's Gorgias*,[8] Olympiodorus recasts several of Socrates' arguments in syllogistic form, often noticing whether the first, second, or third figure syllogisms are being employed. The impression is left of a teacher who wishes to reinforce the teaching of Aristotelian syllogistic, while simultaneously showing that Plato's (or Socrates') theses are grounded in formally valid arguments. There is still much discussion of syllogisms in the *Commentary on Plato's First Alcibiades*,[9] though the same feature can be observed in Proclus' commentary on that work, principally because an important predecessor, possibly Iamblichus, had divided the work according to ten syllogisms that bear the brunt of the argument. This meant that neither Proclus nor Olympiodorus could tackle the *Alcibiades* without attention to how the overall argument was constructed, and to its ability to be understood as an interconnected series of deductions.[10] It may therefore be a surprise to find that the *Commentary on Plato's Phaedo* (10.3), in talking about the syllogistic path of the argument, claims that Plato calls it a *mythos* at 70b6 (διαμυθολογῶμεν). Olympiodorus is seemingly using this as a jibe against the value that the Peripatetics placed upon their syllogistic schemata. Had his commitment to syllogistic changed over time?

It would be no surprise if it had done so. I have recently demonstrated that the terminology of proof and argument undergoes some changes between the supposedly early *Gorgias*-commentary and the late commentary on the *Meteorology* dating from no earlier than 565 CE.[11] I did not include the terminology of elenchus there, first because I expected it to skew the results because it was naturally used more frequently in works involving "Socrates" in his elenctic mode,[12] but secondly because Renaud had already published a significant study of Olympiodoran

---

[8] Its early date (possibly in the late 520s) is generally agreed on the basis on its relative dependence on his own teacher Ammonius rather than on such figures as Proclus and Damascius, but certainly not yet proven. This work is translated into English in Jackson, Lycos, and Tarrant (1998); a recent book on it is Bohle (2020).
[9] The work probably postdates AD 546 if 2.79 alludes to Hephaestus the governor of Alexandria, who arrived in that year; see Watts (2006), p. 254, and Griffin (2015), pp. 166–167. Olympiodorus' career continued for another two decades at least.
[10] On this, see Tarrant (2021a), pp. 218–219.
[11] Olympiodorus refers to the appearance of a comet in March/April of 565 at *in Meteor.* 52.31. On the chronology of Olympiodorus' works see Opsomer (2010) and now Tarrant (2021a).
[12] Olympiodorus was not interested in distinguishing between earlier and later works involving two different styles of Socrates, but he would nevertheless have distinguished arguments that were intended to undermine the views of others from those which were intended to give credence to the speaker's own views.

elenchus. Admittedly, that study had observed that, for Olympiodorus, "from a formal point of view, there is nothing to distinguish Socratic elenchus from philosophic argument in general". It also saw elenchus and midwifery as "two complementary functions of Socratic dialectic".[13] But this still allows that the terminology of elenchus should be found very much more in texts where "Socrates" speaks. The nature of the original texts being commented on would contribute to determining whether this vocabulary was to be found.

The result of my previous study, which had excluded elenchus, was as follows: the terminology of syllogistic and demonstration became less frequent in later works, and that of constructing arguments[14] became more common there. This does not permit one to construct a timeline of Olympiodorus' major works with any great confidence, though the *Phaedo*-commentary was in general statistically closer to the *Meteorology*-commentary that to any other work. Though Olympiodorus seems to have remained confident of what could be learned through argument, the shift in terminology might easily reflect a similar shift away from the certainties associated with Aristotelian demonstration towards greater openness to contrary arguments and to reassessment such as works like the *Gorgias* and *Phaedo* show.[15]

3 Platonic Versus Aristotelian Logic

In his *Prolegomena to Logic*, we find the following passage, which goes some way toward explaining Olympiodorus' apparent preference for Platonic argument:

> Olympiodorus, *Prolegomena* 17.37–18.6:
>
> καὶ ἄξιον θαυμάσαι τόν τε Πλάτωνα καὶ Ἀριστοτέλη, καὶ τοῦ μὲν Ἀριστοτέλους ὡς χωρίσαντος καὶ ἐφευρόντος ἄνευ πράγματος τοὺς κανόνας, Πλάτωνος δὲ χωρὶς κανόνων κεχρημένου τῇ ἀποδείξει· οἱ γὰρ παλαιοὶ ἀποδεῖξαι μὲν ᾔδεισαν, ἀπόδειξιν δὲ ποιῆσαι ... οὐκ ᾔδεισαν, ταὐτὸ πάσχοντες τοῖς κεχρημένοις μὲν ὑποδήμασι, σκυτοτομεῖν δὲ ἀγνοοῦσι. καὶ οὐ διὰ τοῦτο χείρονα τὸν Πλάτωνα τοῦ Ἀριστοτέλους οἰητέον, ἀλλὰ τὸ ἐναντίον καὶ κρείττονα· οὐδὲ γὰρ ἀποδεικνὺς ἐκεῖνος τῆς Ἀριστοτέλους ἀποδεικτικῆς μεθόδου ἐδεήθη, ἀλλὰ τοὐναντίον Ἀριστοτέλης τῆς Πλάτωνος ἀποδείξεως.
>
> One should admire both Plato and Aristotle, Aristotle because he separated the rules off and discovered them in isolation from the matter [itself], and Plato because he employed demon-

---

[13] Renaud (2014), pp. 120 and 126.
[14] The relevant verb is κατασκευάζειν and the noun ἐπιχείρημα (ἐπιχείρησις being rather rare); all terminology in συλλογι- other than the meteorological term συλλογίμαῖα was taken into consideration, as was all cases of the noun ἀπόδειξις.
[15] I think of such passages as *Grg.* 509a and *Phd.* 107a–b.

stration without [treating] rules. For the ancients knew how to demonstrate but did not know how to construct a demonstration, having the same experience as those who use footwear, but are ignorant of shoe-making. And one should not think Plato worse than Aristotle on this account, but the opposite—actually better. For when demonstrating he had no need for the demonstrative method of Aristotle; on the contrary, Aristotle needed Plato's demonstration.

Olympiodorus goes on to compare the independence of Homer from Aristotle's *Poetics* or that of Demosthenes[16] from Hermogenes' writings on the rhetorical art. It would, I believe, be typical of Platonism to believe that there were certain gifted persons endowed with some divine portion (*theia moira*),[17] who had no need for any science rules in their sphere of competence. And if one worries that this might seem to make Plato *unscientifically* rational, then one might reflect on Hermias' statement that "every rational soul really does partake of a divine portion".

4 The Case of Elenchus

The terminology that I had not included in my previous study was that of *elenchus*, for it was obvious that it was going to have a much greater role to play in some commentaries than in others. Vlastos drew on the *Gorgias* in particular when trying to describe the rules that applied to Socratic elenchus,[18] for there is much discussion between Socrates and Polus on this subject. Hence there are five cases where the terminology of elenchus appears in the lemmata of the *Commentary on Plato's Gorgias* (11.9 [x2], 13.8, 18.6, 19.14). The frequency of such terminology in Plato encourages a high frequency in the commentary too. It was also to be expected that the *Commentary on Plato's First Alcibiades* would have much to say about elenchus, even though the term was not used in the Platonic text, and this is partly because Olympiodorus divided the dialogue into three parts, of which the first and largest part was treated as elenctic, while the second was protreptic and the last maieutic (*in Alc.* 11.7–23, 57.5, 142.9). There are also historical considerations: even though Proclus had a more integrated notion of the part played by elenchus in the dialogue, so that elenctic and maieutic can at times be blended

---

16 Note that Olympiodorus admired Demosthenes' rhetoric and regarded him as a pupil of Plato (*in Gorg.* 41.10).
17 See Pl. *Ap.* 33c, *Phdr.* 230a, 244c, *Meno* 99e, 100b, *Ion* 535a–542a, etc.; later Platonists seem less inclined to use the phrase themselves than to quote Platonic passages that use it (Plut. *Mor.* 499b [cf. *Phd.* 58e], 1119b [cf. *Phdr.* 230a], Proc. *in Remp.* 1.184.21–25 [cf. *Ion* 534c]).
18 Vlastos (1983).

(Proc. *in Alc.* 209.11), Proclus too had used the vocabulary of elenchus some 94 times in what survives of his commentary.[19]

Olympiodorus' use of the vocabulary of elenchus conformed broadly with my expectations. It was found more in works that we suppose to be employing the elenctic "Socrates" and less in those that do not. There were no instances of such vocabulary in the *Commentary on Plato's Phaedo*, and rather few in the *Commentary on Aristotle's Categories*, but an intermediate number in that on the *Meteorologica* (see tab. 1):

Commentary	Total words	Elenchus-words	Rate per 10000 words
in Categ.	50220	11	2.19
in Meteor.	113456	66	5.82
in Alc.	50719	74	14.59
in Grg.	64789	75	11.58
in Phd.	19110	0	0

Tab. 1: Rates of elenchus-vocabulary in Olympiodorus' commentaries

It has long been appreciated that Olympiodorus' concept of elenchus in the Platonic commentaries where it occurs is especially appropriate to *Socratic* elenchus.[20] The object is not simply to expose a flaw in the *thinking* of the interlocutor, but to expose something more like a character-trait as well. It is remarkable that in the course of two lines[21] the flaw through which Gorgias succumbed to Socrates' elenchus was described first as a "distorted view" (*diastrophos doxa*) and then as a "distorted character" (*diastrophon ēthos*), strongly suggesting the double role of Socratic elenchus. Even if it would seem to be illegitimate to draw a sharp distinction between intellectual and moral weakness when thinking in a Socratic mode, Vlastos was right to note the importance of *moral* beliefs for the Socratic elenchus. It is the interlocutor himself who is "being exposed", not just a flaw in the way that he has argued. Thus, Olympiodorus can find in Socrates' apparent praise of Alcibiades' qualities at the beginning of the dialogue both censures (*psogoi*) and attempts to expose (*elenkhoi*); thus, his unmixed *elenkhoi* later will be preceded here by blended ones that are less intimidating (*in Alc.* 29.17–20).

---

19 My figures have 14.59 cases per 10,000 words in Olympiodorus' commentary to 13.67 in Proclus'; one must acknowledge that none of Proclus' surviving commentary goes beyond Olympiodorus' elenctic part of the dialogue.
20 See Jackson, Lycos and Tarrant (1998), p. 10, and Renaud (2014).
21 *in Gorg.* 27.2 (Westerink, pp. 146.12–13).

The cases of apparent praise with underlying criticism are enumerated between 30.1 and 32.6 in the *theôria*,[22] but a revealing passage follows, in which Olympiodorus feels the need to explain Alcibiades' confidence and pride in what were merely apparent goods—how good his looks, his family, his friends, his guardian and his wealth were. Though all this entails false beliefs, Olympiodorus insists that every false belief originates in a true one, and has its basis in that, being dependent upon it: "for the truth, through its superabundance of power, colors even the falsehood that is its own contrary, nor can a complete eclipse of the common notions arise".[23] So, the false beliefs entertained by Alcibiades have their roots in intelligible beauty itself, in respect for the nobler cause, in the unity of the one itself, and in the guardian *daemon*, things that Alcibiades has a dim awareness of. Thus, Renaud could speak of Socrates entreating "the young man passionately to recognize the ultimate object of his desire".[24] So, Olympiodorus' concept of Socratic elenchus here is closely connected with the "common notions" (*in Alc.* 32.11), seen as what survives within us of our earlier discarnate vision of transcendent truth. The underlying role played by the common notions in the *Commentary on Plato's Gorgias* has also been emphasized,[25] for the interlocutors were there said to differ in accordance with their ability to grasp common notions accurately.

Given that in these two Platonic commentaries the interconnected nature of the moral and intellectual faults exposed by Socrates is given some emphasis, one may wonder what role elenchus could have in the Aristotelian commentaries. Furthermore, it seems that personal interaction was essential to the process as usually conceived, and, given that Renaud was able to discuss not only Socrates' elenctic strategies as a questioner, but also the requirements for the interlocutor,[26] it is clear that Olympiodorus' focus had been on a concept of elenchus that took place within a dialogic setting. If Aristotelian works gave rise to discussions of elenchus, it would have to be elenchus of a rather different kind. Who would be the subject of this practice there, and whose beliefs are exposed?

There are four cases of such vocabulary in the *Prolegomena to Logic* (4.4, 4.15, 4.19 and 15.10). They involve Platonic, Aristotelian and Galenic arguments against Protagorean skepticism, and an argument against a Stoic position. The meaning comes close here to "refutation", and indeed Gertz translates *elenkhein* in this way, as he does also for *exelenkhein*, which is a more explicit term for "refute"

---

22 See also the *lexis* at 35.1–37.3.
23 See 32.9–11; the translation is my own, though influenced by that of Griffin (2015).
24 Renaud (2014), p. 120.
25 Jackson, Lycos and Tarrant (1998), pp. 10, 42–43, 163–164, etc., and Tarrant (1997); cf. Renaud (2014), p. 120 on the *in Alc.*
26 See Renaud (2014), pp. 120–122.

but rare in Olympiodorus.[27] There are just six cases where Olympiodorus finds a use for the vocabulary of elenchus in the following *Commentary on Aristotle's Categories*,[28] and while the first two discuss how Aristotle would have been open to refutation if he had expressed himself in a different way, in the remaining cases it is Olympiodorus himself who employs the elenchus, using the first person plural to refute (a) those who say that substance and accident are the only two real categories, (b) those who say that all things are relative, and (c) those who say that nothing is relative. It is as if the Alexandrian lecturer were himself in conversation with positions that his auditors might be attracted towards, but it would still be strange to think of such uses as properly "dialogic".

In the *Commentary on Aristotle's Meteorologica* there are fifty occurrences of this terminology, mainly in close clusters.[29] Hence there are three cases on p. 21, thirteen on pp. 52–59, ten on pp. 67–73, eight on pp. 127–128, and eleven on pp. 148–156; this amounts to 95% of all occurrences. The final instance in the 338 pages of the Greek text of this commentary occurs two thirds of the way through the treatment of Book II on p. 175, meaning that all such vocabulary occurs over little more than half of the complete Olympiodoran text. While there is one case of the compound *dielenkhein* on p. 214, all thirteen remaining cases where the verb is prefaced by the strengthening prefix *dia-* occur between pp. 7 and 151.[30] So, at first sight one may wonder whether the disappearance of elenchus terminology indicates that the later pages of the commentary were simply written later.

There does seem to be some important factor at play that has almost confined the *elenchus* terminology to the first half, for this is not the only case of terminology becoming rare in the later pages. Another case where a word nearly disappears is the didactic *isteon* ("one must know"), which had been very common in the early *Commentary on Plato's Gorgias*. This appears fifteen times on pp. 1–184

---

27 Gertz (2018), pp. 234 and 238; the only other case occurs in *in Gorg.*
28 32.13, 18, 55.3, 98.9, 99.14 and 15.
29 We lack a complete study of this commentary; Viano (2006), in a wide-ranging book, tackles that on Book IV of the *Meteorologica*. A related text, discussed in Viano (2021), in an alchemical work *mentioning* Olympiodorus' name in the title. While I doubt that Olympiodorus is anywhere responsible for the choice of diction, since I cannot find there traces of certain expressions that I have come to regard as typical of Olympiodorus, Viano (2021) argues that the title reference to Olympiodorus' *Commentary on On Action* refers only to part of the work (1–7 and parts of 18–27), where the compiler can nevertheless add bits from elsewhere, returning to "copy and/or paraphrase other parts from it" (Viano 2021, pp. 20–21).
30 Note that elsewhere in Olympiodorus, *dielenkhein* is rare, being found twice in *in Categ.*, three times in *in Alc.*, and just one in *in Grg.*, so that it seems to have increased in frequency towards the end of his life.

(all within the commentary on Books I-II) but only three times thereafter. Again, all five references to Ammonius by name occur in the commentary on Book I,[31] whereas he is instead referred to as "the great philosopher" in commentary on the later books.[32] Again, another term for an argument, *epikheirhma*, which appears 146 times up to p. 184, if found only 24 times from p. 185 on (about five times more often per page in the earlier material). In this case it seems to be a late term that decreases in frequency in the later pages, for it had been rare in the Commentary on Plato's Gorgias.

This last point is not the only one that makes a chronological interpretation of the differences, making the vocabulary of the commentary on Books I-II significantly earlier in date, seem strained. It was in the earlier pages at 52.31 that we find the information that establishes a very late date for the commentary, and, if the books of the original had been commented on in their established order, one would have to postulate a later *partial* revision, adding material but leaving the vocabulary little touched. A further point against a chronological interpretation is that, however much the original *Meteorologica* itself might seem such that its books could easily have been compiled in a different order, our surviving *Commentary on Aristotle's Meteorologica* seems to be a single course, given over 51 weeks. It is divided into 51 sessions (πράξεις) rather like the *Commentary on Plato's Gorgias* (50 plus an introduction of comparable length).[33] Hence the chronological interpretation of the linguistic differences in Books III and IV is improbable, and a more satisfactory explanation should be found.

Alternative explanations might appeal to Olympiodorus' sources, his recorder, or just the details of the original material being commented on. Among his sources

---

[31] See 6.23, 51.29, 69.17, 75.25, 118.13; the second and fifth of these also describe Ammonius as "the great philosopher".

[32] See Book II, 153.7, 174.15, 175.14/19, and 188.2–3; Book III, 239.13–14, 241.19, 245.3, 255.23/25, 260.25, 263.20, and 270.16, and Book IV, 298.21, 313.24, 322.25, and 332.6.

[33] One has to acknowledge that there are only 28 lectures on the *Alcibiades* plus a biography of Plato roughly equivalent to one lecture, leaving us 22 lectures short of the figure of 51. Perhaps the lecturer had earlier treated other introductory *topoi*, equivalent to 10 divisions of the anonymous *Prolegomena*, but tackled in greater depth. Again, the *Categories*-commentary plus *Prolegomena* was only 34 lectures. However, the surviving 13 lectures of the *Phaedo*-commentary cover just under 20 Stephanus pages, while the length of the original *Phaedo* was about 61 pages. If the amount treated had been roughly the same in each lecture then the complete commentary would have extended to 40 ordinary lectures, plus additional introductory material and an introduction to the myth (equivalent to lecture 44 in the *Gorgias*-commentary). Given that the most difficult material in the *Phaedo* will be tackled after the extant lectures, and that 107c–115a, expounding Socrates' myth-like vision, is comparable to *Gorgias* 521d–524d (6 lectures of the *in Gorg.!*), there is a good chance that the *Phaedo* would also have needed 51 sessions.

the one that stands out repeatedly is his teacher Ammonius. Ammonius' work on the *Meteorologica* is not extant, but Philoponus' *Commentary on Aristotle's Meteorologica Book I* is a late revision of his transcript of Ammonius' lectures on that text,[34] though we cannot determine whether Philoponus heard him lecture on Books II-IV as well. Ammonius was referred to most frequently in Book III (seven times) and least frequently in Book IV (four times), so his influence seems fairly evenly spread. Alexander of Aphrodisias' commentary on *Meteorologica* is also referred to quite often, but terms for argument there are rare, and the vocabulary of elenchus occurs only once,[35] and it seems irrelevant to Olympiodorus' choice of vocabulary. There is a single reference to Proclus' *Commentary on Plato's Timaeus* (266.37),[36] but it may be incorrect and is in any case irrelevant for our purposes.

While I doubt that the identity of Olympiodorus' recorder would have changed during this course, it is possible that the practices of the recorder did. Assuming that they were hard pressed to keep up with what is being said by the lecturer, recorders might possibly decide to drop unnecessary recurrent words. Hence *isteon hoti* ("one must know that") might easily be dropped after a while as an uninformative mannerism, while Ammonius' name was also dropped from the phrase "Ammonius the great philosopher". But it would be much more difficult to explain why the terminology of elenchus should be omitted regularly if it had been used by the lecturer.

Looking more closely at the relationship of the commentary to the original text we find that in most instances Olympiodorus refers to elenchus when he speaks of Aristotle's attempts to refute the opinions or hypotheses of his predecessors. Given that the usual meaning is something like an "attempted refutation", it is scarcely

---

34 See Verrycken (2010), p. 733.
35 On p. 50, still in *Meteor.*, Book I.
36 As noted by Viano (2006), p. 235, this seems to relate to *in Tim.* 1.43.2–17 on *Tim.* 18b1–3, which reminds us that the guardians are not allowed to possess gold, silver, etc. Proclus is alleged to have related seven metals, lead, "electron" alloy, iron, copper, gold, tin, silver, to the seven planets Saturn, Jupiter, Mars, Venus, Sun, Mercury and Moon respectively, but the original passage mentions only Sun, Moon, Saturn and Ares. Outside Olympiodorus, the full set of seven corresponding metals is given only by a scholiast on Pindar, *Isthmian* 5 (*schol.vet.* 2b): ἑκάστῳ δὲ τῶν ἀστέρων ὕλη τις ἀνάγεται· καὶ Ἡλίῳ μὲν ὁ χρυσός, Σελήνῃ δὲ ὁ ἄργυρος, Ἄρεϊ σίδηρος, Κρόνῳ μόλιβδος, Διῒ ἤλεκτρος, Ἑρμῇ κασσίτερος, Ἀφροδίτῃ χαλκός. This system differs from an account in Celsus (Origen, *Cels.* 6.22.5–20) that assigns different metals to Mercury, Venus, Mars and Jupiter, citing the authority of Mithraic rites. Tarrant (2007), p. 136, n. 190, shows that the details were irrelevant to Proclus, and indeed that he criticizes those who promote it as "sight-lovers". Hence Olympiodorus must have known the full system from elsewhere, possibly from Asclepiodotus' *Timaeus*-commentary to which he also refers (321.28), possibly from alchemical writings.

surprising that the terminology disappears from view in Olympiodorus' discussion of Books III and IV of the *Meteorologica*, since there is very little material about the author's predecessors there, whereas Book I, Chapters 6 and 8, are rich in such material, and Book II, Chapters 2, 3 and 7, discuss the *Phaedo*, Democritus, Empedocles, Anaximenes, and Anaxagoras. This also explains the presence, up to p. 151, of fifteen of the sixteen cases of the stronger verb *dielenkhein* that implies the success of the refutation,[37] mostly involving the same contexts.

While the usual sense of *elenkhein* in this commentary involves subjecting a theory to criticism with the goal of refuting it, we should note here that six instances of the passive verb *elenkhetai* between 128.21 and 131.23 are exceptional[38] and show how a particular phenomenon is most easily "exposed" against a contrasting background—motion against a stationary one, white against a black one. I should also translate the passive at 175.17 in this fashion: "For Aristotle is never *exposed* [or *caught out*] as saying this". So, it is not quite the case that the terminology could only be used in relation to passages where Aristotle tries to refute others or is refuted by them, and there is real doubt whether *elenchus* in these cases should be considered a term for a kind of argument, even though that is usually the case. However, I would argue that it is never "Socratic" in so far as it does not involve an attempt to expose or correct any moral weakness alongside the erroneous beliefs. Indeed, in this commentary it tends to be theory rather than earlier theorists that are subject to elenchus. Another word for arguing against a theory, *anaskeuazein*, had been encountered six times in the early *Commentary on Plato's Gorgias* and once in the *Commentary on Aristotle's Categories*, which may well be fairly early too. Yet it had been dropped in Olympiodorus' later works, perhaps leaving him with little choice but to use *elenkhein* to cover such cases.

Other terms that Olympiodorus favors in the late commentary on Aristotle's *Meteorologica* are *epikheirhma* for an argument that one constructs and *kataskeuazein* for constructing an argument in favor of a theory. He refers to syllogistic only five times and to demonstration (*apodeixis*) only fifteen, much less often than in early works. By this date the terms that Olympiodorus favors are those that avoid passing final judgement on the success or failure of the argument, leaving

---

[37] The single instance thereafter is at 214.30, where others try to refute (or thoroughly expose) Aristotle—unsuccessfully in Olympiodorus' view.

[38] It is interesting to hear from Chiara Militello, who is translating the commentary on Books II and III for *The Ancient Commentators on Aristotle* series, that she too believes that a different sense from usual is employed in this passage, possibly involving the perception of something's being different from something else.

more to the student's own assessment both of the premises and the formal validity of the argument.

5 Conclusion

The changing preference, over Olympiodorus' lifetime, for terms referring to arguments can be explained in various ways. It might just be that he has become more skeptical in later life about various issues, perhaps as a result of many years instructing not only those who are firmly wedded to the Greek culture that Byzantine Alexandria was leaving further behind, but also many persons of Christian persuasion. This may not entail deep doubts about his earlier teaching, but rather enhance his awareness of other valid points of view. However, I prefer to explain the changes educationally, as indicating an increasing belief that the student needs to hear not single proofs but collections of arguments. Employing the whole armory of arguments will allow the listener to use personal judgement about the strengths of the various arguments and the truths that they may collectively point to. A paradigm for such a practice is present in the discussion of the legitimacy of suicide in the first extant lecture of the *Commentary on Plato's Phaedo*. In this respect it seems to me that Olympiodorus becomes progressively Platonist in his attitude. Perhaps it was natural, given these increasingly Platonist leanings, to see even Aristotle as employing the techniques of elenchus against the continuing influence of his Presocratic predecessors.

This contribution began by proposing to examine Olympiodoran use of the terminology of elenchus, with a special focus on commentaries where one would not think that a particularly Socratic type of argument needed to be referred to. Given the long history of these terms in rhetorical literature it is not a huge surprise that a significant number of occurrences do occur in the Aristotelian commentaries, mainly in the earlier part of the commentary on the *Meteorologica*. I very much hope that this might provoke further study of the language and content of this commentary, to see how far other such internal differences occur. If they are commoner than I have observed, then we shall have to ask whether this is due to the lecturer's own shifting practices, to changes of source material, or to changes in the recorder's practice before the end of the commentary on *Meterologica*, Book II.

Bibliography

Bohle, Bettina (2020): *Olympiodors Kommentar zu Platons Gorgias.* (*Stud. z. Lit. u. Erkenntnis, 11*). Heidelberg: Winter.

Filippi, Francesca (Ed. and Trans.) (2017): *Olimpiodoro d'Alessandria: Tutti i Commentari a Platone*. Vols. I and II. Sankt Augustin: Academia Verlag.
Gertz, Sebastian (Trans.) (2018): *Elias and David: Introductions to Philosophy with Olympiodorus: Introduction to Logic. Ancient Commentators on Aristotle*. London: Bloomsbury.
Griffin, Michael (Trans.) (2015): *Olympiodorus: Life of Plato and on Plato's First Alcibiades 1–9. Ancient Commentators on Aristotle*. London: Bloomsbury.
Griffin, Michael (Trans.) (2016): *Olympiodorus: On Plato's First Alcibiades 10–28. Ancient Commentators on Aristotle*. London: Bloomsbury.
Jackson, Robin, Lycos, Kimon, Tarrant, Harold (Trans.) (1998): *Olympiodorus: On Plato's Gorgias*. Leiden: Brill.
Joose, Albert (Ed.) (2021): *Olympiodorus of Alexandria: Exegete, Teacher, Platonic Philosopher*. Leiden: Brill
Militello, Chiara (Trans.) (forthcoming): *Olympiodorus: Commentary on Aristotle Meteorologica Books 2 and 3. Ancient Commentators on Aristotle*. London: Bloomsbury.
Morrow, Glenn and Dillon, John (Trans.) (1987): *Proclus' Commentary on Plato's Parmenides*. Princeton: Princeton University Press.
Motta, Anna (Trans.) (2014): *Prolegomeni alla filosofia di Platone*. Rome: Armando Editore.
Opsomer, Jan (2010): "Olympiodorus". In: Gerson, Lloyd P. (Ed.): *The Cambridge History of Philosophy in Late Antiquity*. Cambridge: Cambridge University Press, pp. 697–710.
Renaud, François (2014). "The Elenctic Strategies of Socrates: The *Alcibiades I* and the Commentary of Olympiodorus". In: Layne, Danielle A., and Tarrant, Harold (Eds.): *The Neoplatonic Socrates*. Philadelphia: University of Pennsylvania Press, pp. 118–126.
Renaud, François and Tarrant, Harold (2015): *The Platonic Alcibiades I: the Dialogue and its Ancient Reception*. Cambridge: Cambridge University Press.
Tarrant, Harold (Trans.) (2007): *Proclus: Commentary on Plato's Timaeus: Vol. I, Proclus on the Socratic State and on Atlantis*. Cambridge: Cambridge University Press.
Tarrant, Harold (2021a): "Formal argument and Olympiodorus' development as a Plato-commentator". In: *History of Philosophy and Logical Analysis* 24, pp. 210–241.
Tarrant, Harold (2021b): "Special kinds of Platonic discourse: Does Olympiodorus have a new approach?". In: Joose, Albert (Ed.): *Olympiodorus of Alexandria: Exegete, Teacher, Platonic Philosopher*. Leiden: Brill, pp. 206–220.
Verrycken, Koenraad (2010): "John Philoponus". In: Gerson, Lloyd P. (Ed.): *The Cambridge History of Philosophy in Late Antiquity*. Cambridge: Cambridge University Press, pp. 733–755.
Viano, Cristina (2006): *La matière des choses: le livre IV des Météorologiques d'Aristôte et son interpretation par Olympiodore*. Paris: Vrin.
Viano, Cristina (2021): "Olympiodorus and Greco-Alexandrian Alchemy". In: Joose, Albert (Ed.): *Olympiodorus of Alexandria: Exegete, Teacher, Platonic Philosopher*. Leiden: Brill, pp. 14–30.
Vlastos, Geoffrey (1983): "The Socratic Elenchus". In: *Oxford Studies in Ancient Philosophy* 1, pp. 27–58.
Watts, Edward Jay (2006): *City and School in Late Antique Athens and Alexandria*. Berkeley and Los Angeles: University of California Press.
Westerink, Leendert G. and Segonds, Alain-Phillippe (1990): *Prolégomènes à la philosophie de Platon*. Jean Trouillard (Trans.). Paris: Les Belles Lettres.

Han Baltussen
Simplicius and Aristotle's Dialectic

1 Preface

The reception of Aristotle's dialectical method as expounded in the *Topics* is not easy to trace. Among the writings of his immediate successor, Theophrastus, we find a work on topics (τὰ πρὸ τῶν τόπων α´, D.L. 5.50[1]) and many other titles that point to continuing study of dialectic (and some criticisms) in the first generation after the school's founder.[2] But the work is the only one with a reference to τοποί in the title, and its form suggests that he reduced the material in Aristotle's long work to a shortened version or introductory work, as did his successor Strato (D.L. 5. 59, τόπων προοίμια). This suspicion of the diminishing theoretical interest in dialectic (and hence its importance) is perhaps confirmed by the much later remark that Theophrastus attempted "to find a universal method" with regard to the four predicables in the *Topics*, and subsumed specific property and genus under "definition", but "separated accident from the rest ... and made the subject less clear".[3]

After Theophrastus and Strato, the practice seems to leave no clear traces until we come to the 1st century, when Cicero claims to have written a *Topica* and implies that it has a connection to Aristotle's *Topics* (Cic. *Topica* 1). But on closer inspection, this work turns out to be neither a "translation" nor even closely connected to the Aristotelian text, except perhaps in a very broad sense.[4] In the 2nd century CE, Alexander of Aphrodisias considered it a worthwhile pastime to study the work,

---

[1] An identical title for Aristotle can be found at D.L. 5.24.
[2] On criticism, see Huby (1989), pp. 71–72. In the D.L. catalogue, we find: "definitions", 5.43; "theses", "on affirmation and denial", 5.44; "first propositions" in eighteen books, "collection of problems", "sophisms", "resolution of syllogisms", 5.45; "on differences", 5.46; "political, ethical, physical, erotic problems"; "theses", 5.49. For the remaining fragments see FHS&G (1992) Vol. I, Nos. 118–121 (Definition), 122–136 (Topics). Note that Simplicius also mentions that "some wrote a book τὰ πρὸ τῶν τόπων" (*in Cat.* 15.28). We may compare the remark at 16.14, where, discussing the ordering of logical works by Adrastos of Aphrodisias (*Topics* before *Categories*), he disagrees with the choice of title: ἄτοπος δὲ ὄντως ἡ πρὸ τῶν τοπίκων ἐπιγραφὴ τοῦ βιβλίου.
[3] See 124 A FHS&G (= Alex. Aphrod. *In Top.* 102b27, Wallies, pp. 55.24–7; full reference in n. 5). For the diminishing interest, cf. Gottschalk (1990), pp. 69–70.
[4] One of his last works written around 44 BCE, Cicero mentions he had a copy of "a certain *Topics* of Aristotle" (*Aristotelis Topica quaedam*) in his library at Tusculum. On the problematic passage and the relation between the works, see Huby (1989). On the long dialectical tradition, see now Spranzi (2011).

as his detailed commentary on the treatise shows.[5] But beyond this commentary, we find little that could be considered a study or imitation of the *Topics*, even if we may assume that philosophers were familiar with the work or at least knew of its existence.[6] But no new commentary of the same scope as Alexander's is known before medieval times. In the 3rd and 4th centuries, Plotinus (c. 205–c. 270 CE) and his followers used Alexander's commentaries (*VPlot.* 14), and dialectic would have been a staple of the education of philosophers anyway. Still, his student and biographer Porphyry (c. 235–c. 305 CE) had ensured that Aristotle's works would become a fixed part of the late Platonist curriculum, which developed into a set program of reading Plato and Aristotle.[7] Thus, dialectic became, so to speak, a "submerged skill" that did not always leave a lot of explicit traces.

The focus of this chapter is one aspect of Aristotle's dialectic which has been under-explored until recently and may throw some light on the approach of the late Platonist philosopher and scholar Simplicius (c. 480–c. 540 CE), in particular his Aristotelian tendencies when it comes to constructing his huge commentaries.[8] I am referring to one of the possible applications of the dialectical method as sketched by Aristotle in his first and eighth books of the *Topics*. In my previous work I have been studying this aspect of Aristotle's methodology, emphasizing the important distinction between *propaedeutic* and *applied* dialectic.[9] At the core of those efforts was an attempt to show how one can take Aristotle's claims for a scientific use of dialectic seriously, so long as we have a proper understanding of the status of propaedeutic dialectic as it is expounded in his *Topics* (school practice and exercises) against the applied form of (evolved) dialectic which goes far beyond this early form, debating skills which have become transformed into an internalized form of dialectic.[10]

---

[5] *Alexandri Aphrodisiensis Aristotelis Topicorum libros octo commentaria* (ed. M. Wallies, Berlin 1901).
[6] Huby (1989, pp. 68–71) discusses several late antique authors writing on dialectic, but they have no direct link to Aristotle: Martianus Cappella and Boethius respond to Cicero—for the latter's commentary on Cicero's *Topica*, his *De topicis differentiis*, see Stump (2009). On Boethius, see Ebbesen (1990); On medieval engagement with dialectic and sophistic refutations, see for example Ebbesen (1981) and Renswoude (2017).
[7] On his *Eisagôgê*, see Barnes (2003).
[8] All page references for the commentaries are to the *CAG* edition.
[9] See especially Baltussen (2000), Chapter 2.
[10] Clarified in Baltussen (2008b) and illustrated in a case study in Baltussen (2015).

2 Simplicius and Aristotle's *Topics*

An investigation into Simplicius' role regarding the reception of ancient dialectic, in particular the Aristotelian version, is at first glance not a promising undertaking. After all, so far as we know, Simplicius did *not* write a commentary on Aristotle's *Topics*.[11] As the former student of Plato, Aristotle displays his familiarity with the debating practices of the Academy in his own work on dialectic, the *Topics*. In this work he compiled, under four new headings, some 300 "commonplaces" (τόποι), while both overlaying the content with his own more systematic organization and reorienting it towards possible applications in training and argumentation (*Top.* A.2; cf. A.14) and notably also for research, as we can see in *Top.* A.2, 101a36–101b2:

> T1 *Furthermore*, [it is useful] in relation to the principles in the several sciences. For it is impossible to discuss them at all from the principles proper to the particular science in hand, seeing that the principles are primitive in relation to everything else: it is through reputable opinions about them that these have to be discussed, and this task belongs *properly and most appropriately* to dialectic; for dialectic is the process of criticism wherein lies the path to the principle of all inquiry.[12]

Moreover, there is no testimony telling us that such a commentary of Simplicius existed or may have been lost.[13] Thus, one may well ask whether Simplicius knew the commentary on the *Topics* written by his most trusted guide in reading Aristotle, the 2nd century Aristotelian commentator Alexander of Aphrodisias. And finally, why would he only paraphrase titbits from the *Topics*, when he clearly does quote from Theophrastus' own *Topics* at *in Cat.* 415.16?

Such questions may well discourage us, but things are not as dire as these comments may suggest. Simplicius was fully aware of the works in the Aristotelian corpus (including the *Topics*, as we shall see shortly) and shows some knowledge of the various treatises on argumentation and reasoning (e.g., *in Cat.* 4.31–5.1 mentions both the *Topics* and the *Sophistic Refutations*[14]). For the latter, we may high-

---

11 As Hadot (2014), p. 283, notes, an attribution to Simplicius of a commentary on the *Sophistic Refutations* was proven to be erroneous.
12 τὴν μέθοδον καὶ πρὸς τὰς κατὰ φιλοσοφίαν ἐπιστήμας "ἐξεταστικὴ γάρ, φησίν, οὖσα πρὸς τὰς ἁπασῶν τῶν ἐπιστημῶν ἀρχὰς ὁδὸν ἔχει" (see also below pp. 448–9). This part of the text has been disputed (but cf. Ar. *Rhet.* 1.2, 1358a21–26); see also Smith (1997) and Baltussen (2022).
13 Cf. Baltussen (2018b).
14 He also uses the adverb σοφιστικῶς to label an argument invalid or fallacious (*in de Cael.* 313.11; *in Phys.* 915.18, 1020.10, 1024.16).

light several passages that refer to Aristotle's work: *in Phys.* 52.15; 70.21; 723.12. The first passage, in which Aristotle called Melissos' argument "more coarse" (μᾶλλον δὲ φορτικόν, 52.8), he points out how one absurdity leads to another (52.15). In the second (70.21) he discusses Porphyry's and Alexander's comments on Melissos, and remarks (70.20–21) that Aristotle himself used the type of argument in the *Soph. El.* on a distinction between reality and appearance ("ἔστι μὲν οὔ, φαίνεται δέ"). At 723.12, we do not find the work's title but a reference to "sophistic arguments" used by Aristotle (καὶ ἐκ τῶν σοφιστικῶν λόγων κατασκευάζει).[15]

In addition, like many late Platonists, his commentaries frequently use formalization of arguments as a way of clarifying Aristotle's reasoning. The most common form for this approach is the syllogism, or at least a syllogistic form of reasoning.[16] It is a feature which reveals the origin of these exegetical works in a didactic context, illustrating how Platonist exegesis was geared towards explaining Plato and Aristotle to students.[17]

This feature of Simplicius' method should not tempt us to create an opposition between dialectic and Aristotle's formal syllogistic writings, the *Analytics*. In his polemic against Philoponus, Simplicius discusses reasoning and refutation in a broad sense, showing he knows the *Topics*, as his remark at *in de Cael.* 238.8–11 Heiberg (*CAG* 7) illustrates:

T2 ... τὰ δὲ λογικά, ἅπερ καὶ διαλεκτικὰ καλοῦσιν, ἀληθῆ μέν ἐστιν καὶ αὐτά, κοινότερα δὲ καὶ δυνάμενα καὶ ἄλλοις ἐφαρμόττειν καὶ ἐξ ἐνδόξων μᾶλλον εἰλημμένα, ὧν τὴν μέθοδον ἐν τοῖς Τοπικοῖς ἐπιγραφομένοις ὁ Ἀριστοτέλης παραδέδωκεν τόπους τὰς κοινότητας καλέσας.[18]

Logical arguments [as opposed to practical ones], which they also call dialectical, are also themselves true, more universal ... and taken from reputable opinions ... for which Aristotle passed on the method in his work entitled *Topics*, calling them the most common standard arguments.

I will return to this passage. Unsurprisingly, another passage in his commentary on the *Categories* reveals his detailed knowledge of Plato's dialectic (*in Cat.* 70.14–20):

---

15 Apart from these cases, those in the previous note, and a few occurrences of σοφιστικός (accusative forms), these 14 are the full extent of the words containing the root σοφιστικ- found in Simplicius (based on TLG).
16 Examples for this habit are so common that there is no need to provide a list. See, e.g., *in de Cael.* 236.26 ff. (clarifying Alexander's formal argument). Cf. Sorabji (1990), "Introduction".
17 On the joint reading (συνανάγνωσις) of Aristotle and Plato in a school context, see Mansfeld (1994).
18 For other comments on the dialectical method (τὴν μέθοδον), see *in Cat.* 70.18; 380.13; cf. *in Phys.* 47.21.

T3 when in dialogue Socrates seems to bring forth the beautiful which is in all things and the equal in all things, it is not because the beautiful within the many is completely undifferentiated (ἀδιάφορον), but because the dialectical method (ἡ διαλεκτικὴ μέθοδος) distinguishes our conception (ἔννοια) which was therefore to mixed up (συμπεφυρμένην) ... and then brings round (περιάγει) our henceforth distinguished conceptions to unity (ἕνωσις), in the place of confusion (σύγχυσις), which is the goal (τέλος) of dialectic, as Socrates showed in the *Republic*.

Such passing remarks hardly amount to a strong endorsement of the method, and out of the 27 occurrences of the term διαλεκτικός in his works, a mere 7 concern points such as these.[19] But on general grounds it is plausible that the training Simplicius himself received included the tools of dialectical reasoning, which have become "second nature" in his own writings.

In what follows, I examine the role of dialectic as an evaluation tool of philosophical opinions in Simplicius. Rather than merely review the occurrences of the word "dialektikos", I will propose that Simplicius implicitly absorbs this evaluative feature into his exegesis because it also serves his tendency to preserve what is useful in the Hellenic tradition.[20] Extending my earlier work on this topic, my argument will proceed in three sections before drawing up a brief conclusion. First, we consider the nature of the dialectical method and its connection to the task of a philosophical exegete. Next, I focus on the writings of Simplicius, and it goes without saying that this paper is highly selective, given that the large corpus defies a quick or comprehensive summary.[21] Third and last, I will reflect on how close Simplicius' approach is to Aristotle's.

3 Method and Exegesis in Simplicius

Let us start with some general points about Simplicius' method, including his explicit reflections on Aristotle's dialectic. He makes several passing comments,

---

19 See, e.g., T2.
20 I made the argument for this special use in Baltussen (2018a).
21 Over 3,000 pages survive of the commentaries on logic, physics, and cosmology: we know of six or seven surviving works: four commentaries on Aristotle, i.e., on the *Categories*, *Physics* and *On the Heavens*, and possibly on the *Metaphysics* (lost; Luna 2001), but the authorship for *in de Anima* is disputed (Steel 1978 and Hadot 2002). In addition, there is a commentary on the *Handbook* ('Ἐγχειρίδιον) of the Stoic Epictetus (c. 55–c.135 CE), and perhaps we can also include a summary version of Theophrastus' *Physics* (mentioned in the *in de An.* 136.29). There is evidence of his study of Plato' *Sophist* (see Gavray 2007), and perhaps the *Phaedo* (Hadot 2001). We have some fragments of a lost commentary on Euclid (Hadot 2001, p. xxxvi, n.4). The commentary on Hippocrates' *Fractures* is spurious, as is the mention of a commentary on Aristotle's *Sophistical Refutations*.

which show him to be, as I have argued elsewhere,[22] a scholarly and philosophical exegete, making full use of the accumulated tools of textual interpretation. Simplicius famously describes the task of the commentator in his commentary on Aristotle's *Categories*, the work considered fundamental to the training of a Platonist since Porphyry. Assuming that this passage *in Cat.* 7.23–32 is sufficiently known I will only discuss its main points. Three principles of exegesis are worth highlighting: (1) the need for great intelligence (μεγαλονοίας, 24), (2) thorough knowledge of the corpus (γεγραμμένων ἔμπειρον, 25) and style of the author (συνηθείας ἐπιστήμονα, 25f.), (3) the requirement to be impartial (ἀδέκαστον, 26).[23] These qualities are supposed to assist in avoiding "misplaced zeal to quarrel over something well stated or to obstinately attempt to prove Aristotle infallible" (26–29). But this basic attitude, while reasonable, also encompasses other approaches: most commentators will also apply "creative exegesis" on many occasions, allowing them to rephrase, stretch or "massage" the text.[24] Ironically, Simplicius often accuses others of *rewriting* the text (μεταγράφει, e.g., said of Alexander at *in Phys.* 526.17) and in his more polemical mood he accuses Philoponus of being "verbose" (διὰ πολυστίχων βιβλίων, *in de Cael.* 25.29).

As we saw in the first quotation (T2), Simplicius knew the *Topics*. The passage is worth revisiting briefly, with some further ones added for context (*in de Cael.* 238.8–10):

> **T2*** τὰ δὲ λογικά, ἅπερ καὶ διαλεκτικὰ καλοῦσιν, ἀληθῆ μέν ἐστιν καὶ αὐτά, κοινότερα δὲ καὶ δυνάμενα καὶ ἄλλοις ἐφαρμόττειν καὶ ἐξ ἐνδόξων μᾶλλον εἰλημμένα, ὧν τὴν μέθοδον <u>ἐν τοῖς Τοπικοῖς</u> ἐπιγραφομένοις ὁ Ἀριστοτέλης παραδέδωκεν τόπους τὰς κοινότητας καλέσας.[25]
>
> Logical arguments [as opposed to practical ones], which they also call dialectical, are also themselves true, more universal ... and taken from reputable opinions ... for which Aristotle passed on the method in his work entitled *Topics*, calling them the most common standard arguments.

---

[22] Baltussen (2018a).

[23] The third point is perhaps inspired by *Cael.* I.10, 279b4–11, where Aristotle ends his discussion on disputes by saying "To give a satisfactory decision as to the truth it is necessary to be rather an arbitrator than a party to the dispute".

[24] I borrow the phrase from Mansfeld (1994), p. 26. Porphyry, *de philosophia ex oraculis* 109.9–110.9 Jacoby, offers an intriguing exegetical principle—which is probably more aspirational than fully adhered to—claiming that he had not changed, added, or subtracted anything from the text (for text and discussion see Baltussen 2018a).

[25] For other comments on the dialectical method (τὴν μέθοδον) see *in Cat.* 70.18; 380.13; *in Phys.* 47.21.

References to the work by title are, however, rare.[26] While they appear in the three major commentaries, they seem perfunctory and generally forego going into detail. A quick survey may illustrate this point to make clear that Simplicius does not seem particularly interested in the work. At *in de Caelo* 238.5–11 (= T2 above), the occurrence of the title (at 238.10) merely hints at the nature of the work. Simplicius speaks about logical arguments and indicates that Aristotle's work entitled *Topics* was concerned with "dialectic" (τὰ δὲ λογικά, ἅπερ καὶ διαλεκτικὰ καλοῦσιν, 8) based on "reputable opinions" on which he based his points of attack in debate (τὴν μέθοδον ἐν τοῖς Τοπικοῖς | ἐπιγραφομένοις ὁ Ἀριστοτέλης παραδέδωκεν τόπους τὰς κοινότητας καλέσας, | καθ' ἃς λαμβάνεται τὰ ἐπιχειρήματα). The mention at 523.25–30 (twice) is more precise, since he first refers to book 1 of the *Topics* repeating Aristotle's claim (discussed above) that the method is useful for philosophical research, while a few lines later he again refers to the work by title, characterizing it as a work that assists in finding premises (τὸ προτάσεις εὑρεῖν) so as to syllogize with ease. Several mentions in the commentary on the *Categories*, including a reference to the fifth book of *Topics* (113.27), concern more general comments on the broad output of Aristotle and the place of various works in his corpus.[27]

A more interesting comment for our purposes is found in *Cat.* 12.3–10, when Simplicius discusses Aristotle's views on what expressions signify, contrasting realities (*pragmata*) with concepts (*noēmata*):

> **T4** This is why Aristotle teaches us what each expression signifies and defines realities in accordance with each category. Moreover, he uses the same division both here, where the goal (*skopos*) is primarily about significant expressions, and in the *Metaphysics*, where he teaches about beings *qua* beings. His procedure is similar in his logical works, for instance the *Topics* and the *Nicomachean Ethics*, and in his other treatises, he uses the same division of the primary genera into these ten. This is because the division is the same everywhere ... (trans. Chase 2014, p.27)

Clearly, here he sees the method in various works as unified, including that of the *Topics*.

It is worth noting that Simplicius also follows certain general techniques of exegesis which originated with individual Platonists before him, such as Iamblichus' idea that an Aristotelian work had a singular purpose (σκόπος), or Syrianus' idea that philosophy could be motivated by the assumption that Socrates was a saviour,

---

[26] I found eight instances (based on a TLG search): *in de Caelo* 238.10; 523.30; *in Cat.* 12.8; 143.4; 415.16; *in Phys.* 47.22; 131.23; 335.20.

[27] *In Cat.* 7.17; 12.8; 15.28–32; 16.1, 3, 14; 113.27 (fifth book of *Topics*); 415.16 (Theophrastus' own *Topics*).

or again Proclus' notion of a division into clarification of meaning and text.[28] But when it comes to dialectic, we will have to consider how he viewed Aristotle's stated position, and whether he saw fit to make use of it himself.

4 Simplicius and Applied Dialectic

As far as applied dialectic is concerned, Simplicius expresses himself only on a few occasions. The few examples I have given so far may serve as evidence for his knowledge of some pivotal passages. But there is more to be explored. For instance, in his discussion of universals, he explains why and how universals are better known than particulars with an appeal to the role of dialectic, presented as Plato's "science of dialectic", but combining it with Aristotle's notion of how certain things are "better known to us" (*in Phys.* 16.20–28; my emphasis):

> **T5** The universal resembles the whole in containing confused within itself the articulation of the many things that compose it, as the parts are contained within the whole: for in animal too is contained the variety of the species of animal without distinction (καὶ γὰρ ἐν τῷ ζῴῳ ἀδιόριστος ἡ τῶν εἰδῶν τοῦ ζῴου διαφορά); and so the universal as compounded and confused is better known to us,[29] and first in knowledge as regards us, just as by nature this too is posterior, if it is a by-product of the individuals. For the simpler things are clearer and better known by nature, as being pure and unmixed; and this is why *the science of dialectic* (ἡ διαλεκτικὴ ἐπιστήμη) is accustomed to examine what each "[thing] itself" is, philosophizing among simple forms, inasmuch as it proceeds in tandem with the nature of beings, according to which simpler things are better known and more manifest than compound ones, and pure things than confused ones (trans. Menn).

Furthermore, in the early part of the commentary on the *Physics*, Simplicius chooses to mention what the role of dialectic was according to Plato and Aristotle during his discussion of the views of early Greek philosophers. It is striking that these occur on pp. 47–51 (*CAG* 9, ed. Diels). The following are the most relevant remarks (trans. Menn; paraphrases mine):[30]

> **T6a** *in Phys.* 47.22–26 Diels, where he acknowledges the usefulness of dialectic in Aristotle, when he quotes the principle of "syllogising from reputable opinions" from *Top.* A.2 (καὶ

---

[28] See Mansfeld (1994) and Golitsis (2008).
[29] Reading κατὰ συγκεχυμένον with the Aldine at 16.23 (Diels has κατὰ τὸ συγκεχυμένον, DEF καὶ τὸ συγκεχυμένον).
[30] The lemma that triggers this discussion is Aristotle, *Phys.* 184b25 (*in Phys.* 46.9–10). The passage in Simplicius actually starts with a quotation from Alexander, who uses the technical term πρόβλημα (46.12).

γὰρ ἐν τοῖς Τοπικοῖς χρήσιμον ἐκείνην εἶπε τὴν μέθοδον καὶ πρὸς τὰς κατὰ φιλοσοφίαν ἐπιστήμας "ἐξεταστικὴ γάρ, φησίν, οὖσα πρὸς τὰς ἁπασῶν τῶν ἐπιστημῶν ἀρχὰς ὁδὸν ἔχει.").[31]

T6b *in Phys.* 49.9–11, where he contrasts the way in which Plato and Aristotle define the dialectician's task, with a special eye for the foundations (principles) for establishing a proof and avoiding an infinite regress (6–9, on Plato and geometric proof from "self-sufficient and indemonstrable principles"; 9–11, Aristotle reasons by syllogistic argument based on reputable opinions).

T6c *in Phys.* p. 51.6–7[32], where he makes a brief statement about what is not the task of the dialectician (οὐδὲ γὰρ τοῦ διαλεκτικοῦ ἂν εἴη διαλέγεσθαι πρὸς τὸν τὸ δίκαιον λέγοντα τὴν στοάν, "nor would it be the task of the dialectician to argue against someone who says that the just is the porch").

Again, these are perhaps passing remarks, but they surely indicate how Aristotle's dialectic is on Simplicius' mind in certain contexts, where reasoning from earlier views forms the core of the debate.

The last of the three passages can be usefully compared to *in Phys.* 71.31–72.2, where dialectic is given a more serious role:

T7 γίνεται δὲ διαλεκτικὴ αὐτῷ ἡ ἐπιχείρησις ἀπὸ τῆς τῶν ὄντων | διαιρέσεως ὁρμηθεῖσα· τῇ γὰρ διαλεκτικῇ [sc. τέχνῃ] χρωμένους ἔστι τὰς τῶν ἐπιστημῶν ἀρχὰς κατασκευάζειν.

And the argument turns out to be dialectical, since it proceeds from the division of beings: for it is possible by making use of dialectic to establish the principles of the sciences (trans. Menn)

It is significant how dialectic is here described as having the task of establishing first principles of knowledge (τὰς τῶν ἐπιστημῶν ἀρχάς). The phrase echoes Aristotle's claim of *Topics* A.2 (=**T2**) and was already found at *in Phys.* 47.25–26 (**T6a** above). These select examples of Simplicius' explicit statements about the important role of dialectic in undertaking philosophical research show clearly how he sees its role as one that can be applied in close agreement with Aristotle.

If this is correct, we may look at another test case in which Simplicius' discussion of elemental principles in *Physics* also demonstrates his tacit approval of dialectical evaluation from reputable opinions. In Simplicius, this aspect comes across as an argument from authority. While his approach is inspired by both his-

---

31 Quotation marks are Diels'. See also *in Phys.* 49.9–10 and *in Phys.* 476.26–28: ἡ γὰρ διαλεκτικὴ ἡ Ἀριστοτέλους κοινή ἐστι μέθοδος περὶ παντὸς ἐξ ἐνδόξων συλλογιζομένη, ὡς αὐτὸς ἀρχόμενος τῶν Τοπικῶν φησι.

32 The lemma that triggers this discussion is Aristotle, *Phys.* 185a5–10, which concerns invalid reasoning (with Heraclitus as an example).

torical and philosophical motives, it is also marked by a keen interest in the original words of the earliest philosophers, and in trying to evaluate which proposed elemental principles are the most plausible.[33] His discussion should be viewed partly in light of the Presocratic ideas (especially *In Phys.* 21.22–22.8, 22.22 ff.) as well as Plato's creation account in the *Timaeus*, since he defends Plato's imperishable heaven arguing that the Creator Craftsman made it indissoluble (*in Cael.* 106.4–108.14 Heiberg).

Interestingly, in this commentary, he shows his awareness of the power of principles (cf. *Top.* A.2), when discussing the proof that elements are finite. After going over several claims regarding the (in)finite nature of the elements (Anaximander: infinite; Leucippus and Democritus: finite), he agrees with Alexander that they must be finite. But the final comment demonstrates how important this conclusion is, because it is given a more general significance (*in de Caelo* 202.31–203.1, trans. Mueller):

> **T8** Thus, he elegantly shows that the slightest oversight in the case of the postulation of the smallest magnitude becomes responsible (αἴτιον) for the greatest errors on account of the very great power (μεγίστην δύναμιν) it has as a principle (ὡς ἀρχήν).

Clearly, then, his point is that even if the argument is about a small thing, it need not be trivial, and Alexander's argument is central for making the case. This variety of arguments from authority is an important characteristic of the *Physics* commentary, with Alexander in a crucial role in this work (he is referred to ca. 700 times), with, as a close second, Theophrastus, Eudemus, and Porphyry.[34] Simplicius is using Alexander's commentary on Aristotle's *Physics* (extant in fragments only) on almost every page.[35] At a minimum, these comments show that Simplicius was fully aware of the claims Aristotle made regarding the usefulness and purpose of dialectic as he had developed it beyond the propaedeutic stage of the Academic practice—the form it also has in the central books of the *Topics*.

Based on these considerations, we may tentatively suggest that Simplicius himself used *applied dialectic* in this sense. The way in which he seems to mirror some of Aristotle's "endoxographical method" can be extended to his overall approach in how he viewed the history of Hellenic philosophy. This is not simply a matter of pointing to his lavish use of quotations. While we know that he uses quotations

---

[33] I here summarized one aspect of a fuller analysis in Baltussen (2015).
[34] For the role of Theophrastus (especially in the first thirty pages or so), see Wiesner (1992), Mansfeld (1987), (1989), and Baltussen (2002) as well as (2008a), Chapter 2.
[35] Cf. Baltussen (2006). Alexander's commentary has been partially reconstructed by Marwan Rashed (2012) on the basis of newly found scholia.

to substantiate paraphrase (*in de Caelo* 140.32–33, 298.21–22), to clarify statements containing technical or other obscure terms he has rephrased (*in de Caelo* 140.25.26–30), or to provide proper evidence (*in Phys.* 331.10), his approach becomes both more interesting and more significant, when we allow for the use of applied dialectic. It is more interesting because it militates against regarding him solely as an uninspiring "conduit" of earlier thinkers, from the Presocratics to Damascius, and it is more significant because his prolixity and presumed scholasticism is no longer merely an unoriginal reflection of an exegetical school tradition that started with Plotinus (c. 204–270 CE).

Simplicius acknowledges the contribution of his predecessors in the exegetical tradition (*In Cat.* 2.30–3.4, 3.10–17) and absorbs many into his own exegetical narrative. He also articulates his "principle of preservation", for which three examples can be given, one for Parmenides, two for Empedocles. In each case he gives a specific reason for quoting more from his sources than he needs for his argument:

> **T9a** At *in Phys.* 144.25–29, he announces he will quote a big chunk from Parmenides' monograph, a move which he clarifies by saying that the book of Parmenides is becoming rare (*spanin*):[36] "In order not to seem too pedantic I will in these memoranda quote the verses of Parmenides ... in order to justify my comments on this matter (διά τε τὴν πίστιν) *and because Parmenides' treatise is becoming quite rare* (καὶ διὰ τὴν σπάνιν τοῦ Παρμενιδείου συγγράμματος)" (*in Phys.* 145.1–146.28; follow 53 lines = 28B7 DK[18]).

> **T9b** Empedocles, where he emphasizes how this philosopher has "handed down" an interesting range of ideas, such as the one and the many, etc. (see *in Phys.* 157.25–7), after which he quotes a total of 79 lines across pages 158–161.[37]

> **T9c** or he offers additional justification for a quotation to support his claims: καὶ ἵνα μὴ δοκῶ τισιν κενὰς μακαρίας ἀναπλάττειν, ὀλίγα τῶν Ἐμπεδοκλέους ἐπῶν παραθήσομαι, "to make sure that I don't seem to be making empty claims, I shall quote a few words from Empedocles' poem" (*in Cael.* 140.30–32, cf. 528.32–33; 6 lines follow, 141.1–6 = 31B17.7–8, 10–14 DK[18]).[38]

Such cases illustrate his sincere efforts to preserve materials for several different reasons. But there are other reasons as well for quoting from the Presocratics. One of particular significance relates to his attempt at *integrating them into a unified philosophical account:* at *in Phys.* 77.9–11, he sets out to discuss Alexander's view on Parmenides and Melissos with the aim to determine how he can show briefly why Alexander is right and suggest we "consider briefly how he too believes it is

---

36 The word also occurs in connection with Archytas (*in Cat.* 352.22–24).
37 34 lines at *in Phys.* 158.1–159.4; next 14 lines at 159.13–26; 11 lines at 160.1–11; two sets of 2 lines at 160.16–17, 20–21; 9 lines at 160.28–161.7 and 4 at 161.16–20; and a further 4 lines at 162.18–22.
38 In DK, line 9 is added.

well said and that the wisdom of old remains irrefutable" (10–11, συντόμως ἴδωμεν πῇ καὶ αὐτὸς εὖ λέγειν δοκεῖ καὶ ἡ παλαιὰ φιλοσοφία μένει ἀνέλεγκτος).

He is also keen for his reader (whoever this may be) to believe that his accounts are based on *genuine sources* of those he mentions: as we just saw, he substantiates paraphrase by offering the relevant text (**T9c** above), while after several quotations from Empedocles, he says at *in Phys.* 331.10 "one could find more of that sort of views to quote from Empedocles' physics" (καὶ πολλὰ ἄν τις εὕροι ἐκ τῶν Ἐμπεδοκλέους Φυσικῶν τοιαῦτα παραθέσθαι, ὥσπερ καὶ τοῦτο [1 line quoted] καὶ μετ' ὀλίγον [1 line quoted]). And for Anaxagoras he also quotes some important lines, now from the "beginning of his book" (*in Phys.* 172.1–3; cf. 175.11–15). One remarkable statement regarding his use of quotations also concerns evidentiary reasoning regarding demonstration, in which he (approvingly it seems) explains that Aristotle adduces "testimonies of his predecessors as agreeing with his demonstrations", not as "recent writers" do, by using the testimonies as demonstrations (*in Phys.* 1318.10–15).[39]

Simplicius' comprehensive harmonizing tendencies will put pressure on his skills of creative exegesis because such an ambitious undertaking (a full synthesis of Greek philosophy from Thales to Plotinus) is hardly an easy task. But this does not automatically mean that he agrees with everything they say. There are good examples in which he critiques earlier commentators.[40] For instance, he criticizes several earlier luminaries among his sources, such as Alexander of Aphrodisias (*in Cael.* 297.3 ff.), "who does not understand Plato's doctrines as Aristotle understood them, nor does he accept that their views are in agreement", and Plotinus, who maintained (*Enn.* 6.1–3) that Aristotle's categories do not apply to the intellectual realm of Forms in the way that Plato's "Great Kinds" could (Being, Sameness, Difference, Motion and Rest—*Soph.* 254cff.) or that only the first four categories (substance, quantity, relation, quality) can be applied to the sensible world (e.g., *in Cat.* 73.25–35). Against Iamblichus, he raises objections to his "intellective theory" (*noera theôria*), criticizing him for inconsistency on the category of place (*in Cat.* 364.7 ff.), arguing that time and place are needed for the intelligible world as well as the sensible world; and he famously had a go at Philoponus in contesting his views on the eternity of the world. Seen from within the Platonist tradition, these moves show how allegiance to a particular school of thought did not mean unconditional acceptance, whereas the presumed authority of Aristotelians was not always successfully integrated into the harmonizing strategy.

---

39 I discussed these passages in Baltussen (2008a), p. 44, in the context of Simplicius' general method.

40 Some of the following examples were also discussed in Baltussen (2008a).

Simplicius also goes significantly further than other Platonists before him in his attempt to harmonize the views of *all* Greek philosophers—even if he claims precedence for this stance in Aristotle himself (*in Phys.* 20.12 ff.), where Aristotle in his endoxographic section sees "agreement" on the question as to whether principles in physics are regarded as opposites by the earliest natural philosophers (a view repeated at *in Phys.* 179.28–29, 182.9–10). In addition, differences of opinion, Simplicius argues, can be useful, for instance at *in Cat.* 1.22–2.3, where he speaks of *aporiai* and solutions (a remote echo of Aristotle, *Cael.* A. 10, 279b6–7, describing *aporiai* as the pathway to solutions[41]), which may relate to the dialectical strategy of viewing a problem from two sides (*in Phys.* 86.2–3). They can be reconciled by close study of the text and by eliminating *apparent* disagreements (*in Phys.* 36.24–31, cf. *in de Cael.* 159.3–9). The significance of agreement among all, most, or the experts was, of course, recognized by Aristotle himself, who saw it as a sign of the truth (*Met.* A.7; *Top.* A.14). Simplicius' so-called harmonizing tendency may have taken its cue from these passages in Aristotle. But his more extreme version to declare Plato and Aristotle in agreement probably has a further, more ideological reason, that is, seeking to unite the Hellenic schools of thought against the Christian worldview.

5 Conclusion

I began this chapter by asking the question to what extent Simplicius was familiar with Aristotle's dialectic, with a particular focus on the claim that dialectic can be useful to finding fundamental starting points in the sciences (**T1–2**). We know that Simplicius had solid knowledge of Plato's dialectic (**T3**). By pointing to the absence of a commentary on Aristotle's *Topics*, I did not mean to say that this signaled a lack of interest on his part, but to flag the difficulty of any attempt at finding out what Simplicius may have thought about the original (though youthful) work on dialectic. We saw that the sparse references to the Aristotelian work show that Simplicius was familiar with it (**T4, T5**), offering passing mention of things dialectical (διαλεκτικός, λογικός, τοπικός), but also referring specifically to book I (*in de Caelo* 523.25; cf. *in Phys.* 476.28) and book V (*in Cat.* 113.27).

An additional premise of the evaluation was the position that we should distinguish between dialectic as an exercise (propaedeutic) and dialectic in its evolved

---

41 On which see, in particular, Quandt (1983). Quandt (his n. 32) notes that Simplicius agrees with his reading at *in de Cael.* 292.18 ff.

and applied form.⁴² The former found its expression in the *Topics*, in which Aristotle's description partly reflects the Academic practice, and partly shows how he imposes his more systematic organization and clarification onto the approximately 300 technical "moves" (τόποι) according to the four "predicables" (genus, difference, accident, and definition). He bookended those central sections with what look like later additions, an introduction that explained its purpose(s) in *Topics* I and a final book with a detailed description of the "role-play" of dialectical debates ("la joute dialectique" in Moraux's words) in *Topics* VIII.⁴³ The latter (applied dialectic) had been further developed by trial and error, and in line with the ambitious (and meta-philosophical) claim in *Topics* A.2, that the method was "useful for finding principles" (**T1**).

Based on select examples we can see how Simplicius seems to work along similar (though not identical) lines when it comes to his commentary on the *endoxographic* passages in Aristotle (**T6a–c, T7, T9a–b**). But the limited presence of explicit references to the work may well signal that the method was now so fully integrated that it did not need to be mentioned: it was literally "taken as read" that a well-trained philosopher was familiar with, and could use, these techniques, especially with regard to principles (**T8**).⁴⁴ What further emerges is that he also expands this approach by including an extra layer of post-Aristotelian exegetical comments, as it were extending the pool of "reputable opinions" by including previous experts in clarifying Aristotle. What he contributes is an impressive feat: he attempts a synthesis of commentators and base texts and attempts to reach an integrated and harmonious narrative. This aspect of the dialectical method, sketched at *Topics* A.2 and implemented by Aristotle in various ways in other treatises, thus played a role in the construction of his commentaries. His method in writing an exegesis of Aristotle, while grounded in Platonist ideology, uses Aristotelian methodology: the former directs his attention to particular themes, while the latter allows him to critically evaluate their views and position them within the Hellenic philosophical tradition.

---

42 A position established in Baltussen (2000), Chapter 2, and further illustrated in Baltussen (2008b, 2015, and 2022).
43 Moraux in Owen (1968). For the usefulness of Books I and VIII, see Robin Smith (1990), who is justified in the decision to write a modern commentary on precisely these two books.
44 This seems a more plausible reading than connecting this relative silence to the old debate on whether "logic" (i.e., all forms of reasoning) is a part of philosophy or not.

Bibliography

Baltussen, Han (2000): *Theophrastus Against the Presocratics and Plato. Peripatetic Dialectic in his De sensibus*. Leiden: Brill.

Baltussen, Han (2006): "Addenda Eudemea". In: *Leeds International Classical Studies* 5. No. 1 (August), pp. 1–28. http://www.leeds.ac.uk/classics/lics/2006/200601.pdf. https://hdl.handle.net/2440/36118, last accessed on December 25, 2022.

Baltussen, Han (2008a): *Philosophy and Exegesis in Simplicius. The Methodology of a Commentator.* London: Duckworth.

Baltussen, Han (2008b): "Dialectic in Dialogue: The Message of Plato's *Protagoras* and Aristotle's *Topics*". In: McKay, Anne (Ed.): *Orality, Literacy, Memory in the Ancient Greco-Roman World.* Leiden: Brill, pp. 201–224.

Baltussen, Han (2015): "Simplicius on Elements and Causes in Greek Philosophy. Critical Appraisal or Philosophical Synthesis?". In: Marmadoro, Anna and Prince, Brian (Eds.): *Creation and Causation in Classical and Late Antiquity*. Cambridge: Cambridge University Press, pp. 187–210.

Baltussen, Han (2018a): "Simplicius and the Commentator's Task: Clarifying Exegeses and Exegetical Techniques". In: Strobel, Bernd (Ed.): *Die Kunst der philosophischen Exegese bei den antiken Platon- und Aristoteles-Kommentatoren*. Berlin and Boston: de Gruyter, pp. 159–183.

Baltussen, Han (2018b): "Simplikios". In: Horn, Christian, Riedwig, Christoph, and Wyrwa, Dietmar (Eds.): *Ueberwegs Grundriss der Geschichte der Philosophie. 14. Neuauflage. Band 5: Die Philosophie der Kaiserzeit und der Spätantike*. Basel: Schwabe, pp. 206–230 (text) and pp. 321–329 (bibliography).

Baltussen, Han (2022): "'Received Opinions' in Aristotle, Theophrastus and Simplicius: Doxography or Endoxography?" In: Lammer, Andreas and Jas, Mareike (Eds.): *Received Opinions. Doxography in Antiquity and the Islamic World*. Leiden: Brill, pp. 151–174.

Barnes, Jonathan (2003): *Porphyry: Introduction*. Oxford: Clarendon Press.

Blumenthal, Henry J. (1996): *Aristotle and Neoplatonism in Late Antiquity. Interpretations of the De anima*. London: Duckworth.

Brunschwig, Pierre (1967): *Aristote. Les Topiques, I–IV.* Vol. I. Paris.

Brunschwig, Pierre (2007): *Aristote. Les Topiques, VI–VIII.* Vol. II. Paris.

Ebbesen, Sten (1981): *Commentators and commentaries on Aristotle's Sophistici elenchi: a study of post-Aristotelian ancient and medieval writings on fallacies*. Leiden: Brill.

Ebbesen, Sten (1990): "Boethius as an Aristotelian Commentator". In: Sorabji, Richard (Ed.). *Aristotle Transformed. The Ancient Commentators and Their Influence*. London: Duckworth, pp. 373–391 [pp. 403–421 in 2nd ed., 2016, London: Bloomsbury].

Gavray, Marc-Antoine (2007): *Simplicius lecteur du Sophiste. Contribution à l'étude de l'exégèse néoplatonicienne*. Paris: Klincksieck.

Golitsis, Pantelis (2008): *Les commentaires de Simplicius et de Jean Philopon à la Physique d'Aristote: tradition et innovation*. Berlin and New York: de Gruyter.

Gottschalk, Hans B. (1990): "The Earliest Aristotelian Commentators". In: Sorabji, Richard (Ed.) (1990): *Aristotle Transformed. The Ancient Commentators and Their Influence*. London: Duckworth [2nd ed., 2016, London: Bloomsbury], pp. 55–81.

Hadot, Ilsetraut (2001): *Simplicius. Commentaire sur les Catégories d'Aristote*. Book III. Chapter 1. Leiden: Brill.

Hadot, Ilsetraut (2002): "Der fortlaufende Philosophische Kommentar". In: Geerlings, Wilhelm and Schulze, Christian (Eds.): *Der Kommentar in Antike und Mittelalter. Beiträge zu seiner Erforschung.* Leiden: Brill, pp. 183–199.

Hadot, Ilsetraut (2014): *Le néoplatonicien Simplicius à la lumière des recherches contemporaines. Un bilan critique.* Sankt Augustin: Academia Verlag.

Huby, Pamela M. (1989): "Cicero's *Topics* and its Peripatetic Sources". In: Fortenbaugh, William W. and Steinmetz, Peter (Eds.): *Cicero's Knowledge of the Peripatos.* New Brunswick and London: Transaction, pp. 61–76.

Luna, Concetta (2001): *Trois Études sur la tradition des commentaires anciens à la Métaphysique d'Aristote.* Leiden: Brill.

Mansfeld, Jaap (1987): "Theophrastus and the Xenophanes Doxography". In: *Mnemosyne* 40, pp. 286–313.

Mansfeld, Jaap (1989): "Gibt es Spuren von Theophrasts *Phys. Op.* bei Cicero". In: Fortenbaugh, William W. and Steinmetz, Peter (Eds.): *Cicero's Knowledge of the Peripatos. RUSCH* IV. New Brunswick and London: Transaction, pp. 133–158.

Mansfeld, Jaap (1990): "Doxography and Dialectic. The 'Sitz im Leben' of the 'Placita'". In: Haase, Wolfgang (Ed.): *Aufstieg und Niedergang der römischen Welt* II.36.4. *Philosophie, Wissenschaften, Technik: Philosophie.* Berlin and New York: de Gruyter, pp. 3056–3229.

Mansfeld, Jaap (1994): *Prolegomena. Questions to be Settled Before the Study of an Author, or a Text.* Leiden: Brill.

Mansfeld, Jaap and Runia, David T. (2020): *Aëtiana V. An Edition of the Reconstructed Text of the Placita with a Commentary and a Collection of Related Texts.* 4 Vols. Leiden: Brill.

Moraux, Paul (1968): "La Joute Dialectique". In: Sorabji, Richard (Ed.) (1990): *Aristotle Transformed. The Ancient Commentators and Their Influence.* London: Duckworth, pp. 55–81 [pp. 61–81 in 2nd ed., 2016, London: Bloomsbury].

Owen, Gwilym E. L. (Ed.) (1968): *Aristotle on Dialectic: The Topics.* Oxford: Clarendon Press.

Quandt, Kenneth (1983): "αἱ γὰρ τῶν ἐναντίων ἀποδείξεις ἀπορίαι περὶ τῶν ἐναντίων εἰσίν: Philosophical Program and Expository Practice in Aristotle". In: *Classical Antiquity*, 2. No. 2: pp. 279–298.

Rashed, Marwan (2011): *Alexandre d'Aphrodise. Commentaire perdu à la Physique d' Aristote [Livres IV–VIII].* Berlin: de Gruyter.

Smith, Robin (1997): *Aristotle. Topics: Books I and VIII.* Oxford: Clarendon Press.

Smith, Robin (1999): "Dialectic and Method in Aristotle". In: Sim, May (Ed.): *From Puzzles to Principles? Essays on Aristotle's Dialectic.* Washington, D.C.: Lexington Books, pp. 39–53.

Sorabji, Richard (Ed.) (1990): *Aristotle Transformed. The Ancient Commentators and Their Influence.* London: Duckworth. [2nd ed.: 2016. London: Bloomsbury].

Spranzi, Marta (2011): *The Art of Dialectic Between Dialogue and Rhetoric: The Aristotelian Tradition.* Amsterdam and Philadelphia: John Benjamins.

Steel, Carlos (1978): *The Changing Self. A study on the soul in later Neoplatonism: Iamblichus, Damascius and Priscianus.* Brussels: Koninklijke Academie voor Wetenschappen.

Stump, Elenore (2009): *Boethius's De topicis differentiis.* Ithaca: Cornell University Press.

Wiesner, Jürgen (1992): "Theophrast und der Beginn des Archereferats von Simplikios Physikommentar". In: *Hermes* 117. No. 3, pp. 288–303.

Van Renswoude, Irene (2017): "The art of disputation: dialogue, dialectic and debate around 800". In: *Early Medieval Europe* 25. No. 1, pp. 38–53.

Graeme Miles
Michael Psellos on Dialectic and Allegory

1 Introduction

As Psellos' *Chronographia* gradually approaches his own times and hence his own direct judgements, the author makes some telling remarks about the wrong (and implicitly the right) ways to do philosophy. At the beginning of Book III, he is recounting the reign of Romanus III who, he says, wished to model his rule on those of "the most philosophical Marcus" and on Augustus.[1] This was, however, only a philosophical facade and not the real thing, partly due to Romanus' own limitations and partly because of the shortcoming of the few learned men (λόγιοι) whom that time produced. These would-be philosophers, Psellos says, had come only to "the antechambers of Aristotelian philosophy" and only mouthed "the Platonic allegories" (σύμβολα), "knowing nothing of their hidden meanings". Nor did they know "what men have discovered about dialectic (διαλεκτική) nor about syllogistic deduction (ἀποδεικτική)".[2] The results of these failings are equally interesting: "since their reasoning (κρίσις) was inaccurate, their judgement (ψῆφος) of [Plato and Aristotle] was false". As a result of this, though they proposed theological problems, the majority of perplexities remained unsolved. "The palace had the outer appearance of philosophy, but this was a mask and a facade. There was no testing or scrutiny of truth".[3]

Psellos, as has often been remarked, presents himself as a new start for philosophy, reviving what had long fallen into disuse.[4] Putting aside, however, both his motivations in self-presentation and the fairness or unfairness of his judgement on these unnamed λόγιοι, I would like to turn to how he implies in this passage that one *should* pursue philosophy. The position of Aristotelian philosophy at the start is what we might expect of Psellos, who as a Platonist inherits from his late-antique sources the view that the study of Aristotle (and especially Aristotelian logic) is preparatory to the higher study of Platonism. This, he implies, is necessary for achieving the accuracy which these former λόγιοι lacked and for reaching the correct appraisal of Plato and Aristotle themselves. His remarks about Platonic allegories (πλατωνικὰ ... σύμβολα) are also important. It is not simply a matter of relating

---

1 *Chronographia* 3.2. For this text, the best edition is Reinsch (2014). All translations are my own.
2 *Chronographia* 3.3.
3 *Chronographia* 3.3.
4 On Psellos' presentation of himself as a new start in philosophy, see Duffy (2002).

these inherited allegories, as Psellos himself does at times, but of genuinely understanding the hidden meanings. This view is consistent with what we can see of Psellos' practice in the texts assembled as his *Philosophica Minora:* not only does he relay ancient allegorical readings, he is also intent on developing his own and on teaching his students to do the same.[5]

The pairing of dialectic and σύμβολα in this programmatic and polemical context emphasizes the importance of both to Psellos' understanding of the work of the philosopher. While his use and discussion of these interpretive and argumentative modes is not monolithic and unchanging across his large and multifarious corpus (and we would hardly expect that it would be), the following chapter will look at some broad tendencies that he demonstrates by examining some apposite examples. Both dialectic and allegory, as we shall see, appear in practical use in Psellos' lectures and writings and as objects of theorizing in themselves. Psellos was well aware of the complexity of the ideas and forms of argument and interpretation which he had inherited. One important difference between the two, as we shall see, lies in the range of objects to which dialectic and allegory can be directed. While dialectic, in the broadest of the senses in which Psellos understands it, takes the aspiring philosopher through almost the entire Plotinian ascent from the sensible to the One or else (following Aristotle) is an argumentative method that can be applied to all subjects, allegory, as a process of interpretive transmutation, can find philosophical content in material which would not, on any normal definition of philosophy, count as philosophical.

2 Logic and Dialectic

The lack of progress in Aristotelian studies which he criticizes in the false λόγιοι is expanded shortly thereafter in the passage quoted above from the *Chronographia*. What they should have achieved, and what by implication Psellos believes that he *has* achieved, is an understanding of dialectic and of the deductive method. These philosophical failings are no doubt serious enough in themselves, but the other failure which follows from this, and which again by implication he believes that he has avoided, is a theological one. Or rather, it is a whole set of failings: on

---

5 Contrast, for example, *Philosophica Minora* 1.40 ("On Hades") where he relays the reading of Porphyry and explicitly does not endorse it as his own with 1.44 ("On the Sphinx") where the reading is his own and is presented as an example to his students in the production of similar types of allegory. The two volumes of *Philosophica Minora* assemble the shorter, philosophical pieces ascribed to Psellos: O'Meara (1989) and Duffy (1992).

major *aporiai*, they failed to reach proper clarity and so to resolve the problems.⁶ The unstated implication is that it is the fully trained philosopher who alone is able to unravel theological perplexities. Here too, we can easily map the general proposal onto Psellos' own practice: his theological writings do indeed approach theological problems using the methods which the brief lines in the *Chronographia* would suggest: close reasoning drawing upon Neoplatonic concepts and, when he deems it appropriate, the application of allegorical reading.⁷

The more famous passage in which Psellos outlines his education presents a compatible picture,⁸ but one which I shall not examine at length just now, both because it is so well known and because it is less directly relevant to my topic. Briefly though: here too, Psellos presents a curriculum from rhetorical study to Aristotelian logic, to Plato and Aristotle themselves, then "back" to the late-antique Platonists: Plotinus, Porphyry and Iamblichus, and then, of course (in the most quoted phrase of all) "into the great harbour of Proclus".⁹ From here, he proceeded to the study of mathematics (apparently adopting Proclus' view that mathematical intermediaries are partway between sensibles and the forms proper),¹⁰ along with the allied fields of music and astronomy. Though the order makes this slightly unclear (since the details of these mathematical fields follow after his briefer statement on the movement beyond this stage), the progression from this point is to the intelligible proper and then to "that which is beyond these and is beyond intellect and is superessential" (τι καὶ ὑπὲρ ταῦτα ὑπέρνουν ἢ ὑπερούσιον).¹¹ What Psellos says that he created for himself, in short, is very much what those allegedly shoddy philosophers of the reign of Romanos failed to achieve: a Christianized version of the late-antique Platonic curriculum with a markedly Proclan character.¹²

Within this broad framework, then, I would like to turn next to the nature and roles of dialectic and allegory as Psellos sees them. As one would imagine the topic is a large one, given the centrality of both to his conception of philosophy (as is evident already in the first passage that I discussed) and due to the sheer mass of material which survives from Psellos, much of it only fairly recently available

---

6 *Chronographia* 3.3.
7 For discussion of some of Psellos' texts on Christian topics, see Criscuolo (1991), Kaldellis (2005), Lauritzen (2008, 2010, 2011, 2012, and 2021), Miles (2017 and forthcoming 2022), Walter (2017), and Diamantopoulos (2021). On the terms *theologia* and *philosophia* in Psellos, see Edwards (2021).
8 *Chronographia* 6.36 ff.
9 *Chronographia* 6.38.
10 For Proclus' views on mathematical intermediaries, see D'Hoine (2018).
11 *Chronographia* 6.38.
12 For the stages of Psellos' use of Proclus in his writings, see Lauritzen (2020). The position of Aristotelian logic in Psellos' approach is the one most typical of his educational surroundings. On the position of logic in Byzantine philosophical studies, see Cacouros (2007).

in good editions and relatively little discussed. I will focus on just a few texts that bear on the broader points.

Firstly, a short essay, edited as *Philosophica Minora* 2.4, discusses dialectic, *eudaimonia* and the beautiful. The motivation for connecting these three topics is that Psellos is closely following (indeed summarizing, sometimes *verbatim*) Plotinus 1.3 ("On Dialectic"), 1.4 ("On Eudaimonia") along with, very briefly, 1.5 ("On Whether Well-Being Increases with Time") and 1.6 ("On Beauty").[13] This is, then, one of the many brief texts where Psellos is intent on relaying the views of ancient thinkers rather than on developing their ideas in his own way or on finding connections between Platonic philosophy and Christianity.[14] As in many of these texts, it is clear that he writes in response to a question (περὶ οὗ ἠρώτησας).[15] As often, the interest in this kind of essay lies in Psellos' selections, omissions and additions to his source material which are, as ever, revealing of Psellos' own views.

Psellos begins with the definition that:

> Dialectic knowledge (ἐπιστήμη) is the capacity or skill (ἕξις) which, by means of reason (λόγος) is able to speak about each thing and how it differs from others, both as many as exist and the kinds of things which do not exist, as many as are lesser than the things which exist and those which are greater than them.[16]

Though he does not quote the opening of Plotinus' essay, where Plotinus speaks about dialectic as the *technē* or method or practice which can lead us up ultimately to the One,[17] this idea does emerge in Psellos' account, when he speaks of proceeding "until it [i.e., dialectic] goes right through the intelligible … until it comes to the first principle" (εἰς ὃ ἂν ἐπ' ἀρχὴν ἔλθῃ).[18]

By way of summary, Psellos adds that "dialectic is not circumscribed by one subject matter, but it pays attention to everything and attempts to speak about everything" (οὐ γὰρ περιώρισται ὑποκειμένῳ ἑνί, ἀλλὰ πᾶσιν ἐπιβάλλει καὶ περὶ πάντων λέγειν ἐπιχειρεῖ).[19] Dialectic, he summarizes, again following Plotinus, is "the

---

13 See the *apparatus fontium* of Duffy's edition for details.
14 Walter rightly stresses the importance of the *Theologica* for future studies of Psellos' philosophy: Walter (2017), pp. 184–185. On the chronology of these lectures, see Kaldellis (2005).
15 *Philosophica Minora* 2.4.1.
16 *Philosophica Minora* 2.4.1–5.
17 *Enneads* 1.3.1.
18 *Philosophica Minora* 2.4.10–11.
19 *Philosophica Minora* 2.4.5–6.

most important part of philosophy" (μέρος φιλοσοφίας τιμιώτατον),[20] which "is not to be considered merely an instrument of the philosopher" (οὐ γὰρ δὴ οἰητέον ὄργανον εἶναι τοῦτο τοῦ φιλοσόφου).[21] Like Plotinus, he goes on to emphasize the living reality of the intelligibles.

What I would like to emphasize at this point are two things: firstly, that we encounter different senses of dialectic in different contexts in Psellos. This is not surprising given the complexity of the traditions which he inherited and the range of interests and topics which we can find in his work. Even in the common context of summarizing the trajectory of philosophical education (as he is in both of the *Chronographia* passages and in this text from the *Philosophica Minora*), the purpose and emphasis of these summaries varies. Secondly, in each of these passages dialectic (variously understood) is credited with an all-embracing character: in the first *Chronographia* passage it was the necessary technical refinement to make correct judgements about the relative importance of Plato and Aristotle, and about difficult theological issues. In *Philosophica Minora* 2.4, it is, in the expansive Plotinian sense, nothing less than the path from sensible knowledge to and ultimately through the intelligible, ready for the final step, beyond even dialectic, to knowledge of the One. (This, incidentally, is never something which Psellos claims, and which indeed he explicitly denies that he possesses.)[22]

Another discussion of dialectic, this time focusing on dialectic in its Aristotelian rather than its Platonic or Plotinian sense, appears in the text edited as *Philosophica Minora* 2.13. Where the previous discussion had closely followed the *Enneads*, this text follows, largely *verbatim*, Alexander of Aphrodisias' *Commentary on Aristotle's Topics* and the *Topics* itself.[23] While texts like *Philosophica Minora* 2.4 and 2.13 cannot, of course, be taken to be statements of Psellos' position in his own person, the selection of these materials for his teaching do reveal the nature of his knowledge of the Platonic and Aristotelian traditions, and what he found valuable to impart to his students. The Plotinian view of dialectic which he reports in 2.4 has, as was observed earlier, a sweeping breadth, taking the aspiring philosopher through each stage of the ascent towards the One. The Aristotelian view of 2.13 possesses a breadth of a different kind: though it is not concerned, as one would expect, with a philosophical ascent of this kind, dialectic is also on this

---

20 *Philosophica Minora* 2.4.15. Psellos alters Plotinus' positive degree of the adjective to the superlative, τίμιον to τιμιώτατον, whether unconsciously or to add greater emphasis to Plotinus' point, or perhaps due to a difference in reading in the text which he employed.
21 *Philosophica Minora* 2.4.15–16.
22 Again, see the famous passage in *Chronographia*, Book 6.36–38.
23 For details of the passages quoted, see O'Meara's *apparatus fontium*, pp. 40–43.

Aristotelian understanding a method applicable not to a single subject or class of subject but, like rhetoric, to all.[24]

No one would argue that these passages give us Psellos at his most original, but they are important for relaying the traditional views of dialectic which he was willing to endorse and to pass on.[25] Returning to his criticisms of the flawed *logioi*,[26] it is clear, conversely, that the failure to properly appreciate and apply dialectic left these would-be philosophers, in Psellos' opinion, without an adequate foundation on which to build. In the following section I turn to discuss the interpretive mode other than dialectic which Psellos mentions in this polemic, and which also had for him a wide range of potential application: allegorical reading. In the final section, some comparisons will be drawn between the two and some general conclusions about their role in Psellos' thinking.

3 Allegory

Psellos' use of allegory is an aspect of his philosophical activity which has attracted some scholarly comment and which I have discussed in other contexts.[27] Readers of Psellos' theological works and the short texts edited as his *Philosophica Minora* will know that he applies an allegorical method inspired by late-antique Platonic models to Christian and non-Christian texts alike. The results and motivations of Psellos' allegorizing are increasingly well understood: while the interpretations at which he arrives in reading Christian texts are orthodox, they are developed using Neoplatonic arguments and concepts.[28] His allegorizing readings also serve the time-honored purpose of finding acceptable meaning in texts which would otherwise by deemed immoral or otherwise unacceptable.[29] What I would like to add in the present discussion, on the topic of the uses and limitations of dialectic and allegory in Psellos' works, is that allegorical reading also allows him to find encoded wisdom, generally of a Platonic kind, in materials from further "down" the so-

---

24 *Philosophica Minora* 2.13.54–63.
25 On the Psellan corpus and the nature of the texts in it, see O'Meara (1998) and Duffy (2006). For a catalogue of Psellos manuscripts, see Moore (2005). For close study of a particular instance of Psellos' "compilation and originality", see Delli (2007).
26 *Chronographia* 3.2–3.3.
27 On Byzantine allegorical reading of Homer, including that of Psellos, see Cesaretti (1991). For analyses of specific allegorical readings by Psellos, see Lauritzen (2012), Miles (2014), and Lozza (2018).
28 See Lauritzen (2010).
29 Lozza (2018).

cial and intellectual hierarchies of his time: in the expressions of popular speech and in popular religious practices. While these "lower" forms do not offer opportunities for the application of dialectic, Psellos believes that they can usefully be interpreted allegorically, permitting in effect a transmutation of these materials into philosophy.

A relatively small group of texts (some 18 pages in Sathas' edition)[30] ascribed to Psellos deal with this kind of "popular" material. Beyond a few lines in Zervos' study (*Un philosophe Néoplatonicien*, now over a century old), they seem to have received no discussion. As Zervos observed already, Psellos does not see his activity here as a discovery of popular wisdom exactly, but rather the uncovering of higher thought (much like his own) which had been concealed in these expressions and practices by some anonymous wise man or wise men of the past.[31] At the beginning of the second of these short texts, on which I shall focus as an example, Psellos announces his intentions: "It is not only serious things (σπουδαῖα) that philosophy takes up, but also things which seem playful and not serious (παίγνια καὶ ἀσπούδατα) to most people". Philosophy aims, he says, to "bring [this material] over by means of reason" (μεταβιβάζειν τῷ λόγῳ) and to "recreate" it (μεταποιοῦσα).[32] He then reinforces his point by reference to a favorite Biblical analogy for allegorical reading, the transformations carried out by Moses before the pharaoh,[33] and then an image to which he is even more attached, that of bitter and sweet water.[34]

Psellos' approach, and the nature of the interpretation at which he arrives, bear a strong resemblance to the practice and to the resulting readings in his discussion of "high" texts, be they Christian or "pagan". There are, however, some important differences between these types of reading to note at the outset. Firstly, while on some occasions at least Psellos is willing to concede that he may be creating rather than discovering meaning through his allegorizing of pagan texts, he never says this about Christian texts, nor does he say this in the case of these popular materials. On the contrary, the implication of the reference to Moses and of the suggestion that the *panēgyris* for Agatha is a kind of mystery rite is that he sees himself as dealing with venerable wisdom. The other important difference

---

30 See the section of Ἑρμηνεῖαι εἰς Κοινολεξίας in Sathas (1876), pp. 525–543. This appears to be the only edition of this text. Despite the manuscript title "Interpretations of Common Expressions" (Ἑρμηνεῖαι εἰς κοινολεξίας) the essay with which I am concerned here deals with common religious practice rather than with language.
31 Zervos (1920), pp. 183–184.
32 Sathas, *Bibliotheca* 5.527.
33 Sathas *Bibliotheca* 5.527. On this image, see Miles (2017), pp. 88–89.
34 On this image, see Duffy (2001).

in the case of the popular materials is the anonymity of the source of the philosophy that he believes is to be found in them. Whereas for his pagan materials, there is generally a named author (for example, Homer) and only occasionally a broader sense of "the ancients", and in the case of his Christian writings, there is always a scriptural or a patristic source, he considers the wisdom of the popular forms to stem from some unnamed "wise man" or philosopher of the remote past.[35]

As in the allegorical studies of "high" materials, Psellos develops an overall interpretation of the rites for Agatha from a cumulative analysis of details.[36] Much of the content of Saint Agatha's story, by contrast, is left implicit, perhaps because it is meant to be known to his audience already. If this text were our only source for Agatha, we would know practically nothing about her: the flesh and blood woman whom we find in the accounts of her martyrdom (for Psellos, most immediately the account in the *Menologion* of Symeon Metaphrastes) is elided from his interpretation. Concerning Agatha herself, he is concerned instead to develop a conceptual schema from just a few, select details, in particular her name. Psellos proposes that Agatha's name is a version of the Good (*Agathon*), altered from the neuter to the feminine in order to convey to women something of the philosophical (i.e., Platonic) teaching on this subject. All activity, he says, is ultimately for the sake of the Good, including that of the women who are loom-workers, weavers, and carders.[37] For these people, it will be necessary to have a single day of the year when attention turns to the Good, just as the rest of the religious calendar serves to turn their minds to particular religious / philosophical subjects. As he summarizes:

> for the philosophical man the Good is not circumscribed (περιγραπτόν), nor does it return in a cycle, for it is with him by nature and does not at all depart, but for women it is desirable if their recollection can revolve back to it in time (διὰ χρόνου), whenever the holy festivals are in the periods and cycles of the year.[38]

One might wonder, given the implication that an elite reader like Psellos himself seems to be required to decipher this material, what benefit it could have to the less learned people to whom it has apparently been entrusted. Psellos does, however, imply that there is some degree of understanding of this encrypted material by the women performing the rites: the more experienced lead the less experi-

---

35 For examples of these types of attributions see Sathas, *Bibliotheca* 5.528.
36 Sathas, *Bibliotheca* 5.530–531.
37 Sathas, *Bibliotheca* 5.529.
38 Sathas, *Bibliotheca* 5.529.

enced participants who are still learning.[39] The way in which the rite is learned and taught is, however, strikingly unlike the type of philosophical learning which we can witness in Psellos' own practice through his surviving writings: the women's carrying out of the rite and their learning within it are described entirely in terms of physical action. They work at the loom to create an image, presumably of Agatha's story, and in doing this, as Psellos says, "they attend to the shadow as if to the truth" (ὡς ἀληθείᾳ προσέχουσι τῇ σκιᾷ).[40] Some of the participants carry out actions depicting the suffering of Agatha, and while the story of Agatha's martyrdom is not related and analyzed by Psellos, it is implicitly present as the object of the rite's mimesis. This is evidently a kind of ritual drama, where the punisher (ἡ τιμωρός) oversees "those who whip" (ταῖς μαστιζούσαις) and "the one who is whipped cries out" (ἡ δὲ μαστιζομένη ἀπολοφύρεται).[41] It is telling that the moment which is the defining one in the hagiographic and indeed in visual sources concerning Agatha, the severing of her breast, is omitted. Perhaps this did not feature in the ritual of the women in Byzantium, though it does feature in contemporary rites for Agatha in Sicily.[42] Rather than these details of human suffering Psellos concentrates on the rite as a representation of abstract truth. This is not to deny that its emotional content is clearly important to its effect. As he writes: "those watching these [women] cry out in lament with them and understand the image of the suffering and fear to suffer the same".[43]

Within this same passage, Psellos summarizes that "they create an image of the truth by means of their figures" (τοῖς σχήμασιν εἰκονίζουσα τὴν ἀλήθειαν).[44] These "figures" (σχήμασιν) appear from the further details in the following paragraph to be dance figures. Though he does not claim (and indeed implicitly denies) that the practitioners of the *panēgyris* understand the full meaning of their actions, there is equally an implication that there is still some value in creating this image of the truth. Moreover, the suffering represented in this "initiation" (τελετή) leads back to joyful things. The last stage, which is apparently a dance in which the women join hands and "turn alternately to each side" (ἐναλλὰξ ἐπὶ θάτερον στρέφονται), represents:

> both the spheres which wander and those which do not (αἱ σφαῖραι πλανώμεναί τε καὶ οὐ πλανώμεναι), [turning] now towards the dawn and now towards the evening. And the rhythm

---

39 Sathas, *Bibliotheca* 5.530–531.
40 Sathas, *Bibliotheca* 5.530.
41 Sathas, *Bibliotheca* 5.531.
42 On the modern feast of Saint Agatha in Sicily, see De Luca and Sanfratello (2019).
43 Sathas, *Bibliotheca* 5.531.
44 Sathas, *Bibliotheca* 5.531.

is inspired by the Muses (ἔμμουσος) and the song (ἐπῳδός) is in accordance with reason, and the steps move rhythmically in relation to the time, and the alternation of the figure is already akin to the time.[45]

Psellos emphasizes, in these lines, both the harmony of the dance with cosmic reality as he sees it and the dance's time-bound nature. This returns to the theme which he had raised earlier: for the women carrying out the *panēgyris* there are points in the year at which to be reminded of the Good, whereas for the philosopher this is always present.

Despite, in other words, the truth and indeed beauty which Psellos finds in the *panēgyris* for Saint Agatha, this embodied and temporal knowledge is treated as lower than the philosophic knowledge to which he is urging his students. It is to this theme, with which his essay began, to which he returns at its conclusion, urging his audience to carry through a Platonic, philosophic ascent.

4 Dialectic and Allegory

Not unlike his late-antique philosophical predecessors, Psellos shows a marked desire to synthesize and render consistent the various intellectual currents surrounding him.

His teaching is not simply constituted by relaying the views of ancient philosophers and theologians (though it does include this) but also by demonstrating and passing on learnable techniques of argument, interpretation, and contemplation. Dialectic features prominently among these, drawing on Aristotelian logic as filtered through late-antique Platonic commentators as well as Plotinus' conception of dialectic, and so does allegory. I began with the pairing of dialectic and allegory at *Chronographia* 3.3 and will now conclude with some general observations on the two.

Dialectic and allegory are both treated by Psellos as capacious in their range of application, as not limited to a single type of appropriate subject matter. Despite his willingness to repeat ancient views which ascribe to dialectic, on occasions at least, the potential to approach anything at all, allegory, it seems, can reach further down the hierarchies of types of knowledge, tradition, and practice, to find philosophical content in what we might call popular culture. The other broad difference (and it is one that I believe holds good for Psellos' practice in general) lies in the different ways in which these two approaches or practices are conceptualized and indeed in the kinds of terms and metaphors employed to speak of them.

---

45 Sathas, *Bibliotheca* 5.531.

While dialectic is a matter of clarifying and making judgement or, in its Plotinian sense, of traversing intelligible terrain (though he makes less of this metaphor than did Plotinus himself), allegory is a practice of both uncovering and of transforming. This double conception of allegory is in large part what allows Psellos to apply allegory to the range of materials, broad in chronological and religious range as well as in class origin, in which he can find philosophical material and allows him to transform it into the kind of discourse which he most values. As a philosopher, his world is, of course, one of argument and reason, but it is also a world of *symbola*, some of which require considerable effort to unearth.

Bibliography

Primary Sources

Criscuolo, Ugo (Ed.) (1991): *Epistola a Giovanni Xiphilino*. Naples: Bibliopolis.
Duffy, John M. (Ed.) (1992): *Michaelis Pselli Philosophica Minora Volume I. Opuscula logica, physica, allegorica, alia*. Berlin: de Gruyter.
O'Meara, Dominic J. (Ed.) (1989): *Michaelis Pselli Philosophica Minora Volume II. Opuscula psychologica, theologica, daemonologica*. Berlin: de Gruyter.
Reinsch, Dieter Roderich (Ed.) (2014): *Michaelis Pselli Chronographia*. Berlin: de Gruyter.
Sathas, Konstantinos (Ed.) (1876): Μεσαιωνική Βιβλιοθήκη. *Bibliotheca Graeca Medii Aevi*. Vol. V. Venice: Phoenix.

Secondary Sources

Cacouros, Michel (2007): "Survie culturelle et rémanence textuelle du Néoplatonisme à Byzance. Éléments généraux—Éléments portant sur la logique". In: D'Ancona, Cristina (Ed.): *The Libraries of the Neoplatonists*. Leiden: Brill, pp. 177–210.
Cesaretti, Paolo (1991): *Allegoristi di Omero a Bisanzio. Ricerche ermeneutiche (XI-XII secolo)*. Milan: Guerini e Associati.
Delli, Eudoxie (2007): "Entre compilation et originalité. Le corps pneumatique dans l'œuvre de Michel Psellos". In D'Ancona, Cristina (Ed.): *The Libraries of the Neoplatonists*. Leiden: Brill, pp. 211–229.
De Luca, Maria Rosa and Sanfratello, Giuseppe (2019): "Shaping Sacred Spaces: The Feast of St Agatha and the Development of its Urban Rituality". In: *Revista Digital de Musicologia* 10, pp. 1–11.
D'Hoine, Pieter (2018): "The Metaphysics of the 'Divided Line' in Proclus: A Sample of Pythagorean Theology". In: *Journal of the History of Philosophy* 56. No. 4, pp. 575–599.
Diamantopoulos, Georgios (2021): "Remarks on Psellos' Attitude Towards the Patristic Exegetical Tradition in his *Theologica*". In: *Theologia Orthodoxa* 66. No. 1, pp. 39–80.

Duffy, John (2001): "Bitter Brine and Sweet Fresh Water. The Anatomy of a Metaphor in Psellos". In: Sode, Claudia and Takács, Sarolta (Eds.): *Novum Millennium: Studies in Byzantine History and Culture Dedicated to Paul Speck, 19 December 1999*. London: Ashgate, pp. 89–96.

Duffy, John (2002): "Hellenic Philosophy in Byzantium and the Lonely Mission of Michael Psellos". In: Ierodiakonou, Katerina (Ed.): *Byzantine Philosophy and its Ancient Sources*. Oxford: Oxford University Press, pp. 139–156.

Duffy, John (2006): "Dealing with the Psellos Corpus: From Allatius to Westerink and the Bibliotheca Teubneriana". In: Barber, Charles and Jenkins, David (Eds.): *Reading Michael Psellos*. Leiden: Brill, pp. 1–12.

Edwards, Mark (2021): "Michael Psellus on Philosophy and Theology". In: *Theologia Orthodoxa* 66. No. 1, pp. 81–100.

Kaldellis, Anthony (2005): "The Date of Psellos' Theological Lectures and Higher Religious Education in Constantinople". In: *Byzantinoslavica* 68, pp. 143–151.

Lauritzen, Frederick (2008): "Psello discepolo di Stetato". In: *Byzantinische Zeitschrift* 101, pp. 715–725.

Lauritzen, Frederick (2010): "L'ortodossia neoplatonica di Psello". In: *Rivista di Studi Bizantini e Neoellenici* 47, pp. 285–291.

Lauritzen, Frederick (2011): "Stethatos' Paradise in Psellos; Ekphrasis of Mt. Olympos (Orat. Min. 36 Littlewood)". In: *Vizantijskij Vremennik* 70, pp. 139–150.

Lauritzen, Frederick (2012): "Psellos the Hesychast: a Neoplatonic reading of the Transfiguration on Mount Tabor". In: *Byzantinoslavica* 70, pp. 167–180.

Lauritzen, Frederick (2020): "A Lifetime with Proclus: Psellos as Reader". In: *Byzantinische Zeitschrift* 113. No. 1, pp. 69–80.

Lauritzen, Frederick (2021): "Psellos' Commentary on the Jesus Prayer". In: *Theologia Orthodoxa* 66. No. 1, pp. 117–134.

Lozza, Giuseppe (2018): "Michele Psello esegeta di Omero fra neoplatonismo e cristianesimo patristico". In: *Koinonia* 42, pp. 443–454.

Miles, Graeme (2014): "Living as a Sphinx: Composite Being and Monstrous Interpreter in the 'Middle Life' of Michael Psellos". In: Kambackovic, Danijela (Ed.): *Conjunctions of Mind, Soul and Body from Plato to the Enlightenment*. Dordrecht: Springer, pp. 292–305.

Miles, Graeme (2017): "Psellos and his Traditions". In: Mariev, Sergei (Ed.): *Byzantine Perspectives on Neoplatonism*. Berlin: de Gruyter, pp. 79–102.

Miles, Graeme (2022): "Psellos on Achieving 'Likeness to God' and Being 'In the Image of God'". In: Anagnostou, Eva and Parry, Ken (Eds.): *The Neoplatonists and their Heirs*. Leiden: Brill. Forthcoming.

Moore, Paul (2005): *Iter Psellianum. A Detailed Listing of Manuscript Sources for All Works Attributed to Michael Psellos, Including a Comprehensive Bibliography*. Toronto: Pontifical Institute of Mediaeval Studies.

O'Meara, Dominic (1998): "Aspects du travail philosophique de Michel Psellus (*Philosophica Minora*, Vol. II)". In: Collatz, Christian-Friedrich, Drummer, Jürgen, Kollesch, Jutta, and Werlitz, Marie-Luise (Eds.): *Dissertatiunculae Criticae. Festschrift für Günther Christian Hansen*. Würzburg: Königshausen und Neumann, pp. 431–439.

Walter, Denis (2017): *Michael Psellos—Christliche Philosophie in Byzanz*. Berlin: de Gruyter.

Zervos, Christian (1920): *Un philosophe néoplatonicien du XIe siècle. Michel Psellos. Sa vie. Son œuvre. Ses luttes philosophiques. Son influence*. Paris: Ernest Leroux.

Index Locorum

Aeschines
In Timarchum
   173   166

Albinus
Introductio in Platonem (*Isagoge*)
   3   352, 400

Alcinous
Didaskalikos or *Handbook of Platonism*
   5   414
   5.156.24–33 (Whittaker)   338
   5.156.34–157.10 (Whittaker) 339
   5.4   415
   5.6   415
   6   400, 415, 427
   6.158–159/12–13 (Whittaker) 359
   6.159.45–160.3 (Whittaker) 344
   6.10.165.5–6 (Whittaker)   339
   6.10.165.5–16 (Whittaker) 339

Alexander Aphrodisiensis
De anima
   100.1   270
De fato
   177.6   260
   196.21   270
   204.4   260
In Aristotelis Metaphysica commentaria
   9.6   252
   23.2–3   252
   41.17   252
   52.6   268
   70.12–71.4   252
   78.2–4   253
   84.7–16; 16–21   99
   84.15   256
   85.7–19   256
   136.8–11   273
   136.11–14   253
   136.15–17   253
   137.15 ff.   252
   138.4–6   253

   164.20–21   268
   169.20 ff.   252
   170.3   252
   171.9–11   252
   172.8–9   253
   172.18–19   252
   172.18–22   251
   173.21–174.4   251
   173.27–174.4   254
   174.1   253
   174.2   256
   176.35   256
   177.8–9   252
   177.8–10   254, 265
   178.19–21   251, 252
   180.31–33   251, 253
   184.7   273
   191.5   273
   193.12–16   273
   195.13–14   273
   206.12–13   251, 273
   210.25–26   252
   218.17   251, 254, 273
   233.1–235.6   255
   236.26   256
   236.26–29   272
   236.28   255
   237.13–238.3   251
   237.15–16   253
   238.2–3   252
   245.33–35   265
   246.13–24   251
   246.14–15   253
   247.2–8   254
   247.22–29   323
   250.3–5   251
   251.7   253
   251.7–9   251
   257.12–16   251
   264.23–27   251
   264.31–34   251, 253
   266.24–25   265
   271.24–275.20   289

273.4–19	292	27.24–31	263
273.22–26	295	28.4	265
273.24–26	294	28.5–6	265
274.1–13	295	28.7–8	266
274.13–20	296	28.12	266
274.18	297	28.16–17	266
274.20–32	298	28.23–29.5	401
275.8–20	300	28.23–29.16	267
275.21–22	302	29.6	270
275.23–31	302	29.19–20	259
275.23–290.21	289	30.9–12	259
276.3–6	298	30.12–13	254
276.3–28	305	32.18–26	272
276.17–27	306, 327	33.1–2	272
276.37–277.9	315	33.19	272
278.21–23	308	40.28–29	256
278.24–279.14	309	55.24–7	441
278.27–30	309	62.16	270
279.32–280.20	321, 322	63.9–19	255
280.30–31	303	65.17–19	316
281.1–10	323	67.3–15	318
282.3–19	327	68.19–21	255
282.19–36	327	70.5–6	255
290.24–292.21	289	83.29–32	263
292.24–293.32	289	83.32	265
293.35–297.6	289	85.6–7	257
297.7–25	289	86.27	340
297.28–300.22	289	87.7	270
300.24–301.25	289	149.22–25	260
344.14–15	265	548.1 and 15	260
374.37–375.9	319	*In priora analytica Aristotelis commentaria*	
In Aristotelis Topicorum libros octo commentaria		1.3 ff.	254
1.7–8	269	1.3–4.29	348
3.18 ff.	270	1.7–2.34	264
3.21–23	263	2.20–22	264
3.23–24	263	3.7	254
4.5	254	3.20–24	257
5.5	270, 271	8.25	269
6.18	271	14.17	270
6.21–7.2	257	262.28–32	261
18.33	260	263.26	261
23.22–25.9	241	326.20–328.5	359
26.12–13	257	*Quaestiones*	
27.3–4	263	1.7	278
27.8–18	259	1.8	278
27.19–20	262	1.10	278
27.21–24	271	1.16	278

2.3	278, 282–284	Aristophanes	
2.25	278	*Nubes*	
3.11	303	1173	61
3.12	270	1485	166
4.5	278		
4.16	278	Aristoteles	
25.21	260	*Analytica posteriora*	
101.18	270	71b9	238, 244
		71b9–10	244
Ambrose of Milan		71b23	233
De viduis		72a5–7	233
13.75–77	336	72b1–3	219
De fide		74a13–16	220
1.13.74	342	74a16–17	220
		74a28–29	244
Ammonius		75b38	238
ap.Olympiodorum *in Gorgiam*		76a4	238
41.9	384	76a27	238
In Aristotelis Analyticorum Priorum librum I commentarium		76a38–b2	235
		77b9–14	218
8.20–9.2	264	77b12–14	243
In Aristotelis Categorias commentarius		77b19	217
93.9–12	388	78a30–b4	219, 220
In Aristotelis librum de interpretatione commentarius		79b24	217–218
		83a33	70
54.24	303	85b5–7	220
		97b14–26	46
Anastasius Sinaita		*Analytica priora*	
Quaestiones et Responsiones		1.30–31 ch.	374
48	343	24a16–17	315
		45b19	359
Anonymous		46a17–22	375
Prolegomena philosophiae Platonicae (Introduction to Platonic Philosophy)		53a2	238
		59b20–24	230
1.1–12	428	61a31	220
1.8–12	428	61b10–15	230
4.15 Westerink	384	61b18, 32	220
22.40	45	62a11	220
26	421	62b12–20	70
Apocalypsis Johannis (Revelation)		62b26	220
1:16	360	63b15	220
19.12	360	66b11, 8	290
		Ars rhetorica (Rhetoric)	
Apuleius		1355b10–11	270
De dogmate Platonis		1355b16–21	240
Book II	45	1358a21–26	443

Categoriae
1a20–22	317
1a25–29	314
1a29–b3	317

De anima
429a27–28	71

De caelo
279b4–11	446
279b6–7	453

De generatione animalium
724a20–30	227
777a32–b16	221

De generatione et corruptione
335b10–14	70

De interpretatione
3, 16b19–22	303
3, 16b22–25	303

De partibus animalium
642a9	70, 75
642a19	385
642b–644a	337

Ethica Eudemia
1216b26–32	373, 376
1216b32–33	241
1216b40–1217a10	244
1217a 7–10	243
1220a16–18	241
1235a4–b18	376
1249b5–6	241

Ethica Nicomachea
1109a34–35	59
1124b28–1125a2	46
1143b11–14	378, 385
1145b2–7	374
1145b20	374

Fragmenta
fr. 13 Rose (*ap.* Synesium *calvit. encom.* 22)	378
fr. 53 Rose (Cicero, *Tusculan Disputations* 3.69)	386

Historia animalium
289a1–3	323

Metaphysica
982b15	253
983a33–b4	379
984a18–19	385
986a2	255
987b14–18	71
987b19	255
987b31–33	70
987b32–33	332
991a9	253
991a21–22	70
991b3–7	70
992b19	255
993a25	253
993a27–b11	371, 377
994a22–b3	227
995a17–19	252
995a25–26	253
995a25 and 34	252
995b2–4	268
995b4	251, 252
996b26–997a15	288
998b9	255
1003a22–37	337
1004a32 and 34	251
1004b18–25	240
1004b19–20	258
1005a22–23	288
1005a23–27	235
1005b12–25	292
1005b14	337
1005b19–20	288
1005b26–27	288
1005b35–1006a11	289
1005b35–1006a28	289
1006a3–4	288
1006a11–18	290
1006a11–28	289
1006a12–13	293
1006a13–16	294
1006a15	290
1006a16–17	300
1006a18–20	294
1006a18–31	306
1006a20–21	300
1006a22–23	303
1006a24–25	296, 297
1006a26	299, 300
1006a27	298
1006a28–b11	301
1006a28–1007b18	289, 301
1006a31–b11	305

1006a31–b34	306	1310a 32	41
1006b11–b34	313	*Sophistici elenchi* (*Sophistical Refutations*)	
1006b28–30	326	ch. 22	98
1006b28–34	315	165a31	240
1006b30–33	327	168a18	217
1006b33–34	288, 327	168a40–b10	244
1007b5–15	314	168b5–8	243
1007b18–1008a2	289	168b8–10	244
1008a2–7	289	168b15	217
1008a7–34	289	169b20–25	212
1008a34–b2	289	169b20–29	212
1008b2–31	289	169b23	215, 216, 232, 233
1008b31–1009a5	289	169b23–24	214–215, 222–223, 228, 231
1017b21–22	319		
1023a26 ff.	227	169b24	214, 217
1026a18	358	169b25–27	232, 242
1033a5–23	227	169b25–29	213
1043b4–32	94	169b26	217, 223–225, 228
1043b7, 10	227	169b26–27	225
1051a24–25	233	169b27	214
1057b8	325	169b27–28	242
1072a26	281	169b28	217
1072b	87	170a23	220, 223
1072b4	281	170a23–24	219
1072b3 f.	281	170a 23–31	218
1074a38–b14	378	170a23–39	218
1074b34	87	170a 24–26	218
1075a 11–15	87	170a25	220
1092a21–35	227	170a25–26	220
1179b33–35	399, 408	170a26	221
Physica		170a31	225
184a15–21	373	170a33–34	232
184b25	448	170a34–39	223
185a5–10	449	170a36–38	218, 223
188b29–30	385	170b1	290
190a21–31	227	171a2–4	290
190b4–5	227	171b4–6	223
207b29	340	171b5	217
209b14–15	75	171b6	214, 223, 228, 230
218a30–b20	381	171b11	241
265b17–266a5	385	171b11–16	237
Politica		171b12	232, 239, 241
1261a6	70	171b12–13	239, 240
1261a16	70	171b16–22	234
1263b29–31	70	171b19	232
1275b20–21	40	171b21	232
1284b19	59	171b29	217, 222

172a6	234	Asclepius	
172a8–9	233, 234	*In Aristotelis Metaphysica*	
172a21–27	223, 224, 226	260.33–34	299
172a25	226		
172a27	217	Athanasius	
172a29	226	*De decretis Nicaenae Synodis (Decr.)*	
172a32	226	27.1–2	349
172a37–38	226		
172a39	226	Aulus Gellius	
172b8	238	*Noctes Atticae (Attic Nights)*	
172b10–11	240	1.26	262
172b35–173a30	50		
175a36	220	Basilius of Caesarea	
Topica		*Homiliae super psalmos*	
100a20–24	374	32.341 a	358
100a25–105b36	337		
100a27–30	240	Cicero	
100b23–101a4	240	*Academica*	
100b24–25	229	II. 42.129	85
101a5	239	II. 91–98	338
101a10	239	*De inventione*	
101a13	239	I.46	271
101a13–17	240	*De Fato*	
101a14	232, 240, 241	28–30	350
101a15–17	241	*De oratore*	
101a27 ff.	269	I, 47	44
101a28–29	259	III, 129	44
101a35	259	*Topica*	
101a35–36	271	1	441
101a36 ff.	218	*Tusculanae Disputationes*	
101a36–37	259	2.3.9	262
101a36–101b2	443		
103b20	316, 317	Clemens Alexandrinus	
103b27	317	*Stromata*	
104b2	257	1.176.3	339, 340
104b14	270	1.177.1	340
105a7–8	256	1.9.39–44	359
105b19–21	267	5.12.81.5–82.1	340, 341
105b 30–31	337	6.10.80.4	339
110a26–27	260	8.17.4 (Havrda)	340
119a30	325	8.19.3 (Havrda)	340
155b23	50		
156a7–13	50	Cornutus	
159a32	258	*Epidrome (Introduction to the Traditions of Greek Theology)*	
161b35	270		
162b8	238	35,75.18–76.5 Lang	381

Damascius
In Philebum
  Sect. 244   399, 409

David
Prolegomena Philosophiae
  lectures 7–8   428
  16.3   428
  31.34   428
  32.1–2   428
  63.32   428

Demetrius
De elocutione
  5, 205–298
Deuteronomy
  23: 1   336

Diogenes Laertius
De vitis clarorum philosophorum
  I, 1–12   381
  II, 30   166
  II, 106   84, 85, 87, 94,
                               166, 337
  II, 107   95, 166
  II,108   95
  II, 118   100
  II, 119   99, 101
  III, 6   84
  III, 58   400
  VI, 39   100
  V, 23   337
  V, 24   441
  V, 43   441
  V, 44   441
  V, 45   261, 441
  V, 46   261, 337, 441
  V, 49   441
  V, 50   441
  V, 59   441
  VII, 39   349
  VII, 41–83   338
  VII, 42   271
  VII, 180   350
  IX, 25   337
  IX, 51=DK 80b 6a   164
  IX, 55   163

Diomedes
Ars grammatica
  299.19–23   422

Dionysius Halicarnassensis
Antiquitates Romanae
  VIII. 71.1.3   299
  X. 40.2.1   299
  X. 41.2.3   299

Elias
Prolegomena Philosophiae
  lecture 5   428
  14.9–10   428

Epictetus
Dissertationes ab Arriano digestae
  1.10.8–9   262
  1.26.1   262
  1.26.13–18   380
  2.1.31–32   380
  2.14.1   262
  3.21.6–7   380

Epiphanius
Adversus Haereses (=AH)
  2.522.7   349
  64.3   335
Epistola ad Hebreos (*Epistle to the Hebrews*) [traditionally attributed to St. Paul but probably not by him]
  4:12   360

Eriugena John Scotus
Periphyseon
  2.559B   361
  3.632D–633 A   361
  4.749 A   361
  4.749AC   361
  4.765C   362
  4.780BC   363
  4.807 A   362

Euclides
Elementa
  Proposition I. 30   236

Euripides
Antiopē
  Frg. 189   164

Eupolis
  Frg. 352   166

Eusebius
Historia Ecclesiastica (HE)
  6.8.1–3   335
  6.14.10   349
  6.19.4–8   349
  6.19.12–14   354
  6.36; 6.37   343
  7.33.3   344
Praeparatio Evangelica
  14.17.1   85

Galenus
De placitis Hippocratis et Platonis
  2.3   351
  4.13   339
  7.200   339
  9.9 (V 796–805)   339
De naturalibus facultatibus (On Natural Capacities)
  35.6   384
De differentia pulsuum (On Types of Pulse)
  8.579   384

Gregorius Nyssenus (Gregory of Nyssa)
De opificio hominis
  5.2   341

Gregory Thaumaturgus
Oratio panegyrica in Origenem (Thanksgiving or Panegyrical Oration to Origen)
  8   346
  13.150   358

Hermias
In Platonis Phaedrum scholia
  248.4–8   407

Herminus
  ap. Porphyrius *in Aristotelis categorias commentarium*
  59.17–33   388

Hesiodus
Opera et dies
  11–12   170

Homerus
Odyssea
  5.259–261   76

Isocrates
Ad Demonicum
  3–4   97
Antidosis
  170   166
  235   166
  268   166
  270   166
  313   166
Helenae Encomium
  1   166
  9   166
In sophistas (Against the sophists)
  1–2   173
Panēgyricus
  3   166

Jerome
De viris illustribus
  2   349
  54   349
  III. 70   343
Epistulae
  33.3; 33.4: *Dialogus adversus Candidum Valentinianum*   343, 349

Lucretius
De Rerum Natura (RN)
  3.1072   354
  5.1185   354

Marinus
Vita Procli
  22   383
Martyrium Iustini
  3–4   335

Matthaios (Matthew)
Kata Matthaion Evangelion (Gospel of Matthew)
   19.12   335, 336

Olympiodorus
In Aristotelis Categorias commentarium
32.13	434
32.18	434
55.3	434
98.9	434
99.14–15	434

In Aristotelis Meteora commentaria
6.23	435
51.29	435
52.31	429
69.17	435
75.25	435
118.13	435
128.21–131.23	437
153.7	435
174.15	435
175.14/19	435
175.17	437
188.2–3	435
214.30	437
239.13–14	435
241.19	435
245.3	435
255.23/25	435
260.25	435
263.20	435
266.37	436
270.16	435
298.21	435
313.24	435
321.28	436
322.25	435
332.6	435

In Platonis Alcibiadem commentarii
1.3–9	428
2.79	429
11.7–23	431
29.17–20	432
30.1	433
32.6	433
35.1–37.3	433
32.9–11	433
32.11	433
57.5	431
142.9	431

In Platonis Gorgiam commentaria
4	391
11.9	431
13.8	431
18.6	431
19.14	431
27.2	432
41.9	384
41.10	431
47	371, 391

In Platonis Phaedonem commentaria
10.3	429

Prolegomena logica (Prolegomena to Logic)
4.4	433
4.15	433
4.19	433
15.10	433
17.37–18.6	430

Origenes
Commentarii in Canticum Canticorum (C. Cant.)
prol. 3.1	357
prol. 3.1–2	357
prol. 4.15	353
2.1.48	335
2.5.21–28	353
2.6.13	355
2.10.7	346
5.2.14	347
3.13.9–17	355
3.14.34	355

Commentarii in Evangelium Iohannis (C. Io)
2.6.51	360
2.7.54–57	360
2.11.65	351
2.177	356
2.179	356
2.180	356
8:42	351
13.5.27–30	333
13.46.303–304	359
20.17.135–140	351

Commentarii in Evangelium Matthaei (C. Mat.)
15.1	336

15.3	335

Contra Celsum
2.20	350
3.77	341
4.48	350
5.22	357
6.6	333
6.9	360
6.10	
6.22.5–20	436
7.12–15	351
7.42	358
7.48	336

De oratione
2.465	345
3.42, 461, 557, 607, 663	345
3.509	345
3.577, 586,663	345
17.16	341

De principiis (Princ.)
6	346
2.8.4–5	345
2.11.1	
2.11.4	354

Epistola ad Amicos Alexandrinos (Letter to Friends in Alexandria)
PG (Patrologia Graeca) 17.625B 343

Fragmenta in Sancti Pauli Epistulam ad Ephesios
35.24	360

Homiliae in Ezechielem (H. Ez.)
13.3	346

Homiliae in Genesim (H. Gen.)
11.3	355
12.5	355
13.3	355
14.3	359

Homiliae in Lucae Evangelium (H. Luc.)
fr. 83.14	353

Philocalia, Anthology of Origen's Works
10.2	356
14.2	358
24	343

Protrepticus ad martyrium
37.20	360

Selecta in Psalmos [Dub.]
144	341

Pamphilus
Apologia pro Origene
7	345
9	354

Parmenides
DK B7.8	61

Patrologia Latina (PL)
26.649.21	
122.1035 A–1036 A	362

Persius
Saturae (Satyrs)
3.66	354

Philo
Quod deterius potiori insidiari soleat (Det.)
176	335

De agricultura
15–16	351

Philodemus
De vitiis
22,30–32

Philoponus, Joannes
In Aristotelis Analytica Priora commentaria
9.18–19	401

Pindarus
Isthmia
5(schol.vet. 2b)	436

Plato
Alcibiades I
114e	384
117e–118a	180
119c–124a	162

Apologia
17b	46
21c–d	182
21d	345
21e	182
22d7–8	155
22e–23a	182
23c	182

23c–e	181	*Euthydemus*	
23d	48	272d	164
24d–28a	49	275d–278e	96
29	352	275d–277c	96
29b	180	275e	97
29d–31c	165	279a–282e	96
32a4–5	60	283a–286b	96
32e1	49	288a–293a	96
33a5–6	181	293b–303a	96
33c	431	*Euthyphro*	
38a	332	5d8–e2	202
Charmides		6d–e	202
157c6–7	67	11c4–5	66
164e7–165a7	178	*Gorgias*	
165b5–c2	111	448d	347
Cratylus		449b	347
385b7–8	68	450c7–e2	202
390c10–11	106	451a–b	49
399d11–e2	72	452a–d	49
406	352	452d–456a	337
421	352	453a	347
421d	162	454a–455a	48, 52
431b2–c2	64	454c4–5	66
431b5–c1	62	454e–455d	41, 42
436c7–d7	68	454e9–455a2	42
439c6–d	68	456c7–8	53
Crito		457c4–458b3	183
46b4–6	62	457d4	53
Epistulae		457e1–3, 4–5	51, 53, 67
312 e	406	458a2–b	111
341a5–7	73	460c	45
341c7	74	463d2	41
342ab	360	463e–466a	41
342a4	74	464b–d	202
342a7–b1	73	464b–466a	37, 202
342b1–2	73	464b3–4	39, 50
342d4	74	465a	48
342e2	74, 109	466a2	51
342e3	74, 109	467b11	50
343a1	74	468c2–5	42
343b8–c2	74	468e–469c	38
343c2–3	60	471e–472c	347
343c8–d1	73	473d3	50
344b3–c1	74, 107	473e	37
344b6	73	473e6	40
360b–c	99	474c	49
617–9	332	475b–c	49

Index Locorum

475d5–e1	38	515b6	53
477e7–8	40	516e–517a	43
479b–e	38	516e9–517a6	43
480c–d	38, 42, 43	517a	40
480c1–3	42	517a–b	45
480c5–7	38	517a5	41
482c	50	517b	39
482c5	50	517b2–c2	40
482e3–4	50	519d5–6	50
487e6–7	67, 68	521d	37, 45
489e1	50	521d–524d	435
490c8–d1	50	521d6–8	39, 41
490d6	50	521d8–e4	40
490d10	50	521e 6–8	38
490e4	50	524a8	52
490e9–491a3	50	524d4	52
492a–c	51	524e–525a	38
492d–494a	52	525b8–c1	39
493a	178	525e5	52
493b–c	52	526a–b	37
493c3–d3	52	526a4	52
494a3, 6	52	526c3	52
494d	50	526d3	52
494d4	50	526d4	52
495a	347	526e1–4	53
495b2	49	526e–527a	38
497a3	50	527b7–c4	44
500e	40	527c1–4	47
500e3–4	50	527c5	52
503a–b	37, 42	527e7	52
503a2–9	41	*Hippias Major*	
504d–e	37, 42	302c12	66
504d5–6	41, 42, 43	*Ion*	
504e1–3	39	534c	431
505a–b	40	535a–542a	431
505c3–4	38, 41	*Laches*	
506b–c	182	188c5	42
507c3–5	50	*Leges*	
508a	51	631	352
508c–509c	38	715e	406
508c1	45	751d	162
509a	430	804b3–4	75
509a5	40	853e10–854a1	74
509a4–7	41	891	352
511c–513c	41	895c11–12	72
515a–b	40	897e 1–2	59
515a3–4	41		

Index Locorum

Lysis		145b6	405
219c4	67	*Phaedo*	
Meno		58e	431
71e1–72a5	202	59a4	61
72a–74b	202	60e1–61a4	68
73a7–b2	132	63c7–8	60, 73
75cd	333, 347	65e4	72
79	352	66b3–4	60, 72
81	352	66b7	61
81a–d	70	66e4–67a2	72
81c–d	202	67a3	72, 73
85b–86b	70, 71	68e	122
86b1	71	70b–c	166, 170
85b7–86d2	69	70b6	429
86e3	69	70c1	60, 70
87c5–6	69	74a5–75a3	71
99e	431	74c1	71
100b	431	76d8–9	70, 71
Parmenides		76e4–5	70
127d–128e	167	77a5	75
128e6–7	132	77b2–3	75
130a7–9	155	77c2–6	75
130b1	131	77d4	75
130b7–9	143	78d1–2	61, 73
135b5–c3	131	79c7	59
135b5–c6	129, 143, 153	80b3	72
135c–136c	167	80b10	72
135c2–3	131	85c9–d1	69
135c8	70	85d1–2	59, 75, 76
135d	166, 170, 268, 418	87a4–c4	75
135d3	62, 131	89c7–8	70
135d3–6	266	89d4	62
135d4–7	70	90b4	62
135d5	60, 70, 400	90b7	60
135d7	400	90c1	61
135e8–136a2	133	91c	384
136a1–2	70	95a–99d	59
136b6–c5	405	96a8	61
136b7–c5	207	96a9–10	60
136c5	70, 207	97b6–7	60, 61
136c6	207	97c5–6	59
136e4	400	98c2–e1	59
137d	406	98e1	59
138a3–6	407	99b2	59
142b5–7	407	99b3	59
142c7–d	407	99c9–d1	57, 59
142d9–143a3	407	99d8–e1	64

99e2–3	59, 73	236d4–5	118
99e5–6	59, 60	236e5	133
99e5–100a4	70	237a	118
99e5–6	59	237a9	118
99e6	57	237a–b	118
100a1	61	237b	112
100a3–4	65, 69	237b2–5	119
100a4–7	68	237b7–c5	120
100a5	65, 67	237d3–5	119
100b1–9	68, 69	237d6–9	113, 120
100b5	64, 68	237d–e	112
100b7–9	70	237e2–238a2	120
101d	418	238b7–c4	113, 120
101d3–5	67	238c6	120
101d–102a	347	238d5–7	120
101e	95	238e4–5	121
101e1	69	238e–239b	122
101e2	61, 70	239c4–5	122
103e5	72	240a–b	115
104e1	71	240b5–7	117, 122
105d3–4	72	240c4–6	122
105e8	75	240d4–e7	117
107a–b	430	241b	122
107b1	74	241d3	118
Phaedrus		241e8	118
227c	115	242b8–d2	120
230a	431	242e3	118
230a3–6	112	242e5–243a1	126
230e7–231a2	114	243a	127
231b7	114	243a4	118
231d–e	114	243c2	118
232b–d	114	243e–57b	416
233a1–2	114	244a	126
233b6–c1	114	244a1	118, 196
233c–d	117	244c	431
233e	114	244c2	120
234 a–b	114	245a1	196
234c	115	247a8–b3	420
234d1	115	247b8	419
234d5	115	248b	416
235a	116	248d	415
235c5–6	116	249d4–5	196
235c–d	116	249e	109
235d6–7	116	252b2	118
235e	116	253c7	118
236b5–8	116	259e–262c	47
236d	118	261a7–9	46, 47

261b2	47	272a–b	123
261c	410	272d–273b	123
261d10–e4	47	273a8–b1	121
261e2	47	273d–274a	123
262b5	46	273d8–274a3	333
262b5–8	113	273e9–274a2	121
262c1–3	113	275c1–2	123
262c8	133	276a5–7	124
262c10–d2	113	276be	332
262d	110	276e5–6	60
264a	113	276e5–277a4	124
264b	113	277b–c	124
264e7	118	277d6–e3	124
265a9–11	196	278a–b	124
265b2	196	278c–d	124
265b3–4	120	279b8	120
265b8	118	*Philebus*	
265c1	118	11c	347
265c6	118	16b–19a	198
265c8	117	16b5–6	195
265c8–d1	122	16c1–2	195
265c8–266c8	190	16c–17a	416
265d–266a	169	16c5–17a5	195
265d3–5	122, 191, 194	17a	347
265d5–7	194	17a6–18d2	195
265 d7	118	18a7–b5	195
265d8–e3	194	19c2–3	59
265e1–3	122, 191, 194	19c1–3	73
265e3	118	27	352
265e3–266b1	194	23b5–26d10	198
265e4–266b1	195	23b7	190, 198
266a3	118	23c9–d8	198
266b	346	50b	347
266b3–4	191	64e5	74
266bc	338	*Politicus*	
266b3–c1	125, 414	257a1–2	192
266b8–c1	191	257a3–5	192
268a	110	257c1	192
269c–272b	45	258b6–7	192
270a	166	258c3	60
270b4–9	123	258d4–5	202
270b–c	45	262b6–8	201
270c6–7	67	262c8–263a3	195
270c10–d7	122	266d5	192
271a5–8	123	268d4	207
271a10–b5	123	278c9–d6	74
271b–272b	47	284b7	192

285d4–7	208	508a	90
285d5–6	206, 207	508b	90
285e4–b2	198	508c	90
286a5–6	198	508e	90
286a7	198	509b	91
286b10	192	509b9	420
287a3–4	206, 207	509c	91
287c	200	509d1–511e5	133
287c3–4	201	510b	91, 105, 418
287c4–5	203	510b4–9	413
287c6	201	510b7	69, 70
288a3ff	203	510c	91
291c3–4	192	510c–511c	413, 415
299b	166	510c2–511d5	148
300c2	59	510c6	68
303b6–c7	193	510d5–511a1	413
		510d2	68
Protagoras		511a–c	413
318a3	261	511b	92, 105, 106
329a	50	511b3–c2	413
329b	164	511b4	105
333a6–8	67	511b8–c2	413
334d	164	511c	92
335a	162	511c1–2	420
336b	50	511c5	105
337ab	347	511d6–e4	65
338e–347b	372	511e2	64, 72
Respublica		517d8–9	71
336	352	523–525	131
359bc	335	526b	168
368	352	528	352
372d	91	531–9	333
409c5–9	60	531c	334
416a–c	180	531d–535a	415
454a	337	531d–e	106
454a6	202	531e	334
454b6–7	106	531e4–5	106
475a–477b	337	532a	105
475d1–476d6	165	532b	334
476a1–8	143	532b–535a	202
476c1–d2	143	532d8	105
476d	105	532e3	415
477a10–478d12	205	533a2	332
477b–478b	105	533a8	105
480a 11–12	165	533a10–c6	148
489a	166	533b	334
506e	89	533b5–c3	143
507b	89		

533c	105, 334	218c6–e6	207
533c5–6	68	218dff.	338
533c7	60, 73, 105, 190	219 a1–2	204
533c7–d1	106	219a7	200
533c8	413	219a8–d2	159
534a1–5	64, 72	219a10–b2	200
534b	106, 334	219b–225d	170
534b3–4	105	219c2–7	159
534b–c	334, 413, 415, 416	219d4–7	160
534b7–d1	165	219d4–e2	159
534c	106	219d5	160
534c1–2	68, 107	219d9–e2	159
534c3	68	221	352
534d	106	221c9–d2	200
534e	334, 418	222a	110
534 e 2–3	399	223c6–7	160
535cd	334	224c10	160
535d1–2	202	224e1	160
536d	334	225a	162
537–539d	418	225a–c	110
537a	335	225a8–b1	162
537c	334	225b13–c9	110
537c1	106	226a–c	175
537c7	106, 202	226b–231b	177
537d3–7	165	226b–c	200
537d5	105	226c	172
539a	97	226d	172, 175
539b	337	226d5–7	174
539bc	332	227a–b	160
539e2	70	227a7–b6	175, 176
540a7–9	69, 70	227a8	195
540c	335	227a–c	174, 178
548c5–7	184	227a7–c6	195
581a9–10	184	227b	175
598b6–8	64, 72	227b6–c6	176
605b	178	228a–231b	110
607b7	60, 70	228b2–4	177
608b4–8	53	228c 1–5	177
Sophista		228c4	177
216c2–217a2	192	228c7–8	177
216c4–5	110	228c10–d2	177
216d1–2	110	228d4	178
217a4–9	192	228e–230d	109
217b1–2	204	229b7–230d5	48
217b2–4	204	229d5–6	203
217c–218a	164	230a	180
217c5–7	132	230a 5–9	178

230a 6–9	111	261a–c	110
230b–d	157, 181, 182	263b3–7	68
230b–e	174, 179, 180	264a8–9	62
230b4–5	177	264a8–b4	148
230b 4–8	181	264c	338
230b–235a	337	264d12–265a2	195
230c–d	179	264e1	196
230c3–d1	181	266aff.	200
230c8	179	266a1–2	196
230c8–d1	181	266a4	196
230d1	183	266d5–8	197
230d1–4	111	267d	338, 346
230d 7–8	177	267e	337
230d8	112	*Symposium*	
230e1–4	155	175e2–3	68
231a	180	204a	180
231a–d	169	212a	117
231a1–3	111	215c7	133
231a7–8	111	*Theaetetus*	
231b5–8	111	142b7–c1	133
232b–e	164	144d1–3	148
231e	165	146a6	133
232b12	164	146b3	133
232d2	164	146c7–d3	134
232d9–e1	163	146e7–8	133
232e3	164	147d4–148b3	193
232e4	164	147d9–e1	193
235b8–c7	195	147e5	193
240d	110	148e1–3	133
248d	177	149a4–151d6	134
250b–258c	337	150b–151d	181
253be	333	150c4–d7	181
253b–254a	169	151c2–5	182
253c7–9, 7–8	193, 198	152a1–183c3	139
253c9	333	152a1–186e12	138, 139
253d–e	198, 333	152a5	134
253d2–3	198	152c5–6	152
253d5–6	198	152c8–d6	140
253d5–e2	198	152d2–3	132
253e	418	153a4	134
253e7	192	154d8–e5	134
254a9	198	155d1–7	133
254b4	192	157c7–9	141
254cff.	452	157d7–9	140
255bff.	416	157d10–12	134
255d5–e7	322	161b9–10	133, 145
259e4–6	416	161c–d	182

161d2–e3	140	185e2	147
161d7–162a2	154	185e3–5	133
161e4–5	134	186a2	144, 149
162a4–5	133, 145	186a2–8	147
164e4–7	145	186a2–e12	148
161e6–7	133	186a2–b1	139, 150
162a4–b5	133	186a4–5	148
165a1–3	133	186a10–b1	138, 142
166d5–167a6	181	186a11	144
169a1–3	145	186b2	143, 144–145
169a6–b4	145	186b2–4	149
171c8–9	145	186b6–7	144
172c–173e	163	186b6–9	149
172c–176a	165	186b11–c5	146
173c–d	163	186c3	144
175b7–d2	155	186c7	151
175 c1–3	165	186c9–10	151
175e	165	186d2–6	152
177c3–5	146	186e4–8	135
177c6	141	186e11–12	138
177c6–179b9	138–141, 149	187a1–6	134
177c6–d5	141	187a3–6	144
177e4–178a3	141	187a7–8	144, 148
177e7	141	187c7	
178a5–7	141	189e4–6	62
178a7–8	141	189e4–190a2	154
178c2–7	141	195b	166
178d4–6	141	195b10	60, 70
179a10	133, 145	195b–c	170
182b6–8	146	201	352
182d8–e6	135, 142	202b3–5	62
183b	133	210b11–c4	134
183b7–c3	137,138, 145, 146	210b–c	181
183c3	146	210c2–4	182
183c4–7	146	210c4–d2	134
183e7–184a2	132	*Timaeus*	
184b3–186a1	139, 148	18b 1–3	436
184b4–186a1	147	21e–25d	379
184b3–186e12	129, 131–132, 134–139, 144	30a2–7	59
		30b	178
185a4	147	46c7	59
185a4–d5	135	47	352
185a9	143	51d3	76
185c5–6	143	52b	268
185c8	151		
185c9	144, 147		
185d6–e2	135, 151		

Plotinus
Enneades
1.3	415, 460
1.3.1	460
1.3.1, 9–10	415
1.3.1, 16–17	415
1.3.3, 5–10	415
1.3.4	416, 421
1.3.4, 1–6	416
1.3.4, 2–7 and 12–15	344
1.3.4, 4–5	416
1.3.4, 6–9	416
1.3.4, 7–8	416
1.3.4, 9–12	416
1.3.4, 14–15	416
1.3.4, 15–18	417
1.3.4, 18–20	417
1.4	460
1.5	460
1.6	460
2.4.7.14–15	340
3.7.1	382
3.7.7	382
4.3.9.1	353
6.1–3	452

Plutarchus
Adversus Colotem
1120d–e	94

De Stoicorum repugnantiis
9	349

Moralia
499b	431
1119b	431

Polybius
Historiae
8.36.6.2	58

Porphyrius
De philosophia ex oraculis
109.9–110.9 Jacoby	446

Scholia ad Odysseam
schol. ad Od. ι 106 (*SSR*, V.A. 189)	91

Vita Plotini
3.35–37	352
13.10–17	352
14	349, 380, 442
19	353

Vita Pythagorae
19	335

Praedestinatus
21	343

Proclus
In Platonis Alcibiadem
10.3–18	384
21.5–8	384
209.11	432

In Platonis Parmenidem commentaria
I 622.18–33	418
I 630.37–633.12	400
I 635.31–638.2	401
I 637.5 ff.	427
I 637.9	401
I 638.14–640.16	401
I 640.17–24	402
I 640.27–37	403
I 641.1–4	402
I 648.1 ff.	418
I 648.2	167
I 648.20–649.8	167
I 649.1–8	418
I 649.32–651.9	169
I 652.29–653.2	168
I 653.1	269
I 653.3–5	168
I 653.5–7	167
I 653.7–18	167
I 653.18–33	419
I 653.18–654.1	169
I 654.1–10	180
I 654.1–14	169
I 654.2	169
I 654.11–13	419
I 654.15–19	168
I 654.17	420
I 654.19–34	168
I 654.34–655.2	168
I 654.34–655.6	180
I 655.2–12	168
I 655.12–656.14	169
I 656.17–26	170

I 656.27–657.18	170	I, 11, p. 49, 3–11	398
I 657.23–27	170	I. 11, p. 50, 2–3	405
I 657.27–658.26	170	I, 11, p. 53, 19–22	398
IV 907.8–18	172, 183	II, 12	406, 407
V 981.1–9	269	II, 12, p. 68, 6–22	406
V 989.10–17	167	II, p. 83, 10–18	405
V 989.10–23	169		
V 989.11–14	169	Psalm(s)	
V 989.16–23	168	103: 24	355
V 990.1–11	170		
V 991.21–992.7	170	Psellus Michael	
V 997	408	*Chronographia*	
V 997.16–1007.34	400	3.2	457
V 1001.13–21	397	3.2–3.3	462
V 1001.37–42	397	3.3	457, 459, 466
VI 1050.5–26	406	6.36 ff.	459
VI 1061.20–1064.12	403	6.36–38	461
VI 1062.10–17	403	6.38	459
VI 1099.27–31	398	Ἑρμηνεῖαι εἰς Κοινολεξίας, in Sathas 1876	
VI 1114.25	398, 406	pp. 525–543	463
VII 1142.10–1143.39	404	*Philosophica Minora*	
VII 1191.34f.	398	1.40	458
In Platonis Rempublicam commentarii		1.44	458
I 184.21–25	431	2.4	460, 461
I 285.5–9	424	2.4.1	460
I 292.6	400	2.4.1–5	460
I 295.19–24	410	2.4.5–6	460
I 383.13–14	358	2.4.10–11	460
In Platonis Timaeum commentaria		2.4.15	461
I, 16.6–12	384	2.4.15–16	461
I, 43.2–17	436	2.13	461
I, 350.8–20	399	2.13.54–63	462
In primum Euclidis elementorum librum commentarii			
		Pseudo-Aristotle	
31.4–17	399	*De Mundo* 6	283
76.8			
98.13–14	407	Pseudo-Galenus	
235.15–236.8	236	*History of Philosophy* 4 (XIX 237) 339	
374.18–21	236		
375.8–12	236	Pseudo-Platonic *Definitions*	
Institutio theologica (*Elements of Theology*)		*Def.* 415b10	66
176, 10–16	408		
Theologia platonica		Quintilianus	
I, 1	358	*Institutio oratoria*	
I, 7, p. 31 and 25–27	398	II, 15, 24–29	45
I.9	401	II, 15,27,28 and 32	45
I, 9, p. 38, 4–7	401	II, 15, 29	45

Sathas
Bibliotheca
5.527	463
5.528	464
5.529	464
5.530	465
5.530–531	464, 465
5.531	465, 466

Seneca
Epistulae morales ad Lucilium (Letters)
49.8	95
90.5	381
108.23	380

Sextus
Sententiae (Sentences)
13, 273	335

Sextus Empiricus
Adversus Mathematicos (AM)
2.6–7	348
7.16	349
7.22	349, 351
7.60	163
8.174–175	270
8.225	352
11.187	349

Pyrrhoniae hypotyposeis
2.3	351

Simplicius
Commentarius in Epicteti enchiridion
pr.87–90	383

In Aristotelis Categorias commentarium
1.22–2.3	453
2.26	387
2.30–3.4	451
3.2–9	384
3.10–17	451
3.13–17	384
4.31–5.1	443
7.17	447
7.23–29	371, 386
7.23–32	446
11.30–13.19	387
12.3–10	447
12.8	447
12.10–13.4	387
15.28	441
15.28–32	447
16.1,3, 14	447
16.14	441
18.28–19.7	254
38,19–24 Kalbfleisch = Boethus fr. 10 Rashed	323
70.14–20	444
70.18	444, 446
73.25–35	452
113.27	447, 453
143.4	447
159.10–15	387, 388
163.28–29	388
352.22–24	451
364.7 ff.	452
380.13	444, 446
415.16	443, 447
438.33–36	384

In Aristotelis De caelo commentaria
25.29	446
106.4–108.14	450
140.25.26–30	451
140.30–32	451
140.32–33	451
141.1–6=31B17.7–8, 10–14 DK	451
159.3–9	453
236.26 ff.	444
238.5–11	447
238.8–10	446
238.8–11	444
238.10	447
292.18 ff.	453
297.3 ff.	452
298.21–22	451
313.11	443
523.25–30	447, 453
523.30	447
528.32–33	451

In Aristotelis libros de anima commentaria
136.29	445

In Aristotelis Physicorum commentaria
3.13	379
6.31	379
16.20–28	448

16.23	448	Socrates	
20.12 ff.	453	*Historia Ecclesiastica (HE)*	
21.22–22.8	450	6.13	349
22.22 ff.	450	SSR (=*Socratis et Socraticorum Reliquiae*)	
36.24–31	453	II. A. 5	84
46.9–10	448	II. A. 8 (*Socratic Letter* 14, 2)	84
46.12	448	II. A. 30	85, 87
47.21	444, 446	II. A. 31	85
47.22	447	II. A. 34	95
47.22–26	448	II. B. 13	95
47.25–26	449	II. O. 6	100
49.9–11	449	II. O. 26	85
51.6–7	449	II. O. 27	99
52.8	444	IV. A. 211	94
52.15	444	V. A. 150	94
70.21	444	V. A. 189	91
71.31–72.2	449	*Sources Chrétiennes (SC)*	
77.9–11	451	222.196–200	359
86.2–3	453	278.158–160	340, 341
131.23	447	290.224–226	351
144.25–29	451	*Stoicorum Veterum Fragmenta (SVF)*	
145.1–146.28=28B7 DK	451	1.75	348
157.25–7	451	2.35	349
158.1–159.4	451	2.37	349
159.13–26	451	2.38	349
160.1–11	451	2.42	349
160.16–17	451	2.44	349
160.20–21	451	2.242	352
160.28–161.7	451	2.248	351
161.16–20	451	2.957	350
162.18–22	451		
172.1–3	452	Strabo	
175.11–15	452	*Geographica*	
179.28–29	453	XIII.1	262
182.9–10	453		
202.31–203.1	450	Straton	
331.10	451, 452	Fr. 19 Wehrli	262
335.20	447		
476.26–28	449, 453	Synesius	
526.17	446	*Dio*	
542.19–22	268	3.5	345
640.12–18	386		
723.12	444	Syrianus	
915.18	443	*In Aristotelis Metaphysica commentaria*	
1020.10	443	2, 18–3, 1	399
1024.16	443	3, 17–24	399
1318.10–15	385, 452	55.38–56.2	399

81, 38–83, 11	399	Vergilius	
119, 28	399, 408	*Georgica*	
In Platonis Parmenidem commentaria		2.490–2	354
VI 1041.22 ff.	403, 404		
VI 1061.20–1064.12	403	Wittgenstein	
VII 1142.10	405	*Tractatus Logico-Philosophicus*	
		2.12	64
Theophrastus			
FHS&G (1992)		Xenophon	
fr. 68.3c and 85 A	293	*Memorabilia*	
vol.1, Nos.		I.1.11	166
118–121 (Definitio)	441	I.2.60	261
122–136 (Topica)	441	I.6.13–14	166
124 A (= Alex. Aphrod. *in Top.* 102b27, p. 55.24–7 Wallies)441		IV.5.11–12	106
		Cynēgeticus	
		ch. 13	166

Index Nominum

Adamantius 343, 349
Adrastos of Aphrodisias 441
Aelius Aristides 44f., 344
Aeschines 85, 166
Agatha 464f.
Alcibiades 27, 117, 359, 384, 428f., 432f., 435
Alcinous 15, 344, 359, 414
Alexander 16f., 24f., 98f., 241, 249-265, 267-273, 277-279, 281-284, 287, 289f., 292-300, 302-312, 314-328, 348, 387, 442, 444, 446, 448, 450f.
Alexander of Aphrodisias 15-16, 98, 249, 277, 287, 347, 349, 441, 452
Ambrose 336, 342
Ammonius 264, 303, 384f., 388, 428f., 435f.
Ammonius Saccas 333, 344
Anastasius 343
Anastasius Sinaita 343
Anaxagoras 73, 88, 437, 452
Anaximander 450
Anaximenes 437
Andronicus of Rhodes 264
Antisthenes 5, 22, 83, 85f., 94, 173
Apollonius of Tyana 284
Apuleius 45
Archytas 451
Aristarchus 380
Aristides the Just 37
Aristippus 83, 85, 94
Aristocles of Messene 85
Aristophanes 5, 166, 173
Aristotle 2f., 5, 8, 13-16, 18-21, 24-28, 40f., 46, 50, 59, 70, 73, 85, 87, 94, 98, 167, 189, 211-214, 216-232, 235-244, 249-254, 256-259, 261-263, 265-273, 277-284, 287-295, 297-319, 321-326, 328, 332, 337, 340f., 348f., 358, 371-380, 384-387, 390, 400, 427f., 430f., 434, 436-438, 441-450, 452-454, 457-459, 461
Athanasius 349
Augustine 342f.
Augustus 457
Aulus Gellius 262

Bardaisan 344
Basil 358
Beryllus in Bostra 343f.
Boethus of Sidon 388
Bryson 99, 236, 240

Callicles 37, 39-43, 48-52, 112
Candidus the Valentinian 343
Carneades 270
Celsus 350f., 358, 360, 436
Chaeremon 354
Chrysippus 262, 348-350
Cicero 13, 44, 85, 262, 271, 338, 350, 386, 441f.
Cimon 40, 42f.
Clement of Alexandria 339
Cornutus 354, 381
Crassus 44
Crates of Thebes 100
Crito 50, 62, 96f.
Ctesippus 97

Damascius 399, 409, 451
Democritus 437, 450
Demosthenes 422f., 431
Diogenes Laertius 5, 12f., 84f., 87, 94f., 99-101, 163f., 166, 261, 264, 271, 337f., 349f., 381, 400
Diogenes of Sinope 100
Dionysodorus 96, 164

Elea 89, 159, 165, 170, 191-193
Empedocles 437, 451f.
Epictetus 262, 380, 445
Epicurus 349
Epiphanius 335, 349
Eriugena 361-363
Euclid 85, 88-89
Eudemus of Rhodes 98
Eusebius 85, 335, 343f., 349, 354
Eustathios from Thessaloniki 58
Euthydemus 96
Eutropius 343
Evagrius 334, 361

Ficino, Marsilio 62, 73, 79

Galen 257, 264, 268, 338 f., 351, 384
Glaucon 53, 91
Gorgias 48
Gregory of Nyssa 341 f., 344, 356
Gregory Thaumaturgus 346, 353, 358
Gual 62, 81

Hephaestus (the governor of Alexandria) 429
Heraclides 73, 343
Heraclitus 3, 135 f., 146, 449
Hermias 407, 410, 422, 431
Hesiod 4, 170 f.
Hierocles 349
Hilary 341
Homer 4, 26, 161, 371, 422 f., 431, 464
Horus 336
Hypatia 335

Iamblichus 332, 358, 383 f., 387, 401, 405, 429, 447, 452, 459
Isocrates 97, 166, 173

Justin 155, 335

Kebes 70

Laius 350
Leucippus 450
Longinus 352–354
Lucretius 354
Lyceum 98, 262
Lysias 108 f., 112–119, 121, 124, 126
Lysis 168

Marcus 343, 457
Martianus Cappella 442
Matthew 336, 355, 388
Maximus of Tyre 344
Megara 12 f., 84
Megethius et Marcus (the two Marcionites) 343
Melissos 444, 451
Melissus 85, 444, 451
Miltiades 40, 42 f.
Moses 463

Nicomachus 354
Numenius 334, 354

Olympiodorus 19, 21, 27, 45, 345, 371, 384, 391, 421, 427–438
Origen of Alexandria 16, 331

Pamphilus 335, 343, 345, 354
Parmenides 85, 143, 167, 331, 451
Pausanias 58
Pericles 5, 40, 42 f.
Persius 354
Phaedrus 47, 106–121, 123–127
Philo 335 f., 341, 351, 354
Philoponus 400, 436, 446, 452
Photius 349
Plotinus 17, 25 f., 333, 340 f., 344, 352 f., 358, 361, 372, 379–383, 387–390, 401, 415–417, 421, 442, 451 f., 459–461, 466 f.
Polos 38, 42, 45, 49 f.
Polus 40, 43, 431
Polyxenus 98 f.
Porphyry 91, 335, 345, 349, 352 f., 358 f., 380, 387, 401, 405, 442, 444, 446, 450, 458 f.
Posidonius 381
Proclus 167–173, 180 f., 183, 269, 344, 397–410, 417–421, 424, 448, 459
Prodicus 170
Protagoras 133–134, 139, 164, 181
Psellos 457–467
Pseudo-Dionysius 361
Pythagoras 335

Quintilian 45

Romanus III 457
Rufinus 349, 354, 357

Sathas 463–466
Seneca 95, 380 f.
Serapis 336
Sextus 270, 335 f.
Sextus Empiricus 163, 270, 338, 348 f., 351 f.
Simmias 70, 76
Simplicius 384–389, 441–454
Solon 379
Stilpo 99–101

Strabo 262
Strato 262, 441
Synesius 345, 378
Syrianus 401–411, 447

Theaetetus 20, 23, 111, 129, 131–140, 142–155, 163–166, 168, 170, 173, 181 f., 192 f., 195, 197, 200 f., 205, 333
Themistius 344
Themistocles 40, 42 f.
Theodorus (the mathematician) 133
Theophrastus 261, 359, 441, 450
Thyphoon 112
Timon of Phlius 96

Virgil 354

Xenophanes 85, 88
Xenophon 83, 106, 166, 261, 344

Zeno 2 f., 5, 7, 85, 131, 337, 348, 397, 400, 403
Zeno of Elea 2, 7, 337
Zeus 52, 108, 121 f., 125

Allan, Donald James 376
Anscombe, Elizabeth 287, 304 f., 314–316
Apelt, Otto 62, 80
Aubenque, Pierre 249 f., 262, 280

Barnes, Jonathan 264, 281, 289, 323, 374–377, 384 f., 442
Berti, Enrico 2, 161, 167, 268, 281 f., 287
Bluck, R.S. 62, 73
Bonitz, Hermann 173
Booth, N.B. 179
Boys-Stones, George 13, 371 f., 378 f., 381, 386
Bredlow, Luis 4, 86
Broadie, Sarah 281 f., 332
Brown, Lesley 191, 195 f., 204, 423
Burckhardt, Jacob 161 f., 170
Burnet, John 37, 58, 60–63
Burnyeat, Myles 74, 135–137, 141, 143 f., 147, 193

Cherniss, Harold 85, 189
Cooper, John M. 39, 47, 53, 110, 136, 138 f., 144, 147, 184, 384

Cordero, Néstor 86
Cornford, Francis MacDonald 68, 111, 138, 140, 143, 158, 163–166, 175, 178 f., 189, 199, 421, 424
Coulson, M.J. Lee 162

DaVia, Carlo 14 f., 373, 375
Davidson, Donald 69, 71
Delcomminette, Sylvain 6, 70, 192, 206, 249
Diès, August 39, 165
Dillon, John 15, 18, 167–169, 269, 344, 358, 373, 383, 397 f., 408, 418 f., 427
Dixsaut, Monique 60, 62 f., 73, 80, 194
Dodds, Eric Robertson 37, 46, 384, 408
Donini, Pierluigi 249, 283
Döring, Klaus 86
Dorter, Kenneth 9, 11 f., 160, 176, 189, 199, 414

Ebert, Theodor 60, 63, 80

Falcon, Andrea 14, 373 f., 376, 380
Ferber, Rafael 12, 21, 57 f., 60, 64, 68 f., 71, 73–76, 91 f., 189, 202
Frede, Dorothea 14, 63, 73, 151, 199, 249, 317, 373–375, 378–380
Futter, Dylan 372, 390 f.

Gallop, David 12, 62–64, 69, 80
Gertz, Sebastian 428, 433 f.
Gill, Mary Louise 129 f., 153, 200 f., 207, 405
Goethe, Johann Wolfgang von 75
Gottlieb, Paula 291
Griffin, Michael 14, 18, 26, 264, 371 f., 389, 428 f., 433
Grube, Georges Maximilien Antoine 53, 57, 61 f., 420
Guthrie, William Keith Chambers 179

Hackforth, Reginald 59, 62 f., 69, 115, 117, 120, 122, 126, 190, 194, 199
Hadot, Pierre 262, 267, 348, 372, 380, 383, 390, 443, 445
Heidegger, Martin 2, 135, 137, 152, 157 f., 160–162, 174
Horn, Christoph 62, 95, 299

Ierodiakonou, Katerina   15, 20, 263, 338, 349, 399
Irwin, Terence   13 f., 41, 287, 374 f.

Jaeger, Werner   87

Kanayama, Yahei   12, 58 f., 62 f., 66, 75
Karbowski, Joseph   14, 216, 244, 373–376
Kerferd, B. George   5, 111, 179, 181
Kupreeva, Inna   15 f., 25, 244, 249 f., 253–255, 272, 278, 284, 287, 292

Laks, André   279–282
Leibniz, Gottfried Wilhelm   22, 67

Madigan, Arthur   250 f., 253 f., 272, 290, 299, 308
Martinelli Tempesta, Stefano   58
McDowell, John   138, 143 f.
McNeill, David N.   377, 387
Menn, Stephen   87 f., 280, 287, 291 f., 372, 448 f.
Mignucci, Mario   287, 306, 311 f., 325 f.
Morrow, Glenn R.   18, 73, 167–169, 269, 397, 418 f., 427
Moss, Jessica   37, 184, 205
Most, Glenn   1, 6 f., 11, 13 f., 16 f., 19–21, 25, 38, 40, 43 f., 48, 50, 65, 75, 91–93, 106, 109, 111, 115, 122, 134, 139, 142, 146, 148, 165, 168 f., 174, 176–178, 180 f., 183, 189, 195, 198, 202, 204, 213, 217, 220, 222 f., 228, 233, 237, 239–241, 253, 259, 267, 270, 273, 279–284, 293, 310, 314, 331–333, 337–339, 342, 344 f., 350, 353, 356, 363, 372, 374–377, 382, 390, 414, 419, 421 f., 435–437, 443 f., 446, 448, 450, 453, 457, 459, 461–464, 467
Mühlenberg, Ekkehard   341

Naess, Arne   62
Narcy, Michel   47, 50, 157, 160 f., 165 f., 179 f., 182, 267
Natorp, Paul   62, 64
Nehamas, Alexander   6, 47, 164, 166, 181, 183
Notomi, Noburu   105 f., 158, 166, 179, 181, 190, 195
Nussbaum, Martha   375

Owen, Gwilym Ellis Lane   375, 386, 454

Pernot, Laurent   37

Reale, Giovanni   61 f., 81, 280
Renaud, François   21, 37, 40, 44, 48–50, 372, 429 f., 432 f.
Robinson, M. Thomas   7, 9–11, 57, 67, 105, 107, 129, 135, 158 f.
Robinson, Richard   7, 9–11, 57, 67, 105, 107, 129, 135, 158 f.
Rose, Lynne E.   61, 378, 386
Rosen, Stanley   115, 175, 179, 310
Ross, William David   62, 65 f., 70 f., 281, 294, 371, 377
Rowett, Catherine   135, 137, 145–147, 152, 155
Rufener, Rudolf   62
Runciman, Walter Garrison   190, 198 f.
Ryle, Gilbert   136, 143, 151

Schleiermacher, Friedrich   62, 79
Sedley, David   1 f., 13, 70, 80, 136–139, 149–151, 333, 349, 380 f., 388
Sferlea, Ovidiu   341
Shorey, Paul   65 f.
Steel, Carlos   349, 397, 405, 407, 418 f., 445
Szlezák, Thomas   39, 74 f., 107, 161, 332
Szram, Mariusz   336

Tarrant, Harold   1, 5, 7 f., 27, 52, 371, 380, 385, 421, 427–429, 432 f., 436
Taylor, Alfred Edward   119, 192, 199, 421, 424
Tredennick, Hugh   62
Trevaskis, John R.   179, 181, 189, 196, 199, 208

Van Campe, Leen   404–406
Van Ophuijsen, Johannes Max   256, 259–263, 265–267, 269, 271, 316, 319
Vlastos, Gregory   7 f., 11, 22, 58, 65 f., 129, 431 f.
von Fritz, Kurt   85 f.

Weedman, Mark   341
Whitaker, C.W.A.   303 f., 315, 323
Wittgenstein, Ludwig   64, 201
Woodruff, Paul   47

Yunis, Harvey   46 f., 118

Zervos, Christian   463

Zeyl, Donald J.   39f., 42–44, 47, 52f.

Index Rerum

Academics 338
accident, accidents 15, 319, 323 f., 326, 339 f., 340, 382, 415, 434, 441, 454
– accidental 322–324, 326 f.
accounts 7, 20, 23, 62, 89, 138, 185, 189, 203, 206, 244, 254, 268, 315, 375, 383, 388, 452, 464
– reasoned 62
admonition (νουθετητικόν) 109, 111, 114, 178
– traditional 85, 92, 111, 185, 232, 359, 423, 462
ἀδολεσχία 60, 70, 170 f., 173
affirmation and negation 303
– two types of negation 321
ἄγνοια 217
Agōn 161
– agōn in relation with *logos* 161
– conventional Athenian *agōn* 162
– verbal contest (ἀγών) 172, 183
agonal age of the history of the Greek civilization 161
– Athen's agonal ethos 162
agōnistikon 163
– agonistic spirit 162
– examples of the agonistic motif in *Odyssey* and *Iliad* 161
agreement (homologia, or harmony) 7, 41, 62, 120, 127, 143, 146, 202, 249, 297, 320, 377, 382, 385, 390, 449, 452 f.
– harmony, *homologia, symphonia* between Aristotle and Plato 26, 65, 67, 162, 371, 376 f., 383, 385 f., 390, 466
ἀγών 161, 162
αἰσχύνη (shame) 50
akrasia 374
– Socrates' denial of 374
Alchemy 284
– alchemical thought 284
– alchemists and chemists 284
– history of Alchemy 284
Alcinous 15, 338–340, 344, 359, 400, 414 f., 417, 427
– *Didaskalikos* 339, 359, 400, 427
– *Handbook of Platonism* 15, 414

alētheia 151, 376
Alexander of Aphrodisias 15 f., 21, 24 f., 98 f., 249, 277, 287, 340, 347, 349, 359, 436, 441, 443, 452, 461
– Alexander's commentaries 442
  – on Aristotle's *Meteorologica* 428, 434 f., 437
  – on Aristotle's *Physics* 450
  – *On De Generatione et Corruptione* 284
– Alexander's school collections, *Mantissa* and *Quaestiones* 308
– Alexander's view on Parmenides and Melissos 451
– *Quaestiones* 16, 25, 250, 260, 277 f., 308, 343
– theory of prime matter 284
Alexandria and Pergamon 379
Alexandrian School 427
Allegorical 28, 336, 360, 459, 462–464
– ancient allegorical readings 458
– Byzantine allegorical reading of Homer 462
allegory, allegoresis 22, 84, 89, 91 f., 458, 462, 466 f.
Ἀλήθεια ἢ Καταβάλλοντες (λόγοι) 163
ἀμαθία 178
ἁμιλλᾶσθαι 162
ἁμιλλητικόν 162
analogizesthai 149
analysis 2, 7, 11 f., 15, 17, 20, 23–25, 27 f., 86, 92, 96, 122 f., 157, 159, 161, 169, 171, 173, 201, 212, 283, 292 f., 325 f., 331, 344, 362 f., 374 f., 399, 410, 415, 417, 419, 450, 464
– analytics 338
anaskeuazein 437
anger against oneself 182–184
another form of pursuit (ἄλλος τρόπος τῆς μεθόδου) 60 f.
antilogic 6, 61, 161, 167, 169–171, 173, 181
– the Protagorean type of antilogic 181
Antisthenes' natural communities 91
– *Sathon* 94
ἄπειρον 340 f.
apokatastasis 350, 363
aporetic 16, 25, 105, 107, 183, 250, 253, 273, 278, 280, 342, 372, 375

- aporetic dialogues   108, 181, 227, 332
- aporetic or zetetic aim   279
aporia, aporiai   25, 250-255, 267, 272 f., 277-280, 282, 288, 391
- aporiai and solutions   453
- aporiai vs. *Problemata*   279
- as the pathway to solutions   453
- multiple solutions for a difficult philosophical problem   278
appropriate   24, 212-215, 217, 219, 221 f., 224, 228, 231-244
- appropriate for the first philosopher   252
- appropriateness to its object   211-213, 217, 229, 244
- explanatory appropriateness   24, 217, 222, 229, 231-235, 242-244
- explanatory appropriateness and primariness   235
argument   1, 3-9, 12 f., 16, 18-20, 23-25, 27, 45, 47, 50, 61-64, 66, 70, 75, 80, 83, 86 f., 89, 94-101, 106 f., 115, 118 f., 125, 131-139, 141-145, 147 f., 150, 152, 163-165, 167-170, 173, 178 f., 181, 191 f., 195, 202, 211-227, 229-238, 240-244, 249-252, 255, 259, 262, 266, 268-273, 277 f., 281, 287-290, 292-302, 304-315, 317, 319, 321, 326-328, 331, 335, 337, 341, 343, 345, 347, 349-353, 356 f., 363, 372, 374, 376, 385, 390 f., 405, 413, 418-421, 427-430, 433, 435-438, 443-447, 449-451, 458, 462, 466 f.
- anti-Megarian argument   90
- Bryson's argument   234-237, 240
- by performance   27, 413, 421
- demonstrative (*meta apodeixeōs*)   15 f., 24, 163, 240, 243, 249, 287, 385, 399, 431
- dialectical   444-446
- elenctic   7
- from authority   390, 449 f.
- from signification   25, 287, 289, 293, 295, 297, 301, 305-307, 312, 314 f., 317, 326, 328
- logical   444, 446 f.
- patterns of sophistic argument   427
argumentation   1-4, 11, 21, 39, 45, 48 f., 51, 95, 193, 218, 256 f., 259 f., 262 f., 273, 348, 400, 443
- peirastic   24, 213-215, 217, 222-232
- philosophical strength in dialectical argumentation   349
- rational   363
- sophistical   24, 211-215, 231-234, 236-244
Aristotelian   4, 13-16, 18, 20, 24, 26 f., 50, 168, 249 f., 253, 258, 260, 263 f., 268-270, 273 f., 277 f., 284, 310, 328, 337, 340, 348, 351, 359, 371 f., 379, 388, 410, 415, 417 f., 427-430, 433, 438, 441-443, 447, 452-454, 457-459, 461 f., 466
- conception of the four elements
  - fire, air, water, earth   284
- corpus   9, 37, 48, 84, 181, 249, 257 f., 270, 282 f., 323, 328, 357, 383, 389, 417, 443, 445-447, 458, 462
- doxographical methodology   372
- logic   15-20, 264-266, 268, 348, 430, 454, 457-459, 466
- syllogistic   15, 19, 224, 235, 337, 417, 444, 449,
- tradition   15, 250, 278, 337, 461
Aristotelianism   16, 20, 25, 249, 277, 282, 381
- as a commentary tradition   381
- as an exegetical tradition   277, 282
Aristotle's philosophy
- account of refutations   217
- account of signification   311, 315
- *Analytics*   19, 219, 224, 230, 233, 235, 238, 243, 243 f., 249, 261, 264, 287
- aporiai   14, 16, 25, 129, 277-283, 374 f., 383, 386 f., 453, 459
- Aristotle's later Neoplatonist exegetes   373
- *Categories*   254, 265, 314, 316-322, 328, 341, 384-389, 400, 434 f., 441, 452
- classification of sophistical fallacies in *Sophistical Refutations* 8   211
- *De Generatione et Corruptione* II.10   282
- dialectic   13-16
- dialectical method   14
- discussion of substance in the *Categories* and *Metaphysics* Z   314
- doxography   102, 374 f., 381
- *Eudemian Ethics*   241, 244, 375
- formal syllogistic writings   444
- introductions to the *Topics* and *Prior* and *Posterior Analytics*   374
- *Metaphysics*   1, 3, 14, 16, 24-26, 61, 227, 233, 235, 240, 249-255, 257 f., 263, 265 f., 268,

270, 272 f., 279–281, 283, 287, 290, 295, 298, 316, 319–321, 328, 358 f., 377–379, 399, 403, 408, 410, 428, 445, 447
- *Metaphysics* 1 and *Physics* 1.5   372
- *Metaphysics* α   377, 386
- *Metaphysics* book *Beta*   279
- *Meteorologica*   27, 282, 432, 434–438
- method in *EN* 1.7   387
- *Nicomachean Ethics*   Organon
- *Physics* 4   375, 381, 386
- Poetics   431
- procedures in natural science   373
- *Rhetorics*   240
- *Sophistical Refutations*   98, 213–215, 237, 242, 337, 445
- theory of argumentation   216
- *Topics* A.2   449, 454
- *Topics*   13–16, 19, 27, 50, 218, 238–241, 254 f., 257–259, 263, 267–273, 292, 316–318, 320, 326, 328, 337, 374, 441–444, 446 f., 450, 453 f., 461
- *Topics* VIII   261, 454
art   1, 5, 13, 17 f., 37–41, 44, 46–51, 60, 71, 109–113, 123, 125, 141, 154, 158–161, 163–165, 170, 172, 176, 179, 202, 263 f., 266, 332, 338, 347, 360 f., 379, 390, 422, 431
- art of arts and the science of sciences' (ἡ τέχνη τῶν τεχνῶν καὶ ἐπιστήμη τῶν ἐπιστημῶν)   359
- cathartic   23, 168 f., 174, 184 f., 421
- concerning the *logoi* (ἡ περὶ τοὺς λόγους τέχνη); „the art of dialectic" (διαλεκτικὴ τέχνη)   60
- of acquisition (τέχνη κτητική)   159 f., 173, 176
- of diairesis   171, 173, 176
- of discriminating, discrimination (διακριτικὴ τέχνη)   160
- of politics   39–41, 45, 53
- which cares for the soul   39
ascent   17, 93, 107, 416 f., 420, 424, 461
Asclepiodotus' *Timaeus*-commentary   436
association   84, 88, 270, 420
- between the sun and the Good   84
- between unity and being   88
astronomy   346, 375, 459
atheists   353
ἄτομον   203, 340

Attic writers   422
Augustine's dialogues/works
- *Contra Academicos*   342
- *De beata vita*   342
- *De immortalitate animae*   342
- *De libero arbitrio*   342
- *De magistro*   342
- *De musica*   342
- *De ordine*   342
- *De quantitate animae*   342
- *Soliloquia*   342
axioms   18 f., 91, 235, 288

babbling (ἀδολεσχία)   165, 170, 418
- babbler (ἀδολέσχης)   165 f., 170, 345
basic qualities hot, cold, dry, wet   283
Beauty   74, 98, 109, 117–119, 121, 125 f., 347, 355, 409 f., 433, 460, 466
Begriffe, conceptos or ideas   62, 79
Being (*ousia*), essence   2–4, 8, 13, 17 f., 20–24, 38, 40–44, 50, 58–67, 69, 73, 75 f., 86–91, 95, 97 f., 101, 107–111, 113 f., 116, 118, 121–124, 127, 129–131, 133–136, 138–155, 168–170, 172, 175, 177 f., 180–183, 185, 189 f., 192, 198, 206 f., 211–214, 216 f., 219 f., 225, 227, 229, 231, 233–238, 240–243, 250–252, 254 f., 257 f., 260, 263–265, 268, 279, 281 f., 284, 288 f., 292–294, 296 f., 299, 301–304, 308, 310, 312–317, 320–324, 328, 333 f., 336, 339 f., 344–346, 350, 353, 360, 362 f., 377, 380, 387–389, 391, 397 f., 401–410, 415, 418 f., 428–430, 432–437, 446–449, 452, 460
- and judgements   151
- and truth   9, 53, 73, 152
- being appropriate to its object (οἰκεῖον τοῦ πράγματος)   212–215, 217, 222, 238
- being, life, and mind   409
- claim that knowledge depends on being   152
- essence (to ti ên einai)   4, 11, 15, 17, 28, 61, 68, 73 f., 105 f., 121, 144, 147, 149, 151, 255, 304, 315 f., 319–321, 334, 362, 409 f., 413, 415
- essential nature of things   344, 382
- existence   19, 42, 47, 49, 70, 85, 87 f., 95, 99, 111, 131, 144, 147, 224, 284, 375, 397, 401, 404, 408–410, 421, 442

- giving an account of being (λόγον διδόναι τοῦ εἶναι) 61, 73
- hierarchy and unity of being 94
- reality 361–363
- superessential 459

beliefs 7–9, 14, 18, 27, 69, 71f., 85, 89, 97, 102, 154, 168, 177, 179, 183–185, 289, 291, 293, 332, 378f., 419, 432f., 437
- justified true 72
- moral 7, 38, 43, 47, 51, 71, 122, 180, 184, 265f., 277, 332, 346, 357, 391, 417, 423, 432f., 437
- tribunal 41

beloved, the 'non-beloved', „non-lover" 109, 112, 114–117, 119f., 122, 126f., 281

benefit 108, 115, 134, 140, 143, 146, 149, 151, 155, 163, 181, 258f., 378, 390, 422, 428, 464

Biblical books 358
- Ecclesiastes 358
- Proverbs 358
- the Song of Songs 348, 356, 358

body, bodies 38, 60, 72f., 93, 115, 122f., 146f., 162, 176, 178f., 181, 194, 202, 262f., 281–283, 321, 345, 353, 383, 423
- celestial spheres 280
- ensouled 283
- heavenly 283, 355, 357
- heavenly movement 282
- simple 3, 7, 27, 42, 58f., 62, 73, 112, 122, 140, 144, 147, 158, 215, 249f., 283, 363, 384, 387, 390, 423, 448

Boethius 442
- *De topicis differentiis* 442

Buddhist Pāli canon (Tittha Sutta) 377
- Udana 6
  - 4 377

by means of questions and answers (ἐρωτῶντες καὶ ἀποκρινόμενοι) 61, 73

Byzantine Alexandria 438

calculations about *being* and *benefit* 150f.
Cartesian dubitatio de omnibus 71
categories 20, 74, 98, 159, 254, 265, 314, 316–322, 325f., 328, 341, 384–389, 400, 427, 434f., 441, 452
- category or genus 387

- first four categories (substance, quantity, relation, quality) and the sensible world 452
- in a subject (in the sense of the *Categories* 2) 314

cause, causes, causation 18, 26, 41, 45, 50, 59, 61, 87, 90, 115f., 172, 177f., 194f., 198, 214, 216, 251–253, 255f., 282f., 296, 319f., 354f., 362, 377, 379, 400–403, 405f., 411, 433
- co-causes (συναίτια) 59
- explanatory power of the four-causal schema itself 379
- formal 61, 86, 107, 176, 190, 226, 234, 320, 379, 430, 438, 444
- material causation 379

Celarent 225–227, 230
Chaldean Oracles 420
chatter (ἀδολεσχικόν) 165
chemistry 284
Christ and Logos 26, 331, 357, 360
Christian 13, 16, 21, 25, 28, 278, 331, 334, 336, 338–340, 343f., 346, 349–351, 354, 357, 359f., 363, 438, 453, 459, 462–464
- imperial pre-Plotinian Platonist, Platonism (commonly, a Christian 'Middle Platonist') 25, 338, 339, 359
- Platonism 25, 338f., 344, 363, 431,
- Platonists 331, 338f., 344, 359
- theories in Plato 360

Chrysippean anapodictical arguments 351
cleansing (καθαρμός) 172, 174f., 383
coincidence of the epistemological and theological realms 362
collection and division 10, 20, 22, 24, 27, 109, 113, 123, 125f., 169, 189f., 193f., 196, 198f., 201–204, 206, 208, 414, 416–421, 424
commentary, commentators 16–18, 21, 26f., 167, 250–258, 261, 263–265, 268–273, 284, 287, 289f., 297f., 302f., 316, 318f., 327f., 336, 345, 348f., 355f., 358f., 371, 381, 383–386, 389, 391, 397f., 400–408, 410, 421, 424, 428–438, 442–448, 450, 453f., 461
- ancient 3, 5, 10, 13f., 19–21, 24, 26, 28, 39, 44, 46, 83, 85–87, 89, 155, 161f., 178, 189, 236, 244, 249, 252, 261, 268, 284, 331, 338, 342, 356, 372f., 378, 380–385, 387, 389–391, 402, 421, 431, 437, 443, 460, 464, 466
- philosophical commentary 390

common items   24, 214, 226, 228 f., 231
common notions, ennoiai   19, 382, 387, 391, 433
– common properties (*ta koina*)   136, 147–149
– *ennoiai* or *prolēpseis*   391
– *koinai ennoiai* (common concepts)   382 f.
communion (κοινωνία)   20, 23, 157, 180, 408
competition   6, 161–163, 170
concept   1, 5, 39, 62, 87 f., 95, 99 f., 105, 107, 114, 136, 143, 145 f., 149, 151, 172, 178, 201, 203, 267, 289, 299, 308, 310, 315, 323, 331, 338, 341, 361, 372, 381, 383, 387–389, 391, 415, 422, 427, 433, 447, 459, 462
– concept formation in Middle Platonism   382
– concept of 'one nature'   322
conceptual   151, 157, 345, 382, 386–388, 464
– conceptual analysis   389
– conceptual language   389
– universal conceptual grammar   383, 387
conclusion   19, 23 f., 26, 37, 50, 68, 91 f., 94–97, 101, 111, 125–127, 131, 135, 137–139, 141, 144, 146, 150 f., 165, 171, 211–213, 215 f., 218–221, 223–229, 235 f., 238, 241 f., 266, 273, 297, 307, 312, 328, 342, 351, 357, 371 f., 377, 390, 404 f., 407, 410, 413, 420, 424, 438, 445, 450, 453, 462, 466
– scientific   211, 215–223, 231, 235, 238, 371
conflict   24, 41, 49, 88, 98, 120, 171, 181 f., 184, 223, 227, 229–232, 242 f., 300
– between Athenian democracy and true politics   41
– between democracy and true rhetoric   41
conjectures   64, 72
– through images   64, 72
– through images of images   64, 72
consensus (ὁμολογία)   7, 26, 68, 166, 264, 279
considering the common items in accordance with the object (κατὰ τὸ πρᾶγμα θεωρῶν τὰ κοινά)   229
consistency   21 f., 43, 58, 65, 67–69, 72, 75, 249, 290, 299, 371
– criterion of consistency (συμφωνία)   65
contemplation   354, 358, 466
contest   4, 53, 162, 350
contradiction   1 f., 5–7, 25, 37, 44, 61, 71 f., 86, 97, 110, 181, 220, 266, 287 f., 290–293, 298 f., 307 f., 312, 324, 376, 386, 400

– denial of the principle of non-contradiction   289
contrast   20, 22, 46, 59 f., 62 f., 66, 98, 116, 129, 133 f., 136 f., 144, 155, 163, 166, 175, 189, 216, 223, 237, 241, 244, 250, 252, 255, 269 f., 281, 308, 310, 316, 352, 390 f., 405, 413, 416, 424, 449, 458, 464
corpora   371
– Chaldaean   371
– Egyptian   371
– Orphic   371
correlation between language and reality   100
creator, craftsman   343, 450
crisis   331
culture   19, 161, 185, 262, 332, 381, 424, 438, 466
– pagan   338, 463 f.
currency   372 f., 375
Cyclops   91
Cynics   335
Cyrenaics   22, 83

Daemon   433
Damascius   26, 372, 399, 409–411, 428 f., 451
– *Commentary on the Philebus*   26, 399, 409–411
Darapti or proto-Darapti   224
debating (ἐριστικόν)   164 f., 173, 342, 349, 442 f.
deducibility   67
deduction   75, 107, 211–214, 222–225, 231, 235, 239, 241, 376, 404, 429, 457
– valid (or sound)   19, 68, 75, 211–213, 223, 225, 293, 429, 438
definition   1, 8, 10, 15, 18, 22 f., 28, 39 f., 61 f., 66, 94, 105 f., 108–110, 113, 121, 125–127, 129, 135, 137 f., 142, 157–162, 164–166, 169, 171–176, 178 f., 181, 183, 185, 189, 194, 203 f., 214, 233, 268, 310, 312, 315 f., 318–321, 324 f., 344 f., 354, 388, 399, 410, 415, 419, 441, 454, 458, 460
– definitional formula   304, 308, 310 f., 317 f., 321–328
demonstration, demonstrating   18–20, 24 f., 27, 50 f., 95, 171, 211, 216–223, 228 f., 231 f., 234–239, 241 f., 244, 251, 256 f., 265, 287, 289–293, 296–299, 302, 304, 306 f., 311, 328, 337, 344, 361, 372, 376, 378, 385, 391, 399, 403 f., 410, 424, 427, 430 f., 437, 452

- apodeixis   27, 376, 385, 391, 437
  - appropriate   212-215, 217, 219, 221f., 231-244
  - elenctic   289-293, 296f., 301-304, 306f.
  - geometrical   145, 235-237
  - opposite   219-222
  - refutative   45, 289
  - scientific   211, 215-223, 231, 235, 242, 255-258, 265f., 337, 399
  - *simpliciter*   133, 244, 292
  - sophistical   211-215, 217, 219, 222f., 228f., 231-234, 236-244
  - undemonstrated sayings and opinions (*anapodeiktois phasesi kai doxais*)   378
desire to learn [*discere cupias*]   355
διαιρεῖν (dividing)   174
diairesis   179, 407
- art of diairesis   175
dialectical
- cosmos   3f., 87, 363, 381, 384, 397f., 401
- debates   165
- inquiry   150-152, 154f., 159f.,190, 192, 204-206, 208
- proof   3, 18, 42, 69, 100, 289-294, 296f., 299, 307, 309f., 313, 336, 349, 376, 384f., 429, 438, 449f.
- pursuit (διαλεκτικὴ μέθοδος)   60, 73, 163, 250, 417
- reductio ad absurdum   3, 290, 293
- rigour   333
- strategy of viewing a problem from two sides   453
dialectic, dialectics, dialectical   1-28, 37-41, 47-53, 57, 60, 70, 83f., 89, 92, 94-97, 105-107, 109f., 121, 123-125, 129f., 132-134, 136, 138, 140, 148, 153-155, 157f., 161, 164, 167-177, 180-182, 184f., 189f., 199, 205, 211, 213, 216, 224, 228-230, 244, 249-251, 253f., 256-259, 261-266, 268-271, 273, 277, 279, 285, 331-335, 337-340, 342-350, 352, 354-363, 374, 397-401, 403f., 406f., 409f., 413-421, 423f., 430, 441-445, 447-450, 453, 457-463, 466f.
- ancient dialectics   278f.
- and allegory   28, 457-459, 466
- applied dialectic   338, 448, 450f., 454
- dialectic diairetic activity   360
- dialectic in the *Gorgias*   48, 53
- dialectics and theology   26, 340, 352, 356f., 359f.
  - dialectics and philosophical theology   363
- genealogy of dialectic   161
- main tasks of dialectic (Aristotle's *Topics* 1.2)   292
- sophistic   5, 12, 22f., 50, 108, 110, 112f., 119, 123, 126, 157, 180-182, 240, 427, 442-444
dialogue   1-3, 6-12, 17f., 20-24, 27, 37-39, 44-49, 52f., 60, 64, 66, 75, 83f., 88, 91, 93, 96f., 101, 105-110, 112, 114, 117, 120-122, 125-127, 129-135, 137-140, 145, 148, 155, 157f., 164, 167, 173f., 176, 178, 180, 182f., 185, 189-196, 199f., 202-207, 268, 279, 285, 299, 331-333, 335, 342-344, 349, 352, 361-363, 383, 389, 398, 401f., 408, 413f., 416-418, 420-422, 424, 431f., 445
- cooperative   96, 101
- instructional (ὑφηγητικοί)   352
- investigative' or 'zetetic' (ζητητικοί)   352
- Socrates' insistence on collaborative dialogue   96
diamartanein   141
dianoia   28, 91, 130, 153, 413, 423
diaporia   253, 267f., 271, 273
- diaporēsantas   374
diarthrōsis (articulation)   383, 388
dichotomy, dichotomies   24, 159, 189, 196, 201-203, 207
διδόναι δίκην   38, 44
differentia specifica   164
- of the sophist   6, 20, 23, 109f., 133, 157-159, 164-166, 169, 171-174, 176, 179f., 182f., 189, 192-196, 199f., 205, 207, 337
Dionysian corpus   362
discernment   266, 331
disciple   38, 111, 334f., 339, 346f., 349, 352f., 362f., 397
discipline   14, 17, 38f., 41, 43, 49, 51, 53, 211, 215f., 223, 226, 232-235, 238, 240, 242, 244, 259, 262, 264f., 271, 288, 333f., 346, 356f., 361f.
- justice as discipline   39
- scientific   216, 242

- theoretical   3, 15, 27, 37, 39, 61, 83, 86, 100 f., 119, 137, 191, 195, 252, 256 f., 268, 277 f., 282, 284, 288, 332, 358, 441
disciplining   37–39, 42–44, 53
- twofold use of the notion of disciplining   39
discourse, speech   3, 6, 8, 28, 41, 45, 47, 49, 93, 97, 105, 108, 118–120, 123 f., 130, 162–164, 265, 268, 311, 348, 358, 467
- forensic or public type of speech   164
- rational discourse   289, 299 f., 305 f., 308, 311, 328
disproportion   177–179
disputation (ἀντιλογικόν)   5 f., 23, 157, 163 f., 166, 170, 386, 400
distinction (διάκρισις)   11, 14, 17, 20, 27, 42, 45, 58, 74 f., 140, 144, 147 f., 150, 171–175, 180, 183, 191, 244, 255 f., 261, 265, 267, 287–289, 298, 312 f., 315 f., 320–322, 324–326, 339, 345 f., 359, 406, 432, 442, 444, 448
- between *propaedeutic* and *applied* dialectic   442
- between reality and appearance   444
divine   22, 87, 107–109, 112–114, 117, 120–122, 125–127, 195–197, 199, 204, 282 f., 333 f., 341, 345, 347 f., 350, 357 f., 360–363, 378 f., 403, 405, 408, 420 f., 424, 431
- henads   403
- portion (*theia moira*)   431
- providence   341, 347
- revelation   46, 126, 345, 359 f.
Division   10, 15, 17 f., 22–24, 27, 37, 50, 111, 125, 157–160, 162–165, 170–176, 189–205, 207, 230, 265, 267, 323, 338 f., 344, 346, 356, 360, 363, 399, 407, 410, 415 f., 427, 435, 447, 449
- and unification   338
- divisions and definitions   169
  - dividing   10, 125, 169, 172, 174–176, 189, 196–198, 203, 338, 410, 419 f.
δημηγορία   42
doctrine   15, 73, 75, 86 f., 105, 131, 140 f., 148, 154, 167, 171, 191, 249, 252 f., 258, 277, 332, 339, 341, 353, 388, 390, 452
- flux   3, 148, 154
Doxa   148, 183, 205, 416, 432
- alēthēs   148
- diastrophos   432

δόξα and πίστις in the *Gorgias*   52
doxazein   144, 148
δόξης ἐκβολή   178
dynamis   105
- *tēn tou dialegesthai dynamin*   130, 143, 153

Egyptian, Celtic and Indian sages   371, 379, 381
Egyptian Christians   336
εἰδέναι   41, 53, 417
εἶδος   106, 255, 260 f., 320, 324, 339 f., 346
εἰκάζω   57, 63
Eleatic, Eleaticism   2, 13, 85 f., 88, 148, 192
- Eleatic influence over Euclid   85
- Plato's parricide in the *Sophist*   86
ἔλεγχος   4, 37, 48, 53, 220, 290, 292, 296
elenctic   179 f.
elements
- (in)finite nature of the elements   450
elenchus 2, 5–8, 22 f., 27, 66, 70–72, 158, 174, 177–185, 227, 427, 429–434, 436–438
- analogy of the elenchus to medicine   181
  - dielenkhein   434, 437
  - elenctic mode   429
- double role of Socratic elenchus   432
- *elenchus* terminology   434
empirical observations   373–375
endoxa   13, 15
ἔνδοξα   216
endoxa   216
ἔνδοξα   270 f.
endoxa   374 f., 377, 387
- approved (ἔνδοξα)   257, 259, 263, 266, 271
- authoritative   374 f., 382, 390
- endoxic approaches   373
- endoxic method   375, 386
- endoxographical method   450
  - *endoxographic* passages   454
- reputable opinions   374, 443 f., 446–449, 454
enquiry
- rational 355
enquiry   168, 354 f.
epagogic   338
Epictetus' Handbook   383
Epicureans   263, 335, 338
- Epicurean prolēpseis   382
epistēmē   134, 373

ἐπιστήμη   1, 41, 46, 52, 68, 73, 331, 333, 349,
    358 f., 448, 460
- τοῦ εὖ λέγειν   46
epoptics   334, 356-359
erga   62 f.
eris   5, 171
- dual meaning of *Eris*   170
eristic, eristics   5 f., 23, 94-97, 101, 157, 161 f.,
    165 f., 170-173, 236 f., 240, 332, 418
- and philosophy   17, 23, 138, 143, 161, 257 f.,
    344, 348, 379, 390
- and dialectic   5-6, 94-97, 161, 164, 167-176,
    180-182
- engaged in eristic (περὶ τὰς ἔριδας
    διατρίβοντας)   173
ἐριστικὴ σοφία   164
Eriugena (Origenian; the last Patristic Platonist)
    26, 357, 361-363
eros   22, 105 f., 108, 120 f., 125, 127
error (τὸ ψεῦδος)   1, 18, 22, 71, 94, 168, 171,
    180 f., 183, 219, 271, 287, 343, 450, 454
ethics   17, 46, 263 f., 266, 338, 346, 348, 351,
    356-359, 373 f., 376, 378, 447
- ethics as a science (*epistēmē*)   373
    - ethical inquiry   373
êthos   432
- diastrophon   432
Eubulides' paradoxes   95 f.
Eucharist   346
Euclid   12 f., 85-88, 90, 95 f., 102, 236, 241, 376,
    445
- Euclid's dialectic   89
    - method   94, 96 f.
    - Euclid's *Elements*   376
    - theoretical position   85
- thesis about the Good-One   88
Eudemian method   376
evil   38, 106
- greatest   38
examination   3, 7-9, 15, 20, 26-28, 71 f., 147,
    173, 178, 182, 212 f., 224, 226-228, 242, 260,
    280, 345, 356, 383, 410
- and discussion   165, 345, 446, 458
- critical   2, 44, 85, 100, 135, 166, 182, 212 f.,
    224, 228, 277, 281 f., 287, 331, 346
    - critical examiner   223 f., 226-228, 230 f., 242
    - cross-examination   8, 71, 111, 127, 230, 380

- or inspection (*episkopein*)   147
- peirastic (πειραστική)   213-215, 217, 222-232
exchanging part (ἀλλακτικόν, μεταβλητικόν)
    159
exegesis   19, 21, 25 f., 249 f., 263, 278, 283 f.,
    336, 346, 352 f., 356, 360, 371 f., 378-381,
    390, 400, 406, 415, 445 f., 452, 454
- and textual edition   283
- creative   19, 26, 362, 371 f., 375, 384, 390, 422,
    446, 452
- of canonical texts   380
- of poetic and philosophical texts   379
- or commentary (*exēgēsis, hermēneusis*)   19,
    378, 390
- philosophical commentary and exegesis   380
exegesis of the Bible
- allegories of the sun, the line and the cave in
    Plato's *Republic*   83, 86, 95
- allegory of the line   90-93
- in Plato's *Republic*   356, 413
- Platonic allegories   457
exegesis of the Bible   336
exegetical   16, 85, 249, 254, 277 f., 297, 321, 353,
    358, 379, 381, 390, 444, 446, 451, 454
- exegetical „inclusion" or harmony   377
- exegetical methodology   371
- exegetical or metaphilosophical views in Aristotle   379
- exegetical posture   372
- exegetical tradition   16, 278, 282, 451
    - school tradition   451
- methodology   11, 28, 174, 331, 373, 398, 442,
    454
exemplarism   334
- exemplaristic   357
exercise (γυμνασία)   6, 13 f., 25 f., 70, 88 f., 94,
    96, 121, 133, 167 f., 170 f., 179, 181, 185, 206 f.,
    260-264, 269, 271, 332, 334, 352, 361-363,
    371 f., 380, 383 f., 389, 397, 400, 418, 421,
    427, 442, 453
- logical   167 f., 398-404
experience   12, 20, 61, 75, 88 f., 93, 99, 126, 146,
    152, 183, 207, 378, 386 f., 389, 431
- those who have experienced and are older (*tōn
    empeirōn kai presbuterōn*)   378
- value of practical experience   378

expert   8, 24, 46, 96, 110, 140–142, 145, 149, 151, 164, 168, 180, 214, 218 f., 221–228, 230, 232, 235 f., 240, 242–244, 387, 453 f.
- ability to predict what *will be* good or beneficial to someone   140
- expert knowledge (*epistēmē*)   134, 142, 145 f., 154, 211, 213, 223–228, 243
- false   211–214, 217–219, 221–230, 232, 234–236
- knowledge of the good and beneficial   141
- scientific expert   218 f., 221
expertise   7, 10, 14, 134, 139–141, 144, 155, 159 f., 165, 172, 211, 215, 228, 230, 233 f., 237 f., 240–242
- scientific expertise   24, 211, 218, 222, 238, 242
  - and dialectical skills   218
explanandum   216, 219, 221 f., 232 f., 238, 244
explanation   23, 38, 40, 52, 61, 84, 88, 90, 105, 131 f., 137, 152, 200, 206, 218 f., 221 f., 224, 235, 238, 263, 283, 289 f., 294–297, 300 f., 303 f., 306, 308, 336, 355, 379, 399 f., 409, 414 f., 435
- scientific   215–223
expositions (exegeseis)   277
exposure   126, 225
- peirastic exposure   211, 225
  - of false pretenders   211
  - of ignorance   177, 179 f., 217, 225
ἐξ ὑποθέσεως ἀναγκαῖον   70, 75

Faith and Reason   346
fallacies   237, 239
- sophistical   237, 239
form (εἶδος), forms   1–4, 8–10, 13, 17–20, 23, 26, 37, 47 f., 50–52, 59, 64 f., 70 f., 86 f., 90, 98 f., 105, 110 f., 115–117, 120–125, 127, 129–132, 135–137, 143, 148, 150, 153, 155, 161, 163 f., 167–169, 171, 174, 177, 180, 185, 189, 191, 193 f., 198 f., 202–204, 214, 221, 231, 240, 249, 255 f., 259, 271, 280, 283 f., 297, 299, 306–308, 312 f., 319–321, 323–326, 332, 335, 337, 339 f., 345 f., 354–357, 362, 378, 381, 389, 398–400, 407–410, 413, 415 f., 418–420, 422–424, 428 f., 441–444, 448–450, 454, 457–459, 463 f.
- enmattered   319 f.
- forms and the One   407

- theory of Forms   3, 61, 97–99, 101
from God to God   361

genus, genera   22, 61, 108, 160, 174–177, 226, 274, 315, 317 f., 339 f., 346, 363, 388 f., 420, 441, 454
geometry   145, 193, 237, 239, 241, 243, 334
gnoseology   96
gnostic   16, 339, 359, 361
God   11, 76, 85, 87 f., 94, 108, 120–123, 126, 196, 333 f., 338–341, 345, 347, 350–352, 354–357, 359, 361–363, 378 f., 398, 403–406, 420
- as „Nous and Logos"   341
- as One   3, 52, 99, 145, 148, 183, 194, 214, 265, 341, 348, 388, 403, 421, 449, 459, 461
- God's greatness   341
- God's infinity   341
- God's Logos (λόγος τοῦ θεοῦ)   360
- One primal God   407
good, the Form of the Good   12 f., 17 f., 23, 37 f., 40–46, 53, 63, 68–71, 74, 84 f., 87–90, 93, 101, 106 f., 111, 114, 116, 118, 120–122, 124–126, 129, 131, 133 f., 136, 139–143, 145 f., 149–151, 155, 162, 171, 173, 183 f., 195, 199, 234, 240, 260, 265, 269, 280, 282, 287, 294, 305, 310–312, 332–334, 356, 359, 376, 378, 381, 384–386, 388 f., 410, 414, 424, 433, 435, 452, 460, 464, 466
- Form of the Good (ἐπέκεινα τῆς οὐσίας)   12, 14, 84, 87, 91 f., 101, 332, 360, 413, 420
- good and the bad, the beautiful and the ugly in the *Theaetetus*   142, 144 f., 149, 171
  - good and beneficial   136, 140–142, 145 f.
- knowability of the Good   90
- relation between the Good and the tangible things   89
- the Good itself   69 f., 334
- the Idea of the Good   334
- the lecture on the Good   101
Gregory of Nyssa's philosophy
- *De hominis opificio*   341
- idea of divine infinity   342
Gregory Thaumaturgus' *Thanksgiving Oratio*   335
gymnastics   162, 179, 259, 335
- psychic gymnastics   179

harmony 26, 65, 67, 162, 371, 376 f., 383, 385 f., 390, 466
- harmony thesis 371
- of ideas 69 f., 75, 331, 371, 451
- of the dance with cosmic reality 466
hedonism 22, 50
heirs of Plato 331
Hellenic philosophical tradition 454
Heraclitean ontology 137–139, 142
Hermogenes' writings 431
Hippocrates' *Fractures* 445
homonymy, heteronymy 50, 98, 309, 313
- heteronyms 323
- homonymously 304, 312
- homonyms 305, 308, 311, 323
- univocal and multivocal (homonymous) expressions 308
horistic 338
human 3, 18, 23, 28, 40, 73–75, 90 f., 93, 107, 109, 113, 121, 124, 131, 134, 142, 146, 155, 157, 167, 171, 185, 195–197, 202, 204, 279, 332, 340, 342, 345, 348, 353–355, 357, 360–362, 377 f., 381, 390, 418, 420, 465
- rationality 347, 357
hunting (θηρευτικόν) 159 f., 335
hypothesis, hypotheses 2, 22, 57, 61, 66–72, 75 f., 108, 126, 252, 397 f., 400–405, 407, 415, 418, 422, 424
- first and second hypotheses of the *Parmenides* 398 f., 410
- hypothesis of ideas 58, 68, 70, 75
- hypothetical 91, 107, 132, 351, 359, 404, 427
  - hypothetical method 61, 95, 208
  - non-hypothetical 19, 68, 92

ideas 1, 3, 19, 62, 64, 69–71, 76, 84–86, 91 f., 99, 167, 169, 171 f., 270, 333, 344, 350, 353, 357, 372, 390, 407, 415–417, 450, 458, 460
- as paradigms 70
ignorance 38, 41–43, 48, 51, 59, 73, 109–111, 116, 122, 124, 134, 177–179, 213, 217 f., 223–225, 243, 345, 360
- avowal of ignorance 40 f., 48 f.
- double ignorance 169, 171, 178, 180
illumination (ἐξέλαμψις) 74 f.
- illumination *as such* 75

image, images 21, 41, 45, 53, 57 f., 63–65, 80, 91, 93, 98, 195, 345, 355, 417, 420, 463, 465
- images or reflections of real things 65
- of Forms 61, 64, 129, 131, 154, 198, 399, 408, 416, 420, 452
- of images 63 f., 93, 413 f.
- of logoi 21, 58, 62 f., 72, 74
- of the truth 71, 124, 207, 218 f., 266, 371, 376 f., 418, 453, 465
  - images of 'the truth of things' 72
Indian philosophy 377
- inclusivism in
  - Indian metaphor, found in early Buddhist and Jain literature 377
  - Indian philosophical and contemplative commentary traditions 377
Induction, inductive 15, 18, 270, 318, 415
- inductive proof 319
- inductive reasoning (*epagōgē*) 158
initiation (τελετή), initiator 405, 465
„in itself", „in another" 398 f., 405–409
instrument (ὄργανον) 16 f., 24, 38, 44, 64, 96, 101, 184, 249, 252, 259, 264 f., 269 f., 356, 360, 461
- transversal 'instrument' (ὄργανον) for all sciences 348
intellect 17–20, 85, 87 f., 94, 339, 341, 345, 351, 354, 361, 386, 388, 399, 403, 408, 416, 420
- beyond intellect 459
intellective theory (*noera theôria*) 452
intellectual 17, 19, 49, 57, 59, 84, 88, 97, 102, 105, 107 f., 114, 121 f., 133 f., 150, 163, 165, 259, 271, 335, 385, 388, 405 f., 408 f., 432 f., 452, 463, 466
- intuition 74, 107, 271, 391, 424
- world / realm 403, 406 f.
intelligible 17 f., 26, 90–93, 98, 167–169, 249, 382, 388 f., 398–400, 402, 404–409, 411, 415 f., 418 f., 423, 428, 433, 452, 460 f., 467
- intelligible proper 459
- intelligible-sensible 355
- universe, world 17, 167–169, 404–405
investigation 3, 7 f., 10, 12, 20, 25, 86, 94, 133, 201, 204, 206, 208, 250, 273, 278, 323, 342, 345 f., 349, 352, 354–356, 358, 371 f., 375, 377, 400, 443
- of names 87, 94, 195, 295, 303, 310, 323

– rational 354–357
– theōria 371, 377
irony 46, 48, 50, 58f., 118, 123, 148, 181
irrational passions 347

Jabir's theory 284
judge (*krinein*) 65, 136, 140–142, 144f., 149–151, 259, 266, 268, 343, 347, 355, 360, 378, 391
judgement (ψῆφος) 52, 86f., 112, 136, 143f., 146, 152f., 376, 386, 391, 437f., 457, 461, 467

καθαρμός and self-knowledge 176
κατηγορεῖν 42
κενεαγορία 60, 70
κενολογεῖν 70
knowledge 1f., 6, 8f., 12f., 16, 18, 20–23, 37, 41, 43, 45, 47f., 52, 68, 72–75, 89, 93–97, 105–107, 109–113, 121–125, 127, 129f., 132–137, 139f., 142–149, 151–155, 157, 159, 167f., 173, 177–179, 181, 183f., 213, 221, 224, 227f., 231, 242, 244, 257, 261, 279, 332, 334f., 337, 339, 348, 354–356, 358, 360–363, 376, 384, 388, 398f., 416, 418, 423f., 427f., 443f., 446, 448, 453, 460f., 466
– by illumination 75
– false claims of knowledge 214, 217
  – false claims of expert knowledge 213, 217, 223, 231
– impossibility of guaranteed knowledge 94
– lack of expert knowledge 217, 223
– modern, or Kantian, conception of knowledge 7, 13, 38f., 46, 61, 190, 295, 304, 390, 420, 454, 465
– pure 72f., 85, 87–90, 92, 96, 133, 176, 343, 403, 422, 448
– scientific 211, 215–223, 228, 244, 265f.
– second-order 59, 227, 255, 264
know yourself 356
κολακεία 42

language (logos) 5, 28, 50, 61f., 87, 94, 107, 168, 178, 375, 387–390, 406f., 420, 438, 463
– and reality 387
– language logic 358
– nature of language 94
– ordinary 389
– use of language 295, 372, 422

layman 118, 134, 140–142, 144f.
legislation 38, 141
lexis 371, 386, 433
life 5, 7, 9, 41, 49, 52f., 72, 75, 93, 98, 107, 116, 119, 126, 134, 142, 162, 164, 170f., 343, 353–355, 380f., 383, 409f., 423, 434, 438
– living and ensouled beings 283
likeness and unlikeness, identity and otherness 147
logic
– and dialectic 15–20, 347–352, 356–360
– and syllogisms 417
– in the service of theology 359, 361, 363
– predicative 351
– propositional 75, 107, 226, 351
– the position of logic in Byzantine philosophical studies 459
logic 15–20, 264–268, 347–352, 356–360, 457–459
Logos-Christ 360
logos, logoi 5, 23, 48, 62, 68f., 105, 108, 118, 129, 133, 157f., 161, 185, 198, 206, 264, 294–297, 299, 337, 343, 346–348, 354, 356f., 360, 389
– false 64, 66–72
– flight into the *logoi* 21, 57, 62, 64, 68f., 75
– *logoi* or arguments 74
– weakness of the logoi 22, 57, 75
λόγος τῆς οὐσίας 61, 73

maker of persuasion (πειθοῦς δημιουργός) 347
materialist views 85
mathematics 17, 20, 23, 91, 134, 148, 168f., 256f., 332, 399, 415, 459
– mathematical 18, 68f., 148, 193, 235, 376, 378, 399, 402, 413, 416, 459
  – disciplines in *Republic* 148
  – *Elements* (Euclid) 376
  – thinking 413
μάθησις 41
medicine 38, 179, 272, 350
– psychic medicine 179
Megarians, Megarian 5f., 12f., 20, 22, 83–86, 88–90, 94, 96–99, 101, 166, 179
– beliefs 85, 89, 97
– claim about the unattainable Good 91

- difference between Platonic and Megarian philosophers   94
  - denial of participation, rejection of participation   101
  - Megarian metaphysics   93
  - position   89
μέγιστον ἀγαθόν   38
μέγιστον κακόν   38
μέγιστον τῶν κακῶν, ἔσχατον κακῶν   38
mercury   284, 436
metals   284, 436
metaphysics   249-255, 358, 358f.
method, methodos   1-3, 6-12, 14-21, 23-28, 51, 58-61, 65-69, 71, 73, 84, 94, 96f., 105-107, 111-113, 119, 122-125, 158, 164, 167-170, 172-176, 179-183, 189f., 193, 195-208, 249f., 252-254, 256f., 259, 261-263, 269, 273, 308, 331, 333f., 344f., 347f., 352f., 356, 363, 371-373, 375-377, 397-401, 408, 413-421, 427, 431, 441, 444-447, 452, 454, 458-460, 462
  - hypothetical   11f., 68f., 91f.
  - indirect method of studying things   65
  - maieutic   172, 182, 279, 431
  - mathematical   18, 68f.
  - methods of modern analytic thought   390
  - nature of the dialectical method and its connection to the task of a philosophical exegete   445
  - the method of hypothesis (ἐξ ὑποθέσεως)   11f., 69, 419
Middle Platonists   13
  - Middle Platonic reception of Stoicism   382
midwifery (μαιεία)   134, 154, 172, 182f., 430
Milesians   379
mimesis   114
  - the mimetic character of flattery   41
  - the rite's mimesis   465
μισολογία   62
Mouseion   379
multiplicity   89f., 99, 125, 127, 310, 403, 405-407
multivocity   308
μῦθος   52, 118f.
mystery rite   463
mythological symbolism   371

name, name (to onoma)   8, 10, 12f., 23, 37, 42, 64, 85, 87f., 94, 99-101, 106f., 120, 162, 166, 170, 172-174, 176, 190, 195, 253, 263f., 272, 278, 295f., 301-305, 307-316, 318, 321-328, 361, 387, 418, 434-436, 464
  - indefinite   235f., 296f., 310, 323
  - negated   321-325
  - the number of 'things' signified by a name   310
nature   1, 3-5, 8, 15, 19, 25, 40, 42, 48, 52, 59, 83, 87, 90, 99f., 106, 109, 112, 119, 122-125, 133, 136, 142-144, 147, 149, 151, 163, 169, 171, 175, 185, 191, 195, 204, 215, 219, 226, 252, 255, 260, 265f., 271, 277, 283f., 287, 292-294, 299, 302-305, 307, 310-312, 315-317, 321f., 326-328, 333, 341f., 348, 357, 362, 371, 377f., 381f., 385, 387-389, 398, 404, 406, 408, 413f., 417, 419, 421, 423f., 430, 433, 445, 447f., 459, 461-464, 466
  - of time   43, 378, 381-383
Neoplatonic, Neoplatonist   16, 18, 20, 26, 358, 371f., 381, 387, 421f., 424, 459, 462
  - commentators of late antiquity   372
  - Neoplatonists' methodological Aristotle   373
  - philosophy of language and philosophy of education   387
  - psychology and pedagogy   372
  - treatment of philosophical authorities   372
Neuplatonismus   371
noeric hebdomad   405
noetic-noeric triads   405
noetic triads   405
Nous   74, 80, 341, 371, 377, 384, 386, 388, 390, 420
  - beyond Nous   341
νοῦς   74, 87f., 306, 384

object, objects
  - in the sensible world   64, 416
Olympiodorus' philosophy
  - Commentary   428-438
    - on Aristotle's *Categories*   432, 434, 437, 446
    - on Plato's *First Alcibiades*   429, 431
    - on Plato's *Gorgias*   429, 431, 433-435, 437
    - on Plato's *Phaedo*   429, 432, 438
    - on the *Meteorology*   429

- Olympiodorus' concept of elenchus   432
  - concept of Socratic elenchus   433
  - Olympiodoran use of the terminology of elenchus   438
ὁμολογία   67 f.
one, One   1-4, 6-19, 21-28, 37, 39-47, 49-53, 57-59, 61-65, 69, 72, 74-76, 80, 84 f., 88 f., 95, 98-101, 105 f., 108-114, 116 f., 119 f., 122-127, 130-139, 141 f., 144-147, 149-151, 153-155, 157-163, 165 f., 168, 171-176, 178-182, 184, 189-195, 197-208, 212-220, 222-227, 229-234, 237-241, 243, 250 f., 253 f., 256, 258-260, 262-269, 271-273, 277-279, 281-284, 289-292, 294-296, 298-303, 305, 307-322, 325-328, 333 f., 336, 339 f., 342 f., 345, 347-349, 353, 356, 361-363, 371 f., 376-380, 382 f., 385-388, 390, 398-411, 414 f., 417-419, 421-424, 427, 430-438, 441-444, 446, 448, 450-452, 457-462, 464-467
- numerically   205, 315 f.
- relationship between monads and the One   410
- the One and Being   403 f.
- the One and the intelligible universe   26, 402, 410
- the One and the universe   398, 403
- the One „in itself" or „in another"   398
- the One is both protological (ἕν = first principle) and theological (ἕν = supreme deity)   358
ontology   3, 18, 88 f., 95 f., 101, 135, 137, 142, 154, 287, 361, 363
- ontological-epistemological co-extension   357
- transformation of ontology into „agathology"   88
opposites   3-5, 88, 110, 170, 220, 453
opposition   11, 45, 48, 60, 163-165, 184, 219 f., 225, 258, 261, 324 f., 337, 347, 444
- between dialectic and rhetoric   48
- *contradictory*   1 f., 44, 47, 91, 164, 166, 176, 179, 181, 184, 213, 215-221, 224 f., 229-231, 242, 256, 260, 291, 300, 302 f., 305, 307, 310, 324-326, 351, 376
- not contradictory   264, 324
Orator   10, 37, 40, 42-47, 51, 163, 166, 424
organon   15 f., 20, 25, 250, 264, 328

Origen's philosophy   346
- ascetic lifestyle   336
- dialectic   331-335, 337-340, 342-350, 352, 354-363
  - taught dialectic   346
- *First Principles* (*De principiis*)   14, 16, 218, 332 f., 346, 353, 355 f., 360, 375, 401, 449
- Origen's allegoresis of the Logos   360
- self-emasculation, self-mutilation   335
- zetetic, heuristic method   16, 249, 279 f., 331, 342, 344 f., 349, 352-356
Orphic   371
- Orphic-Pythagorean tradition   178
ousia (οὐσία)   129, 136, 139, 141 f., 144, 146-151, 316, 320

pagan   16, 28, 331, 336, 338, 343 f., 349, 463 f.
- philosophers   331, 344
- tradition   331
paideia   19, 345, 352, 422-424
- ordinary   424
palaioi, the ancient philosophers   378, 382, 386
paradigms   357
paraphrase, paraphrases   159, 271, 277, 293, 297, 300, 434, 443, 448, 451 f.
παραφροσύνη   177
Parmenides   2, 4, 18, 20, 23, 25-27, 61 f., 69 f., 85 f., 88, 90, 98, 129-134, 138, 140, 143, 150, 153, 155, 166-168, 170, 172, 183, 207, 266, 268 f., 331, 333, 337, 359, 397-410, 417-419, 427, 451
- Parmenidean   4, 69, 167 f., 322, 406, 409, 418 f.
  - dialectic   4, 167-171, 397-401, 403 f., 406 f.
  - *gymnasia*   401
  - paradox of Not-Being in the *Sophist*   322
- Parmenides and Euclid's metaphysical argument   89
paronyms   323
participation   19, 26, 70, 83, 88, 90, 97-99, 299, 361, 404, 411, 419
particulars   10, 189, 194, 198, 448
pedagogical   25 f., 206, 279, 281, 372
- questions   279
- scope for questioning   279
peirastic (πειραστική)   24, 211, 213-215, 217, 222-232, 241 f.

peras and apeiria 409
perception (aisthēsis) 23, 60f., 107, 112, 129, 134–140, 143–145, 148, 150–152, 154, 437
– and judgement 144
– perceptual use of the mind 147
περὶ ἀρχῶν 280
Periphyseon 361–363
persuasion 21, 40–42, 45–47, 50–52, 111, 127, 438
– two kinds of (δύο εἴδη πειθοῦς) 10, 41, 49, 180, 197, 279, 316
petitio principii 294, 303, 305
phainomena or appearances 373
– tithenai ta phainomena 374
Philocalists 343
Philoponus' Commentary on Aristotle's Meteorologica Book I 436
philosophical 1–3, 6, 8, 10–14, 16, 18–21, 25 f., 28, 45, 50, 53, 57, 60, 83 f., 89, 93, 121, 127, 132 f., 155, 161, 174, 178, 190, 193, 198, 201, 249, 254, 259, 261, 263, 266, 269, 277–280, 282, 288, 334, 338, 341–344, 346, 348, 352–354, 356–358, 363, 371–375, 380–383, 386, 389–391, 413, 415, 420 f., 424, 428, 446 f., 449–451, 454, 457 f., 461 f., 464–467
– enquiry 168, 354 f.
– riddles 381
– section 1, 13, 20–22, 24, 37 f., 48, 62, 76, 101, 122, 135–137, 139, 143, 152, 190, 224, 230, 278, 301, 304, 306, 313, 333, 341, 346, 373, 408, 445, 453 f., 462 f.
– „ways of life" 26, 371
philosophise 391
philosophy
– Classical 1, 5, 11, 14, 20 f., 57, 84, 218, 279, 379 f., 384, 419, 423
– division of philosophy, partitions of philosophy 356, 359
– Greek philosophy from Thales to Plotinus 452
– Hellenic philosophy 450
  – history of 4 f., 18–20, 28, 135, 253, 261, 277 f., 337, 339, 373, 378 f., 389, 400, 438, 450
– Ionian natural 61
– of language 64, 94, 99, 359, 383, 387, 389 f., 422
– of the commentators 381

– post-Hellenistic philosophy 270, 371, 381
– relationship between philosophy and rhetoric 44
– the fullest form of philosophy (τῆς τελειοτάτης φιλοσοφίας) 349, 359
– tripartition of philosophy 267, 349, 356
philosophy 1, 3, 5 f., 9, 14–17, 19–21, 23, 26, 44 f., 61, 83 f., 87, 94, 97, 105, 107, 109, 114, 117, 122, 129 f., 134 f., 140, 148, 153–155, 162–166, 249, 252–254, 256 f., 259, 262–268, 274, 278 f., 287, 328, 332–334, 337–339, 342, 345–348, 350, 353–359, 363, 371–373, 378–381, 383 f., 387, 389 f., 410, 421, 427 f., 447, 454, 457–461, 463 f.
phlogiston theory of combustion 284
phōnai 387
φιλονικία, φιλονίκως 53
πίστις 41, 52
Plato 2 f., 5–18, 20–27, 37, 44–46, 48 f., 51, 57–62, 64–68, 70 f., 73, 75 f., 83–87, 89–99, 101 f., 105, 107, 109–112, 117, 119–121, 125 f., 129–133, 135–139, 143, 146, 153–155, 157–159, 162–167, 169, 172–174, 178–181, 183–185, 189 f., 192, 198 f., 202, 205–208, 252, 261 f., 266, 268 f., 279, 322, 331–335, 337–339, 344–350, 352–354, 356, 358–360, 363, 371 f., 379–381, 384–386, 388 f., 398–401, 405 f., 413–415, 417 f., 420–424, 427–431, 435 f., 442–445, 448–450, 452 f., 457, 459, 461
– account of thumos in the Republic 184
– Alcibiades I, Gorgias, Phaedrus 9, 162, 180, 383
– conception of knowledge 8, 134, 138 f.
– Cratylus 10, 17, 64, 106, 162, 173, 427
– creation account in the Timaeus 450
– dialectic 6–12, 89–97, 105–107, 109 f., 123–125, 129 f., 132–134, 136, 138 f.,153–155, 157 f., 161, 164, 167–177, 180–182, 184 f., 189 f., 199, 205, 331–335, 359, 413 f., 417–421
– Euthydemus, Parmenides, Cratylus 5 f., 22, 51, 83, 96, 109, 164, 173, 427
– Gorgias 20 f., 27, 37–41, 43–53, 111 f., 168, 178, 202, 345, 383, 421, 428–432, 435
– „Great Kinds" 452
– Parmenides 129–134, 138, 140, 143, 150, 153, 155, 166–168, 170, 172, 183, 207, 266, 268 f., 359, 397–410, 417–419, 427

- *Phaedrus* 10, 20, 22–24, 27, 44–48, 105–110, 112–121, 123–127, 133, 166, 169, 171, 189–196, 198 f., 202, 204, 338, 383, 407, 410, 414–416, 419–422
- Plato's Socrates of the *Phaedo* 75
- Plato's unwritten doctrines 332 f.
- 'science of dialectic' 448
- *Theaetetus* and *Sophist* 132
- *Timaeus, Meno, Sophist, Republic* 59, 75 f., 132, 178, 268, 333, 363, 423, 436

Platonic
- allegories (σύμβολα) 84, 87, 89, 93, 101, 458
- authorship 384, 445
- criticism of Socratic „intellectualism" 38
- dialectic and the practice of theurgy 420
- dialectician 10, 12, 24, 106, 109, 124 f., 143, 154 f., 165, 173, 180 f., 191, 218, 228 f., 258, 334, 337 f., 340
- dialectic in a theological way 339
- philosopher 3, 13, 16 f., 28, 51, 65, 68, 71, 92, 96, 110–112, 122, 124, 127, 130, 133, 136, 154, 165 f., 169, 171, 179 f., 182, 192 f., 198, 204, 252, 254, 256, 264, 273, 331, 333 f., 337 f., 343, 350, 352–354, 371, 373, 384–386, 413, 424, 428, 435 f., 442, 451, 454, 458 f., 461, 464, 466 f.
- philosophic ascent 466
- Republican dialectic 27, 414, 419, 424
- rhetoric 39–53
- theories in Scripture 360
- tradition 354, 381

Platonic 3, 9, 11, 14–18, 20–22, 25, 27, 37–39, 45, 48, 50 f., 58–61, 64, 66–71, 73, 75 f., 83, 87, 92, 96, 98 f., 101, 105, 127, 148, 164–167, 169, 173, 175, 177, 181, 184, 194, 201, 258, 263, 268, 337–339, 344 f., 348 f., 353–356, 358–361, 379, 381, 383 f., 401, 403, 410, 414, 417–424, 427 f., 430–433, 459–462, 464, 466

Platonism 15 f., 25, 28, 338 f., 344, 348, 352, 363, 414, 431, 457
- imperial 338 f., 359
- imperial pre-Plotinian 25, 338
- middle 6, 11 f., 15, 20, 25, 27, 94, 105, 126, 129, 179, 201, 260, 278, 282, 289, 321, 338 f., 348, 406 f., 414, 418

Platonists 13, 16, 21, 26 f., 268, 331, 335, 338 f., 344, 359, 373, 384, 389–391, 414, 421, 427, 431, 444, 447, 453, 459
- Origen as a Christian Platonist 339
- Platonist exegesis 444
- Platonist ideology 454
- Platonist tradition 452

plausible (πιθανόν) 8, 38, 84, 96, 117, 121, 147, 179, 213, 217, 228, 266–272, 342, 445, 450, 454

Plotinus' philosophy 17
- ascent from the sensible to the One 458
- conception of dialectic 17, 415–417

polemic against the Academy 373

politics 21, 37, 39–41, 43–45, 48–50, 53
- true art of politics, true politics, true political art 39 f.

πολιτικῆς εἴδωλον 41

polyonyms 322 f.

popular 28, 50, 463 f., 466
- culture 466

possession-taking (χειρωτικόν) 159–163, 173

postulate 3, 9, 86, 241, 328, 435

practice of criticism of Homer 380

Praedestinatus 343

pragma, pragmata 215, 236
- the things (τὰ πράγματα) 2, 53, 60, 65, 80, 89, 98, 106, 111, 122, 130 f., 134, 136, 139–141, 144, 148, 154, 174 f., 197 f., 242, 288, 295, 302, 305, 307–309, 311, 318, 324, 327, 344, 362, 377, 388, 416, 460
  - pragmateuesthai 144
  - pragmatic 39, 289, 378
- the thing (*to pragma*) 73 f., 174, 207, 292, 303, 311 f., 316, 318 f., 321, 397, 400

predicate, predicates 89, 98, 237, 254, 265, 269, 299, 302, 311–314, 316–319, 324–327, 397
- in the category of substance 304, 316
- in the non-substantial categories 316

predication 216, 219–222, 228, 312, 314–320, 323–328, 388 f.
- accidental, non-accidental 322–324, 326 f.
- definitional 8, 16, 324 f., 327
- false 217–219, 221–230
- four predicables in the *Topics* 441
- in the same category 318, 320, 325 f.
- non-identifying 312

- the predication of 'what-it-is'   325
- two types of categorial predication   318
- two types of predication   316
premise, premises   1, 7f., 13, 16, 24, 49f., 63–66, 68, 70, 95, 107, 137, 143, 212, 215f., 218, 221, 223–225, 227, 229–235, 237f., 240–242, 244, 252, 255f., 263, 265, 267–270, 351, 403, 405, 410, 417, 438, 447, 453
- common   5, 14, 27, 37, 39, 46, 48, 60, 65, 84, 88, 94, 98, 100, 105f., 113f., 120, 125, 134, 139, 147, 151, 154, 159, 161–163, 166f., 171, 175, 180f., 185, 227, 234f., 258, 261, 266, 270, 278f., 304, 382–388, 415, 430, 434, 438, 444, 446, 461, 463
- explanatory   219–222, 232f., 232f.
  - explanatory role of the premises   220
- mistaken   37, 94, 195, 263, 265, 288, 351, 386
Presocratics   2, 451
primary explanatory factors   217
prime matter   283f.
prime unmoved mover   87, 280f.
principle, principles   2–5, 10–12, 14–17, 19, 24f., 41, 49, 66, 69f., 72, 75, 87, 90f., 106f., 112, 136, 170, 177, 197, 219, 222, 226f., 233–235, 240, 242, 251f., 255f., 259, 280, 282f., 287–290, 292f., 298, 356, 358, 397–402, 404f., 408f., 413, 415, 420, 443, 446, 448–451, 453f., 460
- appropriate   233–235, 240
- basic principles of reasoning   288
- indemonstrable   287, 293, 298, 449
- of all the axioms   298
- of exegesis   446f.
- of knowledge   3, 11f., 20, 42, 48, 69, 73–75, 90, 96, 101, 109, 133, 167, 178, 185, 213, 218, 268, 361f., 398f., 413, 418, 449, 466
- of non-contradiction (PNC)   2, 287f.
- of science   15, 25, 45, 250, 259, 287, 292, 348, 356
- principles belonging to a given scientific expertise   218
- problem of the 'unproved principle' (ἀρχὴ ἀναπόδεικτος)   22, 57, 66
- theory of principles   280f.
- the power of principles   450
- ultimate principles of being   288

Proclus   17f., 21, 23, 27, 157, 167–173, 180f., 183, 236, 269, 344, 349, 358, 383f., 397–408, 410f., 413, 417–422, 424, 427–429, 431f., 436, 459
- allegorical exegesis of the gods   384
- Proclus' Elements of Theology   408
- Proclus on Dialectic in Plato's *Sophist*   170
- Platonic Theology   401, 404–406, 420
- Proclus' notion of a division into clarification of meaning and text   448
- Proclus' performative argument   424
- Proclus' *Timaeus Commentary*   421
- the great harbour of Proclus   459
- tripartite division of Platonic dialectic   419
production and acquisition   160
proem of the *Timaeus*   379
Prolegomena   17, 45, 332, 421, 427f., 430, 433, 435
- to Plato's Philosophy   332, 428
Prologue   343, 345f., 351, 356, 361
property   8, 148, 207, 317, 321, 405, 441
- perceptible   3, 140, 150, 266
proposition   12f., 24, 62–66, 75, 107, 159, 170f., 215–217, 219–221, 224, 226–230, 232, 234–236, 238, 240–242, 292, 296, 351, 373, 377, 385, 397–399, 404f., 407–409, 411, 423, 441
- false   232, 234–236
- true   240–242
Protagoras   5, 133f., 136, 138–143, 145f., 148f., 152, 154, 162–164, 168, 181, 372
- „On wrestling"   163
- „Truth or Refutations"   163
Protagorean   137, 139f., 142, 145, 154, 167, 169, 181, 433
- measure doctrine   137f., 140, 152, 154
  - Platonic, Aristotelian and Galenic arguments against Protagorean scepticism   3, 9, 11, 14–18, 20–22, 25, 27, 37–39, 45, 48, 50f., 58–61, 64, 66–71, 73, 75f., 83, 87, 92, 96, 98f., 101, 105, 127, 148, 164–167, 169, 173, 175, 177, 181, 184, 194, 201, 258, 263, 268, 337–339, 344f., 348f., 353–356, 358–361, 379, 381, 383f., 401, 403, 410, 414, 417–424, 427f., 430–433, 459–462, 464, 466
Proverbs (Biblical book)   358

Psellos 19–21, 28, 457–467
- allegorical reading in Psellos
  - double conception of allegory 467
  - Psellos' use of allegory 462
  - specific allegorical readings by Psellos 462
  - uses and limitations of dialectic and allegory in Psellos' works 462
- catalogue of Psellos manuscripts 462
- *Chronographia* 20, 457–459, 461f., 466
- knowledge of the Platonic and Aristotelian traditions 461
- *Philosophica Minora* 28, 458, 460–462
- theologia and philosophia in Psellos 459
Pseudographemata 24, 211, 237, 239f., 242
psogoi 432
psychagogy 26, 337, 344, 346f.
- rhetorical 46–47
public 40f., 47, 163f., 243, 343f., 355
- assembly 164
- debates 164f.
Purification 23, 157, 175–179, 183
- two types of purification of thinking (καθαρμὸς τῆς διανοίας) 60, 125, 147f., 154f., 390
Pythagoreans 252, 335, 379

quality, qualities 73, 96, 116, 152, 162, 283, 304, 317–319, 387, 452
- four elementary qualities or 'natures' 284
  - hotness, coldness, dryness, and moistness 284
question (ἐρώτησις) 2, 5–9, 16–18, 23f., 27, 40–42, 44, 47–51, 64, 66f., 69, 86, 93–95, 97, 100, 106, 132–137, 139, 141f., 144f., 147–151, 153, 160, 164–166, 168f., 171, 173, 176, 180, 182, 190, 199–201, 204, 207f., 213f., 216f., 222, 224, 227–229, 231–235, 237–242, 244, 255, 257, 259f., 266–268, 272, 279–283, 288, 290–300, 305f., 308–312, 318, 321, 327, 338, 352, 377, 382, 387f., 398f., 406, 413f., 416, 418f., 423, 433, 453, 460

raising difficulties 170f.
raising objections 23, 170
rational 354–357, 362f.
- *unscientifically* rational 431
rationale [*rationes*] 357
rationality 347, 357

realism 71, 375
- early internal 375
reality 59–61, 334, 359, 362f., 389
reason
- critical 331
  - critical personal 331
- give and receive reason (δοῦναί τε καὶ ἀποδέξασθαι λόγον) 334
- reason (λόγος) 69, 75, 334, 346f., 347
reasoning 69, 95, 152, 169, 171, 180, 257, 268, 300f., 312, 314, 322, 334, 351, 357, 413, 443–445, 449, 457, 459
- inductive 159
- reasoning (κρίσις) 457
- scientific 13–15, 18f., 25f., 155, 169, 171, 180, 255–258, 262f., 265f., 268, 337, 371, 399, 442
- reasoning and refutation 444
reductio ad absurdum 3, 290, 293
refutation, refutations 2, 4, 6–9, 23f., 37f., 41, 48–50, 53, 95f., 100, 106, 108–112, 134–141, 144–148, 152, 169, 171, 178, 180–183, 212, 214, 218–222, 224f., 227, 234, 238, 242, 257, 269, 290–293, 296–300, 304, 306f., 328, 351f., 433f., 436f.
- as a method of catharsis 169, 171
- genuine 18, 25, 96, 110, 117, 145, 167, 178, 219–221, 250, 323, 418, 452
- of the thesis that knowledge is perception 23, 131, 134, 138, 146
- refuting *in accordance with its object* (κατὰ τὸ πρᾶγμα ἐλέγχοντες) 214f., 222, 231
- scientific 217–223
- sophistical 211–215
reign of Romanus III 457
relativism 139, 142
religious books 278
rhetoric
- and dialectic 37–53
- as oratory 46
- conventional 38–40
- *forensic* 6, 47, 163, 165, 216
- philosophic 38, 44, 46f., 428, 430, 466
- psychagogic function of rhetoric 347
- rhetoric of belief (πιστευτική) 48, 52

- rhetoric of the *Phaedrus* 48
  - defense of rhetoric as an art (τέχνη) in the *Phaedrus* 45
  - philosophic rhetoric of the *Phaedrus* 46
- true 41-44
- true rhetoric in the *Gorgias* 44
- two kinds or two parts (*partes*) of rhetoric 45
  - two rhetorics or two uses of rhetoric, one juridical or conventional and the other philosophical 45
rhetoric 39-53, 347 f.
rite as a representation of abstract truth 465
ritual 465
- drama 23, 134, 139, 145, 192, 465
- of the women in Byzantium 465
Roman 45, 371, 422
- Empire 371, 381, 422
- law 50, 87, 115, 124, 141, 162 f., 168, 288 f., 291, 378, 406

sages 356, 381
Saint Agatha 465 f.
- Saint Agatha's story 464
  - Agatha's name 464
  - contemporary rites for Agatha in Sicily 465
  - *panēgyris* for Agatha 463
  - rites for Agatha 464
  - the story of Agatha's martyrdom 465
  - the suffering of Agatha 465
Scholastic and neo-Scholastic thinkers 280
science 1, 3, 10, 14-16, 20, 24, 45 f., 59, 69, 91, 105, 129 f., 153 f., 198, 239 f., 249, 251 f., 254-259, 263 f., 266 f., 273, 288, 333 f., 337, 348 f., 356-358, 378, 399, 401, 404, 431, 443, 448 f., 453
Scripture 16, 25, 333, 343-346, 349, 352, 355 f., 360
self-
- self-conceit 169, 171, 180
- self-control 39, 43, 51 f., 109, 115, 117, 122
- self-knowledge 23, 73, 158, 162, 176-178, 181, 184, 342, 361
- self-mastery 116 f., 127
- self-purification 23, 180, 182 f., 185
- self-recollection 158, 185
- self-recovery 158, 183, 185
- self-understanding 185, 424

senate-house 164
senses 66, 98, 122, 135 f., 144, 147, 174, 253, 255, 259, 270, 319 f., 404, 414, 428, 458, 461
- sense-perception 131
sensible world 18, 452
signification 297, 301, 307, 309-312, 314 f., 320, 328, 388-390
- signifying 295, 298-317, 319-326, 328
  - about one 315, 321 f.
  - about one thing 312-317, 321, 323, 325
  - essentialist interpretation of 'signifying one thing' 326
  - one thing 85, 87, 112, 132, 135, 154, 301, 304 f., 307-309, 311-323, 325-327, 408
  - signifying nature and substance (ousia) 316
  - signifying one thing in respect of the subject 326
  - signifying the essence of a thing 323
silence 454
Simplicius 19, 21, 27 f., 254, 268, 323, 371 f., 379, 381, 383-390, 441-454
- Commentaries
  - on the *Categories* 386, 444 f., 447
  - *Physics* Commentary 450
- discussion of elemental principles in *Physics* 449
- polemic against Philoponus 444
- practice of the history of philosophy 372
- role of dialectic as an evaluation tool of philosophical opinions in Simplicius 445
- Simplicius' so-called harmonizing tendency 453
sketch accounts (*hupographas*) 388
skill (ἕξις) 110, 206 f., 215, 335, 442, 452, 460
skopeisthai 139, 142, 149
skopos 384, 386 f., 400, 402 f., 421, 447
Socrates 5-9, 20-24, 27, 37-53, 57-75, 83, 86, 89, 96 f., 106-110, 112 f., 115-127, 131-155, 158, 163-170, 173, 179-183, 190-196, 198, 201, 204 f., 207, 261, 313-318, 321-326, 332-335, 338, 345, 349, 372, 380, 384, 408, 413, 416-419, 429-433, 435, 445, 447
- avowal of ignorance 40 f., 48 f.
- elenchic method 332
- flight into the *logoi* 57, 62, 64, 68 f., 75
- in love 114, 119, 127
- intellectualism 48, 177, 183

- protreptic 96f., 168, 431
- the dialectician 11, 14, 74, 105, 107, 110, 113, 123, 142, 169-171, 173, 180f., 183, 194, 205, 223f., 265, 270, 272, 291, 332, 334, 404, 413, 416, 449
  - Socrates' 'dialectical turn' 61
- use of irony 49f., 110
- use of myth and other extra-logical means 37

Socratic 2, 5, 7-9, 20, 23, 27, 37, 40, 48, 53, 58-60, 66, 70, 76, 83-87, 94, 102, 105, 108, 118, 142, 157f., 164, 166-169, 172f., 176f., 180-184, 227, 260f., 279, 337f., 344, 353, 430-432, 437f.
- antilogic 5f., 61, 161, 167, 169-171, 173, 181
- dialectic, cathartic dialectic 167-177, 180-182, 184f.
- dialectic in the *Gorgias* 48, 53
- Dialectic School 338
- elenchus 6-8, 27, 66, 70-72, 158, 177-185, 429-434, 436-438
- *logoi* 5, 21f., 57f., 60, 62-72, 74f., 133
- maieutic 48, 172, 182, 431
- Megaric School 13, 338, 348
- rhetoric 48-53
- Socratic circle 13, 22, 83-85, 102, 166
- Socratics 5, 12, 83, 97
- Socratic way 353
- Socratism 86
- the Socratic paradox 38, 45, 48

solution (λύσις) 7, 16, 22, 25, 85, 121, 129, 131, 224, 250-252, 261, 267, 271, 273, 277f., 281-285, 308, 322, 355, 357, 428

„Son of Horus" (Origen) 336

„son" of the Good 90

sophist 2f., 5f., 9f., 15, 20, 23f., 48, 51, 86, 97-99, 109-113, 119, 122, 125, 129, 132, 134, 148, 157-160, 162-171, 173, 175, 178-182, 189-202, 204, 206f., 222, 228, 232f., 242f., 272, 332f., 338, 414, 416, 418f., 421, 427, 445
- historical sophists 158
- noble 110-113
- of noble lineage 159, 169, 174

sophistry 111, 134, 165f., 169, 171, 173f., 179, 240, 243
- and eristic 50, 124, 165f.

σοφιστικοὶ ἔλεγχοι 213f.

soul 3, 9, 17, 19, 22f., 28, 39-43, 45-47, 49, 52, 59, 69-75, 87, 105, 108-115, 120-126, 129, 133, 135, 139, 142, 144-151, 167, 169, 176-181, 183f., 263, 320, 346f., 353, 360, 377, 380, 382f., 388f., 403, 415f., 419f., 431
- as the 'place of ideas' (τόπος εἰδῶν) 71
- care of the soul (ἐπιμέλεια ψυχῆς) 9, 176, 202
- inferior level of our soul (τὸ ταπεινὸν τῆς ψυχῆς) 346
- soul's salvation 416

species [εἶδος] 333, 340

spiritual eunuch 336

statement 10f., 40, 58, 68, 91, 95, 154, 164, 182, 199, 265, 291, 296, 298f., 303, 307f., 313, 315f., 326-328, 332, 376, 402f., 406, 409f., 431, 449, 452, 459
- affirmative 303
- negative 303
- statements (ἀποφαντικοὶ λόγοι) 4, 14, 26, 43, 62, 86, 94, 96, 158, 181, 294, 299, 303, 326, 376, 382, 400, 402, 404, 406f., 449, 451, 461
- sworn statement (διωμοσία or ἀντωμοσία) 163

Stoic, Stoics 15-17, 25f., 263f., 335, 337-339, 341, 348-352, 354, 381, 390,
- logic 15-16, 25, 337f., 341, 348-352
- materialistic immanentism 359
- refutations 50, 68, 107, 137-139, 146, 211, 213, 217-221, 223, 341, 345, 427, 442f.

student of *erga*, student of *logoi* 21, 63

study of nature (φύσεως ἱστορία) 57, 59-61

subject 6f., 10, 17, 19, 23, 28, 47, 115, 124, 134, 137, 149, 153f., 157, 161, 163f., 166, 176, 181f., 207, 215, 224, 230, 245, 252, 257, 259, 261, 266-269, 271f., 279, 281f., 288, 312, 314, 316-319, 324-328, 339f., 346f., 373-375, 381, 388, 397f., 401f., 413, 415, 423f., 427, 431, 433, 437, 441, 458, 460, 462, 464, 466
- the nature of the subject 312

sublunary world 282

substance 98, 251, 284, 304f., 311, 314, 316f., 319-321, 324f., 338-340, 379, 408, 434, 452
- of each thing (οὐσία ἑκάστου) 11, 15, 60, 105f., 318-320, 413

substantialist interpretation 304, 317

sulphur 284
συμφωνία 65, 67 f.
supposition (ὑπόθεσις) 125, 154, 356, 400
supreme Godhead 339
syllogisms 14 f., 20, 230, 237, 243 f., 255, 257, 359, 427, 429, 441
– hypothetical 91, 351, 359, 427
syllogistic, syllogistics 15, 19, 224, 337 f., 415, 417, 427, 444
– deduction (ἀποδεικτική) 457
– form of reasoning 444
symbola 467
symphonia 385
synagôgê 415
synonyms, synonymy 312, 322 f.
synoptic 106, 307, 315, 334
Syrianus 18, 21, 26, 255, 373, 397–411, 422, 447
– Commentaries on *Metaphysics* M and N, and B and Γ 373
system 14, 25, 72, 98, 105, 107, 161, 249, 273, 278, 325, 356, 436

τὰ ὄντα 57, 59 f., 66
tautonyms 323
taxonomy 24, 174, 189, 202, 204 f.
technē and epistēmē 134
τερετίσματα 70
Theodorus, Theaetetus' teacher 133 f., 140, 145 f., 148, 154 f., 192 f.
theology 16, 26, 334, 337 f., 352 f., 356–360, 362 f., 408
– affirmative (cataphatic) 362
– negative (apophatic) 362
– Origen's apophatic theology 333
– philosophical theology 25, 331, 337, 342, 347, 352 f., 359
– theology as coextensive with dialectics 362
– theology-epoptics 357
Theophrastus 259, 261 f., 279–281, 292–294, 359, 441, 443, 445, 447, 450
– *Physics* 445
– *On Affirmation and Negation* 293
– Theophrastus' aporetical book 280
– *Topics* 441, 443, 447
theorem 400
theōria 371, 377
theurgical practice 383, 420

things 2–4, 10 f., 14, 16, 18, 23, 39 f., 43, 57, 61, 63–66, 68 f., 72, 80, 87, 89 f., 92 f., 95, 98, 100, 106, 116, 119, 122 f., 126, 131 f., 134, 137–143, 145–147, 150 f., 153 f., 159, 172, 174, 177, 180, 182, 184, 191, 194, 197 f., 214 f., 226, 235, 243, 250, 254, 256, 258 f., 265 f., 272, 281, 292, 295 f., 300–302, 304 f., 308–313, 315–317, 319–321, 327 f., 342, 349, 351, 354 f., 357 f., 362 f., 371, 375–377, 379, 382 f., 388, 399, 401–404, 406, 408, 410, 414–417, 419, 423, 427, 433 f., 443, 445, 448, 453, 460 f., 463, 465
– accidental to something 322
– things signified 302, 308, 311
  – signified by names 311
– things that are in a subject 317
thinking (*dianoein*) 19 f., 151 f., 154, 157, 176
– thinking or calculation 152
– mathematical thinking 413
third man argument 22, 98
thought 1, 3, 7 f., 13, 15, 17, 19–21, 23–25, 28, 43, 51, 63–66, 80, 83–85, 87, 91, 115, 118, 120, 130, 140, 144, 153, 157, 169, 172, 176, 180 f., 183–185, 190, 192, 197, 202, 206, 244, 249, 251, 254, 281–283, 295, 301, 303, 305, 314, 325, 331, 337 f., 341, 348, 350, 353 f., 359 f., 362, 377–379, 400, 402, 409, 413, 417, 419 f., 422, 452 f., 463
– cognitive norms of thought 342
thumos 177, 183 f.
ti esti questions 150 f.
ti kata tinos 316
'to be', the verb 136
tode 327
topoi 14, 46, 48, 435
– introductory 435
– of judiciary rhetoric 48
– of sophistic rhetoric 46
τὸ σκοπεῖν τὰ ὄντα ἐν [τοῖς] ἔργοις 66
tradition 10, 14 f., 25, 28, 45, 185, 250, 278, 284 f., 288, 295, 331, 335, 337, 342, 346, 354, 381, 384, 390 f., 422, 441, 445, 454, 461, 466
transcendence 94, 401 f.
– of the One 26, 74, 213, 223, 225, 254, 278, 308, 359, 397 f., 401–405, 407, 410 f., 433, 461
transcendentalist 359

translation, translations   15, 18, 57, 61–64, 67 f., 73, 79, 92, 110, 135, 157, 163, 167, 191, 199, 230, 237, 243, 249, 260, 277, 281, 290, 320, 362 f., 374, 397, 428, 433, 441, 457
- sentential translations   62
  - mono-sentential   62
  - non-sentential   62
  - poly-sentential and ambiguous   62
tropos, tropikos   376, 415
- 'two conditionals' or 'two propositions' or 'two figures' (διὰ δύο τροπικῶν)   351
- the true' (τὸ ἀληθὲς)   61
truth   1 f., 7 f., 12, 18, 20–22, 26 f., 40, 46, 50 f., 53, 57 f., 61, 64–68, 71 f., 74 f., 80, 90, 95, 105–107, 113, 117, 122–124, 127, 134–136, 143, 147, 151–154, 163, 167–169, 171, 177 f., 207, 219, 221, 226, 238, 240, 253 f., 257, 263, 266, 268 f., 271–273, 280, 289 f., 301, 303, 331, 342 f., 345, 347, 351, 354 f., 360–362, 371, 373–377, 382, 384–386, 388–390, 400, 409 f., 416, 419–421, 428, 433, 438, 446, 457, 465 f.
- the correspondence theory of truth   65
- 'to look at the truth of things' (σκοπεῖν τῶν ὄντων τὴν ἀλήθειαν)   59
- truth of things (τῶν ὄντων τὴν ἀλήθειαν)   22, 57, 63 f., 66, 68, 71, 80
  - 'truth of things' in images   63, 72

ugliness   177–179
universals   15, 71, 317, 448
ὑπόθεσις   68

verbal controversy (διὰ λόγων ἀμφισβητικόν)   170 f., 173
violence (βιαστικόν)   162 f., 360

virtue   18, 37, 52 f., 69, 88, 96, 109, 114 f., 117, 119, 123, 140, 173, 236, 333, 361, 388, 421–424
- cathartic   421
- civic   170, 421 f.
- contemplative   383 f., 421 f.
- ēthikē aretē   383
- natural, ethical, civic, kathartic, and contemplative gradations of the virtues   10, 59, 113, 122, 144, 150, 189, 191, 194, 266, 277, 283, 319, 345, 354, 357, 384, 421, 427, 438, 453
- phusikē aretē   383
- scala virtutum   383
voyage   21, 57–61, 67, 73, 76
- earlier voyage (πρότερος πλοῦς)   59
- first voyage (πρῶτος πλοῦς)   59, 61, 73

wisdom   109, 124–127, 354 f., 378, 381, 463
- practical   15, 84, 94, 358, 377, 399, 403, 444, 446, 458
  - those who have practical wisdom (*phronimōn*)   378
witness   332, 372 f., 376–378, 383, 385, 390, 465
- of very ancient (*palaioi*) historical persons   378
- 'witnesses' of the *phainomena*   379
wolf pretending to be a dog   169, 180

Zeno's paradox   153, 233 f., 236 f.
zētein   385
- zetetic   331, 342, 344 f., 352–356
  - inquiry   342
  - philosophical strategy   331
  - zetetically   357
  - zetetic, or strictly aporetical scope   16, 249, 279 f., 331, 342, 344 f., 349, 352–356

Notes on the Contributors

Lucas Angioni received his PhD in Philosophy from the University of Campinas (2000), where he is a Professor in the Department of Philosophy (tenured since 2003) and runs the research group MESA (Metaphysics and Science in Aristotle, funded by the CNPq). He has been an Academic Visitor in Oxford many times (Hillary Terms in 2007, 2009, 2010, 2012, Trinity Term 2015). He was also a Visiting Scholar in Stanford twice for the winter term (2014, 2015) with Alan Code. Professor Angioni is the author of several books and book chapters in Portuguese (one of them is a monograph on the central books of Aristotle's *Metaphysics: As Noções Aristotélicas de Substância e Essência*, Campinas: Unicamp, 2008). He has written several articles on Aristotle that have appeared in journals such as *Oxford Studies in Ancient Philosophy*, *Apeiron*, *Ancient Philosophy*, *Logical Analysis and History of Philosophy*, *Manuscrito*, *Archai*, *Studia Philosophica Estonica*, *Eirene* (and others). One of his lifetime projects is to translate a significant part of Aristotle's works into Portuguese (*Physics* I-II was released in 2009, while other translations have appeared in provisional form in collections and journals). His current research is organized around Aristotle's theory of scientific explanation (in the *Posterior Analytics*) without excluding secondary focuses on other topics, especially on Aristotle's theory of argumentation, metaphysics, and ethics.

Han Baltussen is the Walter W. Hughes Professor of Classics at the University of Adelaide and a Fellow of the Australian Academy of Humanities. He has held fellowships at the Harvard Center for Hellenic Studies, Washington DC, the Institute for Advanced Study, Princeton. His research focuses on ancient intellectual traditions and their transmission. He has to his name seven books and many articles on ancient philosophers from the Presocratics to the late Platonists. He is currently finalizing a monograph on ancient consolation strategies, a commentary on Theophrastus' doxographical fragments, and a new translation of Eunapius' *Lives of Philosophers and Sophists* (Loeb, Vol. CXXXIV).

Dirk Baltzly, FAHA is Professor of Philosophy at the University of Tasmania. Born and educated in the USA, he holds one PhD in Philosophy from Ohio State University and a second in Classics from Monash University. After a postdoc at Kings College London, he has worked in Australian universities since 1994. He is best known for his translations of philosophical texts from Late Antiquity, including Volumes III–V of *Proclus Commentary on Plato's Timaeus* (Cambridge, 2007–2013), *Hermias Commentary on Plato's Phaedrus*, Volumes I and II (Bloomsbury 2018 and 2022), as well as *Proclus' Commentary on Plato's Republic* (Cambridge 2018 and 2022). He was among the editors of the 700-page reference work *The Brill Companion to the Reception of Plato in Antiquity* (2018) and has authored over 50 articles and book chapters on ancient philosophy. Baltzly has previously held visiting fellowships at the University of London's Institute for Classical Studies and the Institute for Advanced Study at Princeton. In 2022, he was a visiting fellow at the De Wulf–Mansion Centre for Ancient, Medieval and Renaissance Philosophy at KU, Leuven. He is a fellow of the Australian Academy of Humanities. In 2021, he was elected President of the Australasian Association for Philosophy. In addition to ancient philosophy, Baltzly enjoys bushwalking, camping, and photographing the wild places of Tasmania.

Beatriz Bossi is an Associate Professor of Ancient Philosophy at the Universidad Complutense de Madrid. She is the author of *Virtud y Conocimiento en Platón y Aristóteles* (2000) and *Saber Gozar: Estudios sobre el Placer en Platón* (2008) and the co-editor (with Thomas M. Robinson) of *Plato's Sophist Revisited* (de Gruyter, 2013), *Plato's Statesman Revisited* (de Gruyter, 2018), and *Plato's Theaetetus Revis-*

ited (de Gruyter, 2020). She has written widely on Greek philosophy for publications in Argentina, Chile, the USA, Germany, Italy, and Spain, and she is currently President of the International Plato Society.

Silvia Fazzo is a Faculty Member of Università del Piemonte Orientale. She is critically editing and commenting upon Aristotle's Metaphysics, beginning with Book Lambda, to which she devoted her 2012 (edition) and 2014 (commentary) volumes. Dealing with ancient philosophy and its reception, she strongly specializes in the historical and philological reconstruction of Greek theoretical treatises from the Aristotelian tradition. Her approach to texts and problems proved ground breaking in a number of fields and especially on the Metaphysics, whose emergence she has detected step by step in dozens of journal articles spanning over the last two decades (one can see some of her publications at: http://publicationslist.org/silvia.fazzo).[1] Sharing her skills, thoughts and experience is a crucial part of her mission.

Rafael Ferber is a Professor Emeritus at the University of Lucerne and adjunct Professor at the University of Zurich. His publications include *Zenons Paradoxien der Bewegung und die Struktur von Raum und Zeit* (1981, 1995), *Platos Idee des Guten* (1984, 1989), *Die Unwissenheit des Philosophen oder Warum hat Platon die "ungeschriebene Lehre" nicht geschrieben?* (1991, 2007), *Philosophische Grundbegriffe 1* (1994, 2009⁸), *Philosophische Grundbegriffe 2* (2003), and *Platonische Aufsätze* (2020).

Michael J. Griffin is an Associate Professor of Classics and Philosophy at the University of British Columbia in Canada. He is co-editor (with Richard Sorabji) of the *Ancient Commentators on Aristotle* project (Bloomsbury Academic) and the author of *Aristotle's Categories in the Early Roman Empire* (2015) as well as a two-volume annotated translation of Olympiodorus of Alexandria's lectures *On Plato's Alcibiades I* (2014, 2016). His research focuses on Greek Neoplatonism, and he has published on philosophers like Plotinus, Proclus, Ammonius, and Olympiodorus, with additional recent studies in cross-cultural philosophy. He is currently working on a monograph on the Greek Neoplatonic scale of virtues.

Gweltaz Guyomarc'h is an Associate Professor in Ancient Philosophy at the University of Lyon and a member of the Philosophical Research Institute of Lyon (IRPhiL). His work focuses on the Aristotelian tradition, especially Alexander of Aphrodisias. He published *L'Unité de la métaphysique selon Alexandre d'Aphrodise* (Vrin 2015) and translated Alexander's Commentary on *Metaphysics* Beta into French (Vrin 2021) based on a revised Greek text. He has also co-edited several volumes: for example, *Réceptions de la théologie aristotélicienne d'Aristote à Michel d'Ephèse* (Peeters 2017) with Fabienne Baghdassarian and a collection of papers on Aristotle's *De anima* III (Peeters 2020) with Claire Louguet and Charlotte Murgier. He is currently working on Peripatetic physics from Theophrastus to Alexander.

Inna Kupreeva (PhD Toronto) is a Senior Lecturer in Ancient Philosophy at the University of Edinburgh. She is working on the Aristotelian tradition in the philosophy, science, and medicine of the early Roman empire. Her publications include translations of works by John Philoponus for the Ancient Commentators Series as well as articles and chapters on Aristotle, Galen, Alexander of Aphrodisias, and other Peripatetics of the first two centuries of the modern era.

---

1 Last accessed on November 7, 2022.

Jens Kristian Larsen is an Associate Professor at the Norwegian University of Science and Technology. He has published widely on various aspects of Plato's philosophy in journals such as *Ancient Philosophy*, *Archiv für Geschichte der Philosophie*, *Polis*, *Études platoniciennes* and *Plato Journal*. He is a contributing co-editor of the volumes *Phenomenological Interpretations of Ancient Philosophy* (Brill 2021) and *New Perspectives on Platonic Dialectic* (Routledge 2022).

Claudia Mársico is a Full Professor of Philosophy (History of Ancient Philosophy) at the Faculty of Philosophy and Arts, University of Buenos Aires, and the Head of the Section of Studies in Ancient Philosophy at the National Academy of Science of Buenos Aires; she is also the former Head of the Department of Philosophy, Universidad de Buenos Aires (2013–2018) and Senior Researcher of the National Council for Scientific and Technological Research. She is author of more than twenty-five books in the field of Ancient Philosophy, including the first full translation of the extant texts on Socratic philosophies (*Filósofos socráticos, Testimonios y fragmentos. I/Megáricos y Cirenaicos*, Buenos Aires, 2013, and *Filósofos socráticos, Testimonios y fragmentos II/Antístenes, Fedón, Esquines y Simón*, Buenos Aires, 2014). She has also published many articles and chapters in national and international journals as well as books in the fields of ancient philosophy and philosophy of history.

Graeme Miles is a Senior Lecturer in Classics at the University of Tasmania who conducts research on Greek literature (especially of the Roman Era) and philosophy (especially regarding the Platonic tradition). He has published extensively in both areas and has recently published the first two of three volumes of Proclus' *Commentary on Plato's Republic* with Dirk Baltzly and John Finamore. With Professor Han Baltussen, he is also currently completing a new translation of Philostratus' *Lives of the Sophists* and Eunapius' *Lives of the Philosophers and Sophists*.

Melina G. Mouzala is an Assistant Professor (Ancient Philosophy; her research interests include the Platonic and Aristotelian traditions as well as ancient and Byzantine commentaries on Aristotle) at the Department of Philosophy of the University of Patras. She is author of three books in Greek: *Ουσία και Ορισμός. Η Προβληματική της ενότητος εις τα οικεία κεφάλαια των Μετά τα φυσικά του Αριστοτέλους*, Athens 2007: Harmos; *Περί του καθ'ετερότητα μη όντος στον Σοφιστή του Πλάτωνος*, Athens 2007: Harmos; *Ζητήματα Γνωσιολογίας, Οντολογίας και Μεταφυσικής στην φιλοσοφία του Αριστοτέλους, Υπό το φως αρχαίων και βυζαντινών υπομνημάτων* (Athens 2013: Gutenberg). She is also the editor of three collective works in Greek: *Θέματα Οντολογίας-Μεταφυσικής, Φιλοσοφίας των Μαθηματικών και Λογικής, in memoriam Δημ. Μούκανου*, Athens 2018: Gutenberg; *Αυτογνωσία και Επιμέλεια Ψυχής/Εαυτού: Αρχαία Ελληνιστική· Νεοπλατωνική· Βυζαντινή· Νεοελληνική Φιλοσοφία* (Athens 2020: Papazisis); *Αρχαία Φιλοσοφικά Θέματα στην Πατερική και Βυζαντινή Φιλοσοφία* (Athens 2022: Papazisis) She is currently working as the author of a monograph on Plato's *Charmides* (in Greek, forthcoming in 2023, Gutenberg) and as the editor of the volume, *Ancient Greek Dialectic and its Reception* (2023, de Gruyter) and the volume, *Cognition in Ancient Greek Philosophy and its Reception: Interdisciplinary Approaches* (forthcoming).

Anna Pavani is an early-stage researcher in Ancient Philosophy. During her PhD research (2017–2020), she was an MSCA Fellow at a.r.t.e.s. Graduate School for the Humanities at Cologne as well as a Visiting Research Fellow at Brown University (RI-USA). In Cologne, she worked as a teaching assistant, lecturer, and assistant researcher for the Chair of Ancient Philosophy, for the Thomas Institute, and for the Chair of Greek Philology. She has presented at conferences in Berlin, Bochum, Boston, Flagstaff, Leuven, New York, Oxford, Paris, Philadelphia, Prague, and Providence. She has three papers forthcoming about the role of language in Plato's late dialogues.

Ilaria L.E. Ramelli is a Professor, FRHistS, and holds two MAs, a PhD, a Doctorate h.c., a Postdoctorate, and various Habilitations to Ordinarius. She was Professor of Roman History, Senior Visiting Professor of Greek Thought at Harvard and Boston Universities, of Church History at Columbia, and of Religion at Erfurt MWK. She has served as Full Professor of Theology and Endowed Chair at the Angelicum, and twice as a Senior Fellow at Durham University as well as at Princeton, at Sacred Heart University, and at both Corpus Christi and Christ Church in Oxford. She is also a Senior Member of the Centre for the Study of Platonism at Cambridge University, a Humboldt Forschungspreis Fellow at Erfurt MWK, a Senior Fellow at Bonn University (elect), and a Professor of Theology (Durham University, Hon.) and of Patristics and Church History (KUL). Her recent books include *The Christian Doctrine of Apokatastasis* (Brill 2013), *Evagrius' Kephalaia Gnostika* (Brill 2015), *The Role of Religion in Shaping Narrative Forms* (Mohr Siebeck 2015), *Social Justice and the Legitimacy of Slavery* (Oxford University Press 2016), *Evagrius, the Cappadocians, and Neoplatonism* (Peeters 2017), *A Larger Hope?* (Cascade 2019), *Bardaisan of Edessa* (Gorgias 2009 and de Gruyter 2019), *T&T Clark Handbook to the Early Church* (co-edited, T&T Clark Bloomsbury Academic 2021), *Eriugena's Christian Neoplatonism and Its Sources* (ed., Peeters 2021), *Patterns of Women's Leadership in Ancient Christianity* (co-edited, Oxford University Press 2021) and *Lovers of the Soul, Lovers of the Body* (co-edited, Harvard 2022).

François Renaud is a Professor of Philosophy at the Université de Moncton (New Brunswick, Canada). Trained in both classics and philosophy (PhD Tübingen), he has been a visiting scholar at the Universities of Crete, Rome (La Sapienza), and Würzburg as well as a Visiting Professor at Durham University. He has published mostly on Plato, Platonic interpretation in Antiquity and in modern times, and Plato's Socratic legacy. His publications include *Die Resokratisierung Platons: Die platonische Hermeneutik Hans-Georg Gadamers* (Sankt Augustin 1999), *The Platonic* Alcibiades I: *The Dialogue and its Ancient Reception* (Cambridge University Press 2015, paperback 2018), co-authored with Harold Tarrant, *Brill's Companion to the Reception of Plato in Antiquity* (2018), coedited with Harold Tarrant, Dirk Baltzly and Danielle Layne, and *La justice du dialogue et ses limites: Étude du* Gorgias *de Platon* (Les Belles Lettres 2022). His current research mostly deals with the relationship between philosophy and poetry in Plato and the reception of Homer in Antiquity and in modern times.

Harold Tarrant studied at St. Johns College Cambridge and Durham University and now lives near Cambridge; however, most of his working life was at the University of Sydney (1973–1993) and the University of Newcastle Australia (1993–2011), where he retains the title of Professor Emeritus. Most of his work has involved the Platonic tradition from Plato to Late Antiquity. He has received grants from the Australian Research Council. Besides his many articles and book chapters, he has authored several books: *Skepticism or Platonism?* (Cambridge 1985, reprinted in 2007), *Thrasyllan Platonism* (Ithaca 1993) K.R. Jackson, K. Lycos and H. Tarrant, *Olympiodorus on Plato's Gorgias* (co-edited with Robin Jackson and Kimon Lycos, Leiden 1998), *Plato's First Interpreters* (London 2000), *Recollecting Plato's Meno* (London 2005), *Proclus: Commentary on Plato's Timaeus*, Vol. I (Cambridge 2007) and Vol. VI (Cambridge 2017), *From the Old Academy to Later Neo-Platonism* (Aldershot 2011), *The Platonic Alcibiades I: the Dialogue and its Ancient Reception* (with François Renaud, Cambridge 2015), and *The Second Alcibiades: A Dialogue on Prayer and on Ignorance* (Parmenides Publishing, 2023). He has co-edited various multi-authored volumes: *Reading Plato in Antiquity* (with Dirk Baltzly, London 2006), *Alcibiades and the Socratic Lover-Educator* (with Marguerite Johnson, London 2012), *The Neoplatonic Socrates* (with Danielle Layne, University Park 2014), and *Brill's Companion to the Reception of Plato in Antiquity* (with Danielle Layne, Dirk Baltzly, and François Renaud, Leiden 2018).

Sarah Klitenic Wear (BA Classics, The University of Chicago; PhD Classics, Trinity College Dublin) writes on the later Platonic tradition. Her monographs include *Despoiling the Hellenes: Pseudo-Dionysius and the Athenian School of Platonism* (co-written with John M. Dillon, Ashgate 2007) and *The Teachings of Syrianus on Plato's Timaeus and Parmenides (*Brill 2011). In addition, she has written commentaries, such as *Plotinus on Beauty and Reality (Enneads I.6 and V.1): A Greek Reader* (Bolchazy 2017) and *Patience and Salvation in Third Century North Africa: A Christian Latin Reader* (Catholic University of America Press, Catholic Education Press, 2023). She co-edits (with Frederick Lauritzen) the book series *Theandrites: Byzantine Philosophy and Christian Platonism* (Franciscan University Press).

www.ingramcontent.com/pod-product-compliance
Lightning Source LLC
Chambersburg PA
CBHW031719230426
43669CB00007B/186